# QUALITATIVE AND QUANTITATIVE BEHAVIOUR
## OF PLANETARY SYSTEMS

# QUALITATIVE AND QUANTITATIVE BEHAVIOUR OF PLANETARY SYSTEMS

## Proceedings of the Third Alexander von Humboldt Colloquium on Celestial Mechanics

*Edited by*

R. DVORAK

*Institut für Astronomie, Universitätssternwarte, Vienna, Austria*

and

J. HENRARD

*Département de Mathématique, FNDP, Namur, Belgium*

Reprinted from Celestial Mechanics and Dynamical Astronomy
Volume 56, Nos. 1-2, 1993

SPRINGER SCIENCE+BUSINESS MEDIA, B.V.

A C.I.P. Catalogue record for this book is available from the Library of Congress.

ISBN 978-94-010-4898-9     ISBN 978-94-011-2030-2 (eBook)
DOI 10.1007/978-94-011-2030-2

*Printed on acid-free paper*

# TABLE OF CONTENTS

## SESSION ON MISCELLANEOUS PROBLEMS
(*Chairman:* V. Szehebely)

Participants in the Third Alexander von Humboldt Colloquium on Celestial Mechanics

# PREFACE

The success of the two first *Alexander von Humboldt Colloquia on Celestial Mechanics* (March 1984 and March 1988) encouraged us to meet again this year from March 29 to April 4, 1992 in the " Alpengasthof Peter Rosegger"" in the Styrian Alps (Ramsau, Austria).

This time the colloquium was devoted to the "Qualitative and Quantitative behaviour of Planetary Systems". The papers covered a large range of questions of current interest from the behaviour of dust particles to the stability of the Solar System as a whole, without forgetting the motion of Asteroids and their classification into families. KAM theory, chaotic motions, resonances, Lyapunov characteristic exponents, perturbation theory, numerical integration were — of course — on the menu every day, served with sauces and accents from various part of the world: from China, from Brazil, from the United States and from all over Europe.

To be able to organize this — now well established — meeting , we have to thank primarily the munificence of the Austrian Ministry of Science, furthermore the Steiermärkischen Wissenschafts- und Landesfonds and the Österreichische Nationalbank for financial support. Also , the Österreichische Forschungsgemein-schaft and the University of Vienna made it possible to invite participants to this meeting from abroad. Support in form of "Tagungsunterlagen" was given by the Creditanstalt-Bankverein in Vienna and coffee during the whole meeting time was donated by Hornig-Kaffee Graz.

Many thanks are due to the Chairmen (unfortunately we did not have any chairwoman this time) of the sessions. They acted as Editors of the papers presented during "their" session, arranging for competent and fast refereeing so that the papers could be reviewed and, when necessary, corrected in a very short time.

Our model hosts Barbara and Fritz Walcher contributed a lot to the success of the meeting, as they did their best providing pleasant and friendly surroundings. We all are looking forward to 1996 when we plan to meet again at the Alpengasthof Rosegger.

R. Dvorak                                                                                          J. Henrard

# REVIEW OF PLANETARY AND SATELLITE THEORIES

P. K. SEIDELMANN

*U S Naval Observatory*

**Abstract.** Planetary and satellite theories have been historically and are presently intimately related to the available computing capabilities, the accuracy of observational data, and the requirements of the astronomical community. Thus, the development of computers made it possible to replace planetary and lunar general theories with numerical integrations, or special perturbation methods. In turn, the availability of inexpensive small computers and high-speed computers with inexpensive memory stimulated the requirement to change from numerical integration back to general theories, or representative ephemerides, where the ephemerides could be calculated for a given date rather than using a table look-up process. In parallel with this progression, the observational accuracy has improved such that general theories cannot presently achieve the accuracy of the observations, and, in turn, it appears that in some cases the models and methods of numerical integration also need to be improved for the accuracies of the observations.

Planetary and lunar theories were originally developed to be able to predict phenomena, and provide what are now considered low accuracy ephemerides of the bodies. This proceeded to the requirement for high accuracy ephemerides, and the progression of accuracy improvement has led to the discoveries of the variable rotation of the Earth, several planets, and a satellite. By means of mapping techniques, it is now possible to integrate a model of the motion of the entire solar system back for the history of the solar system. The challenges for the future are: Can general planetary and lunar theories with an acceptable number of terms achieve the accuracies of observations? How can numerical integrations more accurately represent the true motions of the solar system? Can regularly available observations be improved in accuracy? What are the meanings and interpretations of stability and chaos with respect to the motions of the bodies of our solar system?

There has been a parallel progress and development of problems in dealing with the motions of artificial satellites. The large number of bodies of various sizes in the limited space around the Earth, subject to the additional forces of drag, radiation pressure, and Earth zonal and tesseral forces, require more accurate theories, improved observational accuracies, and improved prediction capabilities, so that potential collisions may be avoided. This must be accomplished by efficient use of computer capabilities.

**Key words:** Planetary theories – satellite theories – artificial satellites

## 1. Introduction

There is a real solar system as part of the real universe that is moving and changing due to the real existing forces. We attempt to compute, observe, and understand the real thing by means of models and approximations. Unfortunately, our theories and computations are incomplete, and our observations are subject to systematic and random errors.

There is a need to develop new theories for the planets, the moon, and artificial satellites. Technology has advanced such that we have significantly increased computer capabilities, both speed and memory, much better software, and algebraic manipulators. The challenge is to use these improvements to satisfy the requirements for new theories.

Prior to the 1950's planetary and lunar theories were developed from two body models by expansions in terms of small quantities and by orders of the power of the perturbing masses. Theories by Newcomb (1898), Hill (1898), Leverrier

*Celestial Mechanics and Dynamical Astronomy* **56**: 1–12, 1993.
© 1993 *Kluwer Academic Publishers.*

(1858, 1859, 1861), and Brown (1919) were developed by hand and accompanied by tables to facilitate the hand computations of the ephemerides of the bodies. The theories were based on a dynamical reference system, or an equinox of date, so that the constant of precession and the location of the equinox were embedded in the theories. The lunar theory required empirical correction terms to agree with observations. Due to the Moon's proximity to the Earth, the accuracy with which it could be observed, and the complexity of its motion, it presented a special problem.

When Pluto was discovered, it presented a special and continuing problem for general theories. Since its heliocentric distance at perihelion is less than the heliocentric distance to Neptune, but at aphelion, the distance is larger, an expression for the relative positions of Pluto and Neptune could not be developed in terms of the ratio of the distances. Thus, expressions for the perturbations are not convergent. Theories have only been developed by expanding in terms of the resonances (Goodrich and Carpenter, 1966) or by combining a theory and a numerical integration (Sharaf, 1955 and Sharaf and Budnikova, 1964).

The computer provided the capability to perform rapidly the tedious computations of numerical integration. Previously, the method was not practical for long accurate computations by hand. The accuracies of numerical integrations are limited by computer word length, the truncation of the series expansions, the integration method, the accuracies of the initial conditions, and the completeness of the force model. The integrations degraded as a function of time from the epoch, when compared to observations.

Numerical integrations require the use of a space-fixed rectangular coordinate system. The numerically integrated ephemeris, DE200/LE200, covers a 250-year period from 1800 - 2050, and is fit to a combination of optical, radar, lunar-laser ranging and spacecraft observations (Standish, 1982). DE102 is a long-term ephemeris covering 4000 years with the lunar motion adjusted so that it would be consistent with the Improved Lunar Ephemeris, and with a slightly inconsistent equinox (Newhall et al 1983). These represent the most widely used numerically integrated ephemerides.

## 2. Planetary and Lunar Theories

While numerical integrations were the primary means of computing ephemerides, Chapront and the other astronomers at the Bureau des Longitudes undertook the preparation of new general theories. They followed the work of Deprit et al (1971), Henrard (1979), and others who had addressed the main problem of the lunar theory, but they attacked both the planetary and lunar problems. Thus, Bretagnon (1980, 1981, 1982), Simon and Bretagnon (1975, 1978, 1984), and Simon (1983) have developed general theories for the planets. Chapront-Touze and Chapront (1980, 1983, 1988) have developed a complete lunar theory including planetary terms. These theories have been fit to the numerical integration DE200/LE200, as opposed to fitting observations directly. They show good agreement to the accuracy of the theories. The accuracy of these theories is limited by 1) the number of terms included in the theories, and 2) the limit on the order of the expansion, as demonstrated by Carpino, Milani, and Nobili (1987), based on a long-term integration. There are significant higher order terms, whose omission degrades the accuracy of the general theory.

## 3. Planetary Satellite Theories

For satellites, with rapid motion such as Jupiter's Galilean satellites and Mars satellites, numerical integration quickly loses accuracy as a function of time, making it inaccurate for extended periods of time.

An extensive theory for the motions of Phobos and Deimos has been prepared at Mission Control Center in Moscow, primarily for the PHOBOS missions to the satellites of Mars. Those theories reach an accuracy of 2-3 km for Phobos and 6-9 km for Deimos (Ivanov et al 1988). Chapront-Touze (1988 and 1990) has developed semi-analytical theories (ESAPHO and ESADE) for the orbital motions of Phobos and Deimos. Sinclair (1989) solved for orbital parameters of the Martian satellites based on all available positional observations. The secular acceleration of Phobos was well determined and indicates a 40 million year future lifetime.

Theories of the Galilean satellites, such as those generated by Lieske (1977, 1980), Vu (1977), and Thuillot and Vu (1983), can provide the necessary accuracy over an extended time period. Recent observational data (Pascu, 1992) indicate a scale discrepancy for the Galilean satellite theory of Lieske. This could be due to observational inaccuracies or a system mass error.

For the outer satellites of Jupiter, a numerical integration provides better accuracy than theoretical expressions. Rohde (1992) has fit all observations of the outer satellites with a new integration which is included in the Satellite Almanac floppy disk. Bukova and Shikhalev (1986) have investigated high accuracy numerical theories of the outer satellites of Jupiter (VI-XII) and Saturn (VI-IX). The orbital parameters of the small inner satellites of Jupiter, discovered by Voyager, are given by Synnott (1984).

Duriez and Vienne (1991) and Vienne and Duriez (1991) have developed a general theory of the motion of the eight major satellites of Saturn using the methods of Duriez (1979) and Laskar (1984). For Hyperion disturbed by Triton a numerical procedure of analyzing the Lagrange equations was used. New terms larger than 100 km and 1000 km have been found in the elements of Iapetus and Hyperion, respectively. Taylor (1992) has developed a synthetic theory of the perturbations of Hyperion by Titan and added solar perturbations and the motion of the orbital plane. The resulting theory was fit to observations. Elements for the smaller Saturnian satellites have been given by Reitsema (1981) and Synnott et al (1981, 1983).

The co-orbital satellites of Saturn were discovered in 1966, but only understood from ground based and Voyager observations in 1980. Explanations of their motions have been given by Harrington and Seidelmann (1981), Dermott and Murray (1981a,b), Yoder et al (1983) and Broucke (1987). Until recently, these satellites had only been observed from the ground during the ring plane crossings in 1966 and 1980, and during the two Voyager encounters. Recent observations using infrared detectors, when the northern portion of the rings is blocked by the planet, have provided a better understanding of the motions of these objects (Nicholson et al, 1992). The librational satellites of Saturn have been observed since 1980 and data reduction analyses are needed on these satellites.

Laskar and Jacobson (1987) have developed an analytical ephemeris for the first five satellites of Uranus. A complete semi-analytical theory for the motion of the five greatest satellites of Uranus is given by Lazzaro (1991). The orbits of the ten small satellites of Uranus were determined by Owen and Synnott (1987). The orbits of Triton and Nereid were numerically integrated and fit to Earth-based astrometric observations and some Voyager spacecraft observations. In addition, Jacobson (1990) determined the mass, second zonal harmonic, and pole orientation of Neptune. The orbits of the newly discovered satellites of Neptune were determined by Owen et al (1991).

The combination of speckle interferometric observations and mutual event data provided the bases for an orbit solution for Pluto's satellite, Charon, which gave a semi-major axis of $19640 \pm 320$ km and an inclination of $98°.3$. The diameters were determined to be 2305 and 1230 km for Pluto and Charon, respectively. The orbital radius and period imply a system mass of $1.35 \pm 0.07 \times 10^8$ inverse solar mass (Beletic et al 1989).

The Satellite Almanac, prepared by Rohde at the U S Naval Observatory, contains accurate ephemerides and satellite data on a floppy disk for MS/DOS machines.

## 4. Artificial Satellite Theories

Artificial satellite theories present an interesting problem. The catalog of tracked objects now exceeds 7000 objects. Due to the fact that drag changes their orbits, in many cases quickly, the elements of these objects are valid for a rather limited time period. The current observational accuracy for many of these objects is in the kilometer range. Frequent numerical integrations of the orbits of so many objects for short time periods is not computer effective. Where special accuracies are required, numerical integration does fulfill the requirements, but in other cases that's not necessary.

Artificial satellite theories were originally developed by Brouwer (1959), Garfinkel (1959), Vinti (1959, 1961), and Kozai (1959). Brouwer and Hori (1961) expanded the Brouwer theory to include a theoretical evaluation of atmospheric drag. Lyddane (1963) modified the Brouwer theory for cases of small eccentricities and inclinations and Kozai (1962) extended it to second order in short period terms and third order secular terms. Giacaglia (1964) extended Brouwer's theory by including higher order terms in zonal harmonics expansions. Hoots (1981) reformulated Brouwer's Theory.

There have been many more investigations or developments of artificial satellite theories: Liu and Alford (1980), Hoots and France (1987), and Henrard (1976) to mention a few. There have been investigations of adding the tesserals to the force function to improve the accuracy, and of the tesseral problems with resonances by Gedeon (1969) and Wnuk (1988). Simplifications or improved formulations have been developed by Deprit (1981) and Saari (1974). Methods for modelling drag effects on the satellite orbits have been investigated by Mittleman and Jezewski (1982), Segan (1988), Watson, Mistretta and Bonavito (1975), and many others.

However, in the U. S. the primary programs generally used for artificial satellite theories are still based on modifications of the early theories by Brouwer and Kozai. The accuracies are primarily limited by the problem of predicting the effects of drag and the truncations of the force function. A combination of resistance to change and a scientific failure to address the practical realities of the problems also contributes to the problem.

## 5. Representative Ephemerides

With the present memory and cycle times for computers, ephemeris look up on an external storage device may not be desirable. Rather, a representative theory as part of the program can quickly compute the ephemeris data for any date and time, without interpolation.

Thus, it is desirable, either from a general theory or a numerically integrated ephemeris, to develop a representative ephemeris using polynomials, or a combination of a mean orbit and polynomials, in order to achieve efficient and accurate methods of determining positions of the bodies. Chebychev polynomials were utilized for the distribution of development ephemerides (Newhall 1989). The Almanac for Computers (Kaplan et al, 1976), Connaissance, des Temps (1979) and Compact Data for Navigation (Yallop 1981) utilized Chebyshev polynomials or power series expansions to provide efficient methods of computing positions. A combination of a mean orbit and Chebyshev polynomial representation of the differences from the mean orbit, along with a compacting scheme for all the data, was used by Kammeyer (1989) to generate compressed ephemerides. This scheme is being used in the Multiyear Interactive Computer Almanac (MICA) being developed at the U. S. Naval Observatory. Such representative ephemerides can be tailored according to the requirements for accuracy and the period of validity.

## 6. Chaos, Stability and Evolution

Long time period investigations of the motions of the solar system objects are the subject of many papers and cannot be adequately covered in this review. Investigations such as Roy et al (1988), Milani et al (1989), Nobili et al (1989), Wisdom and

Holman (1991), Innanen and Mikkola (1987), and Kovalevsky (1987), to mention only a few, are designed to examine the long-term stability or repeatability of the motions of the solar system bodies and to explain the observed features of those motions. Milani (1987) gives a review of work on the long term changes of the orbital elements of the planets. Laskar (1986, 1989, 1990) has investigated the chaotic behavior of the solar system, except for Pluto. Due to the limited accuracies of the available initial conditions for the bodies and the averaging methods required for the computation, these investigations actually reveal the characteristics of model solar systems which can only approximate the real solar system. The definitions and meaning of the terms chaos, stability and evolution, as they relate to the real solar system, require further consideration.

## 7. Future for General Theories

The future for computers certainly can be anticipated; computer capabilities will improve with time, memories will be larger and cheaper, execution time will be faster, vector and parallel processing will be improved, displays and graphics will be better, and software capabilities, including algebraic manipulation, will provide improved capabilities. However, the increased capabilities of computers won't mean a universal solution to all the problems of celestial mechanics.

For the development of general theories, the challenge is how to provide increased accuracy with reduced expansions. Deprit (1981) has been working on this problem with his transformations for the elimination of the parallax and the relegation of the node. Can these transformations be effectively applied to the lunar theory? Using mapping techniques investigated by Coffey and Deprit (1980), techniques for combining terms, and methods to minimize arithmetic operations, Miller (1989) has developed a LISP program which optimizes a Fortran program for evaluating Fourier series. There have been a number of investigations of the use of elliptic functions and elliptic integrals for planetary theories (Richardson 1982, Williams et al 1987, Williams 1992). The hope is that small divisors can be avoided by this technique and that algebraic manipulators can be developed to integrate the expressions. Message (1987) developed planetary perturbation theory from Lie series, including resonance and critical arguments. Near resonance conditions present difficulties in this development.

Bretagnon (1990) has developed an iterative method for the construction of a general planetary theory and Bretagnon and Simon (1990) have used the iterative method for the Sun-Jupiter-Saturn case. They find long period terms in the semimajor axis which produce long period terms in the mean longitude. The question remains whether the iterative method of Bretagnon is subject to the convergence problems experienced in the iterative method of Seidelmann (1970). Is there a way of overcoming the convergence problem and controlling the number of terms? Are there special mathematical functions to be used in place of Fourier series to reduce the expansion? Can algebraic manipulators be used to develop an efficient compact theory? It is clear that the use of computers in the future will prefer the use of general theories over numerical integrations for realizing positions and velocities. The challenge is to find effective, efficient accurate methods of developing such theories.

## 8. Future for Numerical Integration

Accuracy improvements based on numerical integration methods will be dependent on improved force models and improved observational data. For force models the PPN relativistic expression needs to be expanded to second order. The present first order terms vary in significance and include terms that are smaller than some second order terms. There is a choice between doing a fully parameterized theory, which involves solutions based on the observational data, and a theoretical development based on the complete adoption of the theory of general relativity. In other words, should we stop conducting tests of the general theory of relativity and instead fully incorporate general relativity into the calculations for the force function, observations and their reduction, coordinate systems and their transformations, time scales, and units? The force model also needs to include all the known bodies in the solar system, particularly the large number of minor planets. Consideration must be given to including the Kuiper belt, the Oort cloud, the galaxy, and other forces, which heretofore have been ignored. What else has been neglected, or truncated, to the accuracies of previous computations that should be incorporated for improved accuracies?

For observational data, we have the challenge of whether we can better utilize existing data. Are there systematic errors in the old observations? Can we properly weight the observations to combine the different types of observational data, including spacecraft, radar and optical observations? How can we achieve more accurate observations in the future? The optical interferometer, under development, will achieve much more accurate stellar observations. Unfortunately, it cannot be used to observe extended bodies, like planets. Can it observe the satellites and achieve improved accuracies for the positions of the planets? Can radar observations of planets and satellites be obtained on a systematic basis for ephemerides?

CCD arrays provide the opportunity to observe over wide dynamic ranges and thus bright satellites could be observed with respect to faint stars or quasars. Could we plant transponders on planets and observe the positions of these planets as radio sources by VLBI to a better accuracy? Budgetary considerations make it unlikely that we will achieve a large number of spacecraft observations in the foreseeable future.

## 9. Future for the Artificial Satellite Problem

The artificial satellite problem also presents an interesting challenge. On two missions, space shuttle has maneuvered to avoid a collision with another object. The proposed space station will not be maneuverable, nor can the Hubble Space Telescope be maneuvered. There is a large amount of debris in space and when the damage that a centimeter cubed object at orbital velocities can do is considered, it becomes a significant concern. Today, tracking is only achieved down to half-cubic meter sized material. Thus, the need is to obtain more accurate observations and of smaller objects, increasing the number of objects to be tracked from 7000 to an estimated 50,000. The theory being used for the catalog of elements of artificial satellite orbits is Brouwer's (1959). Can better theories, which will provide elements over longer time periods, even in the presence of drag, be developed? Can we develop means of predicting drag, or for monitoring drag, on a real time basis? There is certainly a need for a greatly improved method of predicting the motions of artificial satellites.

The Global Positioning System (GPS) is on the threshold of becoming operational, along with GLONASS. GPS capabilities were dramatically demonstrated in Desert Storm. Achieving the full accuracy potentials of these systems requires proper incorporation of the general theory of relativity, in addition to resonance terms in the ephemeris, and correct processing of accurate observational data.

## 10. Conclusion

There is a continuous competition between the accuracy of observations and ephemerides. As more accurate observations have been achieved, the requirement for more accurate ephemerides has followed. Today, lunar laser ranging can provide precisions of one centimeter. The ephemeris predictions are not that accurate. The future holds potential for more accurate observations of the planets. Accurate ephemerides will have to follow. The accuracies of general theories are currently limited by our mathematical capabilities, not by the computer equipment, not by the observational accuracy. Rather we need to find better means of developing general theories. Numerical integrations, on the other hand, are limited by the force model and observational data. The artificial satellite situation requires computer efficient improved theories, observations, drag prediction, and orbital predictions. I think it's time we adopt the general theory of relativity for the field of celestial mechanics and continue the improvement process based on observations, theories, and computers.

## References

Beletic, J. W., Goody, R.M., and Tholen, J.D. (1989) "Orbital Elements of Charon from Speckle Interferometry," Icarus 79, 38 - 46.

Bretagnon, P. (1980) "Theorie au deuxieme ordre des planetes interieures," Astron. Astrophys. 84, 329-341.

Bretagnon, P (1981) "Construction d'une theorie des grosses planetes par une methods iterative," Astron. Astrophys. 101, 342-349.

Bretagnon, P. (1982) "Theorie du mouvement de l'ensemble des planetes. Solution VSSOP82," Astron. Astrophys. 114, 278-288.

Bretagnon, P. (1990) "Methode iterative de construction d'une theorie general planetaire," Astron. Astrophys 231, 561-570.

Bretagnon, P. and Simon, J.L. (1990) "Theorie general du couple Jupiter - Saturne par une methode iterative," Astron. Astrophys. 239, 387-398.

Brown, E.W. (1919), Tables of the Motion of the Moon (Yale Univ. Press. New Haven, CT)

Broucke, R. and Konopliv, A. (1987) "Some models for the Motion of the Co-orbital Satellites of Saturn," In Roy, A.E. (ed) Long-Term Dynamical Behaviour of Natural and Artificial N-Body Systems, Kluwer A.P., Dordrecht, 155-169.

Brouwer, D. (1959) "Solution of the Problem of Artificial Satellite Theory Without Drag," Astron J. 64, 378-397.

Brouwer, D. and Hori, G. (1961) "Theoretical Evaluation of Atmospheric Drag Effects in the Motion of an Artificial Satellite," Astron J. 66, 193.

Bykova, L.E. and Shikhalev, V.V. (1986) "Developing High Accuracy Numerical Theories of Motion of Outer Satellites of Jupiter and Saturn," Relativity in Celestial Mechanic and Astrometry, (J. Kovalevsky and V.A. Brumberg eds), Kluwer Academic Publishers, Dordrecht.

Carpino, M., Milani, A, and Nobili, A.M. (1987) "Long-term Numerical Integrations and Synthetic Theories for the Motion of the Outer Planets," Astron. Astrophys. 181 182-194.

Chapront-Touze, M. (1990) "Orbits of the Martian Satellites from ESAPHO and ESADE Theories," Astron. Astrophys. 240, 159- 172.

Chapront-Touze, M. (1988) "ESAPHO: a semi-analytical theory for the orbital motion of Phobos," Astron. Astrophys. 200, 255- 268.

Chapront-Touze, M. and Chapront, J. (1980) "Les perturbations planetaires de la Lune," Astron. Astrophys. 91, 233-248.

Chapront-Touze, M. and Chapront J. (1983). "The Lunar Ephemeris ELP 2000," Astron. Astrophys. 124, 50-62.

Chapront-Touze, M. and Chapront J. (1988) "ELP 200-85: "A Semi-analytical Lunar Ephemeris Adequate for Historical Times," Astron. Astrophys. 190, 342-352.

Coffey, S. and Deprit, A. (1980) "Fast Evaluation of Fourier Series," Astron. Astrophys. 81, 310-315.

Deprit, A. (1981) "The Elimination of the Parallax in Satellite," Celest. Mech. 24, 111.

Deprit, A., Henrard, J. and Rom, A. (1971), "Analytical Lunar Ephemeris: I Definition of the Main Problem," Astron. Astrophys. 10 257.

Deprit, A., Henrard, J. and Rom, A. (1971) "Analytical Lunar Ephemeris: Delaunay's Theory" Astron. J. 76, 269.

Deprit, A., Henrard, J. and Rom, A. (1971) "Analytical Lunar Ephemeris: The Variational Orbit," Astron. J. 76, 273.

Dermott, S.F. and Murray, C.D. (1981) "The Dynamics of Tadpole and Horseshoe Orbits I Theory" Icarus 48, 1-11; "II The Coorbital Satellites of Saturn," Icarus 48, 12-22.

Duriez, L. (1979) Approche d'une Theorie Generale Planetaire en variable elliptiques heliocentriques, These, Lille.

Duriez, L. and Vienne, A (1991) "A general theory of motion for the eight major satellites of Saturn, I. Equations and method of resolution," Astron. Astrophys. 243, 263.

Garfinkel, B. (1959) "The Orbit of Satellite of an Oblate Planet," Astron. J. 64, 378-397.

Gedeon, G.S. (1969) "Tesseral Resonance Effects on Satellite Orbits," Celest. Mech. 1, 167.

Giacaglia, G. (1964) "The Influence of High Order Zonal Harmonies on the Motion of an Artificial Satellite Without Drag," Astron J. 69, 303.

Goodrich, E. F. and Carpenter, L. (1966) "Computation of General Planetary Perturbations for Resonance Cases," Goddard Space Flight Center X643-66-133.

Harrington, R.S. and Seidelmann, P.K. (1981) "The Dynamics of the Saturn Satellites 1980 S1 and 1980 S3," Icarus 47, 97-99.

Henrard, J. (1976) "On the Artificial Satellite Theory," Celest. Mech. 14, 331.

Henrard, J (1979) "A New Solution to the Main Problem of Lunar Theory," Celest. Mech. 19, 337.

Hoots, F.R and France, R.G. (1987) "An Analytic Satellite Theory Using Gravity and a Dynamic Atmosphere," Celest. Mech. 40, 1- 18.

Hoots, F. (1981), "Reformulation of the Brouwer Geopotential Theory for Improved Computational Efficiency," Celest. Mech. 24, 367.

Innanen, K. A. and Mikkola, S. (1987) "Where are the Saturnian Trojans?" In Roy, A. E. (ed) Long-Term Dynamical Behaviour of Natural and Artificial N-Body Systems; Kluwer Academic Publishers, Dordrecht, 21-26.

Ivanov, N. M., Kolyuka, Yu F, Kudryavotsev, S.M. and Tikhonov, V.F. (1988) "The Motion of Phobos: orbit parameters from ground- based astrometry and the Mariner 9 and Viking 1, 2 data:," Pis'me Astron Zh 14, 956-960, Sov. Astron. Lilt 14, 5.

Jacobson, R.A. (1990) "The Orbits of the Satellites of Neptune," Astron. Astrophys. 231, 241-250.

Kammeyer, P. (1989) "Compressed Planetary and Lunar Ephemerides," in Applications of Computer Technology to Dynamical Astronomy, (P.K. Seidelmann and J. Kovalevsky eds) Kluwer Academic Publishers, Dordrecht, 311-316.

Kaplan, G.H., Doggett, L.E. and Seidelmann, P.K. (1976) Almanac for Computers, USNO Circular 155.

Kovalevsky, J. (1987) "Orbital Evolution" In Roy A.E. (ed) Long - Term Dynamical Behaviour of Natural and Artificial N-Body Systems, Kluwer A.P., Dordrecht, 27-46.

Kozai, Y. (1959) "The Motion of a Close Earth Satellite," Astron. J. 64, 367-377.

Kozai, Y. (1962) "Second-order Solution of Artificial Satellite Theory Without Air Drag," Astron. J. 67, 446.

Laskar, J. (1984) "Theorie generale planetaire: elements orbitaux des planetes sur un million d'annes," These de Troisieme cycle, Observatoire de Paris.

Laskar, J. (1986) "Secular Terms of Classical Planetary Theories Using the Results of General Theory," Astron. Astrophys. 157, 59-70.

Laskar, J. (1989) "A Numerical Experiment on the Chaotic Behaviour of the Solar System," Nature 338 237-238.

Laskar, J. (1990) "The Chaotic Motion of the Solar System: A Numerical Estimate of the Size of the Chaotic Zones," Icarus 88, 266-291.

Laskar, J. and Jacobson, R.A. (1987) "An Analytical Ephemeris of the Uranian Satellites," Astron. Astrophys. 188, 212-224.

Lazzaro, D. (1991) "Semi-Analytical Theory for the Motion of Uranus' Satellites," Astron. Astrophys. 250, 253-265.

Leverrier, V. J.J. (1858) Ann Obs Paris 4, 1.

Leverrier, V. J.J. (1859) Ann Obs Paris 5, 1.

Leverrier, V. J.J. (1861a) Ann Obs Paris 6, 1.

Leverrier, V. J.J. (1861b) Ann Obs Paris 6, 185.

Lieske, J.H. (1977) "Theory of Motion of Jupiter's Galilean Satellites," Astron. Astrophys. 56, 333.

Lieske, J.H. (1980) "Improved Ephemerides of the Galilean Satellites," Astron. Astrophys. 82, 340-348.

Liu, J.J.F. and Alford, R.L. (1980) "Semianalytic Theory for a Close-Earth Artificial Satellite," J. Guidance and Control 3, 304-11. (Errata 1981 Guidance and Control 4).

Lyddane, R.H. (1963) "Small Eccentricities or Inclinations in the Brouwer Theory of the Artificial Satellite," Astron J. 68, 555.

Message, P.J. (1987) "Planetary Perturbation Theory from Lie Series, including Resonance and Critical Arguments" In Roy A.E. (ed) Long-Term Dynamical Behaviour of Natural and Artificial N-Body Systems, Kluwer A.P., Dordrecht, 47-72.

Milani, A. (1987) "Secular Perturbations of the Planetary Orbits and their representation as Series," In Roy, A.E. (ed) Long- Term Dynamical Behaviour of Natural and Artificial N-Body Systems, Kluwer A.P., Dordrecht, 73-108.

Milani, A., Nobili, A.M., and Carpino, M. (1989) "Dynamics of Pluto," Icarus 82, 200-217.

Miller, B. R. (1989) "A Program Generator for Efficient Evaluation of Fourier Series," In proceedings of ACM-SIGSAM International Symposium on Symbolic and Algebraic computation; ISSAC-89, p. 199-206, ACM Press.

Mittleman, D. and Jezewski, D. (1982) "An Analytic Solution to the Classical Two Body Problem with Drag," Celest. Mech. 28, 401.

Newcomb, S. (1898) Astronomical Papers American Ephemeris VI (U S Government Printing Office, Washington, DC).

Newhall, XX, Standish, E.M. and Williams, J.G. (1983) "DE102: A Numerically Integrated Ephemeris of the Moon and Planets Spanning Forty-Four Centuries," Astron. Astrophys. 125, 150- 167.

Newhall, XX (1989) "Numerical Representation of Planetary Ephemerides," Celest. Mech 45, 305-310.

Nicholson, P.D., Hamilton, D.P., Mathews, K. and Yoder, C.F. (1992) "New Observations of Saturn's Coorbital Satellites," Icarus in press.

Nobili, A.M., Milani, A., and Carpino, M. (1989) "Fundamental Frequencies and Small Divisors in the Orbits of the Outer Planets," Astron. Astrophys. 210, 313-336.

Owen, W.M., Jr and Synnott, S.P. (1987) "Orbits of the Ten Small Satellites of Uranus," Astron J. 93, 1268.

Owen, W.M. Jr., Vaughan, R.M. and Synnott, S.P. (1991) "Orbits of the Six New Satellites of Neptune," Astron J. 101, 1511- 1515.

Pascu, D. (1992) private communication.

Reitsema, H.J., (1981) "The Libration of the Saturnian Satellite Dione B," Icarus 48, 23-28.

Richardson, D. (1982) "A Third Order Intermediate Orbit for Planetary Theory," Celest. Mech. 26, 187-195.

Rohde, J.R. (1992) "Outer Satellites of Jupiter," unpublished.

Roy, A.E., Walker, I.W., MacDonald, A.J., Williams, I.P., Fox, K., Murray, C.D., Milani, A., Nobili, A.M., Message, P.J., Sinclair, A.T., and Carpino, M. (1988) "Project Longstop" Vistas in Astronomy 32, 95-116.

Saari, D. (1974) "Regularization and the Artificial Earth Satellite Problem," Celest. Mech. 9, 55.

Segan, S. (1988) "Analytical Computation of Atmospheric Drag Effects," Celest. Mech. 41, 381.

Seidelmann, P.K. "An Interactive Method of General Perturbations Programmed for a Computer," Celest. Mech. 2, 134.

Sharaf, Sh. G (1955) "Theory of Motion of the Planet Pluto pt 1" Transactions of the Institute of Theoretical Astronomy, 4 USSR Academy of Sciences Press, Moscow-Leningrad, 1955, NASA Tech Translation F490, Jan 1969.

Sharaf, Sh. G and Budnikova, N.A. (1964) "Theory of Motion of the Planet Pluto pt 2,3,4" Transactions of the Institute of Theoretical Astronomy 10, "Nauka" Press, Moscow-Leningrad 1964, NASA Tech Translation F-491, Jan 1969.

Simon, J.L. (1983) "Theorie du mouvement des quatre grosses planetes. Solution TOP82," Astron. Astrophys. 120, 197-202.

Simon, J.L. and Bretagnon, P. (1975) "Perturbations du premier ordre des quatre grosses planetes. Variations litterales," Astron. Astrophys. 42 259-263.

Simon, J.L. and Bretagnon, P. (1975) "Resultats des perturbations du premier ordre des quatre grosses planetes Variations Litterales," Astron Astrophys. Suppl 22, 107-160.

Simon, J.L. and Bretagnon, P. (1978). "Perturbations du deuxieme ordre des quatre grosses planetes. Variations seculaires du demi grand axe," Astron. Astrophys. 69, 369-372.

Simon, J.L. and Bretagnon, P., (1978). "Resultats des perturbations du deuxieme ordre des quatre grosses planetes," Astron. Astrophys. Suppl. 34, 183-194.

Simon, J.L. and Bretagnon, P. (1984). "Theorie du mouvement de Jupiter et Saturne sur un intervalle de temps de 6000 ans. Solution JASON84," Astron. Astrophys. 138, 169-178.

Sinclair, A.T. (1989) "The Orbits of the Satellites of Mars Determined from Earth-based and Spacecraft Observations," Astron. Astrophys. 220, 321-328.

Standish, E.M. (1982) "The JPL Planetary Ephemerides," Celest. Mech. 26, 191-186.

Synnott, S.P., Peters C.F., Smith, B.A. and Marabito, L.A. (1981) "Orbits of The Small Satellites of Saturn," Science 212, 191.

Synnott, S.P., Terrile, R.J., Jacobson, R.A. and Smith, B.A. (1983) "Orbits of Saturn's F Ring and its Shepherding Satellites" Icarus 53, 156-158.

Synnott, S.P. (1984) "Orbits of the Small Inner Satellites of Jupiter," Icarus 58, 178-181.

Taylor, D.B., (1992) "A Synthetic Theory for the Perturbations of Titan on Hyperion," Astron. Astrophys. (in press).

Thuillot, W. and Vu, D.T. (1983) "Analytical Theory of the Motion of the Galilean Satellites of Jupiter. The First Approximation," In: S. Ferraz Mello and P.E. Nacozy (eds) The Motion of Planets and Natural and Artificial Satellites, Universidade de Sao Paulo, 273.

Vienne, A. and Duriez, L. (1991) "A General Theory of Motion for the Eight Major Satellites of

Saturn, II. Short-period Perturbations," Astron. Astrophys. 246, 619-633.

Vinti, J.P. (1959) "A New Method of Solution for Unretarded Satellite Orbits, J. Res. NBS 63 B; Math Phys 2.

Vinti, J.P. (1961) "Theory of an Accurate Intermediary Orbit for Satellite Astronomy," J. Res NBS 65 b; Math. Math. Phys. 3, 169-201.

Vu, D.T. (1977) "Une Nouvelle Presentation des Ephemerides des Satellites Galileans de Jupiter," Astron. Astrophys. Suppl. 30, 361-367.

Watson, J.S., Mistretta, G.D. and Bonavito, N.L. (1975) "An Analytical Method to Account for Drag in the Vinti Satellite Theory," Celest. Mech 11, 145.

Williams, C.A., Van Flandern, T.C. and Wright, E. (1987) "First Order Planetary Perturbations with Elliptic Functions," Celest. Mech. 40, 367-391.

Williams, C.A. (1992) "A Planetary Theory with Elliptic Functions an Elliptic Integrals Exhibiting No Small Divisors," Chaos meeting in Angra dos Reis.

Wisdom, J. and Holman, M. (1991) "Symplectic Maps for the N-Body Problem," Astron. J. 102, 1528.

Wnuk, E. (1988) "Tesseral Harmonic Perturbations for High Order and Degree Harmonics," Celest. Mech. 44, 179.

Yallop, B.D. (1981) "Compact Data for Navigation and Astronomy 1981-1985," Royal Greenwich Observatory Bulletin No. 185.

Yoder, C.F., Colombo, G., Synnott, S.P. and Yoder, K.A. (1983) "Theory of Motion of Saturn's Coorbiting Satellites," Icarus 53, 431-443.

# THE MYSTERY OF PLUTO'S MASS

## —THE RING HYPOTHESIS—

C. MARCHAL

*DES – ONERA – 92320 CHATILLON – FRANCE*

**Abstract.** After a short presentation of the Pluto-Charon system and the history of its mass deter-minations some first reasons are presented that support the existence of a ring of billions of small satellites about Pluto up to tenths of millions of kilometers.

The stability, the shape and the dimensions of such an heavy ring are discussed.

Finally a general review of advantages and drawbacks of this ring theory is presented as well as the possibilities of detection of the eventual Pluto's ring.

**Key words:** Celestial Mechanics – ring – planet Pluto

## 1. Introduction

The existence of planets Neptune and Pluto was suspected because of unexplained perturbations in the motions of other planets. However, in spite of a very good tentative orbit given by Lowell in 1915, the mass of Pluto given by the planetary perturbations is at least 30 times larger than the mass obtained with the motion of its satellite.

The missing mass has been looked for in a tenth planet, in a binary companion to the Sun etc..., but without much success.

Is it possible that this missing mass is in the vicinity of Pluto ? This planet would be a large aborted planet and its initial cloud of matter would have been unable to condense because of a too fast rotation.

There would remain a large and flat ring of small satellites up to millions of kilometers from the planet...

## 2. The Planet Pluto and its Satellite Charon

A recent series of occultations and eclipses in the Pluto-Charon system has allowed to improve very much our knowledge as indicated in the table I from Burwitz et al (1991).

It seems that Pluto and Charon are in a state of co-rotation and face each other continuously with the same hemisphere, as the near side of the Moon facing the Earth. The "pole of positive rotation" of this bound system, usually called North pole, is oriented towards the ecliptic latitude of $-18°.0$ and the ecliptic longitude of $134°.3$.

The history of our ideas on the mass of Pluto's system is presented in table II from Duncombe and Seidelmann (1980). The general tendency is a very fast decrease, Pluto was very much overestimated.

*Celestial Mechanics and Dynamical Astronomy* **56**: 13–26, 1993.

TABLE I

Physical parameters and orbital elements of the Pluto-Charon system

|  | Pluto | Charon |
|---|---|---|
| Radius (km) | $1151 \pm 20$ | $591 \pm 11$ |
| Distance Pluto-Charon (km) | $19640 \pm 320$ | |
| Mean Density (g/cm3) | $2.032 \pm 0.040$ | |
| Mass ($10^{22}$ kg) | $1.47 \pm 0.07$ | |
| Mass (Earth = 1) | $0.00246 \pm 0.00012$ | |
| Albedo | $0.618 \pm 0.020$ | $0.372 \pm 0.012$ |
| Inclination on Pluto's orbit | $114°.4$ | |
| Sidereal orbital period (days) | $6.387244 \pm 0.000007$ | |

TABLE II

Mass determinations of planet Pluto

| Date | Author | Observations | Mass (Earth = 1) |
|---|---|---|---|
| 1848 | J. Babinet | Neptune | 12 |
| 1899 | H. Lau | Uranus | 9 |
| 1908 | W. Pickering | Uranus | 2 |
| 1909 | B. Gaillot | Uranus, Neptune | 5 |
| 1915 | P. Lowell | Uranus, Neptune | 6.6 |
| 1928 | W. Pickering | Uranus, Neptune | 0.75 |
| 1930 | Discovery of Pluto by Clyde Tombaugh | | |
| 1930 | J. Jackson | Neptune | 1.0 |
| 1931 | Nicholson and Mayall | Neptune | 0.94 |
| 1931 | E. Brown | Uranus | 0.5 |
| 1940 | V. Kourganoff | Uranus | 1.0 |
| 1942 | L. Wylie | Neptune | 0.91 |
| 1951 | Eckert, Brouwer, Clemence | Neptune | 1.0 |
| 1955 | Brouwer | Uranus, Neptune | 0.82 |
| 1960's | Availability of high-speed computing equipments | | |
| 1968 | Duncombe, Klepczynski and Seidelmann | Neptune | 0.18 |
| 1971 | Duncombe, Jackson, Klepczynski and Seidelmann | Neptune | 0.11 |
| 1971 | Ash, Shapiro, Smith | Uranus, Neptune | 0.08 |
| 1976 | Cruikshank, Pilcher, Morrison | Albedo | 0.004 |
| 1978 | Discovery of Charon by Christy and Harrington | | |
| 1978 | Christy, Harrington | Satellite Charon | 0.0025 |

TABLE III
Pluto's orbital elements The true orbit and the two tentative orbits.

|  | True orbit | Lowell (1915) | Pickering (1919) |
|---|---|---|---|
| Semi-major axis | 39.72 A.U. | 43.0 | 55.1 |
| Eccentricity | 0.2523 | 0.202 | 0.31 |
| Inclination | 17°.14 | 22°.1 | 15° |
| Longitude of node | 109°.51 | 92°.0 | 100° |
| Argument of perihelion | 112°.99 | 112°.9 | 180°.1 |
| Longitude 1930.0 | 108°.5 | 102°.7 | 102°.6 |
| Latitude 1930.0 | −0°.3 | 4°.3 | 1°.1 |
| Distance 1930.0 | 41.41 A.U. | 42.96 | 72.15 |

The finest analysis of the motions of Uranus and Neptune have finally led to a Pluto's mass of 0.08 to 0.11 Earth-mass in obvious contradiction with the much smaller mass given in table I.

This situation leads most people to consider that Pluto is not the planet of Lowell and Pickering: it was at the good place only by chance. Many astronomers are or have been looking for a tenth planet that would resolve the contradiction (Seidelmann and Harrington 1988). However, notice the following:

**A)** The research of an unknown tenth planet has led to a systematic survey of the sky. We now know that there is no such planet, up to the magnitude 17, at less than 35° from the ecliptic plane and, up to the magnitude 16, at less than 43°; while tiny Pluto has already the magnitude 15, (Tombaugh 1992). However some small doubts remain for the southern half of the Milky Way.

With the magnitude 16 the planet Neptune would be at about 200 A.U. from the Sun, instead of 30, and the Earth would be at 75 A.U. They would be 25 % further away with the magnitude 17. It becomes difficult to find the suitable missing mass that would explain the motions of Uranus and Neptune.

**B)** If we look for the tentative orbit given by Lowell (table III, from Seidelmann and Harrington 1988 and from Seidelmann et al 1980) we find it really very near to the true Pluto's orbit, while the same comparison for planet Neptune (table IV, from the same authors) gives a good agreement between the two tentative orbits, of Adams and of Leverrier, but these two orbits are both very far from the true Neptune orbit. . .even if they have given a good observed position at the time of discovery.

The Lowell orbit is excellent and if we add that it leads to a perihelion passage at 1991.2 while the true perihelion passage has occurred at 1989.68 we obtain such a number of "coincidences" that it becomes difficult to believe to a pure chance.

In these conditions it is possible to imagine that the mass missing in Pluto and Charon lies in a large and flat ring surrounding the Pluto-Charon system up to a distance of 10 millions of kilometers or even more.

TABLE IV
Neptune's orbital elements The true orbit and the two tentative orbits.

|  | True orbit at 1846.7 | Leverrier | Adams |
|---|---|---|---|
| Semi-major axis | 30.109 A.U. | 36.15 | 37.25 |
| Eccentricity | 0.0090 | 0.10761 | 0.12062 |
| Inclination | 1°.78 | not considered | |
| Longitude of node | 130°.10 | (very small inclinations) | |
| Longitude of perihelion | 45°.97 | 284°.75 | 299°.2 |
| Longitude 1846.7 | 327°.40 | 326°.53 | 329°.6 |
| Latitude 1846.7 | −0°.53 | Very small | Very small |
| Distance 1846.7 | 30.347 A.U. | 33.08 | 33.26 |
| Mass (Earth = 1) | 17.24 | 35.73 | 49.94 |

## 3. The Long-Term Stability of Rings and Discs

The most common objection presented to ring or disc systems of millions of small satellites is their lack of stability.

The close encounters and the collisions lead either to the ejection and the escape of many particles, and thus to a continuous loss of matter, or to the formation of massive bodies, especially at the center of the ring. These bodies will attract the surrounding particles and will finally absorb the system completely. However, we must consider the time-scale of these phenomena.

The natural rings of Jupiter, Saturn, Uranus, Neptune are in a situation very different from this of an eventual Pluto ring; they are within the Roche limit of a large and very massive body that prevent the formation of massive satellites inside the ring.

Most numerical simulations of a gravitational system of many colliding particles use the model of hard spheres with a given coefficient of restitution $K$, also called Newton coefficient, $K$ is between 0 and 1 with $K = 1$ for elastic collisions (no loss of energy) and $K = 0$ for soft shocks.

The case $K = 1$ gives a kind of Brownian motion of constant energy, without accretion of matter and with a slow loss of particles. Many binary systems are formed (some of them with periodic collisions) as well as several multiple systems.

The mean density of the central part of the cloud increases slowly while the outer part looses its matter inward and outward.

The central zone of large density becomes always denser and denser but also smaller and smaller while the dispersion goes on.

Finally, for ordinary initial energies, very few particles remain together.

The picture of motion is very different if the coefficient $K$ is smaller than one and for the very cold matter of a proto-plutonian cloud it seems realistic to adopt $K$ between 0.5 and 0.9.

If then we give to the proto-plutonian cloud a large angular momentum, as suggested by the fast rotation of all four major planets, we obtain the three following phases of motion.

A) The first phase is very short (20 to 200 collisions per particles, according to $K$), it leads to a very flat and almost planar disc, in a plane normal to the angular momentum.

This flat cloud has already lost a good part of its matter, most of its particles are on almost circular orbits about the center of mass and a central body of many aggregated particles exists.

The disc has a differential rotation: all particles rotate in the same direction but their periods are smaller near the center of mass (as in a flat galactic system).

For given initial size and energy the proportion of lost particles in this first phase doesn't vary very much in terms of the angular momentum but the mass of the central body and its rate of rotation are very sensitive functions of the angular momentum and when this one is large the central body is very small and almost at the limit of disruption.

This first phase of motion can be pictured as the motion of particles in the general gravitational field of the cloud with very few individual interactions save the collisions and their loss of energy.

B) The second phase of motion has a much longer duration, it is dominated by the mutual interactions of nearby particles and three major very slow phenomena happen.

B.1) The disc expands very slowly in its plane and, in the average, the mutual interactions lead the outer particles to go outward and the inner particles to go inward.

B.2) The same more or less chaotic interactions eject from time to time some outer particles and, more rarely, throw an inner particle on the central body.

B.3) The collisions lead to the progressive formation of massive satellites that drain a part of the disc between two nearby radius. The initial disc is thus broken into ringlets.

It is exceptional to obtain two or several such massive satellites on the same circular orbit and they usually play the role of tiny shepherd satellites controlling the nearby ringlets that can be much heavier, at least during the first half of this second phase of motion.

The presence of these shepherd satellites forbids sometimes the apparition of a new aggregated satellite inside the nearby ringlets when this new satellite has a too fast rotation.

C) The second phase of motion can be very long in spite of the apparent fragility of the system of ringlets, but this system is not strictly stable and tends to the third and final state through the chaotic motions appearing from time to time in the ringlets after a close encounter or a collision.

These chaotic motions lead, rarely, a new particle to the shepherd satellites that grow very slowly and finally absorb all the ringlets.

The system of satellites thus obtained is not necessarily stable and may have a troubled history until a final state similar to the saturnian system with many more or less large satellites rotating in the same direction about the central body on almost circular orbits with wide enough separations(i.e. large separations with respect to the distance of their collinear equilibrium Lagrangian points).

These pictures of the evolution of the cloud of matter constituting a proto-planet are only tentative, being based on numerical simulation of only 500 to 10 000 particles. The mass effects also must be emphasized when particles of different masses are considered: the lightest particles are the most subject to be ejected while the proportion of heavy particles is larger in the central body.

We thus arrive to the following picture of the Plutonian system.

This system has a large angular momentum and is in the first half of its second phase of motion with many heavy ringlets separated by light shepherd satellites.

Because of the large angular momentum the central mass is small and its rotation at the limit of disruption has provoked its separation into two bodies, Pluto and Charon, that have precisely the mass ratio of 12 to 15 % corresponding to the Poincaré disruption of an homogeneous rotating body (Marchal 1968). Furthermore their state of co-rotation and their present mutual distance correspond accurately to this same phenomenon after the end of tidal effects.

The existence of Charon leads of course to the instability of too near orbits and there is thus a hole in the middle of Pluto's ring with a radius of at least 30 000 km.

## 4. Shape and Dimensions of the Pluto's Ring

In the previous section we have not considered the outer perturbations and we have thus obtained flat discs, however the outer perturbations, and especially the solar perturbations will disrupt this beautiful symmetry.

Let us study the gravitational motions of concentric rings, with or without a central body, all of them being considered as solid bodies and having both a symmetry of revolution about their polar axis and a planar symmetry about their equatorial plane (this type of symmetry is the symmetry of oblate and prolate ellipsoids and is called "spheroidal symmetry". Fig. 1)

This gravitational problem has many integrals of motion.

A) For reasons of symmetry all centers of mass remain together at the same point that will be chosen as origin of coordinates.

B) Let us call:

**B.1)** $I_{jP}$ and $I_{jE}$ the polar and equatorial moments of inertia of the $j^{\text{th}}$ body.

**B.2)** $\vec{\Omega}_j$ the rotation vector of the $j^{\text{th}}$ body

**B.3)** $\vec{u}_j$ the unit vector of the northern polar direction of the $j^{\text{th}}$ body.

**B.4)** $\Omega_{jP}$ and $\vec{\Omega}_{jE}$ the polar and equatorial components of the rotation vector $\vec{\Omega}_j$:

$$\Omega_{jP} = \vec{u}_j \cdot \vec{\Omega}_j \qquad \vec{\Omega}_j = \Omega_{jP} \vec{u}_j + \vec{\Omega}_{jE} \tag{1}$$

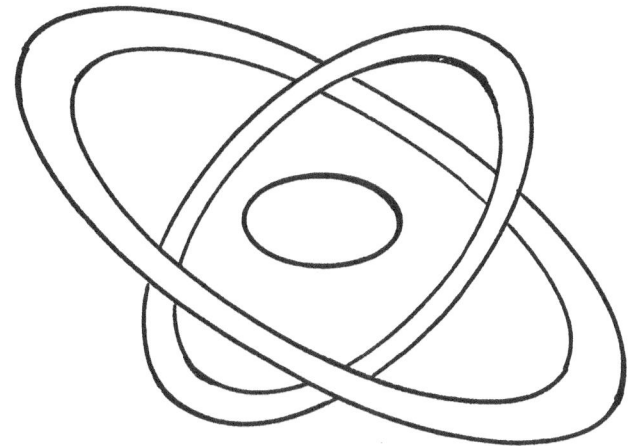

Fig. 1. A central body and several concentric rings with "spheroidal symmetry"

**B.5)** Finally let us call $U$ the gravitational mutual potential of the system of interest:

$$U = G \int dm_1 \, dm_2 / r_{12} \tag{2}$$

with the usual expressions ($G$ is the Cavendish constant, the couple of masses $dm_1$ and $dm_2$, with mutual distance $r_{12}$, must appear one and only one time in the integral).

For given mass distributions and shapes of the different bodies the mutual potential $U$ is only a function of the unit vectors $\vec{u}_j$.

The remaining integrals of motion are then:

The total angular momentum $\vec{c}$:

$$\vec{c} = \sum_j \left( I_{jP} \, \Omega_{jP} \, \vec{u}_j + I_{jE} \, \vec{\Omega}_{jE} \right) \tag{3}$$

The mechanical energy $h$:

$$h = \frac{1}{2} \sum_j \left( I_{jP} \, \Omega_{jP}^2 + I_{jE} \, \Omega_{jE}^2 \right) - U \tag{4}$$

The polar components $\Omega_{jP}$ of all rotation vectors (because the torque on a given body is in its equatorial plane).

If we call $\vec{c}_j$ the angular momentum of the $j^{\text{th}}$ body we obtain the following equations of motion:

$$\vec{c}_j = I_{jP} \, \Omega_{jP} \, \vec{u}_j + I_{jE} \, \vec{\Omega}_{jE} \quad ; \quad \vec{c} = \Sigma \, \vec{c}_j \tag{5}$$

C. MARCHAL

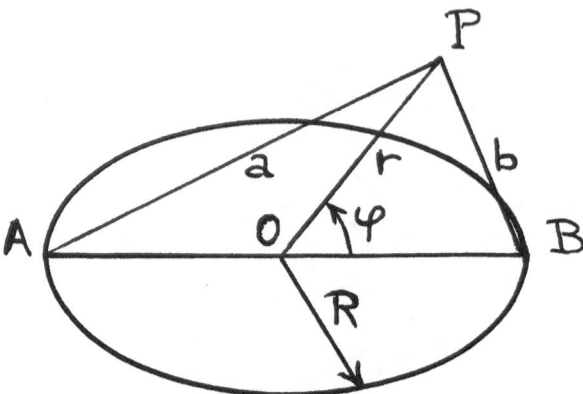

Fig. 2. The potential of a circle of mass $m$ and radius $R$ (with uniform distribution of masses)

$$\begin{cases} \mathrm{d}\vec{c}_j/\mathrm{d}t = \vec{u}_j \times \partial U/\partial \vec{u}_j \\ \mathrm{d}\vec{u}_j/\mathrm{d}t = \vec{\Omega}_j \times \vec{u}_j = \vec{c}_j \times \vec{u}_j/I_{jE} \end{cases} \tag{6}$$

These equations lead of course usually to complex motions, because $U$ is a complex function, however there are very simple cases, the cases in which, for given integrals $\vec{c}$ and the $\Omega_{jP}$, *the mechanical energy $h$ is minimum.*

The motion of all $\vec{u}_j$, $\vec{\Omega}_j$, $\vec{\Omega}_{jE}$ and $\vec{c}_j$ vectors is then a pure and constant rotation $\vec{\alpha}$, the same for all $j$, about the direction of the total angular momentum $\vec{c}$ and each body has a motion of Euler type.

This result can be obtained easily if we consider arbitrary variations of the $\vec{\Omega}_{jE}$ and a small general rotation of the $\vec{u}_j$ (that doesn't modify $U$) and if we study the variations of $\vec{c}$, $h$ and the products $\vec{u}_j \cdot \vec{\Omega}_{jE}$ that must remain zero.

The eventual Pluto's ring would be obtained after a very long evolution with many collisions and many subsequent losses of energy. It is very natural to look for a state in which the mechanical energy $h$ is minimum for given values of the total angular momentum $\vec{c}$ and the rates of rotation $\Omega_{jP}$.

According to the problem of the figure 1 the average effect of solar perturbations can be modelized by a "solar ring" all around Pluto in its orbital plane with the mass and the distance of the Sun, we will also modelize Pluto and Charon by two small rings along their respective orbits in the Pluto-Charon system.

In these conditions the "solar ring" contains practically all the angular momentum $\vec{c}$ and the axis of the constant rotation $\vec{\alpha}$ is normal to Pluto's orbital plane.

The rate of rotation $\alpha$ is extremely small ($+12''$ per century if the Pluto-Charon system is isolated, at most a few degrees per century if a large Pluto's ring exists). Then the problem of the minimization of the mechanical energy $h$ for given $\vec{c}$

and $\Omega_{jP}$ is practically identical to the problem of the maximization of the mutual potential $U$ (including the terms related to the solar ring) for given $\Omega_{jP}$ and given $c_z$; $c_z$ being the component normal to Pluto's orbital plane of the angular momentum $\overrightarrow{c}_R$ of Pluto's system and ring only.

The potential $U_c$ of a thin ring, a circle, of mass $m$ and radius $R$ is well known (Fig. 2), it can be developed easily in terms of the radial distance $r$ and the latitude $\varphi$:

**A)** For $r \leq R$

$$U_c = \frac{Gm}{R} \sum_{q=0}^{\infty} (-1)^q \left(\frac{r}{R}\right)^{2q} \frac{(2q)!}{4^q (q!)^2} P_{2q}(\sin \varphi) \tag{7}$$

that is

$$U_c = \frac{Gm}{R} \left(1 - \frac{r^2}{2R^2} \cdot \frac{3 \sin^2 \varphi - 1}{2} + \right.$$

$$\left. \frac{3r^4}{8R^4} \cdot \frac{35 \sin^4 \varphi - 30 \sin^2 \varphi + 3}{8} - \cdots \right) \tag{8}$$

**B)** For $r \geq R$

$$U_c = \frac{Gm}{r} \sum_{q=0}^{\infty} (-1)^q \left(\frac{R}{r}\right)^{2q} \frac{(2q)!}{4^q (q!)^2} P_{2q}(\sin \varphi) \tag{9}$$

that is

$$U_c = \frac{Gm}{r} \left(1 - \frac{R^2}{2r^2} \cdot \frac{3 \sin^2 \varphi - 1}{2} + \right.$$

$$\left. \frac{3R^4}{8r^4} \cdot \frac{35 \sin^4 \varphi - 30 \sin^2 \varphi + 3}{8} - \cdots \right) \tag{10}$$

These expressions can also be given in terms of the largest and the smallest distances, $a$ and $b$, to the circle:

$$a = \left[R^2 + r^2 + 2Rr \cos \varphi\right]^{\frac{1}{2}} \quad ; \quad b = \left[R^2 + r^2 - 2Rr \cos \varphi\right]^{\frac{1}{2}} \tag{11}$$

$$U_c = \frac{Gm}{2\pi} \int_0^{2\pi} \frac{du}{\sqrt{(a^2 \sin^2 u + b^2 \cos^2 u)}} = \frac{Gm}{L} \tag{12}$$

with:

$$L = \text{``arithmetico-geometrical mean of } a \text{ and } b\text{''} = \mathcal{AG}(a, b)$$

$$\left. \begin{array}{c} \text{that is: } L = \lim_{n \to \infty} a_n = \lim_{n \to \infty} b_n, \quad \text{with:} \\ a_1 = (a + b)/2 \quad ; \quad b_1 = \sqrt{ab} \\ a_2 = (a_1 + b_1)/2 \quad ; \quad b_2 = \sqrt{a_1 b_1} \\ \forall n : a_{n+1} = (a_n + b_n)/2 \quad ; \quad b_{n+1} = \sqrt{a_n b_n} \end{array} \right\} \begin{array}{c} L = \mathcal{AG}(a_1, b_1) \\ = \mathcal{AG}(a_2, b_2) \\ = \text{etc.} \ldots \end{array} \right\} \tag{13}$$

The convergence of these sequences is extremely fast, much faster than in the series (7) and (9).

We can also use:

$$\pi(4a^2 - b^2)/8a \cdot \ln(4a/b) < L < \pi a/2 \ln(4a/b) \tag{14}$$

that is especially useful when $b/a$ is small.

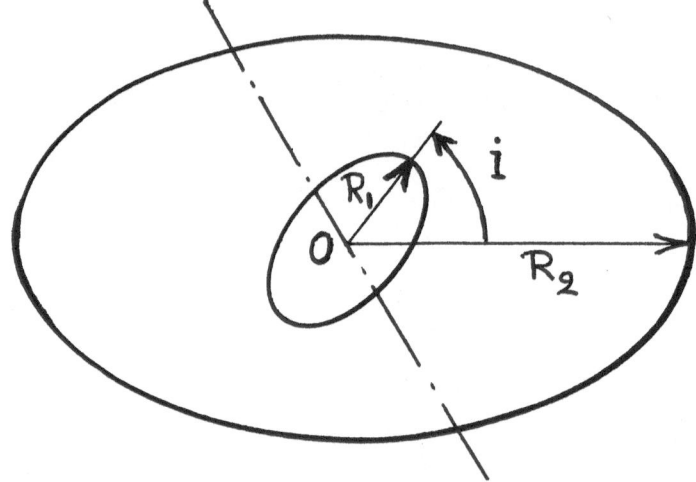

Fig. 3. Mutual potential of two concentric circles

These expressions lead to the mutual potentiel $U_{c_1 c_2}$ of two concentric circles of masses $m_1$ and $m_2$, radius $R_1$ and $R_2$ and mutual inclination $i$.

If $R_1 \leq R_2$:

$$U_{c_1 c_2} = \frac{Gm_1 m_2}{R_2} \sum_{q=0}^{\infty} \left(\frac{R_1}{R_2}\right)^{2q} \cdot \frac{[(2q)!]^2}{16^q (q!)^4} \cdot P_{2q}(\cos i) \tag{15}$$

that is

$$U_{c_1 c_2} = \frac{Gm_1 m_2}{R_2} \left(1 + \frac{R_1^2}{4R_2^2} \cdot \frac{3\cos^2 i - 1}{2} + \right.$$
$$\left. \frac{9R_1^4}{64 R_2^4} \cdot \frac{35\cos^4 i - 30\cos^2 i + 3}{8} - \ldots \right) \tag{16}$$

It is then possible to look for the shape of the Pluto's ring that, for given masses, radius and rates of rotation $\Omega_{jP}$, and for given $c_z$ the normal component of $\vec{c}_R$, gives the maximum to the value of the mutual potentiel $U$.

The influence of the "solar ring" and that of the Pluton-Charon system are equivalent at about 550 000 km to 600 000 km from Pluto and lead to the following

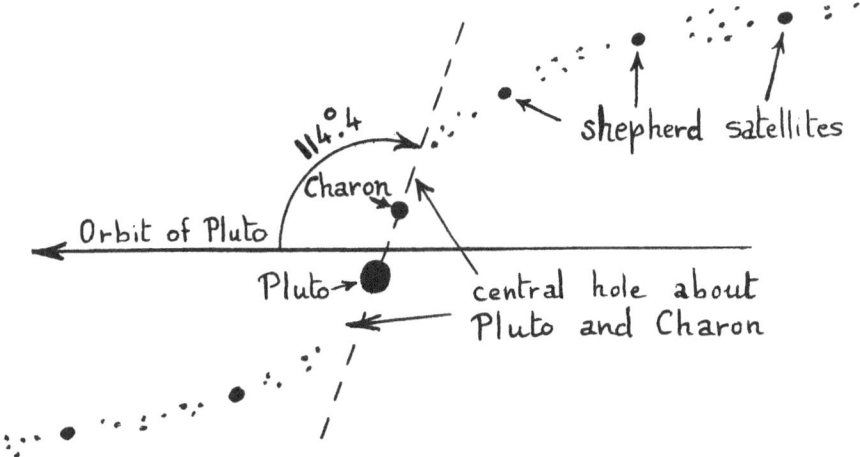

Fig. 4. Section of Pluto's ring by a plane normal to the general line of nodes

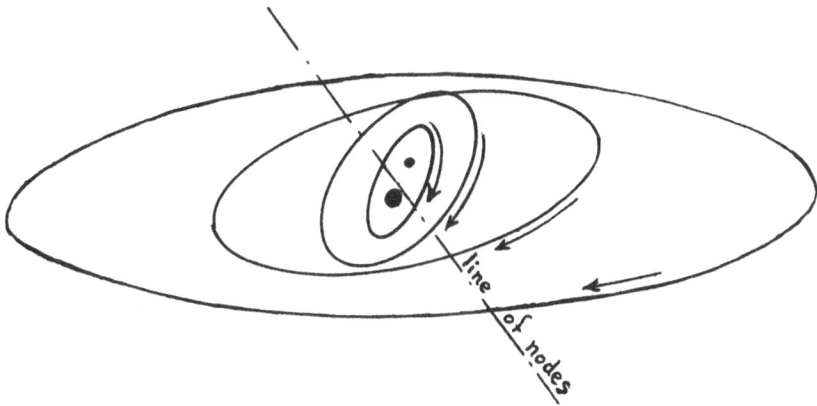

Fig. 5. Pluto, Charon and the successive ringlets with the same line of nodes and with an inclination on Pluto's orbital plane increasing from 114°.4 (Pluto-Charon system) to about 180°. At the end of 1987, the descending node of the general line of nodes was pointing towards the Sun.

approximated picture with all ringlets having the same line of nodes and an inclination increasing with their radius from 114°.4 (Pluto-Charon inclination) to about 180° (Figs. 4 and 5).

The question of the size of the ring is difficult, we hope that its total mass be about 8 to 11 % of Earth-mass but we know that it cannot suffer too much collisions.

With this total mass the average distance of Lagrangian equilibrium points is about 26 millions of kilometers and thus a conservative possibility is a ring with a radius of about 10 millions of kilometers.

The large inclination of the Pluto-Charon system has the useful effect of giving some thickness to the ring (Fig. 4) and we can consider a thickness-to-width ratio of 1 %.

In these conditions the volume of the ring domain would be about 5p28 m$^3$ while the volume of matter inside the ring (with the average Pluto-Charon density of 2 g/cm$^3$) would be about 3p20 m$^3$. The corresponding coefficient of occupation is 3p20/5p28 that is 6n9 (n is for "negative power of ten") and thus the average distance between "neighbours" of kilometric size would be about 500 km.

This distance is large but not large enough to avoid numerous collisions (0.1 to 0.2 collision per millenium and per particle), collisions that will lead to the condensation of the ring into large satellites.

Fortunately the ring has a retrograde rotation and retrograde satellites orbits are much more stable at much larger distances than prograde orbits (Hénon 1969). Nothing forbids the extension of the ring up to 100 millions of kilometers if necessary since no planet approach Pluto at less than 1.4 billions of kilometers. The outer ringlets will then be ellipses centered at Pluto and slightly elongated along Pluto's orbit.

The coefficient of occupation falls then to one over one hundred billions and the ring can now have easily a very long duration.

Let us add a final remark. The particles escaping from the ring will accumulate essentially along Pluto's orbit (a phenomenon that is common for destroyed comets). But the three-to-two resonance between Neptune and Pluto, the resonance that forbids Pluto from approaching Neptune in spite of the proximity of their orbits, will have a similar effect on escaped particles: most of them will remain confined to a small arc about Pluto and must then be counted into the mass of Pluto for the perturbations of Uranus and Neptune.

## 5. Discussion, Observability

Beside the unexplained perturbations of Uranus and Neptune and the apparent absence of a tenth planet, we can notice the following favourable arguments for the existence of a large and heavy Pluto's ring.

A) All four major planets have rings and a very fast rotation, and hence a very large angular momentum.

B) The outer regions of the Solar System seems to have undergone very large perturbations:

- The axis of Uranus has an inclination of 98°
- Triton, the major satellite of Neptune, has a retrograde motion and Nereid has an eccentricity of 0.75, i.e. by far the largest eccentricity of all satellites.
- The orbit of Pluto crosses that of Neptune.
- The inclination of Pluto on its orbit is 114°.4.

In these conditions, it is possible that the proto-plutonian cloud of matter had a too fast and retrograde rotation. The existence of the proportionally very big satellite

Charon and its co-rotation with Pluto, seemingly the product of the disruption of a too rapidly rotating planet, are supplementary favourable arguments.

The retrograde motion and the large inclination of the Pluto-Charon system are two major arguments for a very long duration of a Pluto's ring since large retrograde satellite orbits are much more stable than prograde ones and since the thickness given to the ring by large inclinations reduces very much the coefficient of occupation.

The confinement of escaping particles in the vicinity of Pluto by the three-to-two Neptune to Pluto resonance is not a negligible argument but, in spite of all these favourable elements, the stability of an eventual large Pluto's ring remains weak and only the exceptional isolation of Pluto, always at more than 1.4 billions of kilometers from other planets allows to hope for the existence of a large Pluto's ring.

How can we detect that ring in the vicinity of Pluto and probably near its orbital plane ? There are at least the four following possibilities.

  – If the shepherd satellites have grown up to about 50 km they will be visible (magnitude about 24) when the very large telescope of Chile will be in service and/or when the Hubble space telescope will be corrected.
  – These shepherd satellites will perhaps also provide some star occultations.
  – When a space probe will be sent to Pluto it will undergo the attraction of its ring.
  – The precession of Charon orbit, $+12''$ per century if Pluto and Charon are isolated, can be as large as a few degrees per century if Pluto has a large and heavy ring.

## 6. Conclusion

The hypothesis of existence of a large Pluto's ring, heavy enough to explain Uranus and Neptune perturbations, is very daring and leads to a picture very different from the light Saturnian rings.

The main weakness of this idea is the weak stability of such a large ring. Only the extreme isolation of Pluto allows a ring of long duration neither escaped into space nor condensed into large satellites. Fortunately many arguments of all kinds support this daring idea.

The position of the ring is probably not far from Pluto's orbital plane but its dimensions are difficult to estimate, its radius can be some tenths of millions of kilometers.

The ring is not far from the limit of detection and the next progress will allow to check its existence.

## References

Burwitz V., Reinsch K., Pakull M.W., Bouchet P.: 1991, "New Aspects of the Binary Planet Pluto-Charon",*The Messenger – El Mensajero*, **66**, 23–26

Duncombe R.L. and Seidelmann P.K.: 1980, "A History of the Determination of Pluto's Mass", *Icarus*, 44, 12–18

Hénon M.: 1969, "Numerical Exploration of the Restricted Problem. V. Hill's case: Periodic Orbits and Their Stability"*Astronomy and Astrophysics*, 1, 223–238.

Marchal C.: 1968, "Figures d'équilibre séculairement stables des masses fluides hétérogènes en rotation", *Bulletin Astronomique. Série 3*, 3, 341–360.

Seidelmann P.K., Kaplan G.H., Pulkinner K.F., Santoro E.J. and Van Flandern T.C.: 1980, "Ephemeris of Pluto"*Icarus*, 44, 19–28.

Seidelmann P.K. and Harrington R.S.: 1988, "Planet X-The Current Status", Proceedings of the second Alexander von Humboldt colloquium on Celestial Mechanics: *"Long term evolution of planetary systems"* Ramsau, Austria, March 13–19.

Seidelmann P.K.: 1988, "Round table discussion about Planet X", Proceedings of the second Alexander von Humboldt colloquium on Celestial Mechanics: *"Long term evolution of planetary systems"* Ramsau, Austria, March 13–19.

Tombaugh C.W. *(1980) "Some early vexing optical and mechanical Problems of the 13 Inch "Pluto" Telescope"*, *Icarus*, 44, 2–6.

Tombaugh C.W. *(1992) "La dernière planète"*, *Ciel et Espace*, 266, 32–38.

# RECENT PROGRESS IN THE THEORY AND APPLICATION OF SYMPLECTIC INTEGRATORS

HARUO YOSHIDA

*National Astronomical Observatory, Mitaka, Tokyo 181, Japan*
*e-mail : yoshida@C1.mtk.nao.ac.jp*

**Abstract.** In this paper various aspect of symplectic integrators are reviewed. Symplectic integrators are numerical integration methods for Hamiltonian systems which are designed to conserve the symplectic structure exactly as the original flow. There are explicit symplectic schemes for systems of the form $H = T(p) + V(q)$, and implicit schemes for general Hamiltonian systems. As a general property, symplectic integrators conserve the energy quite well and therefore an artificial damping (excitation) caused by the accumulation of the local truncation error cannot occur. Symplectic integrators have been applied to the Kepler problem, the motion of minor bodies in the solar system and the long-term evolution of outer planets.

**Key words:** Numerical integration methods – long time evolution – symplectic mapping

## 1. Introduction

### 1.1. MOTIVATION

For a given system of differential equations,

$$\frac{dz}{dt} = f(z), \tag{1}$$

we try to get an approximate solution from $z$ at $t = 0$ to $z'$ at $t = \tau$ in the form $z' = \psi(z, \tau)$, where $\tau$ is called the step size and assumed to be small. The most primitive one is called the Euler method which makes use of the mapping

$$z' = \psi(z, \tau) = z + \tau f(z). \tag{2}$$

This Euler method has the 1st order accuracy, since (2) agrees with the Taylor expansion of the true solution,

$$z' = z + \tau f(z) + \frac{\tau^2}{2} f'(z)f(z) + \frac{\tau^3}{6}\left(f''(z)f(z) + f'(z)^2\right)f(z) + \cdots \tag{3}$$

up to the 1st order of $\tau$. We call a mapping $z' = \psi(z, \tau)$ an n-th order integration scheme (integration method, integrator), if it agrees with the Taylor expansion (3) up to the order of $\tau^n$. The well-known classical Runge-Kutta method, which avoids the evaluation of higher derivatives of $f(z)$, is the 4th order in this sense.

When we apply these conventional integration methods to Hamiltonian systems, there occur an artificial excitation or damping which comes from the integration method itself. For example, for the one-dimensional harmonic oscillator with the Hamiltonian

$$H = (1/2)(p^2 + q^2), \tag{4}$$

*Celestial Mechanics and Dynamical Astronomy* **56**: 27–43, 1993.
© 1993 *Kluwer Academic Publishers.*

we know the exact solution

$$\begin{pmatrix} q(\tau) \\ p(\tau) \end{pmatrix} = \begin{pmatrix} \cos\tau & \sin\tau \\ -\sin\tau & \cos\tau \end{pmatrix} \begin{pmatrix} q(0) \\ p(0) \end{pmatrix}. \tag{5}$$

On the other hands, the Euler method (2) approximates (5) as

$$\begin{pmatrix} q' \\ p' \end{pmatrix} = \begin{pmatrix} 1 & \tau \\ -\tau & 1 \end{pmatrix} \begin{pmatrix} q \\ p \end{pmatrix}. \tag{6}$$

One finds easily that at each step, the value of the energy is multiplied by $(1 + \tau^2)$ with the Euler method (6), i.e.,

$$(p'^2 + q'^2) = (1 + \tau^2)(p^2 + q^2), \tag{7}$$

which leads to an indefinite increase of the energy. When we use the 4th order Runge-Kutta method we find, on the contrary, an artificial damping,

$$(p'^2 + q'^2) = (1 - \frac{1}{72}\tau^6 + ....)(p^2 + q^2). \tag{8}$$

This artificial excitation or damping makes the result of long-time integration quite unreliable. Therefore it is desirable to use some special integration scheme for Hamiltonian systems.

For autonomous Hamiltonian systems

$$\frac{dq}{dt} = \frac{\partial H}{\partial p}, \quad \frac{dp}{dt} = -\frac{\partial H}{\partial q}, \tag{9}$$

in general, we know that (i) the value of energy (Hamiltonian) is conserved, and that (ii) the mapping from $(q, p)$ at $t = 0$ to $(q', p')$ at $t = \tau$ along the solution is symplectic (canonical),

$$dp \wedge dq = dp' \wedge dq', \tag{10}$$

i.e., the symplectic structure is conserved. The Euler method and Runge-Kutta method violate not only the conservation of energy but also the conservation of symplectic structure.

Then it is quite natural to search a numerical integration scheme which keeps the above two properties, (i) $H = const.$ and (ii) $dp \wedge dq = const..$. Unfortunately, according to Ge and Marsden (1988), there cannot exist such an integration scheme for non-integrable Hamiltonian systems in general. They claims that if such a scheme $z' = \psi(z, \tau)$ exists then it should coincides with the exact solution up to a reparametrization of the independent variable $\tau$.

Now as a compromise, one may search schemes which keep one of the conservation properties. As for the Hamiltonian-conserving methods, there have been a lot of works by now and they will not be discussed in this paper. See Itoh and Abe (1988, 1989) and references therein. A scheme which conserves the symplectic structure (10) exactly is called a symplectic integration method (symplectic integrator) which will be reviewed from now on in detail.

## 1.2. HISTORY OF SYMPLECTIC INTEGRATORS

The research on symplectic integrators originates from three independent groups. The first one is by Feng and his collaborators in Beijing who developed implicit symplectic schemes based on the generating function. See Feng and Qin (1987) for a review. This implicit method was later applied by Channell and Scovel (1990) extensively.

The second group of research starts with Sanz-Serna (1988) and Lasagni (1988). They found a condition for implicit Runge-Kutta methods to be symplectic. As a result, the family of Gauss-Legendre Runge-Kutta methods (Dekker and Verwer, 1984) are shown to be symplectic. The simplest one is known as the implicit midpoint rule.

As for the third group, Ruth(1983) developed an idea of explicit symplectic schemes for Hamiltonian of the form $H = T(p) + V(q)$. Along the line of Ruth(1983), higher order integrators were presented by Forest(1987), Neri(1988) and Yoshida(1990) later on.

In the next two sections, these implicit and explicit schemes will be explained in more detail. For a general review of symplectic integrators, see MacKay (1991), Sanz-Serna (1991) and Scovel (1991).

One important topic which will not be described in this paper is related to multistep methods. In fact, Eirola and Sanz-Serna (1990) proposed a multistep method which conserves the symplectic structure. On the other hand, Quinlan and Tremaine (1990) worked out with the symmetric multistep method which shows a good behavior on the conservation of energy, and might be understood as a kind of symplectic method. For this method, see also Kinoshita and Nakai (1991), and Quinlan and Toomre (1991).

## 2. Implicit Schemes

## 2.1. GENERATING FUNCTION METHODS

For an arbitrary function of mixed variables $W = W(q, p')$, the mapping $(q, p) \rightarrow (q', p')$ defined implicitly by the relations

$$p = \frac{\partial W}{\partial q}, \quad q' = \frac{\partial W}{\partial p'}, \tag{11}$$

is symplectic (canonical transformation). The function $W$ is called the generating function (of Von-Zeipel type). If we take

$$W = qp' + \tau H(q, p'), \tag{12}$$

then (11) implies

$$q' = q + \tau \frac{\partial H}{\partial p'}, \quad p' = p - \tau \frac{\partial H}{\partial q}, \tag{13}$$

which gives a 1st order implicit symplectic integrator. For the 2nd order integrator, take

$$W(q, p') = qp' + \tau H(q, p') + (\tau^2/2) H_{p'} H_q. \tag{14}$$

and the implicit scheme

$$q' = q + \tau H_{p'} + (\tau^2/2)(H_{p'p'} H_q + H_{p'} H_{p'q}),$$

$$p' = p - \tau H_q - (\tau^2/2)(H_{p'q} H_q + H_{p'} H_{qq}). \tag{15}$$

follows. Higher order integrators ($n > 2$) are similarly obtained by choosing the generating function $W(q, p')$ properly so that the mapping $(q, p) \to (q', p')$ agrees with the Taylor expansion of the solution up to the order of $\tau^n$.

Feng and Qin (1987) derived the generating functions $W(q, p')$ and the corresponding symplectic schemes up to the 4th order, and Channell and Scovel (1990), up to the 6th order. Since it becomes very difficult to write down the computer program by hand for these higher order integrators, Channell and Scovel (1990) developed a preprocessor to generate the FORTRAN source code.

## 2.2. IMPLICIT RUNGE-KUTTA METHODS

For the system of differential equations (1) in general, an s-stage Runge-Kutta method (generally implicit), which is a natural generalization of the classical 4-th order one, is defined as follows. First, vectors $k_i$ are determined by solving the simultaneous algebraic equations

$$k_i = f(z + \tau \sum_{j=1}^{s} a_{ij} k_j), \tag{16}$$

($i = 1, ..., s$), then the mapping $z \to z'$ is

$$z' = z + \tau \sum_{j=1}^{s} b_j k_j. \tag{17}$$

Here $a_{ij}$ and $b_j$ are scalar constants which characterize the scheme. The so-called Butcher table which lists $a_{ij}$ and $b_j$ as an $s + 1$ by $s$ matrix is often used to specify a given Runge-Kutta method. Note that if $a_{ij} = 0$ for $i \leq j$, then the scheme is explicit. These constants $a_{ij}$ and $b_j$ are determined by the order conditions which are derived by the postulate that the mapping $z \to z'$ should agree with the Taylor expansion of the solution up to the desired order of $\tau$.

Suppose now that (1) is Hamiltonian. For general Runge-Kutta methods, the mapping (17) is not symplectic. Sanz-Serna(1988) and Lasagni(1988) found that if the constants satisfies the conditions

$$M_{ij} := b_i a_{ij} + b_j a_{ji} - b_i b_j = 0, \quad (1 \leq i, j \leq s) \tag{18}$$

identically, then the mapping is symplectic. Notice that for (18) to be satisfied, the scheme must be implicit.

The simplest solution (1-stage, $s = 1$) which satisfies the conditions (18) is given by

$$a_{11} = \frac{1}{2}, \quad b_1 = 1, \tag{19}$$

and we have the scheme,

$$k_1 = f(z + \frac{\tau}{2}k_1), \quad z' = z + \tau k_1 \tag{20}$$

or, more concisely,

$$z' = z + \tau f(\frac{z + z'}{2}), \tag{21}$$

which is known as the implicit midpoint rule, and has order 2. For the 2-stage method we have, for example,

$$(a_{ij}) = \begin{pmatrix} \frac{1}{4} & \frac{1}{4} - \frac{\sqrt{3}}{6} \\ \frac{1}{4} + \frac{\sqrt{3}}{6} & \frac{1}{4} \end{pmatrix}, \quad (b_j) = (\frac{1}{2} \quad \frac{1}{2}) \tag{22}$$

and the order is 4. Pullin and Saffman (1991) applied this 4th order implicit Runge-Kutta method to the motion of four vortex motion successfully. These schemes are simplest examples of the family of Gauss-Legendre method and the s-stage Gauss-Legendre method has order 2s (Dekker and Verwer, 1984). This family has a good stability property. For further references in this direction of research, see Sanz-Serna(1991), Sanz-Serna and Abia (1991), and Saito et al. (1992).

### 3. Explicit Schemes

3.1. RUTH(1983)

For Hamiltonian systems of the form

$$H = T(p) + V(q), \tag{23}$$

there exist explicit symplectic algorithms. As for the 1st order one, a small change of the Euler method (2) makes it exactly symplectic. In fact, take

$$q' = q + \tau \left(\frac{\partial T}{\partial p}\right)_{p=p}, \quad p' = p - \tau \left(\frac{\partial V}{\partial q}\right)_{q=q'}. \tag{24}$$

This mapping is symplectic because it is composed of two symplectic mappings $(q, p) \rightarrow (q', p)$ and $(q', p) \rightarrow (q', p')$.

The idea to construct higher order schemes is simply to approximate the original Hamiltonian flow by a composition of trivial symplectic mappings,

$$S_T(c_i\tau): \quad q' = q + c_i\tau \left(\frac{\partial T}{\partial p}\right), \quad p' = p, \tag{25}$$

and

$$S_V(d_i\tau): \quad q' = q, \quad p' = p - d_i\tau \left(\frac{\partial V}{\partial q}\right), \tag{26}$$

repeatedly. Here numerical coefficients $(c_i, d_i)$, $(i = 1, ...k)$ are determined so that the composed mapping $(q, p) \rightarrow (q', p')$ coincides with the Taylor expansion of the solution up to the order of $\tau^n$. Thus an n-th order explicit symplectic integrator is obtained. For example, a second order scheme $(q, p) \rightarrow (q', p')$ is attained by

$$q* = q + \frac{\tau}{2}\left(\frac{\partial T}{\partial p}\right)_{p=p}, \quad p' = p - \tau\left(\frac{\partial V}{\partial q}\right)_{q=q*}, \quad q' = q + \frac{\tau}{2}\left(\frac{\partial T}{\partial p}\right)_{p=p'} \tag{27}$$

which corresponds to the choice $c_1 = c_2 = 1/2, d_1 = 1, d_2 = 0$, and this scheme has been known as the leap-frog method. Ruth(1983) first derived the algebraic equations of $(c_i, d_i)$ for the 3rd order integrator and obtained a solution,

$$c_1 = \frac{7}{24}, c_2 = \frac{3}{4}, c_3 = -\frac{1}{24}, d_1 = \frac{2}{3}, d_2 = -\frac{2}{3}, d_3 = 1. \tag{28}$$

Candy and Rozmus (1991) directly extended the idea of Ruth to obtain the coefficients of the 4th order integrator, and demonstrated the advantage of symplectic schemes in various examples.

## 3.2. NERI(1987)

The above problem to derive explicit symplectic integrator was reformulated by Neri(1987) in terms of Lie algebraic language. First rewrite the Hamilton equation in the form

$$\frac{dz}{dt} = \{z, H(z)\}, \tag{29}$$

where braces stand for the Poisson bracket, $\{F, G\} = F_q G_p - F_p G_q$. If we introduce a differential operator $D_G$ by $D_G F := \{F, G\}$, then (29) is written as $\mathfrak{E} = D_H z$, so the formal solution, or the exact time evolution of $z(t)$ from $t = 0$ to $t = \tau$ is given by

$$z(\tau) = [\exp(\tau D_H)]z(0). \tag{30}$$

For a Hamiltonian of the form (23), $D_H = D_T + D_V$ and we have the formal solution

$$z(\tau) = \exp[\tau(A + B)]z(0). \tag{31}$$

where $A := D_T$ and $B := D_V$, and these operators $A$ and $B$ do not commute in general.

Suppose $(c_i, d_i)$, $(i = 1, 2, ..., k)$ is a set of real numbers which satisfies the equality

$$\exp[\tau(A + B)] = \prod_{i=1}^{k} \exp(c_i\tau A)\exp(d_i\tau B) + o(\tau^{n+1}), \tag{32}$$

for a given integer $n$, which corresponds to the order of integrator. Now consider a mapping from $z = z(0)$ to $z' = z(\tau)$ given by

$$z' = \left[ \prod_{i=1}^{k} \exp(c_i \tau A) \exp(d_i \tau B) \right] z. \qquad (33)$$

This mapping is symplectic because it is just a product of elementary symplectic mappings, and approximates the exact solution (31) up to the order of $\tau^n$. Furthermore (33) is explicitly computable. In fact (33) gives the succession of the mappings

$$q_i = q_{i-1} + \tau c_i \left( \frac{\partial T}{\partial p} \right)_{p=p_{i-1}}, \quad p_i = p_{i-1} - \tau d_i \left( \frac{\partial V}{\partial q} \right)_{q=q_i}, \qquad (34)$$

for $i = 1$ to $i = k$, with $(q_0, p_0) = z$ and $(q_k, p_k) = z'$. An n-th order symplectic integrator (integration scheme) is thus realized. For example when $n = 1$, a trivial solution is $c_1 = d_1 = 1$, $(k = 1)$, which corresponds to the identity

$$\exp[\tau(A + B)] = \exp(\tau A) \exp(\tau B) + o(\tau^2). \qquad (35)$$

and gives the 1st order symplectic integrator (24). When $n = 2$, we find easily a solution $c_1 = c_2 = 1/2, d_1 = 1, d_2 = 0$, $(k = 2)$, which comes from

$$\exp[\tau(A + B)] = \exp(\frac{\tau}{2} A) \exp(\tau B) \exp(\frac{\tau}{2} A) + o(\tau^3). \qquad (36)$$

and implies the 2nd order integrator (27). Forest(1987), Forest and Ruth(1990), Candy and Rozmus(1991) obtained a 4th order integrator in a rather straightforward way with the result,

$$c_1 = c_4 = \frac{1}{2(2 - 2^{1/3})}, \quad c_2 = c_3 = \frac{1 - 2^{1/3}}{2(2 - 2^{1/3})},$$

$$d_1 = d_3 = \frac{1}{2 - 2^{1/3}}, \quad d_2 = \frac{-2^{1/3}}{2 - 2^{1/3}}, \quad d_4 = 0. \qquad (37)$$

Notice that this 4th order integrator requires the evaluation of force function only 3-times in a step, although the classical 4th order Runge-Kutta method needs 4-times.

## 3.3. YOSHIDA(1990)

Yoshida(1990) first noticed that the 4th order integrator found by Forest (1987) is composed of the 2nd order ones. With use of the notation

$$S_2(\tau) := \exp(\frac{\tau}{2} A) \exp(\tau B) \exp(\frac{\tau}{2} A), \qquad (38)$$

the 4th order integrator $S_4(\tau)$ can be written as

$$S_4(\tau) = S_2(x_1\tau)S_2(x_0\tau)S_2(x_1\tau) \tag{39}$$

where

$$x_0 = \frac{-2^{1/3}}{2 - 2^{1/3}}, \quad x_1 = \frac{1}{2 - 2^{1/3}}. \tag{40}$$

In fact, $x_0$ and $x_1$ are determined as the solution of algebraic equations,

$$x_0 + 2x_1 = 1, \quad x_0^3 + 2x_1^3 = 0, \tag{41}$$

and this interpretation gives the simplest derivation of the 4th order integrator (37). Forest et al.(1991) noticed that this idea of composition can be used also to obtain higher order implicit integrators in the previous section.

The above fact was found independently, and *even earlier*, by Suzuki (1990) from a completely different motivation which shares the same mathematical problem (32). Suzuki (1991) further proved the very strong statement that there cannot exist any solution of (32) with all positive $(c_i, d_i)$ when $n \geq 3$. Therefore, the presence of some negative numbers in (28) and (37) is unavoidable. For the 6th order integrator, Yoshida (1990) put

$$S_6(\tau) = S_2(w_3\tau)S_2(w_2\tau)S_2(w_1\tau)S_2(w_0\tau)S_2(w_1\tau)S_2(w_2\tau)S_2(w_3\tau) \tag{42}$$

and three sets of $(w_0, w_1, w_2, w_3)$ were obtained numerically. Five sets for the 8th order integrator were also found. This construction of higher order integrators with minimum number of force evaluation was recently generalized to arbitrary orders by Suzuki (1992).

For other researches on finding explicit symplectic integrators, see Abia and Sanz-Serna (1990), McLachlan and Atela (1991) and Okunbor and Skeel (1991, 1992, 1992a).

## 4. General Property

### 4.1. CONSERVATION OF ENERGY

The original Hamiltonian flow conserves the Hamiltonian (Energy) exactly and it is desirable that numerical integration methods respect this fact.

For the family of symplectic Runge-Kutta methods with condition (18), all the quadratic integrals are conserved exactly (Sanz-Serna, 1988). Thus for linear systems with a quadratic Hamiltonian, the symplectic Runge-Kutta scheme conserves the Hamiltonian and the symplectic structure at the same time. This is possible because linear Hamiltonian systems are integrable. For general non-integrable systems one cannot expect the conservation of energy exactly at each step (Ge and Marsden, 1988). Nevertheless there is an advantage for symplectic schemes.

Let us apply the 1st order explicit symplectic scheme (24) to the one-dimensional harmonic oscillator (4). Then we have the symplectic mapping

$$\begin{pmatrix} q' \\ p' \end{pmatrix} = \begin{pmatrix} 1 & \tau \\ -\tau & 1 - \tau^2 \end{pmatrix} \begin{pmatrix} q \\ p \end{pmatrix}. \tag{43}$$

One finds easily that although the value of energy is not conserved exactly by the iteration of symplectic mapping (43), the error has no secular increase and it is bounded of the order of $\tau$ for moderate small values of $\tau$.

For the one-dimensional harmonic oscillator case, this phenomenon is explained by the existence of a conserved quantity (integral of motion) of the mapping (43), which has the expression

$$\frac{1}{2}(p^2 + q^2) + \frac{\tau}{2}pq = const. \tag{44}$$

If one starts with the initial condition $(q, p) = (1, 0)$ with a fixed small value of $\tau$, the points obtained by iterating the mapping (43) must lie on an ellipse in the $(q, p)$ plane, $q^2 + p^2 + \tau pq = 1$, which differs from the trajectory of the exact solution, $q^2 + p^2 = 1$, only of the order of $\tau$ permanently. Thus the error of the energy caused by the local truncation error cannot grow. Indeed, we have a more general statement;

THEOREM 1. *The symplectic mapping (24) exactly describes the time-$\tau$ evolution of an associated Hamiltonian system $\tilde{H}$, which is close to the original Hamiltonian (23) and has the expression of a formal power series in $\tau$,*

$$\tilde{H} = H + \tau H_1 + \tau^2 H_2 + \tau^3 H_3 + .... \tag{45}$$

*where $H$ is the original Hamiltonian (23), and*

$$H_1 = \frac{1}{2}H_p H_q, \quad H_2 = \frac{1}{12}(H_{pp}H_q^2 + H_{qq}H_p^2), \quad H_3 = \frac{1}{12}H_{pp}H_{qq}H_p H_q, \quad ... \tag{46}$$

*In particular, (24) conserves $\tilde{H}$ in (45) exactly.*

As far as the author knows, this fact was first mentioned by Dragt and Finn (1976) without any motivation on numerical integration methods. See also Dragt et al.(1988). The series (45) reflects on the Baker-Campbell-Hausdorff (BCH) formula (Varadarajan, 1974) for the product of two exponential functions of non-commuting operators $X$ and $Y$;

$$\exp X \exp Y = \exp Z \tag{47}$$

with

$$Z = X + Y + \frac{1}{2}[X, Y] + \frac{1}{12}([X, [X, Y]] + [Y, [Y, X]]) + \frac{1}{24}[X, [Y, [Y, X]]] + ... \tag{48}$$

and $[X, Y] := XY - YX$, etc. For the 1st order symplectic integrator (24), one can apply BCH formula (47) to obtain

$$(\exp \tau D_T)(\exp \tau D_V) = \exp \tau D_{\tilde{H}} .\qquad(49)$$

where

$$\tilde{H} = T+V+\frac{\tau}{2}\{V,T\}+\frac{\tau^2}{12}(\{\{T,V\},V\} + \{\{V,T\},T\})+\frac{\tau^3}{24}\{\{\{T,V\},V\},T\}+\ldots$$

$$= H + \frac{\tau}{2}H_p H_q + \frac{\tau^2}{12}(H_{pp}H_q^2 + H_{qq}H_p^2) + \frac{\tau^3}{12}H_{pp}H_{qq}H_p H_q + \ldots \qquad(50)$$

which proves the theorem. For the 2nd order integrator (27), one obtains

$$(\exp \frac{\tau}{2}D_T)(\exp \tau D_V)(\exp \frac{\tau}{2}D_T) = \exp \tau D_{\tilde{H}_{2nd}}, \qquad(51)$$

where

$$\tilde{H}_{2nd} = T + V + \tau^2 \left( \frac{1}{12}\{\{T,V\},V\} - \frac{1}{24}\{\{V,T\},T\} \right) + o(\tau^4)$$

$$= H + \tau^2 \left( \frac{1}{12}H_q^2 H_{pp} - \frac{1}{24}H_p^2 H_{qq} \right) + o(\tau^4). \qquad(52)$$

In general, for an n-th order integrator, one has the associated Hamiltonian

$$\tilde{H}_{nth} = H + \tau^n H_n + o(\tau^{n+1}), \qquad(53)$$

so that the error of the energy remains of the order of $\tau^n$. Notice, however, that the rigorous convergence of the series (45), (52), (53) are not guaranteed for nonlinear systems in general. Some related works are found in Auerbach and Friedman (1991), and Friedman and Auerbach (1991).

As for the implicit schemes by the generating function method, i.e., for (13) and (15), a similar argument is possible. A near-identity symplectic mapping defined by the generating function, (12) or (14), can be written into the explicit form

$$z' = (\exp \tau D_{\tilde{H}})z, \qquad(54)$$

by the equivalence of von-Zeipel and Hori perturbation methods (Mersman 1971, Giacaglia 1972). Thus $\tilde{H}$, which is again a formal power series in $\tau$, is conserved.

For the symplectic Runge-Kutta method with condition (18), the present author does not know any similar statement how to write down the associated Hamiltonian $\tilde{H}$ from the original Hamiltonian $H$.

# 5. Application to Specific Problems

## 5.1. KEPLER PROBLEM

Kinoshita et al.(1991) applied the 4th order explicit symplectic integrator (SI4) to the Kepler problem

$$H = \frac{1}{2}p^2 - \frac{1}{r}, \tag{55}$$

with the eccentricity, $e = 0.1$, and compared with the results obtained by the classical 4th order Runge-Kutta method (RK4).

For a short time (10 orbital periods), the errors in the semi-major axis $\triangle a$ and in the eccentricity $\triangle e$ by SI4 are much bigger than those by RK4. The superiority of SI4 over RK4 can be seen after a long time (2000 periods), since $\triangle a$ and $\triangle e$ by RK4 grow secularly in time. The errors in the inclination $\triangle i$ and the longitude of node $\triangle \Omega$ come from the round-off error only, since these elements are conserved exactly. As for the error in the mean anomaly $\triangle l$, RK4 allows the quadratic increase in time although SI4 allows only the linear growth. This difference becomes significant for a long time integration. There is one problem with the error in the argument of pericenter $\triangle \omega$. Both integrators give linear growth of $\triangle \omega$, and the error by SI4 is much larger than that by RK4. RK4 is too good by some unknown reason for $\triangle \omega$, since the 6th order symplectic integrator (SI6) and the 6th order explicit Runge-Kutta method (RK6) give almost the same error in $\triangle \omega$ (Kinoshita and Nakai, 1991).

Gladman et al.(1991) also applied SI4 to the Kepler problem and obtained similar results.

## 5.2. SOLAR SYSTEM

The idea of explicit symplectic integrator can be applied also to the system of the form

$$H = H_0(q, p) + H_1(q, p) \tag{56}$$

where $H_0$ and $H_1$ are integrable in the absence of other parts. A Hamiltonian of the form, $H = T(p) + V(q)$, is the simplest case of this situation. Since $H_0$ and $H_1$ are integrable, $\exp \tau D_{H_0}$ and $\exp \tau D_{H_1}$ can be computed without error by introducing the action-angle variables, $(I_0, \theta_0)$ and $(I_1, \theta_1)$ for each part. Sometimes, however, the change of variables between these two sets of action-angle variables is really time-consuming (Kinoshita et al. 1991) and may decrease the advantage of this treatment.

Wisdom and Holman (1991) write the Hamiltonian of the n-body problem (outer planets)

$$H = \sum_{i=0}^{n-1} \frac{p_i^2}{2m_i} - \sum_{i<j} \frac{Gm_i m_j}{r_{ij}}, \tag{57}$$

into the form

$$H = H_0 + H_1 = H_{Kepler} + H_{Interaction} \qquad (58)$$

using the Jacobi coordinates and applied the 2nd order symplectic integrator (27) in the sense above with the step size, $\tau = 1$ year to integrate for $10^9$ years. For the computation of $\exp \tau D_{H_0}$, the $f$ and $g$ functions of Gauss were used. The authors claimed that they reproduced all the principal results of Sussman and Wisdom (1988) and confirmed that the motion of Pluto is chaotic with much fewer CPU time.

Gladman and Duncan (1990) simulated the evolution of test particles in the outer solar system using the 4th order explicit symplectic integrator (37) to explain the apparent absence of a large number of minor bodies between the giant planets.

## 6. Miscellaneous Problems

### 6.1. VARIABLE TIME STEP

When integrating, for example, a very eccentric orbit in the Kepler problem, one often uses a variable time step to obtain better accuracy. What happens if symplectic integrators are used with a variable step ? The result is, unfortunately, a decrease in efficiency, and the error of energy starts to increase without any bound like the result by a traditional integration method.

Let us define the new independent variable $s$, for example, by

$$adt = rds \qquad (59)$$

where $a$ is the semi-major axis of elliptic orbit. This introduction of new variable $s$ implies that when $r < a$ (i.e., near the pericenter where the motion is fast), $ds$ is bigger that $dt$ and more steps are used than the average. On the other hand when $r > a$ (i.e., near the apocenter where the motion is slow), fewer steps are used to integrating the orbit.

If one compares the constant step and the variable step using RK4, there is an obvious advantage in the variable step method (Figure 1). The error of energy (although it grows linearly) decreases one order using the variable step while total number of steps are kept constant. On th other hand, with use of symplectic integrator (SI4), one finds the secular grows in the error of energy which did not exist with the constant step integration. The argument in the previous section to ensure the lack of secular increase in the error of energy has assumed a constant step size $\tau$, or just an iteration of mapping, and is not valid when the step size is changed. Thus one must say that there is no advantage to use a variable step when integrating by known symplectic methods.

More detailed analysis on this subject can be found in Calvo and Sanz-Serna (1991, 1991a). See also Gladman et al.(1991).

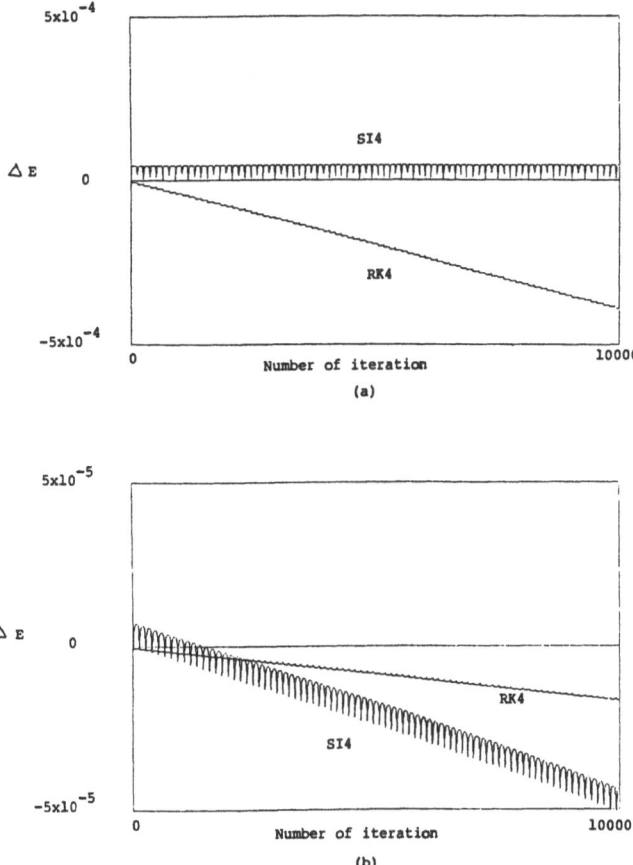

Fig. 1. (a) : The error of energy in the Kepler problem (55) by RK4 and by SI4 $(a = 1, e = 0.5)$. The constant step size is $\tau = 0.05$ and the number of iteration is 10000 (80 orbital periods). The error growth by RK4 is almost linear and periodic by SI4. (b) : The same couple as in (a) using the variable step size (59). The scale of ordinate is 1/10 of that of (a). The energy growth by SI4 is no more periodic.

## 6.2. LARGE TIME STEP AND CHAOS

With a constant time step, symplectic integrators do not produce any secular growth in the error of energy. This fact allows one to use a relatively large step size. Of course, if the step size is too large so that the series (45), (52), (52) really diverge (or, far from convergent) the associated Hamiltonian $\tilde{H}$ never represents the numerical solution and the good conservation of energy is no more guaranteed.

Take an example of the simple pendulum

$$H = \frac{1}{2}p^2 + \cos q. \tag{60}$$

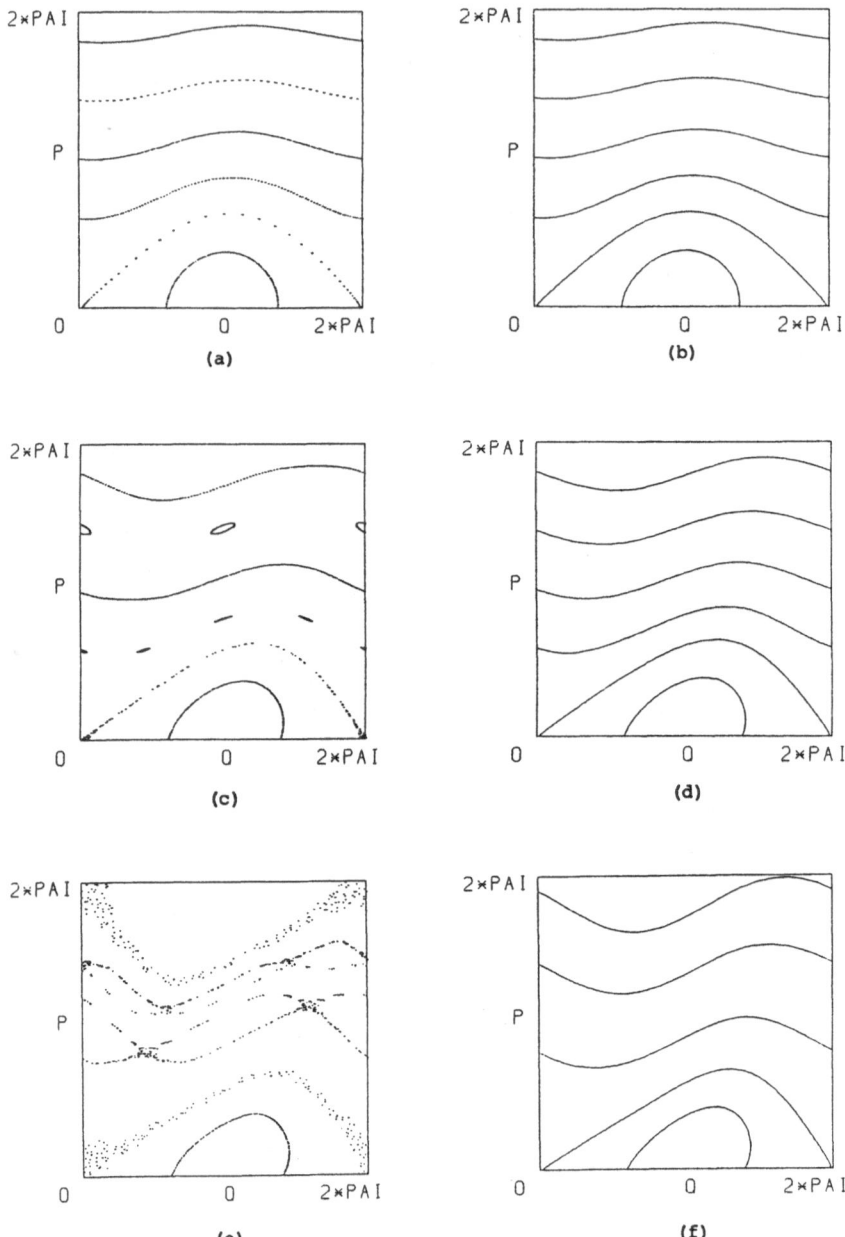

Fig. 2. (a) : Points generated by the iteration of symplectic mapping (61) when $\tau = 0.2$.
(b) : Contour lines of $\tilde{H} = const.$ given by (63) when $\tau = 0.2$. (c) : Same as (a) when
$\tau = 0.7$. (d) : Same as (b) when $\tau = 0.7$. (e) : Same as (a) when $\tau = 1.0$. (f) : Same as (b)
when $\tau = 1.0$.

The 1st order explicit symplectic integrator, the dual of (24), gives the mapping

$$p' = p + \tau \sin q, \quad q' = q + \tau p', \tag{61}$$

which is equivalent (unless $\tau = 0$) to the so-called standard mapping (Greene 1979)

$$P' = P + k \sin Q, \quad Q' = Q + P'. \tag{62}$$

In fact, (60) and (61) are related by the change of scale $\tau p = P$, $q = Q$ and $\tau^2 = k$. The associated Hamiltonian (45) has the expression

$$\tilde{H} = \frac{1}{2}p^2 + \cos q + \frac{\tau}{2}p \sin q + \frac{\tau^2}{12}(\sin^2 q - p^2 \cos q) - \frac{\tau^3}{12}p \sin q \cos q + o(\tau^4). \tag{63}$$

Figure 2 shows the comparison between the points generated by the iteration of the mapping (61) and the contour lines of $\tilde{H}(q,p) = const.$ given by (63) for the values $\tau = 0.2$, $\tau = 0.7$, and $\tau = 1.0$. To draw contour lines, only the terms of up to the order $\tau^3$ are used in (63) and higher terms are neglected. The good agreement of two pictures when $\tau = 0.2$ indicates (but not proves) that the series (63) *practically* converges for this value of $\tau$ and in the range of $(q, p)$ shown. When $\tau = 0.7$ there are already some discrepancy between the iterated points and the contour lines $\tilde{H} = const.$ This discrepancy becomes larger when $\tau = 1.0$. For this value of $\tau$, the series (63) never represents the conserved quantity of the mapping (61) except for the small region around the elliptic fixed point $(q, p) = (\pi, 0)$. For most initial conditions, the iterated points show chaotic behaviors, which cannot be approximated by the perturbation series (63) at all.

These phenomena reminds us of a famous result of Gustavson on the Hénon-Heiles Hamiltonian (Gustavson 1966). He showed that the set of points (obtained by the Poincaré map) on the surface of section has a good agreement with the truncated formal integral obtained by the perturbation series for small energy (small perturbation) but not for large energy, where the Poincaré map becomes chaotic.

## References

Abia, L. and Sanz-Serna, J.M.: 1990, 'Partitioned Runge-Kutta methods for separable Hamiltonian problems', Universidad de Valladolid, Applied Mathematics and Computation Reports **1990/8**

Auerbach, S.P. and Friedman, A.: 1991, 'Long-time behaviour of numerically computed orbits : small and intermediate time step analysis of one-dimensional systems', *J. Comp. Phys.* **93**, 189-223

Calvo, M.P. and Sanz-Serna, J.M: 1991, 'Variable steps for symplectic integrators', Universidad de Valladolid, Applied Mathematics and Computation Reports **1991/3**

Calvo, M.P. and Sanz-Serna, J.M: 1991a, 'The development of variable-step symplectic integrators, with application to the two-body problem', Universidad de Valladolid, Applied Mathematics and Computation Reports **1991/9**

Candy, J. and Rozmus, W.: 1991, 'A symplectic integration algorithm for separable Hamiltonian functions', *J. Comp. Phys.* **92**, 230-256

Channell, P.J. and Scovel, J.C.: 1990, 'Symplectic integration of Hamiltonian systems', *Nonlinearity* **3**, 231-259

Dekker, K. and Verwer, J.G.: 1984, *Stability of Runge-Kutta methods for stiff nonlinear differential equations*, North-Holland

Dragt, A.J. and Finn, J.M.: 1976, 'Lie series and invariant functions for analytic symplectic maps', *J. Math. Phys.* **17**, 2215-2227

Dragt, A.J., Neri, F., Rangarajan, G., Douglas, D.R., Healy, L.M. and Ryne, R.D.: 1988, 'Lie algebraic treatment of linear and nonlinear beam dynamics', *Ann. Rev. Nucl. Part. Sci.* **38**, 455-496

Eirola, T. and Sanz-Serna, J.M.: 1990, 'Conservation of integrals and symplectic structure in the integration of differential equations by multistep methods', Universidad de Valladolid, Applied Mathematics and Computation Reports **1990/9**

Feng, K. and Qin, M.: 1987, 'The symplectic methods for the computation of Hamiltonian equations', *Lecture Note in Math.* **1297**, 1-37

Forest, E.: 1987, 'Canonical integrators as tracking codes', SSC Central Design Group Technical Report SSC-**138**

Forest, E., Brengtsson, J. and Reusch M.F.: 1991, 'Application of the Yoshida-Ruth techniques to implicit integration and multi-map explicit integration', *Phys. Lett. A* **158**, 99-101

Forest, E. and Ruth, R.D.: 1990, 'Fourth-order symplectic integration', *Physica D* **43**, 105-117

Friedman, A. and Auerbach, S.P.: 1991, 'Numerically induced stochasticity', *J. Comp. Phys.* **93**, 171-188

Ge, Z. and Marsden, J.E.: 1988, 'Lie-Poisson Hamilton-Jacobi theory and Lie-Poisson integrators', *Phys. Lett. A* **133**, 134-139

Giacaglia, G.E.O.: 1972, *Perturbation methods in non-linear systems*, Springer

Gladman, B. and Duncan, M.: 1990, 'On the fates of minor bodies in the outer solar system', *Astron. J.* **100**, 1680-1693

Gladman, B., Duncan, M. and Candy, J.: 1991, 'Symplectic integrators for long-term integrations in celestial mechanics', *Celest. Mech.* **52**, 221-240

Greene, J.M.: 1979, 'A method for determining a stochastic transition', *J. Math. Phys.* **20**, 1183-1201

Gustavson, F.: 1966, 'On constructing formal integrals of a Hamiltonian system near an equilibrium points', *Astron. J.* **71**, 670-686

Itoh, T. and Abe, K.: 1988, 'Hamiltonian-conserving discrete canonical equations based on variational difference quotients', *J. Comp. Phys.* **77**, 85-102

Itoh, T. and Abe, K.: 1989, 'Discrete Lagrange's equations and canonical equations based on the principle of least action', *Applied Math. Comp.* **29**, 161-183

Kinoshita, H. and Nakai, H.: 1991, 'New methods for long-time numerical integration of planetary orbits', National Astronomical Observatory of Japan, preprint

Kinoshita, H., Yoshida, H. and Nakai, H.: 1991, 'Symplectic integrators and their application to dynamical astronomy', *Celest. Mech.* **50**, 59-71

Lasagni, F.: 1988, 'Canonical Runge-Kutta methods', *ZAMP* **39**, 952-953

MacKay, R.S.: 1991, 'Some aspects of the dynamics and numerics of Hamiltonian systems', University of Warwick, preprint

McLachlan, R. and Atela, P.: 1991, 'The accuracy of symplectic integrators', *Nonlinearity* **5**, 541-562

Mersman, W.A.: 1971, 'Explicit recursive algorithms for the construction of equivalent canonical transformations', *Celest. Mech.* **3**, 384-389

Neri, F.: 1987, 'Lie algebras and canonical integration', Dept. of Physics, University of Maryland, preprint

Okunbor, D. and Skeel, R.D.: 1991, 'Explicit canonical methods for Hamiltonian systems', Dept. of Computer Sci., Univ. of Illinois at Urbana-Champaign, preprint

Okunbor, D. and Skeel, R.D.: 1992, 'An explicit Runge-Kutta-Nystrom method is canonical if and only if its adjoint is explicit', *SIAM J. Numer. Anal.* **29**, -

Okunbor, D. and Skeel, R.D.: 1992a, 'Canonical Runge-Kutta-Nystrom methods of order 5 and 6', Dept. of Computer Sci., Univ. of Illinois at Urbana-Champaign, preprint

Pullin, D.I. and Saffman, P.G.: 1991, 'Long time symplectic integration, the example of four-vortex motion', *Proc. R. Soc. London A* **432**, 481-494

Quinlan, G.D. and Toomre, A.: 1991, 'Resonant instabilities in symmetric multistep methods', University of Toronto, preprint

Quinlan, G.D. and Tremaine, S.: 1990, 'Symmetric multistep methods for the numerical integration of planetary orbits', *Astron. J.* **100**, 1694-1700

Ruth, R.D.: 1983, 'A canonical integration technique', *IEEE Trans. Nucl. Sci.* NS-30, 2669-2671

Saito, S., Sugiura, H. and Mitsui, T.: 1992, 'Family of symplectic implicit Runge-Kutta formulae', Dept. of information engineering, Nagoya university, preprint

Sanz-Serna, J.M.: 1988, 'Runge-Kutta schemes for Hamiltonian systems', *BIT* **28**, 877-883

Sanz-Serna, J.M.: 1991, 'Symplectic integrators for Hamiltonian problems : an overview', Universidad de Valladolid, Applied Mathematics and Computation Reports **1991/6**

Sanz-Serna, J.M. and Abia, L.: 1991, 'Order conditions for canonical Runge-Kutta schemes', *SIAM J. Numer. Anal.* **28**, 1081-1096

Scovel, C: 1991, 'Symplectic numerical integration of Hamiltonian systems', in *The geometry of Hamiltonian systems*, Ratiu, T. ed., Springer, 463-496

Sussman, G.J. and Wisdom, J.: 1988, 'Numerical evidence that the motion of Pluto is chaotic', *Science* **241**, 433-437

Suzuki, M.: 1990, 'Fractal decomposition of exponential operators with applications to many-body theories and Monte Carlo simulations', *Phys. Lett. A* **146**, 319-323

Suzuki, M.: 1991, 'General theory of fractal path integrals with applications to many-body theories and statistical physics', *J. Math. Phys.* **32**, 400-407

Suzuki, M.: 1992, 'General theory of higher-order decomposition of exponential operators and symplectic integrators', Dept. of Physics, University of Tokyo, preprint

Varadarajan, V.S.: 1974, *Lie groups, Lie algebras and their representation*, Prentice-Hall

Wisdom, J. and Holman, M.: 1991, 'Symplectic maps for the N-body problem', *Astron. J.* **102**, 1528-1538

Yoshida, H.: 1990, 'Construction of higher order symplectic integrators', *Phys. Lett. A* **150**, 262-268

# STABLE PLANETARY ORBITS AROUND ONE COMPONENT IN
# NEARBY BINARY STARS. II

DANIEL BENEST

*O.C.A. Observatoire de Nice, B.P. 229, F - 06304 NICE Cedex 4, FRANCE*

**Abstract.** Numerical simulations are made within the framework of the plane restricted three-body problem, in order to find out if stable orbits for planets around one of the two components in double stars can exist. For any given set of initial parameters (the mass ratio of the two stars and the eccentricity of their orbit around each other), the phase-space of initial positions and velocities is systematically explored.

In previous works, systematic exploration of the circular model as well as studies of more realistic (elliptic) cases such as Sun-Jupiter and the nearby α Centauri and Sirius systems, large stable planetary orbits were found to exist around both components of the binary, up to distances from each star of the order or more than half the binary's periastron separation.

The first results presented here for the η Coronae Borealis system confirm the previous studies.

**Key words:** Restricted problem – stability – planets of double stars

## 1. Introduction

Cosmogonical theories as well as recent observations allow us to expect the existence of planets around many stars. On an other hand, double star systems, or binaries, are established to be more abundant than single stars (such as the Sun), at least in the Solar Neighborhood. We are then faced with two questions; the cosmogonic problem (can planets form easily in binaries or not ?), which we will not deal with here, and the following dynamical problem: assuming that the planets can form, does long-term stability for planetary orbits in double star systems exist ? The earliest paper I know which refers explicitly to a "Planetary Orbit in a Binary" was written by Pavanini (1907); since then, an increasing number of numerical and semi-numerical studies has been published (see e.g. Benest, 1988a, hereafter referred as Paper I, which contains a large list of references).

## 2. Method

We consider the "classical" Plane Restricted Problem of Three Bodies to explore the possible motion of a planet in the gravitational field of a double star, the mass of the planet, of course, being considered negligible with respect to the masses of the two stars (for more detail, see Paper I). Among the known possible types of motion, the two most considered in the literature are the S- and P-types, i.e. referring to a planet revolving either around one –and only one, hereafter called the primary– of the two components of the binary (see e.g. Rabl and Dvorak, 1988), or far around the binary (see e.g. Dvorak, 1984, 1986; Dvorak *et al.*, 1989; Szebehely, 1980, 1992; Szebehely and McKenzie, 1981). Here, we consider only the S-type.

The reduced mass of the primary star is denoted by $\mu$; its value ranges from 0 to 1. $e$ is the eccentricity of the binary; for $e = 0$, we have the "circular" case; when

*Celestial Mechanics and Dynamical Astronomy* **56**: 45–50, 1993.
© 1993 *Kluwer Academic Publishers.*

$e \neq 0$, we are considering the more complicated but more realistic "elliptic" case. $\mu$ and $e$ are the only two parameters of the Plane Restricted Problem. As usual, the equations for the dimensionless relative coordinates are written in a rotating-pulsating frame $(X,Y)$, with respect to which the two stars are at rest; let them be on the X-axis, the primary at $X = 0$ and the second star at $X = -1$. We then explore for every set $(\mu,e)$ systematically the phase-space of initial conditions for the planet $(X_0, Y_0, \dot{X}_0 = U_0, \dot{Y}_0 = V_0)$; using Poincaré surfaces of section, it can be shown (see e.g. Benest, 1978) that we may limit ourselves to $X_0 > 0$, $Y_0 = 0$ and $U_0 = 0$, with any value for $V_0$; therefore, an orbit is represented biunivocally by a point in the $(X_0,V_0)$ plane, in which we can easily recognize the region(s) corresponding to stable orbits. We regard an orbit as stable if there is neither escape nor collision during 100 revolutions of the binary.

A systematic exploration of the circular case (Benest, 1974, 1975, 1976) has shown that stable planetary orbits exist up to distances of their primary of the order of the binary's separation.

## 3. The $\alpha$ Centauri and Sirius Systems

For the exploration of the elliptic case, we take values for $\mu$ and $e$ which correspond to well-known nearby double –and sometimes multiple– systems.

The two first systems studied were $\alpha$ Centauri ($e = 0.52$, $\mu = 0.45$ for $\alpha$ Cen B and $\mu = 0.55$ for $\alpha$ Cen A; cf. Benest, 1988b) and Sirius ($e = 0.592$, $\mu = 1/3$ for Sirius B and $\mu = 2/3$ for Sirius A; Benest, 1989). The most important result is that we could confirm existence of very wide planetary orbits for both of these systems, even up to distances of the order or more of half the periastron separation between the two stars. Moreover, nearly circular orbits (i.e. for which the eccentricity remains under 0.1 during the integration time) were found around $\alpha$ Cen A as well as $\alpha$ Cen B up to distances of the order of a quarter of the periastron separation; this points out the fascinating possibility of stable planetary systems possibly existing around each of these stars, with planets even in the "habitable zone" (Benest, 1991).

## 4. The $\eta$ Coronae Borealis System

The study of the $\eta$ Coronae Borealis system has been recently undertaken; the masses are similar to those in the $\alpha$ Centauri system ($\mu = 0.45$ and 0.55), but the eccentricity is less: $e = 0.28$. First results show globally the same behaviour than for the latter (Figures 1 and 2): stable planetary orbits exist up to distances of their primary of the order between half and three quarter of the periastron separation of the binary, and nearly circular orbits (as defined in section 3) exist up to distances of the order of a quarter of the periastron separation.

More orbits have to be computed for more definitive statements about the boundaries of the stability zone, and the set of the nearly circular orbits.

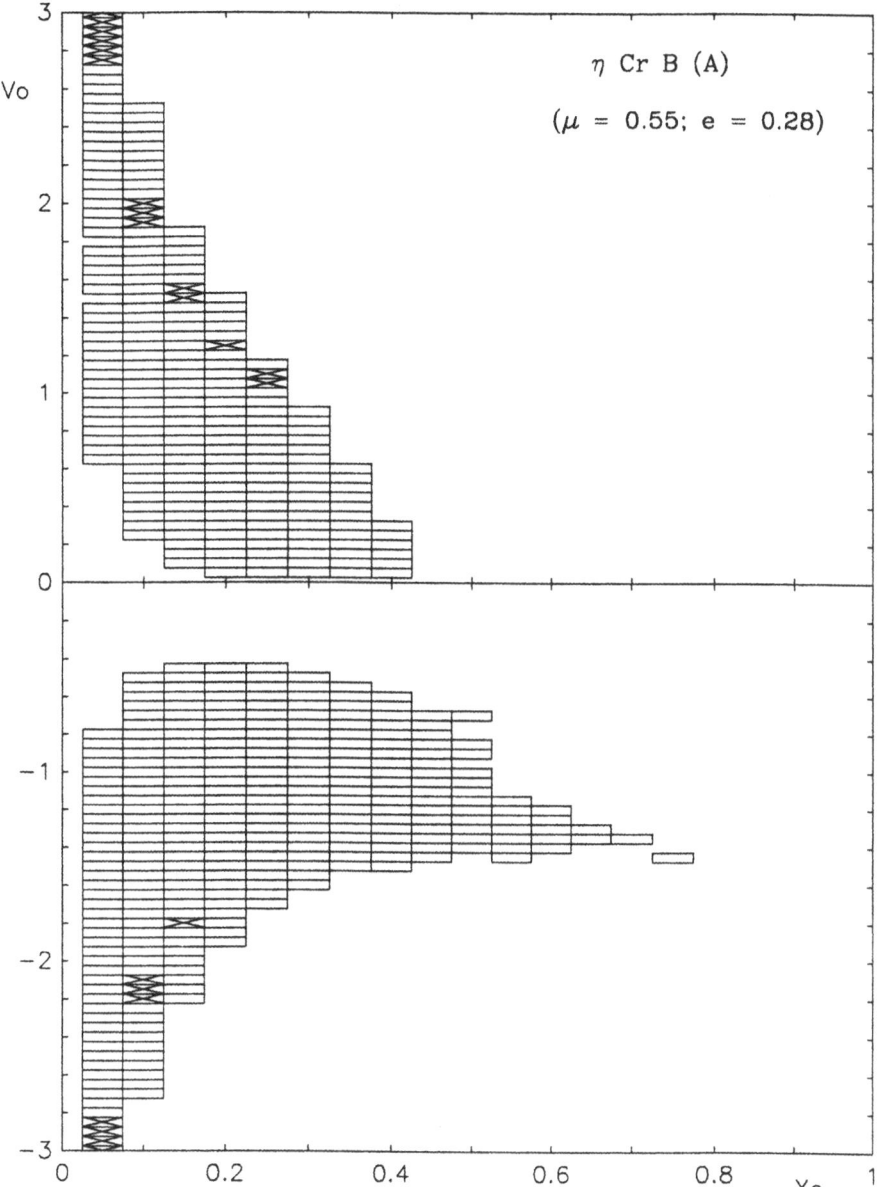

Fig. 1. Stable regions (hatchings) in the $(X_0, V_0)$ plane for planetary orbits around the heavier of the two stars in the $\eta$ Coronae Borealis binary system; crosses: nearly circular orbits; grid: $\Delta X_0 = 0.05$, $\Delta V_0 = 0.05$; 2420 orbits computed.

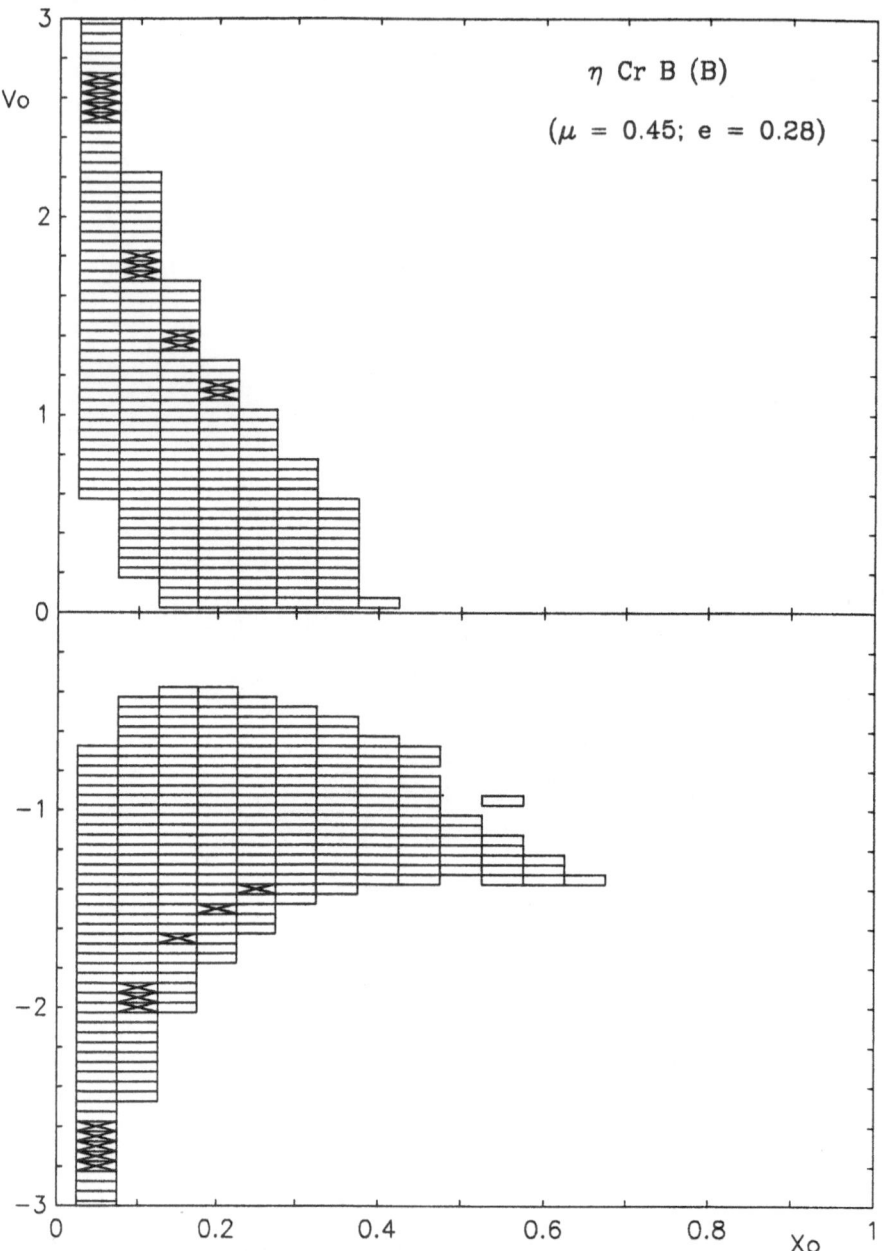

Fig. 2. Stable regions in the $(X_0, V_0)$ plane for planetary orbits around the lighter of the two stars in the $\eta$ Coronae Borealis binary system; same notations as for Figure 1.

## 5. Discussion and Prospect

Each binary studied confirms the main result obtained in the circular case, i.e. the existence of stable potential planetary orbits around one component of double stars up to distance of their parent star of the order or more of half the periastron separation of the binary. This systematic exploration of the elliptic plane restricted three-body problem will be continued for many other nearby binaries; our final goal is to establish the limiting value of $e$ beyond which no such stable planetary orbits exist.

During our study, we sometimes found orbits near the boundary of the stability zone showing some stochastic behaviour. A project for the future is to compute the Lyapunov Characteristic Exponents –these are a quantitative measure of chaos (Gonczi and Froeschlé, 1981; Froeschlé, 1984)– for all the orbits in order to establish, if possible, a relation between the stability as defined here and the stochasticity.

Such calculations, e.g. to get all the data for Figure 1, require very long computation times. In order to cut down on arithmetic, and in anticipation of introducing the computation of the Lyapunov Exponents, I intend to run a parallel version of the model on the "Connection Machine"; this version is being tested.

Finally, we must keep in mind that our study establishes from a dynamical point of view the existence of stable orbits for planets around each star in a binary system. This does not of course imply that such orbits are actually occupied by planets; this is a separate cosmogonic question: is the formation of planet(s) possible, and under which conditions, during or immediately after the formation of the binary system ? This still open question will be solved –we hope in a not too far future– by direct observations, and possibly supported by physically realistic numerical simulations of binary formation (cf. e.g. Tscharnuter, 1989).

## References

Benest, D. : 1974/1975/1976, 'Effects of the Mass Ratio on the Existence of Retrograde Satellites in the Circular Plane Restricted Problem. I/II/III.', *Astron. Astrophys.*, **32**, 39-46/**45**, 353-363/**53**, 231-236.

Benest, D. : 1978, 'Orbites de satellite dans le Problème Restreint.', Thesis, Obs. Nice.

Benest, D. : 1988a, 'Stable Planetary Orbits around one Component in Nearby Binary Stars.', *Celes. Mech.*, **43**, 47-53.

Benest, D. : 1988b, 'Planetary Orbits in the Elliptic Restricted Problem I. The $\alpha$ Centauri System.', *Astron. Astrophys.*, **206**, 143-146.

Benest, D. : 1989, 'Planetary Orbits in the Elliptic Restricted Problem II. The Sirius System.', *Astron. Astrophys.*, **223**, 361-364.

Benest, D. : 1991, 'Habitable Planetary Orbits around $\alpha$ Centauri and other binaries.', *Lecture Notes in Physics*, **390**, 44-47.

Dvorak, R. : 1984, 'Numerical Experiment on Planetary Orbits in Double Stars.', *Celes. Mech.*, **34**, 369-378.

Dvorak, R. : 1986, 'Critical Orbits in the Elliptic Restricted Three-Body Problem.', *Astron. Astrophys.*, **167**, 379-386.

Dvorak, R., Froeschlé, Ch., Froeschlé, C. : 1989, 'Stability of Outer Planetary Orbits (P-Types) in Binaries.', *Astron. Astrophys.*, **226**, 335-342.

Froeschlé, C. : 1984, 'The Lyapunov Exponents. Application to Celestial Mechanics.', *Celes. Mech.*, **34**, 95-115.

Gonczi, R., Froeschlé, C. : 1981, 'The Lyapunov Characteristic Exponents as Indicators of Stochasticity in the Three Body Restricted Problem.', *Celes. Mech.*, **25**, 271-280.

Pavanini, G. : 1907, 'Sopra una nuova categoria di soluzioni periodiche nel problema dei 3 corpi.', *Annali di Matematica*, Serie III, **13**, 179-202.

Rabl, G., Dvorak, R. : 1988, 'Satellite-type Planetary Orbits in Double Stars: A Numerical Approach.', *Astron. Astrophys.*, **191**, 385-391

Szebehely, V. : 1980, 'Stability of Planetary Orbits in Binary Systems.', *Celes. Mech.*, **22**, 7-12.

Szebehely, V. : 1992, 'On the Stability of Outer Planetary Orbits around Binary Stars.', *Celes. Mech.*, this volume.

Szebehely, V., McKenzie, R. : 1981, 'Stability of Outer Planetary Systems.', *Celes. Mech.*, **23**, 3-7.

Tscharnuter, W.M.: 1989, 'Collapse Calculations.', in *E.S.O. Workshop on Low Mass Star Formation and Pre-Main Sequence Objects*, pp. 75-87.

# STABILITY OF OUTER PLANETARY ORBITS AROUND
# BINARY STARS: A COMPARISON OF HILL'S AND
# LAPLACE'S STABILITY CRITERIA

A. KUBALA and D. BLACK
*Lunar and Planetary Institute*
*Houston, Texas, USA 77058*

and

V. SZEBEHELY
*The University of Texas*
*Austin, Texas, USA 78712*

**Abstract.** A comparison is made between the stability criteria of Hill and that of Laplace to determine the stability of outer planetary orbits encircling binary stars. The restricted, analytically determined results of Hill's method by Szebehely and co-workers and the general, numerically integrated results of Laplace's method by Graziani and Black are compared for varying values of the mass parameter $\mu = m_2/(m_1 + m_2)$. For $0 \leq \mu \leq 0.15$, the closest orbit (lower limit of radius) an outer planet in a binary system can have and still remain stable is determined by Hill's stability criterion. For $\mu > 0.15$, the critical radius is determined by Laplace's stability criterion. It appears that the Graziani-Black stability criterion describes the critical orbit within a few percent for all values of $\mu$.

**Key words:** Planetary orbits – stability.

## 1. Introduction

Studying the orbital stability of three-body systems is essential to various fields in astronomy, astrophysics, and aerospace engineering, including such problems as triple-star systems, perturbed satellite orbits, and planetary orbits. Although we still have no definitive evidence of a planetary system other than our own, the search for other planetary systems has gained much support in recent years. Orbital stability studies and three-body calculations can help to constrain and possibly predict the location and type of orbits we may expect to observe in such systems.

This paper is concerned with the stability of outer planetary orbits in binary star systems, where an "outer" planet is one with an orbit that encloses both primaries. Our focus is on binaries because the majority of stars occur in binary systems. The basic question concerning the stability of outer planets can be formulated as follows: for a binary system with given masses ($m_1 \geq m_2$), what is the smallest radius that an outer planet in an approximately circular orbit can have for stability? If the outer planet is far enough from the binary, the perturbation of the binary will not disturb the planet's orbit, and it can be considered stable. If, on the other hand, the size of the outer planetary orbit is below the limit for stability, the binary would disrupt the planet's orbit. It should be noted that if the planetary orbit is very far from the binary, its orbit might be perturbed by other members of the

*Celestial Mechanics and Dynamical Astronomy* **56**: 51–68, 1993.
© 1993 *Kluwer Academic Publishers.*

galactic system, and the binary could lose the planet. This type of instability is not considered in the present paper.

Comparison of various stability investigations is a highly sensitive procedure as the results depend strongly on the definitions used for stability. In 1984, a review (Szebehely, 1984) resulted in approximately 50 different definitions and terminologies used for stability. This number today would be close to 100, indicating the increased interest in stability research.

The two concepts to be compared in this paper are associated with concepts due to Hill and to Laplace. Because many variations of their original formulations and definitions exist today, we will clarify and explain the exact approaches being compared here. Hill's stability criterion uses the concept of zero-velocity surfaces. The stability of an outer planet depends on the value of its actual Jacobian constant as compared to the critical value of the Jacobian constant at the equilibrium point (Szebehely, 1980; Szebehely and McKenzie, 1982). Stability according to Laplace means that no secular trends develop in the orbital elements of any of the bodies during the evolution of the system (Graziani and Black, 1981; Black, 1982; Pendelton and Black, 1983). The difference between these two definitions is considerable in the outer planet case. Hill's stability requirement is still satisfied if the planet escapes from the binary as long as it stays outside Hill's limiting zero velocity curve.

Laplace's concept, on the other hand, considers the escape of the outer planet as a form of instability.

We will compare the results of stability analyses for planar motions in this paper, but both approaches can be generalized to three-dimensional configurations. We will also assume circular motions for the binary. This assumption can also be generalized in the Laplace approach, but it may present some conceptual difficulties for Hill's method which is based on the existence of the Jacobian integral (which does not exist if the members of the binary move on elliptic orbits).

## 2. Review of Previous Work

In 1982, a comparison was made (Black, 1982) between numerically integrated, general three-body problem (TBP) results and an analytically determined set of results for the restricted problem. The analytic results, based on Hill's method (Szebehely, 1980; Szebehely and McKenzie, 1981), are shown in Figure 1 where $\mu = m_2/(m_1 + m_2)$ and $r$ is the distance between the planet and the center of mass of the binary. We are here using the following definitions: an "outer" planet is one whose orbit encircles the binary, "inner" planet refers to a planet that orbits the larger of the primaries, and "satellite" refers to a planet orbiting the smaller primary.

The results of a study using Laplace's method (Graziani and Black, 1981; Black, 1982) are shown in Figure 2 as a solid line. This line, the Graziani-Black (GB) stability criterion, denotes the division between stable (below the line) and unstable

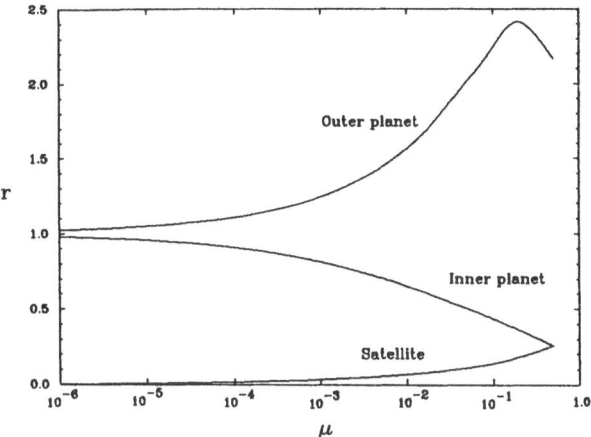

Fig. 1.  Hill's critical orbital radii as functions of $\mu$ for outer planets, inner planets, and satellites.

(above the line) orbits. In this approach,

$$\mu_B = \frac{m_2 + m_3}{2m_1} \quad \text{where} \; m_1 \geq m_2 \geq m_3 \; \text{and} \; \Delta = \frac{2(R_3 - R_2)}{R_3 + R_2},$$

$$\text{where} \; R_2 = \overline{m_1 m_2} \quad \text{and} \; R_3 = \overline{m_1 m_3}.$$

When the data from the restricted, analytical (Hill) approach for the so-called "inner", "outer" and "satellite" configurations are expressed in the variables $\mu_B$ and $\Delta$, they can be graphed with the GB line for comparison. These results are also shown in Figure 2. Note how the orbits that define three separate lines in Figure 1 are all consistent with the single line (the GB criterion) when expressed in these variables.

The results show remarkable good quantitative agreement for all twelve orders of magnitude ($10^{-6}$ to $10^{6}$) in $\mu_B$. The largest disagreement between the two sets of results is for an outer planet with $\mu = \mu_B = 0.5$. The limiting radius (lower limit) for an outer planet will be derived as a function of $\mu$ using both methods in the next section, and Section 4 will present a comparison of these results.

For the elliptic problem see several papers by Benest (1976–1988) and by Dvorak (1984, 1986). Regarding families of periodic orbits and their stability properties see Froeschlé (1971) and Hadjidemetriou (1976).

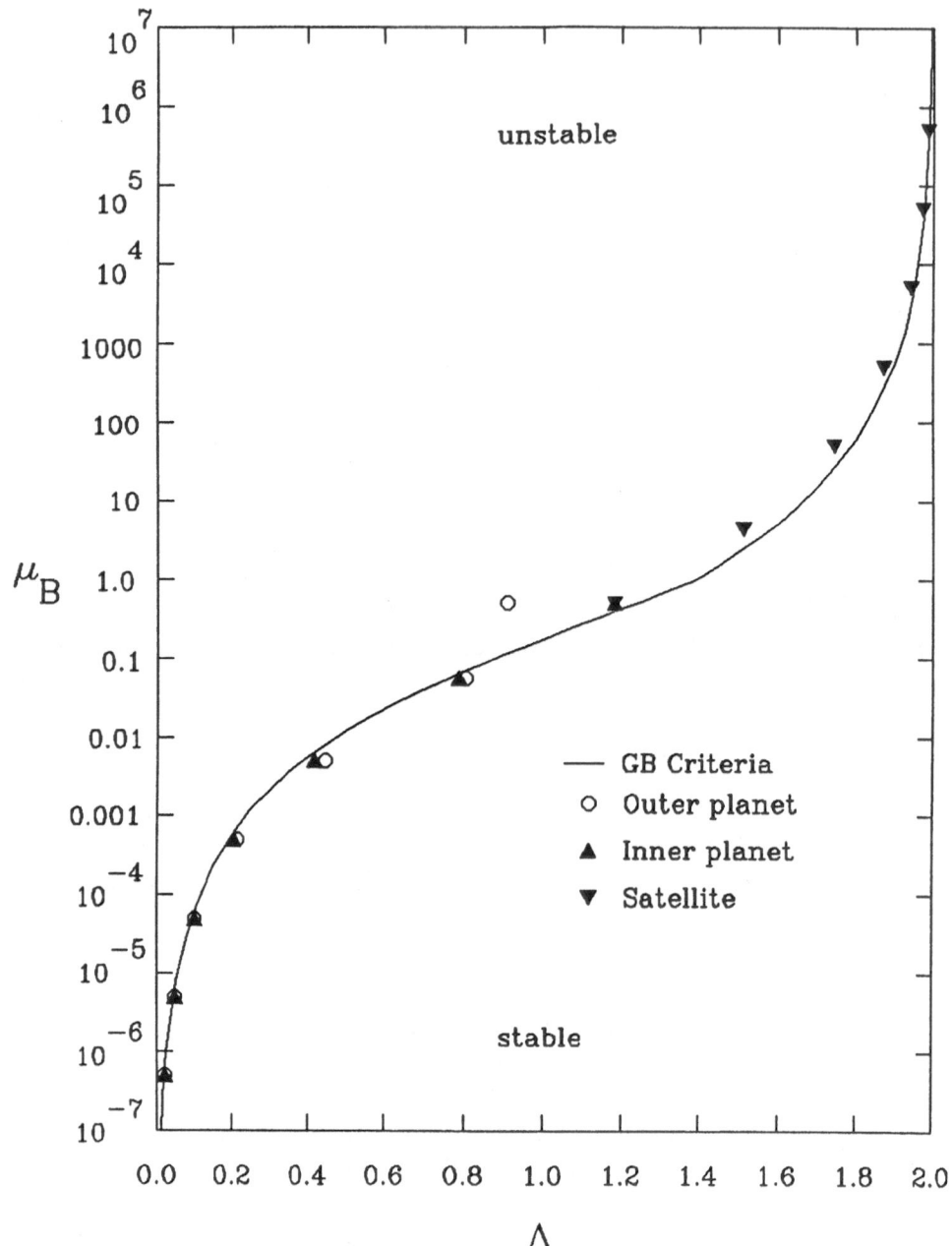

Fig. 2. The Graziani-Black stability criterion compared with data from Hill's method.

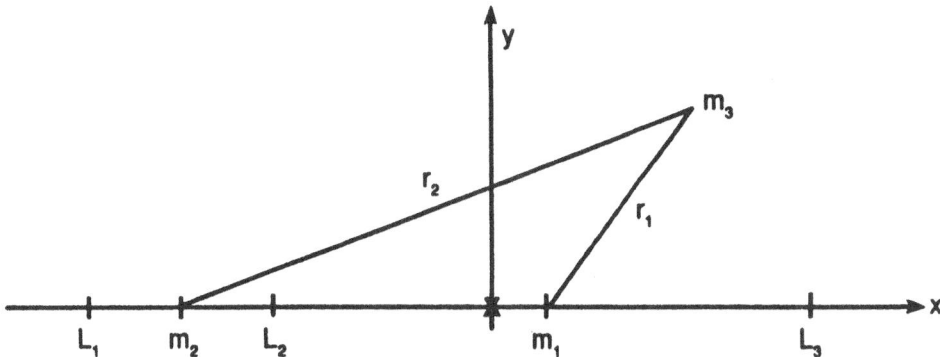

Fig. 3. General arrangement of the masses and of the collinear equilibrium points in the restricted problem of three bodies.

### 3A. Hill's Stability Criterion of Outer Planetary Orbits Around Binaries

Hill's (1878) stability method is related to the curves of zero velocity of the restricted three-body problem. At a given value of the mass-parameter $\mu$ the locations of the equilibrium points are evaluated. The locations of these points are the solutions of the equations

$$\frac{\partial \Omega}{\partial x} = \frac{\partial \Omega}{\partial y} = 0, \tag{1}$$

where

$$\Omega = \frac{1}{2} \left[(1 - \mu)r_1^2 + \mu r_2^2\right] + \frac{1 - \mu}{r_1} + \frac{\mu}{r_2}, \tag{2}$$

$x$ and $y$ are the rotating or synodic coordinates of the third body ($m_3 \ll m_2 \leq m_1$), and $r_1$ and $r_2$ are the distances between the primaries and the third body. The origin of the synodic system is located at the center of mass of the primaries. The primary with the larger mass is located at the right side of the origin of the system at $x = \mu$, and the primary with the smaller mass is located at $x = \mu - 1$. Figure 3 shows the general arrangement, including the collinear equilibrium points $L_1$, $L_2$ and $L_3$. For the notation and the arrangement of the participating masses, see Szebehely (1967).

The locations of the collinear equilibrium points are obtained as the roots of the equation

$$\left(\frac{\partial \Omega}{\partial x}\right)_{y=0} = 0. \tag{3}$$

The abscissa of $L_1$ is $x_1$, and the corresponding distances from the primaries are $r_1 = \mu - x_1$ and $r_2 = \mu - x_1 - 1$. The equation for $x_1$ becomes

$$F(x_1, \mu) = x_1 + \frac{1 - \mu}{(x_1 - \mu)^2} + \frac{\mu}{(x_1 - \mu + 1)^2} = 0. \tag{4}$$

This is a fifth order algebraic equation for $x_1$. The solution, therefore, must be found by iteration. The result is an approximately established relation, $x_1 = f(\mu)$, and is shown on the lower part of Figure 4.

For outer planetary orbits, Hill's stability requirement is that the Jacobian constant of the motion of the third body $(C)$ should be larger than (or equal to) the Jacobian constant at $L_1$. The Jacobian constant of the third body is given by

$$C = 2\Omega - v^2, \tag{5}$$

where the function $\Omega(x, y)$ and the velocity are to be evaluated at any point on the orbit.

The approximately circular motion of the third body in the fixed system requires the dimensionless velocity $v_a = 1/\sqrt{r}$ if the motion is counter-clockwise and $v_a = -1/\sqrt{r}$ is the motion is clockwise. The rotating system is revolving in the counter-clockwise direction and its velocity at a distance $r$ from the origin is $r$ since the value of the angular velocity is one $(v_s = r)$. The velocity of $m_3$ relative to the rotating system is

$$v = v_{\rm rel} = v_a - v_s = \pm\frac{1}{\sqrt{r}} - r. \tag{6}$$

If $r > 1$, the velocity of the system at $r$ is larger than the absolute circular velocity of $m_3$ since $v_s = r > v_a = \pm 1/\sqrt{r}$, and, therefore, $v$ is always negative and the motion is retrograde in the rotating system. In what follows, we select $v_a = 1/\sqrt{r}$, and our orbits will be direct (counter-clockwise) in the fixed system, but retrograde in the rotating system since we study outer planetary orbits for which $r > 1$.

Using Eqns. (5) and (6), the Jacobian constant of the third body can be expressed as a function of $\mu$ and $r$

$$C = 2\Omega(r) + 2\sqrt{r} - r^2 - \frac{1}{r}. \tag{7}$$

The Jacobian constant at $L_1$ is obtained from Eqn. (5):

$$C_1 = 2\Omega[x_1(\mu), 0]. \tag{8}$$

The function $C_1(\mu)$ is shown on the upper part of Figure 4.

For a given value of $\mu$, the relation $C = C_1$ will contain the unknown $r$ in the form of a seventh order algebraic equation, which, solved by iteration, results in an approximate value $r(\mu)$. The approximation at this point assumes that the

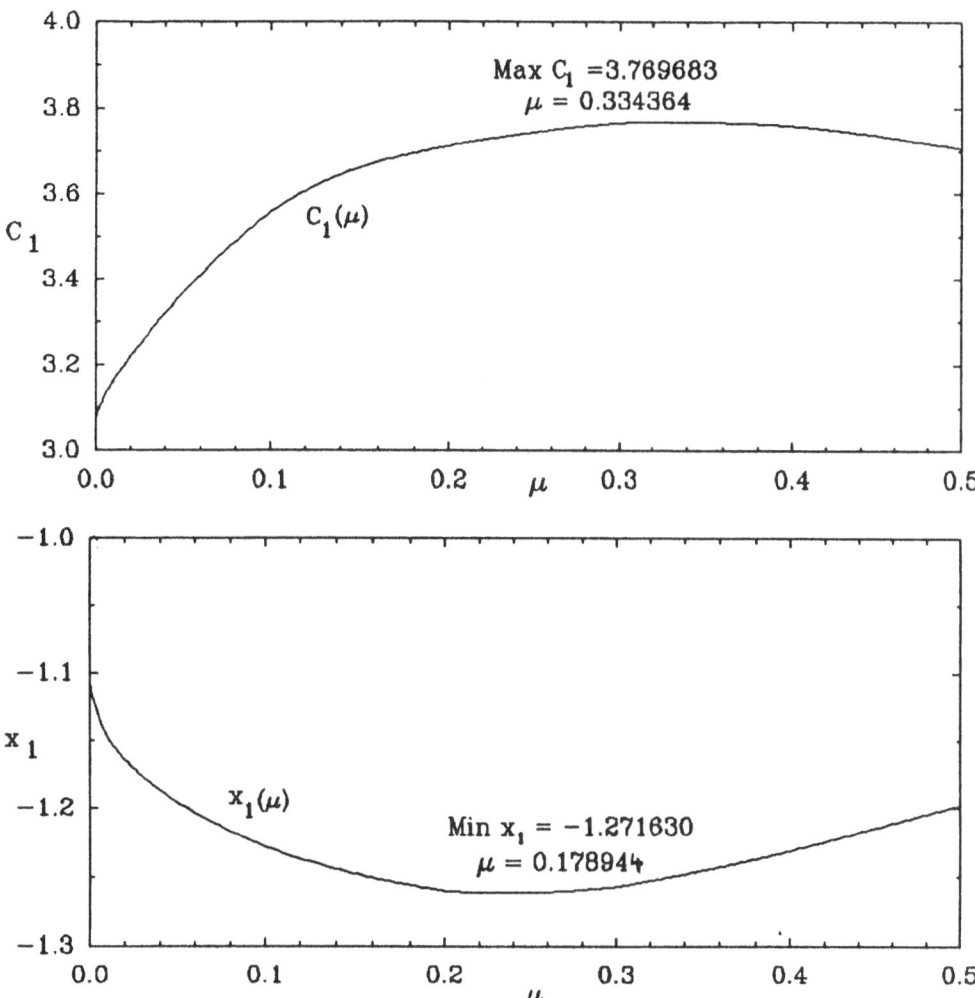

Fig. 4. Representation of the non-monotonic nature of the functions $C_1(\mu)$ and $x_1(\mu)$, where $C_1(\mu)$ is the value of the Jacobian constant at the equilibrium point $L_1$ and $x_1(\mu)$ is the location of $L_1$ relative to the center of mass of the binary.

total mass of the primaries is located at the origin and uses the circular velocity, or, in other words, using a circular approximation for Poincaré's periodic orbit of "premiere sorte."

The function $\Omega(r)$ can be evaluated two different ways. If the radius of the approximately circular orbit is $r$ (as measured from the origin), then this circle intersects the $x$ axis at two points. When $r_1 = r - \mu$ and $r_2 = r - \mu + 1$, the intersection is on the positive $x$ axis, and when $r_1 = r + \mu$ and $r_2 = r + \mu - 1$, the intersection is located on the negative $x$ axis. In the first case $r_2 = 1 + r_1$, while in the second case $r_1 = 1 + r_2$. The corresponding values of the Jacobian constant are denoted by $C_R$ and $C_L$ (subscript $R$ for right hand side, $+x$ axis, and $L$ for left hand side, $-x$ axis).

From Eqns. (7) and (2), we have

$$C_R = \mu(1 - \mu) + \frac{2\mu}{r + 1 - \mu} + \frac{2(1 - \mu)}{r - \mu} - \frac{1}{r} + 2\sqrt{r}$$

$$\text{and} \tag{9}$$

$$C_L = \mu(1 - \mu) + \frac{2\mu}{r - 1 + \mu} + \frac{2(1 - \mu)}{r + \mu} - \frac{1}{r} + 2\sqrt{r}.$$

For given values of $r$ and $\mu$ the above values of the Jacobian constants differ by

$$\Delta C = C_R - C_L = \frac{4\mu(1 - \mu)(2\mu - 1)}{(r^2 - \mu^2)[r^2 - (1 - \mu)^2]} \, .$$

This difference is zero when $\mu = 0$, $\mu = 0.5$ or when $r \to \infty$, as expected. Because of this difference, the solution of the previously mentioned equation $C = C_1$ depends on the use of $C_R$ and $C_L$ in place of $C$.

For a given value of $\mu$, the computational process is to obtain first $x_1$ and from this the corresponding value of $C_1$. Then from $C_R = C_1$ or $C_L = C_1$ the value of $r$ is computed. Because for a given value of $\mu$ the functions $C_R(r, \mu)$ and $C_L(r, \mu)$ are not equal, the solutions for $r$ will be different. At $\mu = 0.1$, for instance, $C_1 = 3.5566844$ and the value of $r$ using $C_1 = C_R$ becomes $r_R = 2.2729$. For the same value of $\mu$, using the relation $C_1 = C_L$, the value of $r$ is $r_L = 2.2431$.

Table I shows the values of $r_R$, $r_L$, $C_1$, $\Delta r$ and $r_{ave}$ for various values of $\mu$. Here $\Delta r = r_R - r_L$ and $r_{ave} = (1/2)(r_R + r_L)$. The difference between $r_R$ and $r_L$ is below 1.8%.

No closed form analytic solution exists for our problem. The multiple roots of the algebraic equations present no serious difficulties because only real roots are of interest, and we ignore any solutions that do not satisfy $x_1 < 0$ and $r > 1$. Furthermore, when more than one root satisfies the condition $r > 1$, we select the largest root. As a consequence of approximations and because no closed form analytic solutions exist, the functional relations $x_1(\mu)$, $C_1(\mu)$ and $r(\mu)$ are known only approximately. The establishment of the qualitative properties of these relations and the extreme values of the above listed functions, $\min x_1(\mu)$, $\max C_1(\mu)$, will be shown in Appendix part 1.

TABLE I
Comparison of the Values of Hill's Critical Radii.

| $\mu$ | $r_R$ | $r_L$ | $C_1$ | $\Delta r$ | $r_{ave}$ |
|---|---|---|---|---|---|
| 0.5 | 2.1652 | 2.1652 | 3.7068 | 0.0000 | 2.1652 |
| 0.4 | 2.3311 | 2.3136 | 3.7589 | 0.0176 | 2.3224 |
| 0.3 | 2.4329 | 2.4077 | 3.7664 | 0.0252 | 2.4203 |
| 0.2 | 2.4409 | 2.4124 | 3.7124 | 0.0285 | 2.4267 |
| 0.1 | 2.2729 | 2.2431 | 3.5567 | 0.0298 | 2.2580 |
| $10^{-2}$ | 1.5911 | 1.5632 | 3.1642 | 0.0279 | 1.5772 |
| $10^{-3}$ | 1.2577 | 1.2385 | 3.0396 | 0.0192 | 1.2481 |
| $10^{-4}$ | 1.1153 | 1.1045 | 3.0090 | 0.0108 | 1.1099 |
| $10^{-5}$ | 1.0526 | 1.0471 | 3.0020 | 0.0055 | 1.0499 |
| $10^{-6}$ | 1.0242 | 1.0215 | 3.0004 | 0.0027 | 1.0229 |
| 0 | 1.0000 | 1.0000 | 3.0000 | 0.0000 | 1.0000 |

## 3B. Laplace's Stability Criterion of Outer Planetary Orbits Around Binaries

According to Laplace, a planetary system is dynamically unstable if (1) the general arrangement shows secular changes or the motion becomes erratic, (2) the order of the participating bodies changes (orbit crossing), or (3) collisions or escapes occur. Orbits for a wide range of masses and separations were numerically integrated (Graziani and Black, 1981) and subsequently extended (Black, 1982) in an attempt to determine a general stability criterion using a Laplacian definition of stability. The orbits were integrated for $10^4$ revolutions of the longest period orbit, and the total energy and angular momentum were held constant to at least one part in $10^{10}$. The general arrangement of the masses and separations is shown in Figure 5.

The basic result of this study was that a planetary system is dynamically unstable if

$$\mu_B = \frac{m_2 + m_3}{2m_1} \geq \mu_{cr} = 0.175 \frac{\Delta^3}{(2 - \Delta)^{3/2}} \quad \text{for } \mu_B \leq 1$$

$$\text{and } \mu_{cr} = 0.083 \frac{\Delta^3}{(2 - \Delta)^3} \quad \text{for } \mu_B \geq 1 ,$$

(10)

where

$$\Delta = \frac{2(R_3 - R_2)}{R_3 + R_2} .$$

(11)

This result is the Graziani-Black (GB) stability criterion and was shown in Figure 2. For a given set of masses, $\mu_B$ can be calculated using the first part of

A. KUBALA ET AL.

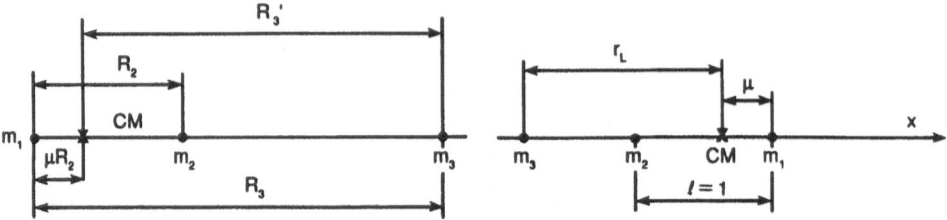

Fig. 5.  Comparison of the general arrangement of masses and separations for the general TBP (GB approach – on left) and restricted TBP (Szebehely approach – on right) and the variables used in each.

Eqn. (10). Eqn. (10) can then be used to determine $\Delta$ by iteration or by using the inverse relation

$$\Delta = \frac{-\mu_{cr}^{2/3} + \mu_{cr}^{1/3}\sqrt{\mu_{cr}^{2/3} + \alpha}}{\beta} \tag{12}$$

where $\alpha = 8 \times 0.175^{2/3} = 2.5029299$ and $\beta = \alpha/4 = 0.6257325$. Eqn. (11) can be written as

$$\Delta = 2\left[\frac{\dfrac{R_3}{R_2} - 1}{\dfrac{R_3}{R_2} + 1}\right] \quad \text{or} \quad \frac{R_3}{R_2} = \frac{2 + \Delta}{2 - \Delta} \tag{13}$$

to determine the ratio $R_3/R_2$.

The distance between the center of mass and $m_3$ is $R_3' = R_3 - R_2\mu$ as shown in Figure 5. Because the distance between the primaries ($R_2$) corresponds to $\ell$ in the circular restricted problem,

$$\frac{R_3'}{\ell} = \frac{R_3}{\ell} - \mu = r_L' \tag{14}$$

using the notation of Section 3A.

An assumption must be made at this point to keep these results consistent with those in the previous section. We need to restrict our discussion to the restricted three-body problem. If $m_3 = 0$, the relationship between the previously defined mass-parameter ($\mu$) and $\mu_B$ becomes

$$\mu_B = \frac{1}{2}\frac{\mu}{1 - \mu}, \quad \text{or} \quad \mu = \frac{2\mu_B}{1 + 2\mu_B}, \tag{15}$$

where

$$\mu_B = \frac{m_2}{2m_1}, \quad \mu = \frac{m_2}{m_1 + m_2}$$

and $\mu_B \leq \mu$. According to these definitions both mass-parameters satisfy the inequality $0 \leq (\mu_B, \mu) \leq 1/2$. The lower limit corresponds to $m_2 = 0$ and the upper limit to $m_1 = m_2$. Table II shows the values of $\mu_B$, $\Delta$, $R_3/R_2$ and $r_L'$ for various values of $\mu$ using the GB (Laplacian) approach.

TABLE II
Laplacian Critical Radii.

| $\mu$ | $\mu_B$ | $\Delta$ | $R_3/R_2$ | $r'_L$ |
|-------|---------|----------|-----------|--------|
| 0.00 | 0.00000 | 0.00 | 1.000 | 1.000 |
| 0.01 | 0.00505 | 0.39 | 1.484 | 1.474 |
| 0.05 | 0.0263 | 0.62 | 1.898 | 1.848 |
| 0.1 | 0.055 | 0.76 | 2.226 | 2.126 |
| 0.2 | 0.125 | 0.93 | 2.740 | 2.540 |
| 0.3 | 0.214 | 1.05 | 3.210 | 2.910 |
| 0.4 | 0.333 | 1.15 | 3.684 | 3.284 |
| 0.5 | 0.5 | 1.23 | 4.195 | 3.695 |

## 4. Comparison of Results Using Hill's and Laplace's Stability Criteria

Hill's approach gives the critical radius $r_R \cong r_L \cong r_{\text{ave}}$ as a function of $\mu$, using the arrangement of

$$r_L = \overline{Om_3} \quad \text{and} \quad \ell = \overline{m_1 m_2} = 1 .$$

The Laplacian approach is based on Eqns. (10) and (11), using the arrangement of

$$R_2 = \overline{m_1 m_2} \quad \text{and} \quad R_3 = \overline{m_1 m_3} .$$

The two systems are related to each other by Eqns. (14) and (15). Figure 6 shows the functions $r_L(\mu)$ and $r'_L(\mu)$.

Note that in Figure 6 the value $r_L$ is used (not $r_R$) for the Hill result. In the GB system the primary with the larger mass $(m_1)$ is located on the left side of the center of mass, while in the restricted problem $m_1$ is on the right side of the center of mass of the primaries (see Figure 5). In the GB system the order of masses from left to right is $m_1$, $CM$, $m_2$, $m_3$. In the restricted problem the order is $m_2$, $CM$, $m_1$ (again from left to right). In order for the two systems to have the same relative orders and separations, $m_3$ should be placed on the left side of $m_2$. This way the arrangement in the restricted problem is $m_3$, $m_2$, $CM$, $m_1$ and the distance between $CM$ and $m_3$ becomes $r_L$.

Also note that in Figure 6, $r'_L$ (the Laplacian stability result) varies approximately linearly with the mass-parameter $\mu$ in the range $0.1 \le \mu \le 0.5$. This is not the case for $\mu_B$, where there is no valid linear approximation. It can be shown either by curve-fitting, by least-squares technique, or by analytical expansions, that

$$\frac{R_3}{R_2} \cong A\mu + B, \tag{16}$$

where $A = 4.92250$ and $B = 1.73375$.

As a consequence, an analytical approximation (in the range of $0.1 \le \mu \le 0.5$) can also be established for the dependence of $R_3/R_2$ on $\mu_B$:

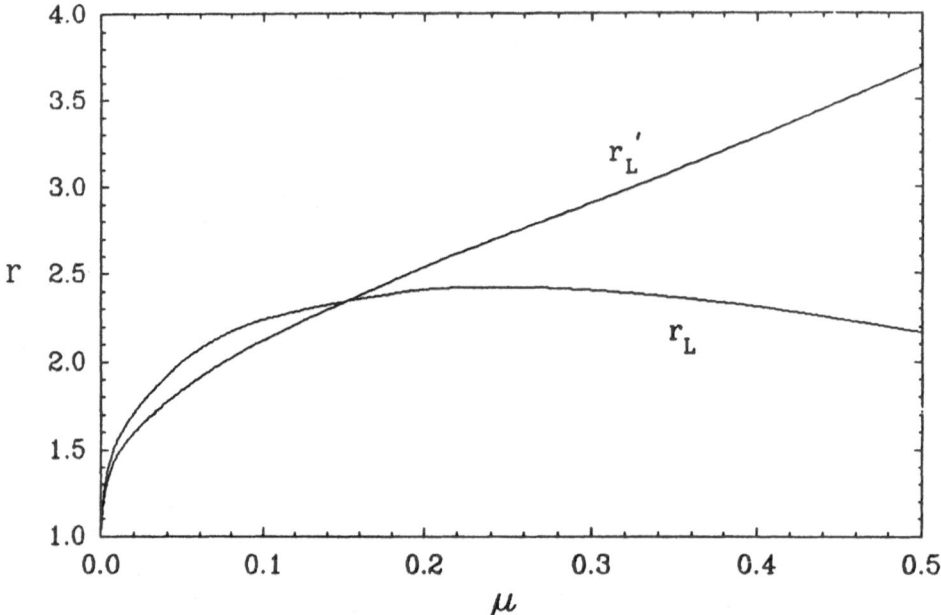

Fig. 6. The radius of the critical orbit (minimum distance) of an outer planet from the center of mass of the binary. Comparison of $r'_L$ (Laplace/GB approach) and $r_L$ (Hill/Szebehely approach).

$$\frac{R_3}{R_2} \cong A \, \frac{2\mu_B}{1 + 2\mu_B} + B \, . \tag{17}$$

If this is substituted in Eqn. (11) we have

$$\Delta = C \, \frac{\mu_B + D}{\mu_B + E} \tag{18}$$

where $C = 1.47755$, $D = 0.064862$ and $E = 0.178531$. Note that the range of validity of these approximations is $0.055 \leq \mu_B \leq 0.5$, since $\mu = 0.1$ corresponds to $\mu_B = 0.055$.

Eqns. (16)–(18) simplify the computation of the quantity $\Delta$ which is related to $\mu_{cr}$ by Eqn. (12). Using this linear approximation, we can compute $\mu_B$ for a given binary and from Eqn. (17) evaluate $R_3/R_2$ or $R'_3 = R_3 - R_2\mu = r'_L$.

Figure 7 shows the locations of the masses of the binary and the limiting radii of the outer planetary orbits ($r_L$ and $r'_L$) for various values of the mass-parameter. This figure also shows the location of the first collinear equilibrium point ($L_1$). As we move down from the top of the figure, we observe that $L_1$ and $r_L$ move to the left, i.e., they move away from the center of mass of the binary. The maximum distances are reached approximately at $\mu \cong 0.2$. As $\mu$ increases further, i.e., we move down in Figure 7, the trend reverses and both distances ($\overline{L_1 O}$ and $r_L$) decrease. At the same time $r'_L$, representing the GB results, shows a monotonic increase with $\mu$.

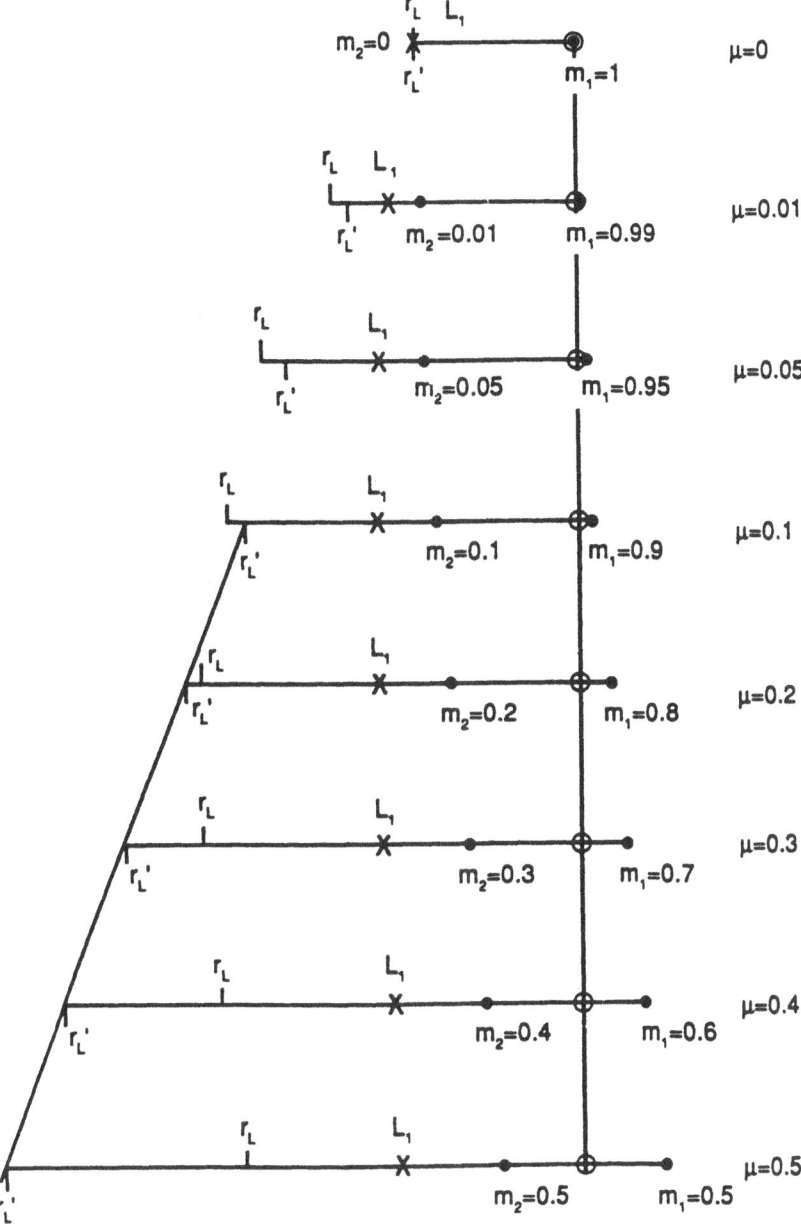

Fig. 7. Summary of the location of the masses, $L_1$, and the radii of the critical outer planet orbits of Hill $(r_L)$ and of Laplace $(r'_L)$.

## 5. Conclusion

Figures 6 and 7 show that $r_L(\mu)$ is not a monotonically increasing function (neither is $x_1(\mu)$ or $C_1(\mu)$ as shown in the Appendix). It has a maximum value of $r_L = 2.4249$ that occurs at $\mu = 0.2439$ (as shown in Appendix part 1). The function $r'_L(\mu)$, however, is monotonically increasing with $\mu$ and is approximately linear for $0.1 \leq \mu \leq 0.5$.

These figures also show that $r_L > r'_L$ for $0 \leq \mu \leq 0.15$ and that $r'_L > r_L$ for $\mu > 0.15$. As we are concerned with outer planets, we are looking for the smallest radius (lower limit) a planet can have and still be stable. Therefore, we should use whichever $r$ is larger ($r_L$ or $r'_L$) so that both stability criteria are satisfied. For $0 \leq \mu \leq 0.15$, the Hill stability criterion is the larger, and $r_{CR} = r_L$ should be used. For $\mu > 0.15$, the Laplace stability criterion is larger, so $r_{CR} = r'_L$ should be used. When $\mu \leq 0.15$, the difference between the critical orbit size as given by the two criteria is less than 7 percent, whereas the difference is as great as 44 percent for $\mu > 0.15$. It appears that the Graziani-Black stability criterion describes the critical orbit within a few percent for all values of $\mu$.

## Appendix

PART 1. THE DETERMINATION OF min $x_1(\mu)$.

Introducing $r_2 = \overline{L_1 m_2} = \mu - x_1 - 1$ as the new variable, Eqn. (4) can be written as

$$F(r_2, \mu) = \mu - 1 - r_2 + \frac{1 - \mu}{(1 + r_2)^2} + \frac{\mu}{r_2^2} = 0, \tag{A.1}$$

or

$$\mu = \frac{r_2^3(r_2^2 + 3r_2 + 3)}{r_2^4 + 2r_2^3 + r_2^2 + 2r_2 + 1} = g(r_2). \tag{A.2}$$

The condition for extremum, using the new variable $r_2$, becomes $dr_2/d\mu = 1$, since

$$x_1 = \mu - 1 - r_2 \text{ and } \frac{dx_1}{d\mu} = 1 - \frac{dr_2}{d\mu} = 1 - \frac{1}{g'},$$

where $g' = dg/dr_2$. The condition for extremum becomes

$$\frac{dg(r_2)}{dr_2} = 1,$$

or

$$6r_2^5 + 15r_2^4 + 16r_2^3 + 3r_2^2 - 4r_2 - 1 = 0.$$

The only positive root of this equation is $r_2 = 0.450574$. The corresponding value of $\mu$ is obtained from Eqn. (A.2) as $\mu = 0.178944$ and, therefore, $x_1$ becomes $-1.271630$.

The second derivative of the function $x_1(\mu)$ is

$$\frac{d^2 x_1}{d\mu^2} = \frac{-d^2 r_2}{d\mu^2} = \frac{g''}{(g')^3},$$

where $g' = dg/dr_2 = 1$ and $g'' = d^2 g/d^2 r_2$ is to be evaluated at $r_2 = 0.450574$. Using Eqn. (A.2) we have

$$g'' = \left[\frac{r_2}{N}\right]^2 (r_2^6 + 4r_2^5 + 6r_2^4 + 14r_2^3 + 26r_2^2 + 24r_2 + 9),$$

where

$$N = r_2^4 + 2r_2^3 + r_2^2 + 2r_2 + 1.$$

This form of $g''$ shows that $g'' > 0$ for any positive value of $r_2$ and consequently $g'' > 0$ when $r_2 = 0.450574$. Therefore, the above $x_1$ value corresponds to a minimum of the $x_1 = x_1(\mu)$ function.


PART 2. DETERMINATION OF max $C_1(\mu)$.

The value of the Jacobian constant at $x_1$ is a function of the mass-parameter as shown by Eqn. (8). On the $x$ axis, the function $\Omega$ becomes $\Omega(x, \mu)$ and

$$\frac{d\Omega(x, \mu)}{d\mu} = \frac{\partial \Omega}{\partial \mu} + \frac{\partial \Omega}{\partial x} \frac{dx}{d\mu}.$$

If the above equation is evaluated at the first collinear equilibrium point $(x = x_1)$, we have

$$\frac{d\Omega(x_1, \mu)}{d\mu} = \frac{\partial \Omega}{\partial \mu},$$

since at $x = x_1$, $\partial \Omega / \partial x = 0$. In order to find the extremum of $C_1$ as a function of $\mu$ we evaluate $\partial \Omega / \partial \mu = 0$ from the expression given by Eqn. (2) and obtain

$$\frac{\partial \Omega}{\partial \mu} = \frac{r_2^2 - r_1^2}{2} + \frac{1}{r_2} - \frac{1}{r_1} = 0.$$

Since $y = 0$ and $L_1$ is located left of $m_2$, we have $r_1 = 1 + r_2$, and the condition for the extremum of $C_1$ becomes

$$\frac{2 - r_2(1 - r_2)(1 + 2r_2)}{2r_2(1 + r_2)} = 0.$$

The only positive root of the cubic expression in the numerator is $r_1 = 0.583156$. The corresponding value of the mass-parameter is obtained from Eqn. (A.2) as $\mu = 0.334364$ and $C_1 = 3.769683$.

PART 3. THE DETERMINATION OF max $r(\mu)$.

In parts 1 and 2, it was shown that neither the location of $L_1$ nor the value of the Jacobian constant at this point are monotonically increasing functions of the mass-parameter, but have extreme values at $\mu = 0.178944$ and at $\mu = 0.334364$ respectively. It should not be surprising, therefore, that the critical radius of the outer planetary orbit also shows an extremum. In what follows, we show that the approximate maximum value of the critical outer planetary radius is 2.4386 and it occurs at $\mu = 0.2426$.

An analytical approach to finding the values of max $r_L$ and $r_R$ is outlined first. This method is similar to those outlined in parts 1 and 2, where the minimum value of the function $x_1(\mu)$ and the maximum value of $C_1(\mu)$ were established.

The equation relating $\mu$ with $r_R$ or $r_L$ is Eqn. (9). Using $C_R$ and equating the corresponding Eqn. (9) with $C_1$ we have

$$C_1(\mu) = \mu(1 - \mu) + \frac{2\mu}{r + 1 - \mu} + \frac{2(1 - \mu)}{r - \mu} - \frac{1}{r} + 2\sqrt{r} , \qquad (A.3)$$

where $r = r_R$. For a given value of $\mu$, this equation when solved for $r$, will give the corresponding $r_R$ value. The same method, but a slightly different equation can be used to obtain $r_L$, therefore, in the following only $r_R$ will be discussed in detail. Eqn. (A.3) can be written in the form of a seventh degree polynomial in $r$:

$$H(r, \mu) = \sum_{i=0}^{7} f_i(\mu) r^i = 0 . \qquad (A.4)$$

From Eqn. (A.4) we have

$$\frac{dH}{d\mu} = \frac{\partial H}{\partial \mu} + \frac{\partial H}{\partial r} \frac{dr}{d\mu} = 0 .$$

To evaluate the extreme value of $r$, we use $dr/d\mu = 0$, so $\partial H/\partial \mu = 0$ is one of the conditions needed. From this, $\mu = h(r)$, or its inverse, can be obtained at least in principle. The other relation to be satisfied is $H(r, \mu) = 0$ or $H[(r, h(r))] = 0$. This gives the critical value of $r$, and $\mu = h(r)$ gives the corresponding value of the mass-parameter.

To evaluate the nature of the critical value, we have

$$\frac{d^2 r}{d\mu^2} = \frac{-h''}{(h')^3} ,$$

where $h' = dh/dr$. The condition for maximum is that $h''/(h')^3$ be positive.

It should be observed that the above method represents a theoretical foundation only, since once again no closed form analytical representation exists for $\mu = h(r)$, and the same applies to the solution of Eqn. (A.3) for $r$. The practical (and approximate) approach is described in the following.

TABLE A.1
Critical Radii Near the Maximum.

| $\mu$ | $r_R$ | $r_L$ |
|------|---------|---------|
| 0.20 | 2.44086 | 2.41239 |
| 0.22 | 2.44940 | 2.42138 |
| 0.24 | 2.45226 | 2.42478 |
| 0.26 | 2.45011 | 2.42326 |
| 0.28 | 2.44350 | 2.41741 |

Eqn. (A.3) can be solved for $r$ by iteration for a given value of $\mu$. The results of this procedure are shown in Table A.1. The functions $r_R = R(\mu)$ and $r_L = L(\mu)$ can be approximated near the value $\mu = 0.24$ by polynomials. These approximations can then be used to give the maximum values with the corresponding values of $\mu$. Table A.1 shows the values of $r_R$ and $r_L$ in the vicinity of $\mu = 0.24$. After curve fitting and differentiating, we obtain

$$\max r_R = 2.4523 \text{ at } \mu = 0.2414$$
$$\text{and}$$
$$\max r_L = 2.4249 \text{ at } \mu = 0.2439 .$$

The average values are max $r_{ave} = 2.4386$ and $\mu_{ave} = 0.2426$.

## Acknowledgements

A portion of this research was done at the Lunar and Planetary Institute which is operated by the Universities Space Research Association under contract No. NASW-4574 with the National Aeronautics and Space Administration. This paper is Lunar and Planetary Institute Contribution No. 804. V. Szebehely wishes to acknowledge the support of the R.B. Curran Centennial Chair of The University of Texas.

## References

Benest, D.: 1988."Planetary Orbits in the Elliptic Restricted Problem," *Astronomy and Astrophysics*, **206**, 143–146, See also **13**, p. 157, (1971); **32**, p. 39, (1974); **45**, p. 353, (1975) and **53**, p. 231, (1976).

Black, D. C.: 1982, "A Simple Criterion for Determining the Dynamical Stability of Three-Body Systems", *The Astronomical Journal*, **87**, 1333–1337.

Dvorak, R.: 1984, "Numerical Experiments on Planetary Orbits in Double Stars", *Celest. Mech.*, **34**, 369–378. See also *Astronomy and Astrophysics*, **167**, 379, (1986); *Bull. AAS*, 842, (1986), with Froeschlé, C.

Graziani, F. and Black, D. C.: 1981, "Orbital Stability Constraints on the Nature of Planetary Systems", *The Astrophysical Journal*, **251**, 337–341.

Froeschlé, C.: 1971, *Astrophysics and Space Science*, **37**, 87.

Hadjidemetriou, J.: 1976, "Families of Periodic Planetary Type Orbits in the Three-Body Problem and their Stability", *Astrophysics and Space Science*, **40**, 201–224.

Hill, G. W.: 1878, "Researches in the Lunar Theory," *Am. J. Math.*, **1**, pp. 5, 129 and 245.

Pendelton, Y. J. and Black, D. C.: 1983, "Further Studies on Criteria for the Onset of Dynamical Instability in General Three-Body Systems," *The Astronomical Journal*, **88**, 1415–1419.

Szebehely, V.: 1967, *Theory of Orbits*, Academic Press.

Szebehely, V.: 1980, "Stability of Planetary Orbits in Binary Systems", *Celest. Mech.*, **22**, 7–12.

Szebehely, V. and McKenzie, R.: 1981, "Stability of Outer Planetary Systems", *Celest. Mech.*, **23**, 3–7.

Szebehely, V.: 1984, "Review of Concepts of Stability", *Celest. Mech.*, **34**, 49–64.

# AN EXCESS MOTION OF THE ASCENDING NODE OF MERCURY IN
# THE OBSERVATIONS USED BY LE VERRIER

TAKESHI INOUE

*Kyoto Sangyo University, Kamigamo, Kyoto 603, Japan*

We have reexamined the twenty-one observations of second and third contacts during transits of Mercury across on the disk of the Sun — the same Le Verrier used himself (Le Verrier 1859) — to check the reality of the excess of the observed motion of the longitude of the ascending node of Mercury's orbit over the theoretically predicted on. To this, we numbered these observations chronologically from 1 to 13 for the transits on November and from 14 to 21 for those on May. The results are as follows :

(1) An initial analysis showed that three of the observations (Nos. 2, 8 and 21) might have been erroneously recorded. If the signs of observations Nos. 2 and 8 are changed and if the value for the observation No. 21 is changed from $-1.''03$ to $-2.''52$, the sum of the squares of the residuals is reduced by almost 60%.

(2) Leaving out these observations altogether, as well as observations Nos. 10, 17 and 18 – which leaves us 15 observations to analyze – the sum of the squares of the residuals is drastically reduced to 5% of its original value (but now, there are of course also fewer observations contributing to this sum), and we obtain an excess motion of $16.''7 \pm 3.''4$ (s.e.) per century. This shows, that the appropriate choice of observations will indeed produce an estimate for the excess motion of the node which exceeds its formal standard error by a factor of 5.

(3) Eliminating eight additional observations and thus utilizing only the seven observations, viz. Nos. 1, 9, 13, 14, 16, 19 and 20, the least squares adjustment of the remaining seven condition equations in the six adjustment unknowns produces an estimate for the excess motion of the ascending node of Mercury's orbit. This time, it gives $15.''2 \pm 0.''1$ per century, with the sum of the squares of the residuals now reduced to the order of $10^{-5}$, even with only even summands contributing, clearly an unrealistic result in view of the precision of observations attained in Le Verrier's time.

The only conclusion one can draw from the data is thus that they do not contribute to a decision as to whether the actual motion of the ascending node of the orbit of Mercury exceeds that predicted by the theory.

We express our cordial thanks to Professor Eichhorn for his keen criticism and also kind advice without which we could not achieve the present study.

## References

Le Verrier, U.-J.: 1859, *Annales de l'Observatoire Impérial de Paris*, V, 76.

# NUMERICAL RESULTS TO THE SITNIKOV-PROBLEM

R.DVORAK

*Institute of Astronomy*
*University of Vienna*
*Türkenschanzstrasse 17*
*A-1180 Vienna*
*e-mail: DVORAK@AVIA.UNA.AC.AT*

**Abstract.** We present numerical results of the so-called Sitnikov-problem, a special case of the three-dimensional elliptic restricted three-body problem. Here the two primaries have equal masses and the third body moves perpendicular to the plane of the primaries' orbit through their barycenter. The circular problem is integrable through elliptic integrals; the elliptic case offers a surprisingly great variety of motions which are until now not very well known. Very interesting work was done by J.Moser in connection with the original Sitnikov-paper itself, but the results are only valid for special types of orbits. As the perturbation approach needs to have small parameters in the system we took in our experiments as initial conditions for the work moderate eccentricities for the primaries' orbit ($0.33 \leq e_{primaries} \leq 0.66$) and also a range of initial conditions for the distance of the $3^{rd}$ body ( = the planet) from very close to the primaries orbital plane of motion up to distance 2 times the semi-major axes of their orbit. To visualize the complexity of motions we present some special orbits and show also the development of Poincaré surfaces of section with the eccentricity as a parameter. Finally a table shows the structure of phase space for these moderately chosen eccentricities.

**Key words:** Sitnikov-Problem - surfaces of section - chaotic motion

## 1. Introduction

A very interesting dynamical problem is known under the name Sitnikov- problem which is cited quite often as a model case for the appearance of chaotic motion. The dynamical model was first described by Pavanini (1907): a massless body is moving in the z-direction perpendicular to the plane of two equally massive primary bodies, which move on Keplerian orbits around their center of gravitation. The circular problem (where the primaries move on circles) was discussed in details by McMillan (1913) where he showed the integrability of the equations of motion with the aid of elliptic integrals which has been rediscussed in detail by K.Stumpff(1965). This is also evident from the fact that in this form it can be regarded as a special case of the Two-Body Fixed Center Problem, which is known to be integrable since Euler (1760).

Much more interesting is the case, where the primaries move in eccentric orbits. Then we can observe periodic orbits, quasi periodic orbits and unbounded motions and additionally the recently rediscovered chaotic motions (already Poincaré mentioned such orbits ,1892)

As first rough definition one can say that two originally very close orbits separate from each other hyperbolically. It is interesting to note that the whole complexity of phase space is already present for very small eccentricities $e \sim 0.0001$ of the primaries' orbit, although it is so close to the integrable circular problem. (J. Liu and S.Sun, 1991). First qualitative results were derived by Sitnikov (1960) himself for special orbits and later by J. Moser (1973). C.Marchal (1990) discussed

*Celestial Mechanics and Dynamical Astronomy* **56**: 71–80, 1993.
© 1993 *Kluwer Academic Publishers.*

the problem also in a qualitative way . H.Juranek (1991) developed a $1^{st}$ order perturbation theory valid for small eccentricities and small oscillations, while J.Hagel (1992) used a perturbation method up to the $3^{rd}$ order in the eccentricities. J.Hagel and T.Trenkler (1992) adapted a technique to find integrals of motion for all eccentricity values of the primaries' orbit but only for small oscillation and derived interesting qualitative results.

We were interested in the structure of phase space for cases not yet studied well; therefore we did the numerical experiments in the range of $(0.33 \leq e_{primaries} \leq 0.66)$. In what concerns the other initial conditions they will be precised later.

## 2. The Formulation of the Problem

Let us know give the equations of motion:

$$\ddot{z} + \frac{z}{\sqrt{r^2 + z^2}^3} = 0 \tag{1}$$

$$r(v) = \frac{1 - e^2}{2(1 + e \cos v)} \tag{2}$$

where z is the distance from the plane of the primaries' orbit; the distance r of the barycenter to one primary varies with the time t according to Keplers $1^{st}$ law ($v$ is the true anomaly).

We used for integration a modified equation developed by K.Wodnar (1991) where T is defined through the following equations:

$$T := \frac{z}{2r} = z \frac{1 + e \cos v}{1 - e^2} \tag{3}$$

$$\frac{dT}{dv} := T' = \dot{z} \frac{\sqrt{1 - e^2}}{1 + e \cos v} - z \frac{e \sin v}{1 - e^2} \tag{4}$$

$$z = T \frac{1 - e^2}{1 + e \cos v} \tag{5}$$

$$\dot{z} = \frac{1}{\sqrt{1 - e^2}} [Te \sin v + T'(1 + e \cos v)] \tag{6}$$

The T-value has the following geometrical meaning: it is half of the tangent of the angle of view of the planet seen from one of the primaries. Finally we are lead to a differential equation of the following form:

$$T'' + \frac{1/\sqrt{1/4 + T^2}^3 + e \cos v}{1 + e \cos v} \cdot T = 0 \tag{7}$$

This equation of motion was integrated with a Lie-integration method with variable step length (e.g. A.Hanslmeier and R. Dvorak, 1984). The time scale was 1000 orbits of the primaries for most of the cases for; then had enough points in the Poincaré surfaces of section, which was defined as T versus T' for every pericentric position of the primaries. In exceptional cases, when we discovered a fractal structure of islands we increased the integration time up to 5000 revolutions of the primaries.

After some test calculations with different initial conditions we fixed the following ones:

- the true anomaly $v_{ini} = 0°$, that means the starting point was always when the primaries are at their pericenter.
- we always set $T' = 0$ and varied only T, which corresponds to a variation of the planet's distance to the barycenter.
- as mentioned above $e_{primaries}$ was chosen between 0.33 and 0.66

It should be kept in mind that this is only a necessary restriction because of limiting computer time available at the moment. Nevertheless it is hoped that we have found the main structures of phase space for moderate eccentricities and the motions not too far away from the primaries' orbital plane.

An appropriate method to find out the structure of phase space is to plot the different surfaces of section and compare them for various values of the eccentricities and the initial T-value. The method was introduced for numerical experiments by Hénon and Heiles (1964) for a simple model of a galactic potential and it is still the most powerful tool to present such numerical experiments. A more rigorous way to determine regions of chaotic motions in phase space is calculate the Liapunov characteristic exponents (e.g. Froeschlé 1984). But this is still a very "expensive" (from the point of view of computer-time) procedure and therefore it was kept for a future project on the same topic.

## 3. The Numerical Results

Before we discuss the global results we want to give some interesting details: the dependance of the location of the periodic orbits (=PO) on the parameter of the system (eccentricity of the primaries) , the decay of periodic orbits in form of bifurcations and the onset of chaos close to a separatrix.

### 3.1. THE CHANGE OF POs WITH THE PARAMETER

Quite well known is the pitchfork-diagram studied extensively at first in the logistic equation. The phenomenon of splitting of 1 PO into 3 is shown in fig.1 for invariant curves surrounding stable POs.

We can see in fig.1 the invariant curves in the SOS starting from the point T=1.13 (T'=0) for 6 different e-values. It is interesting to see the shift of the location of the PO (generally at the center of the island) outwards to greater T-values with

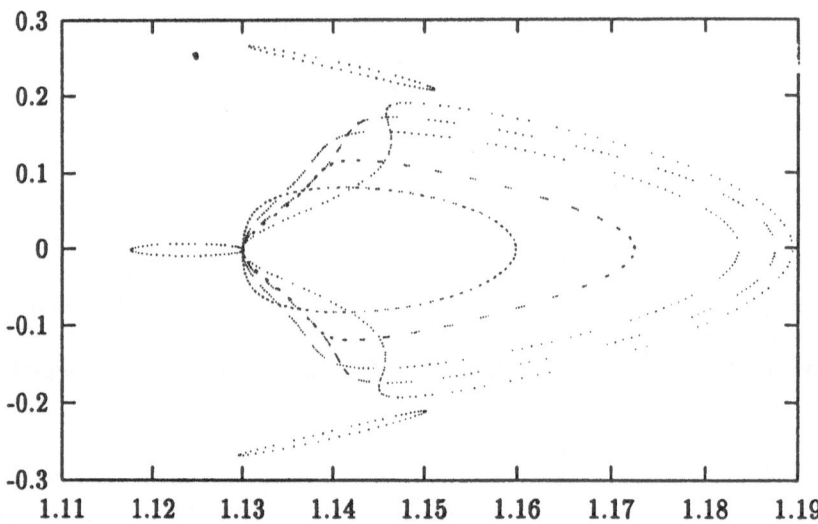

Fig. 1. Decay of a 1 PO to 3 PO with an increasing value of the parameter of the Sit-nikov-system

increasing e-values. Then we suddenly observe a splitting from a 1 PO to a 3 PO (islands around them). The most interesting orbit very close to the separatrix is shown in fig.2. The accumulation of points close to a hyperbolic point (an unstable PO) is also visible on this graph. The starting point was the same as in fig.1, the eccentricity was chosen in between the eccentricity where we derived the last one and the one where we derived 3 islands in the surface of section.

As another example we show the decay of an island around a PO into a chain of islands. We fixed the initial conditions for T ($T = 1.05$, $T'$ is always zero in our experiments) and varied the parameter e again. Fig.3a shows a well defined invariant curve (e=0.23) which decays for e=0.28 int o 7 islands (fig.3b). Then for e=0.33 we see even 17 islands replacing the one from fig.3a. Fig.3d shows the 2 small islands on the left bottom corner of fig.3c on a smaller scale.

### 3.2. MOTION ON A SEPARATRIX

Sometimes the initial conditions were chosen such that the motion is close to a separatrix, a curve connecting the hyperbolic fixed-points (unstable POs) in the surface of section. On fig.4 we see such an example where the consecutive points of intersection are surrounding tiny islands lying around stable POs. It is evident that the very complex shape on the SOS is difficult to derive with any analytical method.

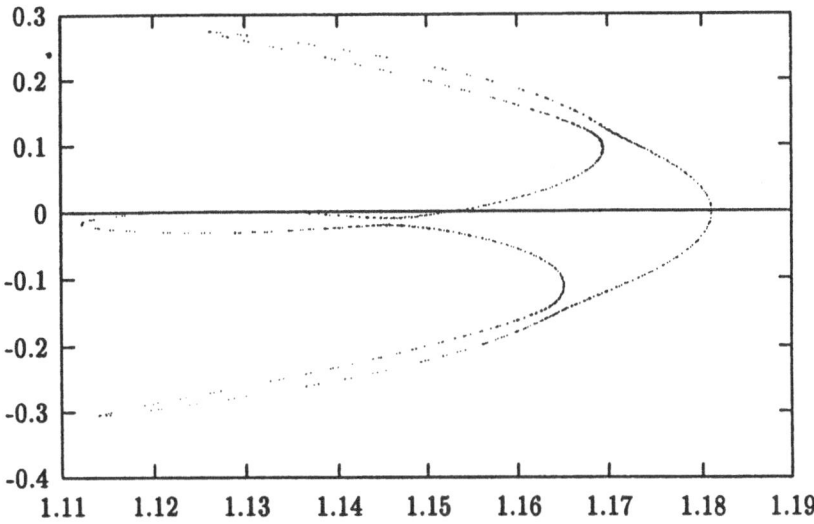

Fig. 2. motion close to the separatrix surrounding the 3 PO

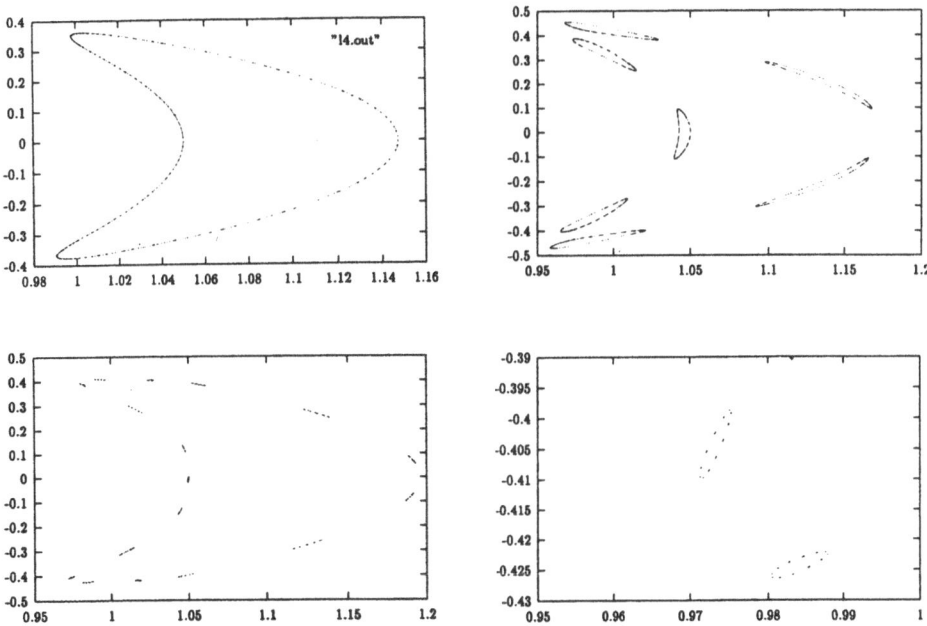

Fig. 3. The decay of an island: fig.3a (top,left) invariant curve for e=0.23; fig.3b(top, right): 7 invariant curves for e=0.28; fig.3c(bottom,left):17 invariant curves for e=0.33; fig.3d(bottom,right) 2 islands from the former graph on a smaller scale; all the plots show surfaces of section $T$ - versus - $T'$

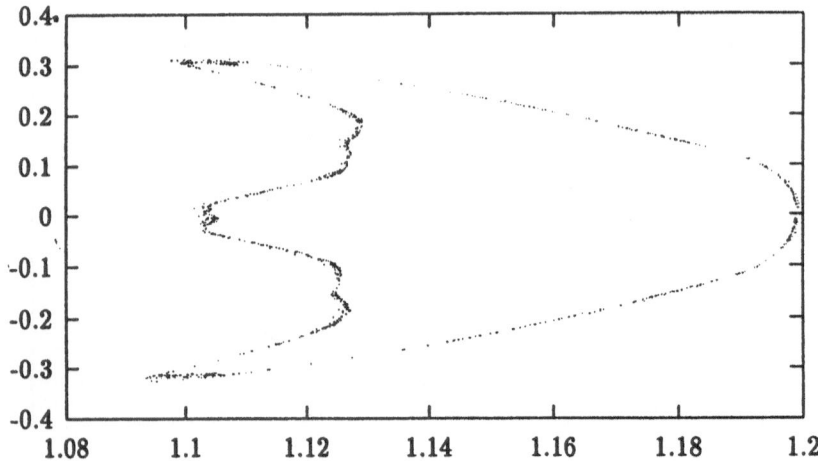

Fig. 4. chaotic motion "on" the separatrix close to multiple periodic orbits

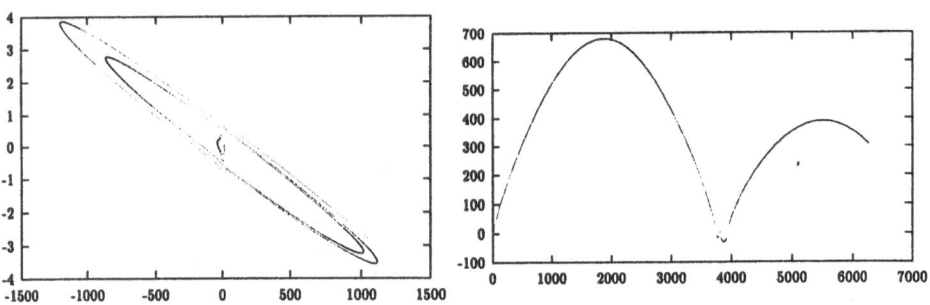

Fig. 5. typical chaotic motion for e=0.66, (left side surface of section T versus T', right side z versus the true anamoly v

## 3.3. A HIGHLY CHAOTIC MOTION

Varying just a little bit the initial conditions of the one orbit "on" the separatrix we have a full chaotic orbit which is shown in fig.5a in the surface of section and fig. 5b in a plot of T (the distance from the barycenter of the massless body) versus the time scaled in $2\pi$ corresponding to one whole orbit of the primary bodies.

It is evident, that the moments of the passage of the third body through the barycenter are very important. The time interval of such events is more or less periodic on an invariant curve and it is practically undistinguishable from a motion in a thin chaotic layer. Great differences of such time intervals can occur for

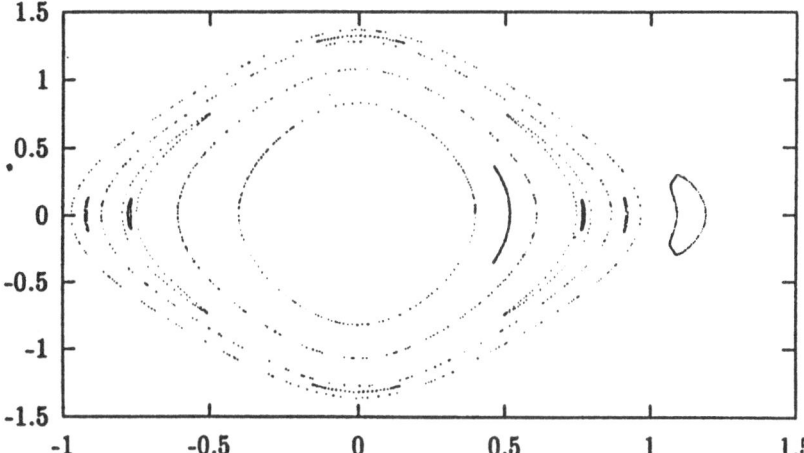

Fig. 6. Overall picture of the Poincare surface of section for e=0.37; T is plotted versus the velocity T'

motions in the global stochastic zone. One example is shown in fig.5b, where we can recognize small variations in T after quite a large one, and then again a very large one. This kind of motion is excluded in Sitnikov (1960), but is in principle included in Moser's book (1973). Although the qualitative picture of the variety of orbits which exist in the Sitnikov problem is quite interesting and complete no method of finding such special initial conditions is given explicitly.

### 3.4. GLOBAL RESULTS

In table I we have plotted the main characteristics of the motions for the initial conditions of the primaries for the eccentricities e of the primaries for which the experiments have been undertaken. The interval in the T-direction was $\Delta T = 0.01$ from T=0.4 to T=1.1, covering the most interesting parts of phase space. Note that we started always in the pericenter position of the primaries. A careful inspection of the Poincaré surface of section lead to the following results of the structure of phase space given in table 1. We marked there the qualitatively different orbits in the following way

- "o" orbit on an invariant curve
- "a" orbit on an invariant curve with some accumulation of points, indicating that the motion is close to a separatrix
- "n" marks the number of islands of the respective orbit
- "s" states that the motion is (on or) very close to a separatrix
- "k" orbit has a chaotic character
- "*" 10 and more islands

## TABLE I
### Results of the systematic research of the Sitnikov-Problem

| $T$ / $e$ | 0.4 | 0.5 | 0.6 | 0.7 | 0.8 | 0.9 |
|---|---|---|---|---|---|---|
| 0.33 | 0000000000 | 1000000a00 | 0a00222220 | 0a00111110 | 0000000000 | 044000669k |
| 0.37 | 0000000000 | 001000000 | 0000000000 | 0000a22222 | 2000000000 | 0044?5?0kk |
| 0.41 | 0000000000 | 0001000000 | 0000000000 | 0000002222 | 2220000000 | 0044000 6kk |
| 0.44 | 0000000000 | 0000100000 | 0000004000 | 0a00002222 | 2222200000 | 300444s6kk |
| 0.47 | 0000000000 | 0000010000 | 0000000400 | 000000a222 | 2222220000 | 030a44skkk |
| 0.51 | 0000000000 | 0000001100 | 0000000000 | 0000000022 | 22222222so | 003*k44kkk |
| 0.55 | 0000000000 | 0000a00111 | 0000000000 | 000a00000s* | 2222222222 | 2d8k3kk4kk |
| 0.58 | 0000000000 | 0000000001 | 1000000000 | 000000000s | 2222222222 | 2sssk3kkkk |
| 0.61 | 0000a00000 | a000a00000 | 1110000006 | 0004000000 | k2*2222222 | 222kkkkkkk |
| 0.64 | 0000000000 | 0000000000 | 0111100000 | 0000440030 | kkkkk22222 | 22222kkkkk |
| 0.66 | 0000000000 | 0000000000 | 0111110000 | 0000044403 | kkkkk2222 | 2220*kkkk |

## TABLE II
### The last invariant curve

| e | 0.33 | 0.37 | 0.41 | 0.44 | 0.47 | 0.51 | 0.55 | 0.61 | 0.64 | 0.66 |
|---|---|---|---|---|---|---|---|---|---|---|
| T | 0.99 | 0.98 | 0.98 | 0.97 | 0.92 | 0.91 | 0.92 | 0.80 | 0.80 | 0.80 |

At first sight we can see that invariant curves exist for greater values of T when the eccentricity e is smaller. As a consequence from low eccentricities on towards higher ones the appearance of the $1^{st}$ island is shifted more and more outwards; also the area (meaning here number of initial conditions) where islands can be observed is becoming more extended with increasing initial T values. The onset of global chaos is in contrary shifted more and more towards smaller T-values. But we shou ld keep in mond that the transformation from $T$ to $z$ and vice versa given in eqs. (3) - (5).

In table 2 we listed the last island for the specific e-value; from here on global chaos can arise. But it is also visible from fig.1 that in this zone of chaoticity there are still regions of regular motions visible through (sometimes very strange formed) islands. Their existence is due to high order resonances.

### 3.5. CONCLUSIONS

What can be said from the systematic numerical study of the Sitnikov problem? First of all we emphasize that this problem of Celestial Mechanics is the most simple problem after the integrable two body problem and the integrable Two-body-Fixed center problem. In this sense it can be regraded as a generic problem! In fig.6 we see a complete picture of the SOS for e=0.33. From that results we can deduce the following: Close to the linear problem, for very small oscillations around the barycenter, we can observe closed invariant curves which were expected to exist because of the KAM-theorem. These closed curves exist up to a certain value of the initial conditions and then they break and unbounded and chaotic motion is possible. But already in the domain of closed invariant curves we can observe islands which exist around stable periodic orbits of the problem. It is known since years (e.g. Henon and Heiles, 1964 Contopoulos, 1968) that in between such island we will have hyperbolic points - separatrices and sometimes only very thin layers of chaotic motion (e.g. Lichtenberg and Lieberman, 1983). Nevertheless this motion is bounded and can never lead to an escape: it is therefore quite important having determined the "$1^{st}$ chaotic orbit" which will be close to the last "KAM-Torus". Well visible in fig.6 is the island in the chaotic sea which is also due to motion close to a stable periodic resonant orbit. The structure of such islands is very complicated as on sees from fig. 4.

Finally it has to be said that this first systematic numerical study of the Sitnikov problem has to be extended to smaller $\Delta e$ and sometimes - in interesting areas to a smaller $\Delta T$. Another point is that for some orbits in the chaotic zone we should calculate also the Liapunov-characteristic exponent; this tool is especially important to determine the zone of the onset of global chaotic motions.

But we emphasize, that the purpose of this paper was to show for the first time explicitly the great variety of possible orbits in the Sitnikov-Problem.

## Acknowledgements

We have to thank Prof. Sun Yi-Sui from the Nanking University, Dr. Hagel from CERN in Geneva and Mr. Wodnar from the Vienna University for many fruitful discussions and most valuable advices.

## References

Contopoulos,G. 1967: 'Resonance Phenomena and the Non-Applicability of the "Third" Integral', *Contr. Astr. Dep. Univ. Thessaloniki*, 31

Euler, L. 1760: *Mém. de l'Acad. de Berlin*

Hagel, J., 1992: 'An New Analytical Approach to the Sitnikov Problem', *Celest.Mech.*, 53, 267–292

Hagel, J., Trenkler T.: 1992, 'A Computer aided analysis of the Sitnikov Problem', *Celest.Mech.*, this volume

Hanslmeier A., Dvorak, R. 1984:' Numerical Integration with Lie- series', *Astron.Astrophys.*, 132, 203

Hénon, M., Heiles, C.: 1964: 'The applicability of the third integral of motion, some numerical experiments', *Astron.J.*, 69, 73

Juranek, H.: 1990, *Zum Sitnikov Problem: Störungsrechnung erster Ordnung*, Diploma Thesis, University of Vienna

Lichtenberg, A.J. and Liebermann, M.A.: 1983, *Regular and Stochastic Motion*, Applied Mathematical Sciences, Volume 38, Springer, New York

Mac Millan, W. D.: 1913, 'An integrable case in the restricted problem of three bodies', *Astron.J.*, 27, 11

Marchal, C., 1990: *The Three Body Problem*, Studies in Astronautics, 4, Elsevier, Amsterdam, p.419ff

Moser, J., 1973: *Stable and Random Motions in Dynamical Systems*, Annals of Mathematics Studies Number 77, Princeton University Press and University of Tokio Press, Princeton, New Jersey

Pavanini, G.: 1907, 'Sopra una nuova categoria di soluzioni periodiche nel problema dei tre corpi', *Annali di Mathematica*, Serie III, Tomo XIII

Poincaré, H. ,1892: *Les méthodes nouvelles de la Mécanique Céleste*, Gauthier-Villars

Sitnikov, K.: 1960, 'Existence of oscillating motions for the three-body problem', *Dokl. Akad. Nauk. USSR*, **133**, 303–306

Stumpff, K., 1965: *Himmelsmechanik*, Band II, VEB Deutscher Verlag der Wissenschaften, Berlin

Lin, J.and Sun, Yi-Sui: 1991, *Celest.Mech.*, **49**, 285

Wodnar, K.: 1991, 'New Formulations of the Sitnikov Problem', in Archie E.Roy (ed): Predictability, Stability, and Chaos in N-Body Dynamical Systems, Plenum Press, p 457

# A COMPUTER AIDED ANALYSIS OF

# THE SITNIKOV PROBLEM

JOHANNES HAGEL

*CERN SL/AP, CH - 1211 Geneva 23, SWITZERLAND*

and

THOMAS TRENKLER

*TU-Graz, Petersgasse 16, A-8010 Graz, AUSTRIA*

**Abstract.** We deal with the problem of a zero mass body oscillating perpendicular to a plane in which two heavy bodies of equal mass orbit each other on Keplerian ellipses. The zero mass body intersects the primaries plane at the systems barycenter. This problem is commonly known as the **Sitnikov Problem**. In this work we are looking for a first integral related to the oscillatory motion of the zero mass body. This is done by first expressing the equation of motion by a second order polynomial differential equation using a Chebyshev approximation techniques. Next we search for an autonomous mapping of the canonical variables over one period of the primaries. For that we discretize the time dependent coefficient functions in a certain number of Dirac Delta Functions and we concatenate the elementary mappings related to the single Delta Function Pulses. Finally for the so obtained polynomial mapping we look for an integral also in polynomial form. The invariant curves in the two dimensional phase space of the canonical variables are investigated as function of the primaries eccentricity and their initial phase. In addition we present a detailed analysis of the linearized Sitnikov Problem which is valid for infinitesimally small oscillation amplitudes of the zero mass body. All computations are performed automatically by the FORTRAN program SALOME which has been designed for stability considerations in high energy particle accelerators.

**Key words:** Sitnikov problem – perturbation theory – symplectic mapping

## 1. Introduction

The Sitnikov Problem represents one of the most simple cases of the elliptic restricted three body problem. A massless body $m_3$ moves on an axis perpendicular to a plane in which two heavy equal masses move on two confocal ellipses with a certain eccentricity $e$ according to Keplers laws. The massless body intersects the primaries plane at the common barycenter of the primaries. The problem can be described by a nonlinear, one dimensional ordinary and explicitly time dependent second order differential equation. For bounded solutions with moderate amplitudes this equation can be expressed in a polynomial form as has been shown in (Hagel, 1991). Using the true anomaly of the primaries ellipses as the independent variable instead of the time (Wodnar, 1991) we may write the equation in a finite closed form which is not possible when using Kepler's equation to express the actual distance of the heavy masses as function of the time $t$.

In the second section of the present paper we set up the equation of motion following (Wodnar, 1991) and approximate the non polynomial part of his equation by a Chebyshev polynomial in a certain limited amplitude range.

In section 3 we deal with the linearized problem in case of very small amplitudes of bounded oscillatory solutions which have been proven to exist in (Sitnikov,

1960). In this case the equation of motion reduces to a simple Hill's equation which can be treated using Floquet's theory.

Section 4 is devoted to the full nonlinear polynomial equation. First the transformation of Courant and Snyder (1958) which transforms the linear part to a harmonic oscillator is applied. Then the remaining time dependent nonlinear part is discretized in equivalent Dirac Delta Function pulses. This symplectic discretization is used to construct a polynomial mapping of the canonical coordinates over one primaries revolution. Finally from the polynomial mapping we generate a first integral of the system i.e. a scalar function of the canonical coordinates taken at an integer number of primaries revolutions which remains constant under the mapping.

In section 5 the just obtained approximate first integral (the polynomial is truncated at a certain finite order) is used to predict the geometry of the phase trajectories for various initial conditions. In addition a relation for the oscillation frequency of the third mass as function of the initial amplitude and primaries eccentricity is derived.

## 2. Equation of Motion

The second order ordinary differential equation of motion for the third body oscillating perpendicular to the primaries plane is given in dimensionless coordinates (Moser, 1972) and is:

$$\ddot{z} + \frac{z}{(r^2(t) + z^2)^{3/2}} = 0 \tag{1}$$

where $z$ is the distance of $m_3$ from the primaries plane as function of the time $t$ and $r(t)$ is half the distance of the primaries given by

$$r(t) = \frac{1}{2}[1 - e \cos E(t)] \tag{2}$$

The function $E(t)$ is the solution of the transcendental Keplerian Equation

$$E(t) = t + e \sin E(t) \tag{3}$$

as given in (Stumpff, 1965). By $e$ we denote the eccentricity of the primaries elliptic orbits. This formulation has the disadvantage that the transcendental equation (3) for $E(t)$ cannot be solved in explicit form and hence a finite closed form for $r(t)$ in Eq. (2) for $e \neq 0$ cannot be given.

A much better formulation of the Sitnikov Problem has been found and introduced by Wodnar (1991). His original idea consists in taking the true anomaly $\varphi$ as the new independent variable using the fact that the primaries ellipses in terms of $\varphi$ are given by the simple finite form

$$r(\varphi) = \frac{p}{2(1 + e \cos \varphi)} \quad ; \quad p = a(1 - e^2) \tag{4}$$

In addition Wodnar uses a transformation of the dependent variable as

$$T = \frac{z}{2\,r(\varphi)} \tag{5}$$

i.e. he takes as the new dependent variable half the tangent of the angle under which the massless body is seen from the primaries positions. The equation for $T$ then becomes (Wodnar, 1991)

$$T'' + \frac{e\cos\varphi + (\frac{1}{4} + T^2)^{-3/2}}{1 + e\cos\varphi} T = 0 \quad ; \quad T'' = \frac{d^2 T}{d\varphi^2} \tag{6}$$

and will be the basis of further analysis.

## 2.1. POLYNOMIAL DIFFERENTIAL EQUATION

The method to construct integrals of the dynamical system described in Eq. (6) is limited to polynomial differential equations of the form

$$T'' + g_1(\varphi)T + \sum_{k=2}^{M} g_k(\varphi)T^k = 0 \tag{7}$$

Thus the first step consists in transforming Eq. (6) to this type. Since in this paper we are only interested in studying bounded oscillatory type solutions of (7) $M$ may be chosen finite (equal to 13 in our case) in order to provide sufficiently accurate results for $|T| < 0.8$. One method to obtain a form (7) out of Eq. (6) consists in expanding the term

$$P(T) = (\frac{1}{4} + T^2)^{-3/2} \tag{8}$$

into its Taylor Series as

$$P(T) = 8 - 48\,T^2 + 240\,T^4 - 1120\,T^6 + 5040\,T^8 - 22176\,T^{10}$$
$$+ 96096\,T^{12} + O(T^{14}) \tag{9}$$

However there exists a serious problem in doing so. Since the holomorphic function $P(T)$ in Eq. (8) has a pole in the complex plane at $T_0 = \pm\frac{1}{2}i$, the radius of convergence of the series (9) is given by $T_{lim} = 0.5$ and values of $T$ beyond this limit cannot be treated. This is why we use a different kind of expansion namely the representation of (8) in terms of Chebyshev polynomials. To find it we use the well known Remez algorithm (Bronstein, 1975). The result to order 12 with coefficients rounded to integer numbers is

$$P(T) = 8 - 47T^2 + 203T^4 - 616T^6 + 1168T^8 - 1206T^{10} + 512T^{12} \tag{10}$$

and this approximation is valid in the range $-0.8 \le T \le 0.8$ In Fig. 1 we compare the function (8) with its Chebyshev-approximation and its Taylor-series expansion

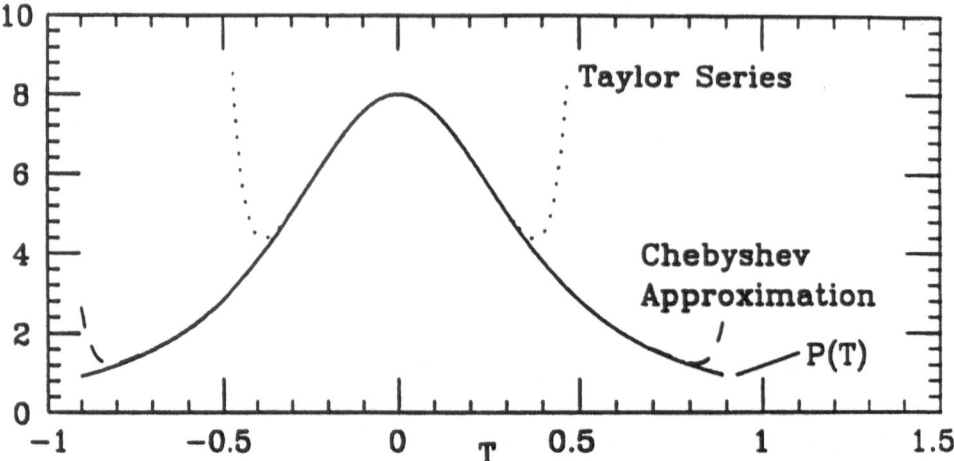

Fig. 1. Representation of $(1/4 + T^2)^{(-3/2)}$ by its Taylor series (9) and the Chebyshev approximation (10)

and we observe the tremendous advantage of the Chebyshev-approximation. The full line corresponds to the function itself, the dots represent the Taylor series to order 12, (see Eq. (9)) while the dashes show the Chebyshev approximation (10) to the same order.

Thus the Chebyshev representation for Wodnars equation (6) in the interval $|T| = [0, 0.8]$ becomes (with $A_{k-1}$ defined in Eq. (10)):

$$T'' + \frac{8 + e \cos \varphi}{1 + e \cos \varphi} T + \sum_{k=2}^{13} \frac{A_{k-1}}{1 + e \cos \varphi} T^k = 0 \quad A_{odd \ k} = 0 \tag{11}$$

## 3. The Linearized Problem

In this section we deal with the limiting case of very small values of $T$ under which condition we may approximate the polynomial equation of motion (7) by taking only into account the linear part i.e.

$$T'' + g_1(\varphi)T = 0 \longrightarrow T'' + \frac{8 + e \cos \varphi}{1 + e \cos \varphi} T = 0 \tag{12}$$

Since $g_1(\varphi)$ is periodic with period $2\pi$, Eq. (12) is of Hill's type. There exists a complete analytical theory of this equation known as **Floquet Theory** (see for example Liechtenberg, 1982). The main statements of this theory are:

−   The solution of Eq. (12) can be written as
$$T(\varphi) = w(\varphi)[a \cos \psi(\varphi) + b \sin \psi(\varphi)] \tag{13}$$
where $a, b$ are arbitrary real constants, $w(\varphi)$ is a periodic function with period $2\pi$ satisfying the equation
$$\frac{d^2 w}{d\varphi^2} + g_1(\varphi)w - \frac{1}{w^3} = 0 \tag{14}$$

and the phase function $\psi(\varphi)$ is related to $w(\varphi)$ by

$$\psi(\varphi) = \int_0^\varphi \frac{d\varphi'}{w^2(\varphi')} \quad ; \quad \psi(0) = 0 \tag{15}$$

— The solution and its first derivative can be written in vectorial form as

$$\begin{pmatrix} T(\varphi) \\ T'(\varphi) \end{pmatrix} = \begin{pmatrix} m_1(\varphi) & m_2(\varphi) \\ m_3(\varphi) & m_4(\varphi) \end{pmatrix} \begin{pmatrix} T(0) \\ T'(0) \end{pmatrix} = M(\varphi) \begin{pmatrix} T(0) \\ T'(0) \end{pmatrix} \tag{16}$$

where the determinant of the matrix is equal to unity and the elements of the matrix are given by

$$m_1(\varphi) = c(\varphi) \quad m_2(\varphi) = s(\varphi) \quad m_3(\varphi) = c'(\varphi) \quad m_4(\varphi) = c'(\varphi) \tag{17}$$

The functions $c(\varphi)$ and $s(\varphi)$ are solutions of Eq. (19) with initial conditions

$$c(0) = 1 \quad c'(0) = 0 \quad s(0) = 0 \quad s'(0) = 1 \tag{18}$$

— The solution of (12) is bounded if the absolute value of the trace of the matrix

$$R = M(\varphi = 2\pi) \tag{19}$$

is less or equal to 2, otherwise the solution is unbounded.

The matrices $M(\varphi)$ and $R = M(\varphi = 2\pi)$ can be easily expressed in terms of the amplitude and phase functions $w(\varphi)$ and are given e.g. in (Courant and Snyder, 1952).

The so called characteristic exponent $\mu$ of is defined by

$$\mu = \psi(2\pi) \quad ; \quad \cos\mu = \frac{1}{2}TrR \tag{20}$$

For $-1 < cos\mu < 1$ the solution is bounded while for $|cos\mu| > 1$ (imaginary $\mu$) it is unbounded. The amplitude function $w(\varphi)$ can be expressed in terms of the elements $m_1(\varphi)$ and $m_2(\varphi)$ as:

$$w(\varphi) = \sqrt{\frac{m_2^2(\varphi)}{w_0^2} + w_0^2 m_1^2(\varphi)} \tag{21}$$

The initial value $w_0 = w(0)$ is

$$w_0 = \sqrt{\frac{r_2}{\sin\mu}} = \sqrt{\frac{r_2}{\sqrt{1 - \cos^2\mu}}} = \sqrt{\frac{r_2}{\sqrt{1 - \frac{1}{4}(r_1 + r_2)^2}}} \tag{22}$$

All these formulae imply that $w'(0) = w_0' = 0$ which for (12) is always the case if $g_1(-\varphi) = g_1(\varphi)$ as can be demonstrated easily by inspecting Eq. (14) for $w(\varphi)$. Inspecting Eq. (12) we realize that this condition is indeed fulfilled. Equations (21) - (22) have been derived in (Courant and Snyder, 1952) with the only difference that in their paper they use $\beta(\varphi) = w^2(\varphi)$ instead of $w(\phi)$.

TABLE I

Matrix $R$, amplitude function and average frequency as function of $e$ for the linearized Sitnikov Problem

| $e$ | $r_1$ | $r_2$ | $r_3$ | $r_4$ | $w(0)$ | $\frac{w(\pi)}{w(0)}$ | $TrR$ | $Q$ |
|---|---|---|---|---|---|---|---|---|
| -0.99 | 0.9325 | -0.0135 | 9.6365 | 0.9325 | 0.1936 | 3.5687 | 1.8650 | 4.9381 |
| -0.80 | -0.5498 | 0.1373 | -5.0817 | -0.5438 | 0.4054 | 1.6663 | 1.0995 | 3.3426 |
| -0.60 | 0.9563 | 0.0675 | -1.2706 | 0.9563 | 0.4798 | 1.3737 | 1.9125 | 3.0472 |
| -0.40 | 0.8534 | -0.1455 | 1.8667 | 0.8534 | 0.5297 | 1.2120 | 1.6117 | 2.8992 |
| -0.20 | 0.5777 | -0.2606 | 2.5563 | 0.5777 | 0.5651 | 1.0960 | 1.1554 | 2.8480 |
| 0.00 | $\cos 2\pi\sqrt{8}$ | $\frac{\sin 2\pi\sqrt{8}}{\sqrt{8}}$ | $-8r_2$ | $\cos 2\pi\sqrt{8}$ | $8^{-1/4}$ | 1.0000 | 0.9461 | $\sqrt{8}$ |
| 0.20 | 0.5777 | -0.3131 | 2.1280 | 0.5777 | 0.6193 | 0.9124 | 1.1554 | 2.8480 |
| 0.40 | 0.8534 | -0.2139 | 1.2703 | 0.8534 | 0.6406 | 0.8249 | 1.7068 | 2.9127 |
| 0.60 | 0.9563 | 0.1271 | -0.6733 | 0.9563 | 0.6591 | 0.7279 | 1.9125 | 3.0472 |
| 0.80 | -0.5498 | 0.3812 | -1.8303 | -0.5450 | 0.6756 | 0.6002 | -1.0995 | 3.3426 |
| 0.99 | 0.9325 | -0.1007 | 0.4458 | 0.9325 | 0.6894 | 0.2801 | 1.8650 | 4.9381 |

## 3.1. AMPLITUDE FUNCTION, FREQUENCY AND STABILITY

The method we chose to compute the matrices $M(\varphi)$ and $R$ consists in numerically computing the cosine and sinelike functions $s(\varphi)$ and $c(\varphi)$ defined in (17)- (18) and constructing $M(\varphi)$ according to (16). The numeric method is the usual Runge-Kutta 4-th order method with a discretization of $\Delta\varphi = 2\pi/200$ which proved to lead to sufficiently accurate results when (-0.99 < e < 0.99). The solution of Hill's equation is oscillatory (if bounded) but non periodic in general as can be seen from the definition of the phase function in Eq. (15). However the average frequency (which is half the number of passages of the third body through the primaries plane within one primaries revolution) can be identified by the lowest order Fourier coefficient of $1/w^2(\varphi)$. This follows from (13) and (15).

$$Q = \frac{1}{2\pi}\int_0^{2\pi}\frac{d\varphi'}{w^2(\varphi')} = \frac{\psi(2\pi)}{2\pi} = \frac{\mu}{2\pi} \tag{23}$$

In Table I we present the results obtained as function of the primaries eccentricity $e$. The second to fourth columns contain the matrix elements of $R$. The fifth and sixth column show the values of the amplitude function at $\varphi = 0$ and $\varphi = \pi$. In the last two columns we find the trace of $R$ (indicating stability) as well as the average frequency $Q$ of $T(\varphi)$.

The meaning of negative eccentricities $e$ becomes clear from the definition of the primaries orbits in Eq. (2). Obviously $e > 0$ means that the primaries start at their closest distance while for $e < 0$ they are located in their largest possible distance when $t = \varphi = 0$. We see that while the average frequency $Q$ as function of $e$ behaves exactly symmetrical w.r.t. $e = 0$ while the amplitude functions do not so. Let us restrict to the case

$$T(0) = T_0 \quad ; \quad T'(0) = 0 \tag{24}$$

Then from the general solution (13) we conclude

$$T(\varphi) = T_0 \frac{w(\varphi)}{w_0} cos\psi(\varphi) \tag{25}$$

Thus from the values for $w(\pi)/w(0)$ in Table I we arrive at the following two important results concerning the linearized motion:

- For $e > 0$ the solution $T(\varphi)$ subject to the the initial conditions (24) is bounded between the positive and negative value of the initial conditions, i.e. $-T_0 < T(\varphi) < T_0$ (The maximum of $w$ occurs indeed at $\varphi = \pi$ since $w(\varphi)$ is symmetric around $\pi$ as can be seen by inspection of (14) and $g_1(\varphi)$ in (12)
- For $e < 0$ the maximum of the solution is larger than the initial value of $T$. The factor as function of $e$ is listed in Table I
- The average frequency $Q$ is an increasing symmetric function of the primaries eccentricity

In (Hagel, 1992) a formula for $Q$ as function of $e$ up to the second power in $e$ has been derived:

$$Q(e) = \sqrt{8} \left[ 1 + \frac{24}{121} e^2 + O(e^4) \right] \tag{26}$$

This relation reproduces the $Q$ values in Table I for $|e| < 0.6$. In Fig. 2 we show the graphs of $w(\varphi)/w(0)$ for various values of $e$ where $w(\varphi)$ has been computed from Eq. (21).

In Fig. 3 the trace of $R$ is plotted against $e$ for the interval $-0.99 < e < 0.99$ and we see that for the entire interval it remains bounded within $-2 < Tr(R) < +2$ thus indicating bounded motion for $T(\varphi)$ as stated above. Note that the non-existence of "instable islands" inside a stable parameter regime is atypical for a general Hill's equation (like Mathieu's equation) and is worth further investigations. Additional and more accurate computations which have been performed by the authors give a strong numerical evidence that this stable behaviour persists at least up to $|e| < 0.99999$. However up to now we could not find a rigorous proof.

## 4. Nonlinear Motion

In this section we investigate the polynomial differential equation (11). The goal is to derive a first integral of the nonlinear equation which makes it possible to predict the phase space dynamics in regions where the integral exists. As first step we apply the "Courant and Snyder Transformation" which transforms the linear time dependent part of (12) to a simple harmonic oscillator with frequency $Q$. Then we discretize the remaining time dependent nonlinear part of the equation in Dirac Delta Function pulses and construct a mapping over one of the primaries period

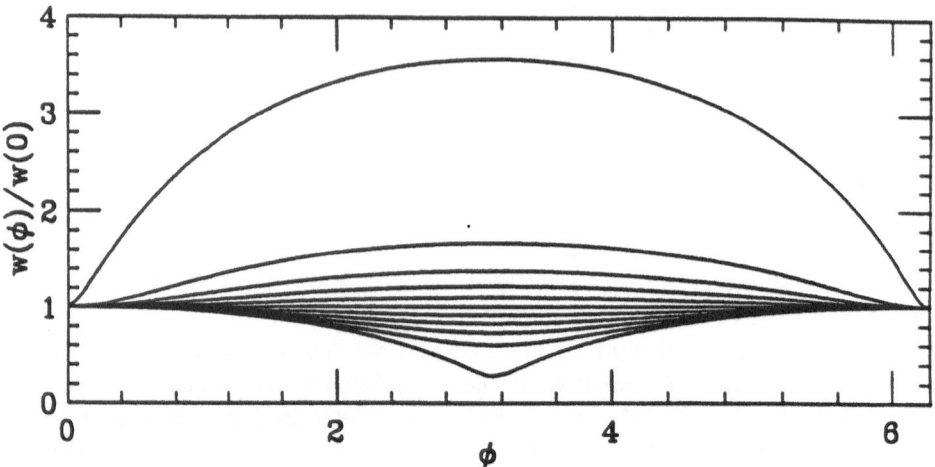

Fig. 2. Amplitude functions $w(\varphi)/w(0)$ for various primaries eccentricities

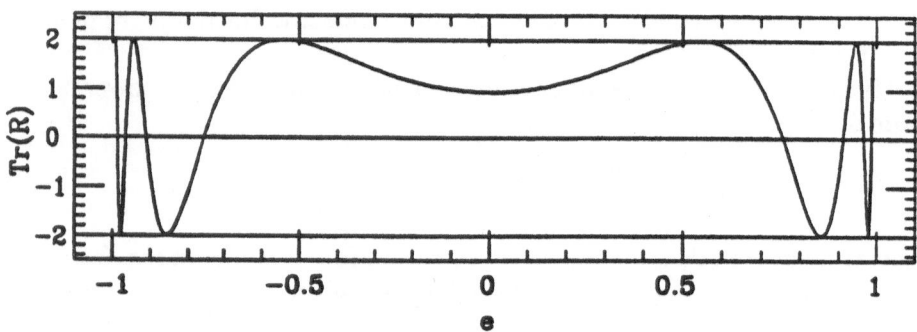

Fig. 3. Trace of the matrix $R$ as function of $e$

for the discretized equation. From this (polynomial) mapping we then construct the first integral i.e. a function of the coordinate and its derivative which remains constant under the mapping. As in the previous section all necessary computations are carried out by the FORTRAN code SALOME.

### 4.1. COURANT AND SNYDER TRANSFORMATION

The polynomial differential equation (11)

$$T'' + g_1(\varphi)T + \sum_{k=2}^{M} g_k(\varphi)T^k = 0 \tag{27}$$

can be transformed to a more simple form by application of a transformation to the dependent and the independent variables $T$ and $\varphi$:

$$y = \frac{T}{w} \quad ; \quad \tau = \frac{1}{Q} \int_0^\varphi \frac{d\varphi'}{w^2(\varphi')} \tag{28}$$

where $w(\varphi)$ is the linear amplitude function treated in section 3. Since in the linearized theory $T = w(\varphi)(a cos\psi + b sin\psi)$, $y = T/w$ must follow a harmonic oscillator equation if we choose $\tau = \psi/Q$ as the new independent variable (see also the definition of $\psi$ in Eq. (15)). All the detailed derivations concerning this transformation are given in (Courant and Snyder, 1952). The application to the Sitnikov Problem is described in (Hagel, 1992). The resulting equation for $y(\tau)$ is

$$y'' + Q^2 y + Q^2 \sum_{k=2}^{M} g_k(\varphi(\tau))w^{k+3}(\varphi(\tau))y^k = 0 \quad ; \quad y'' = \frac{d^2 y}{d\tau^2} \tag{29}$$

with (see Eq. (11))

$$g_k(\varphi(\tau)) = \frac{A_{k-1}}{1 + e \cos \varphi(\tau)} \tag{30}$$

>From Eq. (15) we realize that $\varphi = 0$ implies $\tau = 0$ and $\varphi = 2\pi$ implies $\tau = 2\pi$ since

$$\tau(\varphi = 2\pi) = \frac{1}{Q}\psi(2\pi) = \frac{\mu}{Q} = 2\pi \tag{31}$$

## 4.2. DISCRETIZATION IN DIRAC DELTA FUNCTIONS

In order to construct a mapping and a first integral for the $y$ equation (29) we look for a symplectic discretization of this equation. By this we mean that we search an elementary mapping over a short interval of the "time" $\tau$ relating the coordinates $y$ and $v = y'$ at the entrance of the time interval to the coordinates at the exit of the same interval. In addition this mapping should be exactly symplectic and it must represent the exact solution $y(\tau)$ within that interval up to a certain order in the interval length $\Delta\tau$. Equation (29) can be derived from a Hamiltonian function

$$H(y, \varphi, \tau) = \frac{1}{2}(Q^2 y^2 + v^2) + \sum_{k=2}^{M} f_k(\tau)\frac{y^{k+1}}{k+1} \quad ; \quad v = y' \tag{32}$$

where

$$f_k(\tau) = Q^2 g_k(\varphi(\tau))w^{k+3}(\varphi(\tau)) \tag{33}$$

Now the discretization consists in replacing the continuous function $f(\tau)$ within the interval $(\tau_j - \Delta\tau/2, \tau_j + \Delta\tau/2)$ by a Dirac Delta Function pulse located at the center of the interval.

$$f_k(\tau) = \sum_{j=1}^{N} \epsilon_{kj}\delta(\tau - \tau_j) \quad ; \quad \tau_j = (j - \frac{1}{2})\Delta\tau \quad ; \quad \Delta\tau = \frac{2\pi}{N} \tag{34}$$

$N$ is the number of Delta Function pulses within one primaries period $2\pi$. The
factors $\epsilon_{kj}$ are chosen as the definite integral

$$\epsilon_{kj} = \int_{\tau_j - \frac{\Delta\tau}{2}}^{\tau_j + \frac{\Delta\tau}{2}} f_k(\tau)d\tau \tag{35}$$

which means that the action of all the coefficient functions $f_k(\tau)$ is concentrated
to one point at the center of the interval.

## 4.3. MAPPING OVER ONE PRIMARIES PERIOD

We first construct an exact mapping which relates the canonical variables $y$ and
$v = y'$ (see the Hamiltonian (32)) over one discretization interval containing a
Delta pulse in its center as described above. Thus we start with the discretized
equation of motion

$$y'' + Q^2 y + \sum_{k=2}^{M} \sum_{j=1}^{N} \epsilon_{kj} \delta(\tau - \tau_j) y^k = 0 \tag{36}$$

with $\tau_j$ and $\epsilon_{kj}$ as defined in Eqs. (34) and (35). The mapping of the phase vector
$(y_0, y_0')$ at the entrance of the j-th interval into the vector $(y_1, y_1')$ at the exit of the
same interval can be split into three exactly solvable parts

- Rotation of $(y_0, y_0')$ over half the interval length $\Delta\tau$ according to the linear
  equation
  $y'' + Q^2 y = 0$ (all nonlinearities are concentrated in the center of the interval).

$$\begin{pmatrix} y_a \\ v_a \end{pmatrix} = \begin{pmatrix} \cos(Q\Delta\tau/2) & \sin(Q\Delta\tau/2)/Q \\ -Q\sin(Q\Delta\tau/2) & \cos(Q\Delta\tau/2) \end{pmatrix} \begin{pmatrix} y_0 \\ v_0 \end{pmatrix} \tag{37}$$

- Executing the nonlinear kicks at the $\tau = \tau_j$. In order to obtain a continuous
  solution $y(\tau)$ the kick is only applied to the slope $y' = v$. The kicks are
  obtained by integrating Eq. (29) once with respect to $\tau$ in the interval $(\tau_j - \epsilon, \tau_j + \epsilon)$ with $\epsilon \to 0$.

$$\begin{pmatrix} y_a \\ v_b \end{pmatrix} = \begin{pmatrix} y_a \\ v_a \end{pmatrix} - \begin{pmatrix} 0 \\ 1 \end{pmatrix} \sum_{k=2}^{M} \epsilon_{kj} y_a^k \tag{38}$$

Here we used the definition of the Dirac Delta Function as

$$\int_{\tau_j - \epsilon}^{\tau_j + \epsilon} \delta(\tau - \tau_j) = 1 \tag{39}$$

- Rotation of the vector $(y_a, v_b)$ over the second half interval according to the
  linear part of the equation.

$$\begin{pmatrix} y_1 \\ v_1 \end{pmatrix} = \begin{pmatrix} \cos(Q\Delta\tau/2) & \sin(Q\Delta\tau/2)/Q \\ -Q\sin(Q\Delta\tau/2) & \cos(Q\Delta\tau/2) \end{pmatrix} \begin{pmatrix} y_a \\ v_b \end{pmatrix} \tag{40}$$

Combining these three operations and evaluating all matrix products results in the following polynomial mapping over one discretization interval:

$$\begin{pmatrix} y_1 \\ v_1 \end{pmatrix} = \begin{pmatrix} \cos(Q\Delta\tau) & \sin(Q\Delta\tau)/Q \\ -Q\sin(Q\Delta\tau) & \cos(Q\Delta\tau) \end{pmatrix} \begin{pmatrix} y_0 \\ v_0 \end{pmatrix} - \tag{41}$$

$$\begin{pmatrix} \sin(Q\Delta\tau/2)/Q \\ \cos(Q\Delta\tau/2) \end{pmatrix} \sum_{k=2}^{M} \epsilon_{kj}[y_0\cos(Q\Delta\tau/2) + v_0\sin(Q\Delta\tau/2)/Q]^k$$

This mapping is symplectic since the determinant of its Jacobian matrix is exactly equal to unity. In addition it is an important fact that **the mapping (41) represents the exact solution of the non discretized equation (29) up to order** $(\Delta\tau)^2$. In order to prove this statement we first write the solution of Eq. (29) as Taylor series at $\tau = \Delta\tau$

$$y_1 = y_0 + v_0\Delta\tau - \frac{1}{2}\left[Q^2 y_0 + \sum_{k=2}^{M} f_k(0)y_0^k\right](\Delta\tau)^2 \tag{42}$$

$$v_1 = v_0 - \left[Q^2 y_0 + \sum_{k=2}^{M} f_k(0)y_0^k\right]\Delta\tau \tag{43}$$

$$-\frac{1}{2}\left[Q^2 v_0 + \sum_{k=2}^{M} f_k'(0)y_0^k + k f_k(0)y_0^{k-1} v_0\right](\Delta\tau)^2$$

For simplicity we restrict to the first interval $j = 1$ when considering $\epsilon_{kj}$ in (35). From (34) and (35) we then obtain

$$\epsilon_{k1} = \int_0^{\Delta\tau} f_k(\tau)d\tau = \int_0^{\Delta\tau}\left[f_k(0) + f_k'(0)\tau + \frac{1}{2}f_k''(0)\tau^2 + O(\tau^3)\right]d\tau \tag{44}$$

$$= f_k(0)\Delta\tau + \frac{1}{2}f_k'(0)(\Delta\tau)^2 + O(\Delta\tau^3)$$

Inserting (44) into (41) and expanding the remaining trigonometric functions w.r.t. $\Delta\tau$ leads exactly to the Taylor representation (42) and (43). Thus **the mapping (41) represents a symplectic numeric integrator of order 2 in the time step.** The basic ideas of this method have been developed in (Jimenez and Vazquez, 1990). In order to produce a mapping valid over one period of the primaries ($\tau = (0-2\pi)$) we have to apply the polynomial mapping (41) $N$ times to the initial vector $(y_0, v_0)$ where as above $N$ represents the number of discretization intervals for one period. To do this exactly for a reasonable number of discretization intervals (we used 1000) is impossible because the order of the polynomials occurring in (41) increases rapidly with the number of applications of the mapping. From (41) we see that after one application the order of the polynomial map in $y_0$ and $v_0$ is $M$. After $N$ applications of the mapping the maximum order will be

$$Ord(N) = M^N \tag{45}$$

i.e. for $M = 13$ (see chapter 3) and N=1000 we would expect a number of terms equal to $13^{1000}$ !!! A possible solution of this problem consists in using the fact that we are only interested in bounded solutions of Eq. (29) which means that the maximum values for $y$ and $y' = v$ are finite. Thus we may truncate the concatenated mappings to a certain order in the coordinates and so keep the number of terms for the resulting mapping constant. This is done in the program SALOME (Hagel and Trenkler, 1992). A listing of the part of the FORTRAN 77 code constructing the concatenated and truncated mapping can be found in this paper.

## 4.4. INTEGRAL OF THE MOTION

Following Trenkler (1991) we write the just obtained polynomial mapping in the form

$$\zeta(\xi) = R\xi + \sum_{\sigma=2}^{N} \sum_{n=0}^{\sigma} z(\sigma)_n y^{\sigma-n} v^n \tag{46}$$

where $N$ is the order of the truncated mapping, $\xi$ is the two dimensional vector of the phase coordinates $y$ and $v$, $R$ is the matrix occurring in (41) with $\Delta\tau = 2\pi$ related to the linear part of Eq. (36) and $z(\sigma)$ is the vectorial coefficient of the powers in $y$ and $v$. For this mapping we now search a scalar polynomial function of the coordinates $\xi$ which under the mapping remains constant, i.e.

$$\eta(y, v) = \eta(\xi) = \sum_{\sigma=2}^{\infty} \sum_{n=0}^{\sigma} \phi(\sigma)_n y^{\sigma-n} v^n = const. \implies \eta(\zeta(\xi)) = \eta(\xi) \tag{47}$$

It is convenient to interpret $\phi(\sigma)_n$ as the n-th component of a $\sigma + 1$ dimensional vector $\phi(\sigma)$. Inserting the above condition (47) into the mapping (46) and reordering coefficients gives (after lengthy but straightforward calculations)

$$\sum_{\sigma=2}^{\infty} \sum_{n=0}^{\sigma} \left[ \sum_{i=2}^{\sigma} \alpha(\sigma, i)\phi(i) \right]_n y^{\sigma-n} v^n = \sum_{\sigma=2}^{\infty} \sum_{n=0}^{\sigma} \phi(\sigma)_n y^{\sigma-n} v^n \tag{48}$$

where the $\alpha(\sigma, i)$ are the $(\sigma + 1) \times (i + 1)$ matrices with coefficients

$$\alpha(\sigma, i)_{n,m} = \sum_{j,k} z_y^{(i-m)}(\sigma - j)_{n-k} z_v^{(m)}(j)_k \tag{49}$$

with

$$max\{0, n - \sigma + j\} \le k \le min\{j, n\} \tag{50}$$

$$max\{m, \sigma - N(i - m)\} \le j \le min\{mN, \sigma - i + m\} \tag{51}$$

$N$ denotes the order of the mapping and the $z_\#^{(m)}(\sigma)_n$ are defined as the coefficients of $y^{\sigma-n} v^n$ in the $m$-th power of the mapping component $z_\#(\xi)$ (# = y or v). As we require $\eta$ to exist in a whole domain around the origin, we can compare coefficients in Eq. (48) and obtain

$$\phi(\sigma) = \sum_{i=2}^{\sigma} \alpha(\sigma, i)\phi(i) \tag{52}$$

which is a linear algebraic system of equations for the unknown vectorial integral coefficients $\phi(\sigma)$. A different formulation of the above equation (52) is

$$[I - \alpha(\sigma, \sigma)]\phi(\sigma) = \sum_{i=2}^{\sigma-1} \alpha(\sigma, i)\phi(i) \tag{53}$$

where $I$ is the $\sigma + 1 \times \sigma + 1$ unity matrix. Then follows that

$$\phi(\sigma) = [I - \alpha(\sigma, \sigma)]^{-1} \sum_{i=2}^{\sigma-1} \alpha(\sigma, i)\phi(i) \tag{54}$$

In this formulation the $\phi(\sigma)$ which are the integral coefficients of $\eta$ to the order $\sigma$ can be computed in a recursive way: Knowing the vectors $\phi(2), \phi(3), ...., \phi(\sigma-1)$ Eq. (54) directly gives $\phi(\sigma)$. The linear problem $y'' + Q^2 y = 0$ has the exact integral of second order polynomial form

$$\eta(y, v) = Q^2 y^2 + v^2 \implies \phi(2) = \begin{pmatrix} Q^2 \\ 0 \\ 1 \end{pmatrix} \tag{55}$$

Then follows $\phi(3)$ from Eq. (54) as

$$\phi(3) = [I - \alpha(3,3)]^{-1}\alpha(3,2)\phi(2) \tag{56}$$

Since $\phi(2)$ is a three dimensional vector and $\alpha(3,2)$ is a $4 \times 3$ matrix and $I - \alpha(3,3)$ is a $4 \times 4$ matrix, $\phi(3)$ is a four dimensional vector containing the integral coefficients of $y^3$, $y^2 v$, $y v^2$ and $v^3$. A method to compute fully analytically the inverse of $I - \alpha(\sigma, \sigma)$ is described in detail in (Trenkler, 1991) and (Hagel and Trenkler, 1992). As in the previous sections all necessary computations are done by the program code SALOME which carries out the iteration (54) and evaluates the integral coefficients $\phi(\sigma)$ to the desired order.

## 5. Approximate First Integral for the Sitnikov Problem

In table II we list the components of the vectors $\phi(\sigma)$ (i.e. the coefficients of $y^{\sigma-n} v^n$ in the approximate first integral $\eta(y, v)$) which have been computed by SALOME up to order 6 for various values of the primaries eccentricity.

TABLE II
Integral coefficients for various eccentricities.

| e | 0.0 | 0.2 | 0.4 | 0.6 | 0.8 | 0.99 |
|---|---|---|---|---|---|---|
| $y^2$ | 8.000 | 8.111 | 8.484 | 9.286 | 11.173 | 24.641 |
| $v^2$ | 1.000 | 1.000 | 1.000 | 1.000 | 1.000 | 1.000 |
| $y^4$ | -8.310 | -8.729 | -9.282 | -10.129 | -11.812 | -22.700 |
| $y^3v$ | - | - | - | - | - | - |
| $y^2v^2$ | - | 0.061 | 0.136 | 0.233 | 0.374 | 0.701 |
| $yv^3$ | - | - | - | - | - | - |
| $v^4$ | - | 0.004 | 0.008 | 0.013 | 0.017 | 0.014 |
| $y^6$ | 8.463 | 9.113 | 9.808 | 10.614 | 11.996 | 16.589 |
| $y^5v$ | - | - | - | - | - | - |
| $y^4v^2$ | - | -0.16 | -0.334 | -0.532 | -0.764 | -1.280 |
| $y^3v^3$ | - | - | - | - | - | - |
| $y^2v^4$ | - | -0.011 | -0.020 | -0.026 | -0.027 | -0.010 |
| $yv^5$ | - | - | - | - | - | - |
| $v^6$ | - | - | - | - | - | - |

Only coefficients whose absolute value is larger that $10^{-3}$ have been taken into account. So for example when $e = 0.6$ the first integral reads as

$$\eta(y, v) = 9.286y^2 + v^2 - 10.129y^4 + 0.233y^2v^2 + 0.013v^4 + 10.614y^6$$
$$-0.532y^4v^2 - 0.026x^2v^4 + O(y^m v^n \, m + n \geq 8) \tag{57}$$

Of course the meaning of $y$ and $v$ is $y(\tau = 2k\pi)$ and $v(\tau = 2k\pi)$ i.e the values of $y$ and $v = y'$ taken after an integer number of primaries revolutions. Note that for $e = 0$ only the terms $y^n$ contribute to the first integral which in this case becomes exactly equal to the Hamiltonian (32) as can be seen by inspecting Eq. (30) and (33) with $e = 0$. In Tables III and IV we show the residual variations of $\eta(y, v)$ for $e = 0.6$ and $e = 0.99$ in percents of their value at $\tau = 0$ as function of the initial amplitude $T(0)$ $(y(0) = T(0)/w_0)$ , $v(0) = T(0) = 0)$.

The integral variations have been obtained by inserting the values $y(2k\pi)$ and $v(2k\pi)$ resulting from a numeric integration of Eq. (29) into the analytical integral expressions taken from Table II. For the numeric integration we used the symplectic integrator described in section 4. The integration time was taken over 400 primaries periods.

As can be seen the procedure is semiconvergent. There exists a certain order up to which the integral properties of $\eta(y, v)$ improve strongly with respect to the lowest (second) order approximation. Above this order the approximations start to diverge. In Fig. 4 we show the phase curves $(y, v/Q)$ represented by the first integral $\eta(y, v) = const.$ for $e = 0.7$ and initial conditions $T(0) = 0.05 - 0.60$. Fig. 5 represents the same curves obtained by numeric integration over 200 primaries

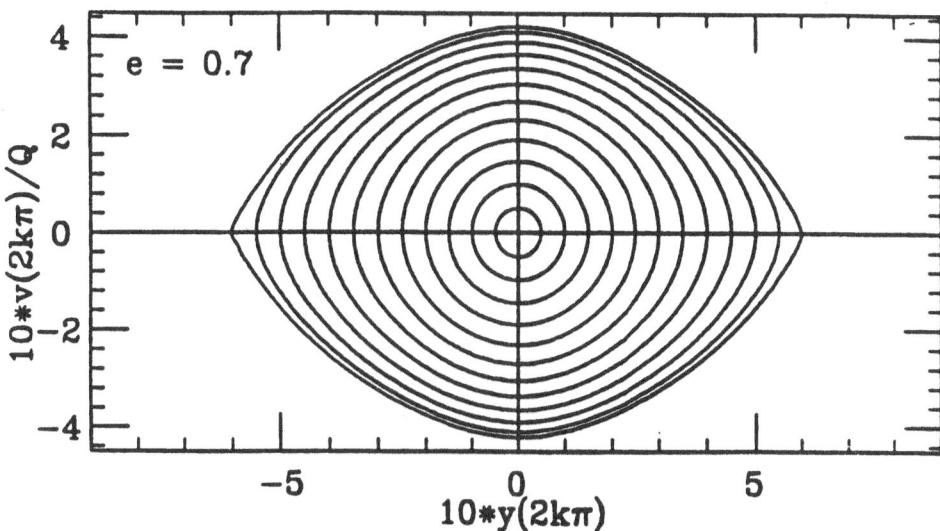

Fig. 4. Analytic phase curves for the Sitnikov problem with $e = 0.7$ and $T(0) = 0.05 - 0.60$ $(y(0) = T(0)/w_0$ , $T'(0) = 0)$

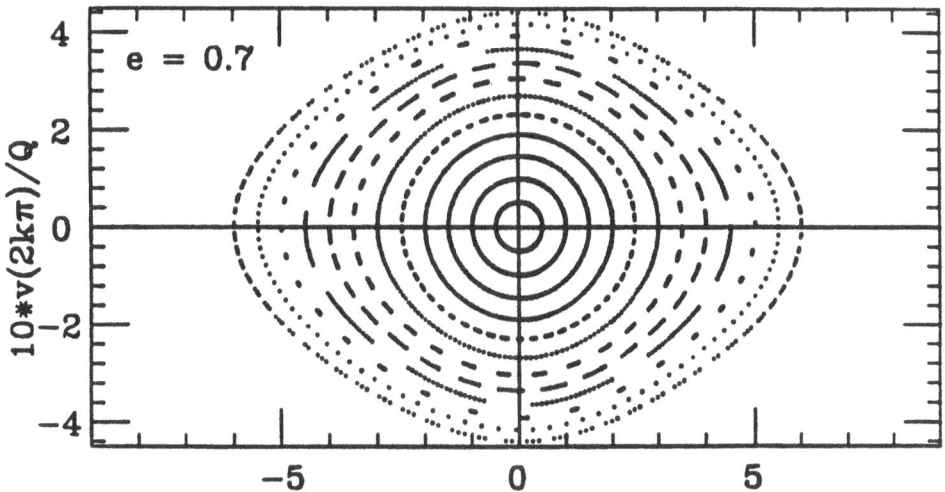

Fig. 5. Numerically obtained phase curves for the Sitnikov problem with $e = 0.7$ and $T(0) = 0.05 - 0.60$ $(y(0) = T(0)/w_0$ , $T'(0) = 0)$

TABLE III

Residual variations in percents of the approximate first integral for $e = 0.6$ as function of the initial amplitude and the integral order.

| Order | $T_0 = 0.05$ | $T_0 = 0.2$ | $T_0 = 0.4$ |
|-------|--------------|-------------|-------------|
| 2     | 0.691        | 9.955       | 29.812      |
| 4     | 0.005        | 1.081       | 20.911      |
| 6     | 0.000        | 0.056       | 4.567       |
| 8     | 0.000        | 0.039       | 2.170       |
| 10    | 0.000        | 0.168       | 44.779      |
| 12    | 0.000        | 0.695       | 143.672     |

TABLE IV

Residual variations in percents of the approximate first integral for $e = 0.99$ as function of the initial amplitude and the integral order.

| Order | $T_0 = 0.05$ | $T_0 = 0.2$ | $T_0 = 0.4$ |
|-------|--------------|-------------|-------------|
| 2     | 0.664        | 9.575       | 28.488      |
| 4     | 0.002        | 0.654       | 12.483      |
| 6     | 0.000        | 0.100       | 1.048       |
| 8     | 0.000        | 0.467       | 25.456      |
| 10    | 0.000        | 2.398       | 145.802     |
| 12    | 0.000        | 11.221      | 111.736     |

periods.

As can be seen the agreement between the analytic and numeric results is very good for all initial conditions $0 < T(0) < 0.7$. For the numeric integration the symplectic integrator described in subsection 4.3 has been used.

## 5.1. FREQUENCY AS FUNCTION OF INITIAL AMPLITUDE

For small values of $T$ (small values for $y(\tau)$) the first integral approaches the one of the harmonic oscillator $y'' + Q^2 y = 0$:

$$\eta_0(y, v) = Q^2 y^2 + v^2 = const. \tag{58}$$

(see Eq. (29)). Consequently in the plane $(y, v/Q)$ we see closed curves with nearly circular shapes when $y$ is small. Any distortion from this circular shape is due to the nonlinear terms in Eq. (29). Inspecting Figs. 4 and 5 we observe that the main effect of the nonlinear contributions is to distort the circular shape of the phase curves to ellipses type objects. Now it is well known that an ellipse can be transformed back to a circle by a one parameter transformation applied to one of the directions of the two main axes. Let $p$ be the crossing point of the phase curves in Figs. 4 and 5 with the $v/Q$ axis. Then the transformation

$$\bar{y} = \frac{p}{y_0} y \tag{59}$$

will generate a circle in the $(\bar{y}, v/Q)$ plane provided the distorted phase curves are exactly elliptic. In this transformed phase space the first integral is again of the harmonic oscillator type, i.e.

$$Q^2 \bar{y}^2 + v^2 = const. \tag{60}$$

Differentiating this equation once w.r.t. to $\tau$ and using $v = y'$ results in

$$\bar{y}'' + \bar{Q}^2 \bar{y} = 0 \quad ; \quad \bar{Q} = \frac{p}{y_0} Q \tag{61}$$

which means that the frequency changes due to the nonlinear terms in the equation of motion. According to the general form (47) of the first integral and the definition of $p$ we find that $p$ is solution of the algebraic equation

$$\sum_{\sigma=2}^{N} \phi(\sigma)_\sigma (Qp)^\sigma = C \tag{62}$$

with

$$C = \sum_{\sigma=2}^{N} \phi(\sigma)_0 y_0^\sigma \quad v_0 = y_0' = 0 \tag{63}$$

Table V shows the dependence of the shifted frequency $\bar{Q}$ on the primaries eccentricity $e$ and the initial amplitude $T(0)$ when $T'(0) = 0$. As can be seen the frequency $\bar{Q}$ decreases with the initial amplitude for all values of $e$. For $T_0 \to 0$ $\bar{Q}$ increases quadratically with $e$ while for larger values of the initial amplitude the frequency decreases with the primaries eccentricity. These results have been found also in (Hagel, 1992) and are analytically described for $|z_0| \leq 0.2$ and $|e| \leq 0.4$ by the relation:

$$\bar{Q}(e, z_0) = \sqrt{8} \left[ 1 + \frac{21}{124} e^2 - z_0^2 \left( \frac{9}{4} + \frac{108}{31} e + \frac{36639}{7688} e^2 \right) \right] \tag{64}$$

$$z_0 = (1 - e) T_0$$

TABLE V

Amplitude dependent frequency $\bar{Q}(e, T_0)$ for the Sitnikov Problem computed from Eqs. (61) - (63)

| $T(0)$ | $e = 0$ | $e = 0.2$ | $e = 0.4$ | $e = 0.6$ | $e = 0.8$ | $e = 0.99$ |
|--------|---------|-----------|-----------|-----------|-----------|------------|
| 0.00 | $\sqrt{8}$ | 2.84802 | 2.91273 | 3.04723 | 3.34258 | 4.96398 |
| 0.10 | 2.77598 | 2.80674 | 2.86849 | 3.00643 | 3.29348 | 4.89341 |
| 0.20 | 2.67461 | 2.69700 | 2.76219 | 2.89060 | 3.17505 | 4.73265 |
| 0.30 | 2.51413 | 2.53423 | 2.60368 | 2.72519 | 2.99919 | 4.60023 |
| 0.40 | 2.30821 | 2.34725 | 2.42091 | 2.52520 | 2.81878 | 4.50321 |

## 6. Conclusions

Using a new method to construct approximate first integrals in near integrable dynamical systems we investigated the Sitnikov problem for low amplitude bounded oscillatory solutions in the full range of primary eccentricities $-0.99 \leq e \leq +0.99$. We found that near integrals in polynomial form can be obtained for sufficiently small oscillation amplitudes in the entire interval of eccentricities. In addition we derived a relation for the nonlinear frequency of the oscillatory solution as function of $e$ and $T_0$. All the necessary computations have been performed by a computer program generally designed for automatic construction of quasi integrals in nonlinear, weakly non integrable systems.

## References

Courant E. D., Livingston M.S. and Snyder H.S.: 1952, *Phys. Rev.*, **88**.

Goldstein H.: 1951, *Classical Mechanics*, Addison-Wesley, Reading.

Hagel J.:1992, 'A new analytic approach to the Sitnikov Problem', *Celest. Mech.* **53**, 267–292.

Lichtenberg A.J. and Liebermann M.A.: 1983, *Regular and stochastic motion*, Springer Verlag.

Mac Millan, W.D.: 1913, 'An integrable case in the restricted problem of three bodies', *Astron.J.*, **27**,11.

Moser J.: 1973, 'Stable and Random Motion in Dynamical Systems', *Annals of Mathematics studies*, **77**.

Moser J.: 1978, *Mathematical Intellegencer*, **1**, 65.

Sitnikov K.: 1960 , 'Existence of oscillating motion for the three-body problem', *Dokl. Akad. Nauk. USSR*, **133**, 303–306.

Stumpff K.: 1965, *Himmelsmechanik, Band II*, VEB, Berlin.

Jie Liu and Yi-Sui Sun: 1990, 'On the Sitnikov Problem', *Celest. Mech.*, **49**, 285–302.

K. Wodnar: 1990, *New formulations of the Sitnikov Problem*, Preprint, Institute of Astronomy, Vienna, Austria, (to be submitted to Celest. Mech.)

# THE ORIGINAL SITNIKOV ARTICLE – NEW INSIGHTS

KARL WODNAR

*Institute of Astronomy, University of Vienna*
*Türkenschanzstraße 17, 1180 VIENNA, Austria*
*e-mail: WODNAR@AVIA.UNA.AC.AT*

**Key words:** Restricted problem – Sitnikov problem – oscillatory motion

In my work (Wodnar, 1992) the extensive proof of Sitnikov's brilliant article on "The Existence of Oscillatory Motions in the Three Body Problem" of March 1960 is reexamined, because of several shortcomings found in the paper. For the first time Sitnikov presented the following nontrivial result: There exists a class of orbits in the three body problem which are unbounded, but nevertheless lead back to the origin of the coordinate system used, an infinite number of times (oscillatory motion). In my treatment only the restricted problem will be dealt with, although Sitnikov gives a generalization to a finite mass of the third body.

The Sitnikov system consists of two primary bodies $m_1$ and $m_2$ of equal mass, moving around each other on Keplerian ellipses and an infinitesimally small 'planet' $m_3$, the motion of which is confined to the line perpendicular to the primaries' plane through the barycenter.

The essential parameter determining the system is the primaries' eccentricity $\varepsilon$. For zero $\varepsilon$ we encounter complete integrability, for sufficiently small $\varepsilon$ values and energies the system performs quasiperiodic motions. For the majority of $\varepsilon$ values in $]0, 1[$ and energies sufficiently close to escape energy, strong chaoticity in the sense of sequence shift embedding on infinitely many symbols occurs. For a numerical analysis of the chaotic behaviour in dependence on $\varepsilon$ see Dvorak (1992).

Sitnikov's result already included part of the sequence shift description, but was confined to monotonically increasing sequences with a lower bound. The generalisation to arbitrary sequences just with a certain lower bound depending on $\varepsilon$ was found by Alekseev. This result was presented by Moser under the more general aspects of Smale's horseshoe theorem and with a generalisation of the latter to infinitely many symbols, going back to ideas of Conley (see Moser, 1972). Nevertheless Sitnikov's derivations gave the first positive answer to the problem of the existence of oscillatory motions in the three body problem posed by Chazy in 1922 (see Arnold, 1985). In addition Sitnikov's proof technique is different to all other related methods and thus may provide informations of how to tackle yet unsolved problems, for which the complete demonstration of the hypotheses of the horseshoe theorem yet fails. This holds especially for Celestial Mechanics problems, because the specific form of the acceleration term plays a central role in Sitnikov's method.

The corrections and improvements of Sitnikov's article made in my work are the following (I assume familiarity with Sitnikov's paper of 1960):

1.) *Numerous* serious printing errors obscuring the meaning of important statements

*Celestial Mechanics and Dynamical Astronomy* **56**: 99–101, 1993.
© 1993 *Kluwer Academic Publishers.*

are corrected and not at all necessary ambiguous notation is replaced by a consistent and more systematic one.

2.) Although I am aware that scientific papers should be written concisely, in the case of Sitnikov's paper one encounters such a short summary of the complicated proof, skipping essential construction steps, that this paper with all other syntactic as well as semantic shortcommings, serves rather as a rough description of the strategy of the proof, than a self- consistent scientific paper, far from being well understandable, as several experts also confirmed. Thus I carried out all steps in detail and reconstructed also large parts of the proof *not* contained in the original paper, like for instance the lengthy integral estimations leading to the validity of inequalities (1) in Sitnikov's article. In this part of the proof and also in many other steps, I found serious shortcommings in argumentation, which was the main reason to give a detailed description of the proof and

3.) discussion as well as correction of problematic and sometimes even wrong statements made in the paper. For instance:

a) In the beginning there is an estimation of the acceleration term of the equation of motion, to establish the existence of a minimum fall height of the small body from barycenter, to fulfill part of the preliminaries of Lemma 2, as well as a lower bound condition on the distance of $m_3$ from barycenter, at pericenter and apocenter times closest to its primary plane passage. Keeping the dependence on the quantity $S$ one step further in the chain of inequalities, saves complicated velocity estimations, which are – by the way – not contained in Sitnikov's article, and thus shortens the proof.

b) In stating the range of validity of inequalities (1) there is a problem of interpretation of the statement: "...We shall show that [inequalities (1)] for the time $t$ up to which $z(t_1 - t)$ increases monotonically and *also*, in the case of the first inequality, for $t = nT$, where $n$ is an integer....". The latter indication of times $t = nT$ is redundant, because these times are already included, from which the question arises, why it is made. On the other hand, after some closer investigation, it turned out that there are periodically imposed gaps (with primary period T) in the range of validity of the first inequality, such that replacing the word 'also' by 'only', which gives quite a different meaning, would turn the statement true. Nevertheless I have found more generally the validity in at least a range $t \in [nT, nT + T/2]$, which makes further argumentation a lot easier. Anyhow either this condition or Sitnikov's condition *must* be taken into account also in further proof steps, which is not the case in the original paper. In my work this point is completely clarified by a detailed analysis of the integral estimations mentioned, *and* a more convenient and also more informative condition is fully included in the application of inequalities (1), which required devising additional proof steps.

c) At some further stage of the proof of Lemma 2, a positive lower bound for the change $\alpha$ in velocity, from last pericenter before crossing the $z = 0$ plane of $m_3$ to first apocenter afterwards, should be used, being universal to all orbits under consideration. In Sitnikov's paper instead, the respective $\alpha$ belonging to an individual orbit is used as an argument of a function which needs to be a lower bound for $S$, at the very final argument to prove Lemma 2. This function goes beyond all bounds, as $\alpha$ approaches zero. But as $\alpha$ still depends on $S$, there is no a priori guarantee that an $S$ can be found, larger than the lower bound of $S$ depending on $\alpha$. This circle conclusion is removed by proving the existence of a positive lower bound for *all* speed differences $\alpha$, no more dependent on $S$, using compactness arguments on an appropriate large set of orbits including all relevant cases.

After all criticism it remains to state that Sitnikov demonstrated an ingenious qualitative method, that in my work is extracted out of a somehow inappropriate presentation. I give a corrected version of the proof and obtain – to my knowledge for the first time – a closed chain of arguments using program-subprogram structure. At the final stage of the proof, I introduce also a new kind of symbolic notation, to characterize the relationship between particular classes of orbits and the corresponding sets of initial conditions.

## References

Arnol'd, V.I.: 1985, *Dynamical Systems III*, Springer Berlin
Dvorak, R.: 1992, "Numerical Results to the Sitnikov problem", *this Volume*
Moser, J.: 1973, *Stable and Random Motion in Dynamical Systems*, Princeton University Press
Sitnikov, K.: 1960, "The Existence of Oscillatory Motions in the Three-Body Problem", *Translation from Doklady Akademii Nauk SSSR*, Vol. **133**, No. **2**, 647-650.
Wodnar, K.: 1992, "The Original Sitnikov Article – New Insights", *Masters Thesis*, University of Vienna

# PROPER ELEMENTS: WHAT ARE THEY ?

ANNE LEMAITRE

*Department of Mathematics - FUNDP*
*8, Rempart de la Vierge - B5000 Namur - Belgium*

**Abstract.** The general ideas about the calculation of proper elements are given here, followed by a comparison of the hypotheses and simplifications made in three different theories: YKM theory, composition of the contributions of Yuasa, Kneževié and Milani , W theory corresponding to Williams's work and HLM theory, paper of Lemaitre and Morbidelli, based on Henrard's semi numerical method. Some short numerical comparisons conclude the paper.

**Key words:** Asteroids – proper elements – averaging method

## 1. Introduction

It is evident now that the understanding of the dynamics of the asteroidal belt could really give important information about the formation of the Solar System and its history. Unfortunately, this dynamics is not easy to model up. The asteroids are in the middle of the Solar System and are drastically perturbed by the main planets; they are numerous and can interact or collide among themselves. The eccentricities and inclinations are relatively high which makes series expansions in these parameters quite problematic. The asteroidal belt is situated in a region between Mars and Jupiter, characterized by a large number of resonances in mean motion (especially with Jupiter) and of secular resonances of first, second or even higher orders, the influence of which is difficult to analyze, each of them requiring a "zoom" and a personal model.

There are some obvious groupings of asteroids in the complicated observed distribution of the planetesimals. They could be the fragments of a same parent body. Of course such families built by celestial mechanicians with the help of dynamical properties have to be checked by chemical or physical data, which could, in some cases, show up the weakness of the hypothesis.

The calculation of proper elements is one of the important parts of the dynamical analysis concerning the asteroidal families. Let us assume that we are able to calculate for each asteroid three invariants of the motion, called the proper elements, depending only on the initial conditions, and staying constant for very long times. We can then imagine that planetesimals with very close proper elements could have been, in the past, physically very close to each other and perhaps fragments of the same important body.

It is obvious that there is an enormous work besides the calculation of the proper elements, which is the classification of the results (when are proper elements "close enough" to be candidates to a same father ?) and the understanding of the fragmentation mechanism (are the fragments coming from a unique catastrophic event or are they fragments of fragments of successive impacts ?). Let us cite here two important contributions dealing with the classification problems: Zappala et al (1990) and Bendjoya et al (1991).

Let us come back to the topic of this paper which concerns the calculation of the proper elements themselves where there are also many interrogation marks. First of all, they are not real invariants of motion but "quasi invariants" because they are results of two averaging processes, the first one over short periodic terms, the second one over long periodic terms. Secondly, how to be confident in results which use secular theories for the main planets on times spans much larger that their reliability ? How can we believe in results obtained by averaging methods of first or second orders using non convergent series expansions ? How can we calculate proper elements of asteroids like Pallas, with very large inclination, or in a region of resonances ?

As an answer to all those questions we can say that it is impossible, nowadays, to give precise values to the proper elements of all the asteroids. However, in many regions, with small eccentricities and inclinations, outside of resonances, not too close to the main planets, the results of three different theories are completely in agreement. These are the three theories, the principles of which we shall develop here. Technical details are published in papers which we shall refer to.

## 2. The Basic Principles

In the three theories for calculating proper elements, the general scheme is the same and can be summarized in six steps:

### 2.1. THE INITIAL PROBLEM

The motion of the asteroid is modeled as a restricted three body problem (or N body problem) Sun - Planet - Asteroid. If we visualize this by the Hamiltonian $\mathcal{H}$ we can write:

$$\mathcal{H} = \mathcal{H}_0 + m_j \mathcal{H}_1 \tag{1}$$

the first term corresponding to the two body problem Sun - Asteroid and the second one, proportional to $m_j$, corresponding to the perturbation due to the $j$th planet. If we include several perturbing planets the second term is just replaced by a summation over the selected planets.

At that stage the problem is guided by the six Lagrange equations (or their corresponding canonical form in a Hamiltonian context) giving the time variations of $a$, the semi major axis, $e$ the eccentricity, $i$ the inclination, $\omega$ the argument of the pericenter , $\Omega$ the longitude of the ascending node and $\lambda$ the mean longitude. We shall also use in some expressions, $\varpi$, the longitude of the pericenter. The right hand sides of those equations contain the corresponding keplerian elements of the perturbing planet (they are designated by a subscript $j$) and are then functions of time. The functions (hamiltonian or perturbing) can be developed in power series of the eccentricity and of the inclination, or kept as close forms of all the keplerian elements.

In any case, at this stage we can distinguish between two types of angles: short periodic angles (namely $\lambda$ and $\lambda_j$) , with periods of a few years and long periodic angles (the arguments of pericenters and the nodes) with periods of a few thousands years.

## 2.2. THE ELIMINATION OF THE SHORT PERIODIC TERMS

The second step consists in eliminating the first types of angles, the short periodic angles to obtain what is usually called the mean elements. This is performed by a classical averaging method, which ends up with a Hamiltonian function (or a perturbing function) independent of the short periodic angles up to a given order, i.e. up to a certain power of the (small) parameter $m_j$. This averaging can be performed completely analytically or numerically, and is truncated at the first or the second order, creating then important differences between the theories . In a schematic way this is done through a transformation:

$$(a, e, i, \omega, \Omega, \lambda) \Longrightarrow (\bar{a}, \bar{e}, \bar{i}, \bar{\omega}, \bar{\Omega}, \bar{\lambda})$$

from osculating elements to mean elements in such a wy that the differential equations in mean elements do not depend any more on the mean longitudes. If we write the mean hamiltonian function $\bar{\mathcal{H}}$, we get:

$$\bar{\mathcal{H}} = \bar{\mathcal{H}}_0 + m_j \bar{\mathcal{H}}_1 + m_j^2 \bar{\mathcal{H}}_2 + O(m_j^3) \tag{2}$$

the first term being now a constant , generally omitted, the last term being absent in the first order theories. Again this function can be presented as a power series of the eccentricity and of the inclination, of the asteroid and (or) of the perturbing planet. The elements varying with time are then the mean eccentricity, $\bar{e}$, the mean inclination $\bar{i}$ , the mean argument of pericenter (for the asteroid and the planets) and the mean nodes. The semi major axis, thanks to the elimination of the longitudes is a constant (up to the order of truncation of the averaging process, i.e its time derivative is proportional to $m_j^2$ for a first order method and to $m_j^3$ for a second order method). The mean semi major axis is then the first quasi invariant of the motion and consequently the first proper element.

## 2.3. THE SEPARATION OF $K$ INTO $K_0$ AND $K_1$

The next step is the preparation of the mean function for the second averaging, over the long periodic terms. It consists in separating $\bar{\mathcal{H}}$ into two parts, a main one and a perturbation. In other words we have to choose a new small parameter $\epsilon$ in the mean problem so to write:

$$\bar{\mathcal{H}} - \bar{\mathcal{H}}_0 = K$$
$$= K_0 + \epsilon K_1 \tag{3}$$

The choice of $\epsilon$ is linked to some hypothesis concerning the order of magnitude of the eccentricity (or the inclination) of the asteroid and of the main planets. We have two solutions: either we consider that the eccentricity and the inclination of the asteroid are as small as those of the planets, or we consider that they are significantly higher. In the first case, we reject in the perturbation $K_1$ all the terms of degree larger than 3 in the eccentricities or inclinations. This is a compromise between the two desirable properties of having a simple, easily integrable, first approximation $K_0$ and a small enough perturbation $K_1$. In the second case, we keep in $K_0$ whatever does not depend on $e_j$ or $i_j$. In any case, we can find in both $K_0$ and $K_1$ terms proportional to $m_j^2$ if the first averaging process has been performed up to the second order.

This step is fundamental because it can lead to completely different topologies for the $K_0$ phase space and for high inclinations the results can be drastically different..

## 2.4. THE ELIMINATION OF THE LONG PERIODIC TERMS

The next step is the averaging over the long periodic terms, namely the pericenter arguments and the longitudes of the nodes. First , for the sake of simplifying this account, let us replace the longitude of the pericenter and the node of the perturbing planet $j$ by a linear functions of time associated with the classical secular frequencies i.e.

$$\dot{\varpi}_j = \nu_j$$
$$\dot{\Omega}_j = s_j \tag{4}$$

These quantities receive different numerical values in different theories. In the three theories that we shall describe this second averaging process is performed up to the first order ; it means that the final Hamiltonian function can be described as a function of $\bar{\bar{e}}$ the averaged mean eccentricity and $\bar{\bar{i}}$ the averaged mean inclination (usually called proper eccentricity and proper inclination) up to the first order, i.e. with an error proportional to $\epsilon^2$. We can write in other words:

$$\bar{K} = \bar{K}(\bar{\bar{e}}, \bar{\bar{i}})$$
$$= \bar{K}_0 + \epsilon \bar{K}_1 + O(\epsilon^2) \tag{5}$$

The two basic frequencies of the problem, called $g$ and $s$ are defined through the partial derivatives of $\bar{K}$. ($g$ is the frequency of the averaged mean longitude of pericenter and $s$ the frequency of the averaged mean longitude of the node, these two angles being generally called the proper angles). Schematically, we can define a transformation:

$$(\bar{a}, \bar{e}, \bar{i}, \bar{\omega}, \bar{\Omega}, \bar{\lambda}) \Longrightarrow (\bar{\bar{a}}, \bar{\bar{e}}, \bar{\bar{i}}, \bar{\omega}, \bar{\bar{\Omega}}, \bar{\bar{\lambda}})$$

such that the differential equations depend only on the proper elements.

## 2.5. THE ITERATIVE PROCESS

The second average can be improved by the introduction of an iterative process; let us recall that the computation of the transformation from mean to proper elements involves an approximation of the proper frequencies (which are computed only at the end of the calculation). This is the reason why it is worth iterating the process to use better and better approximations of these frequencies; this can be written down as the search for a fixed point of a transformation.

## 2.6. THE OUTPUTS

The outputs can be very different from one theory to the other; the questions to answer here are : what is the best output for the families, and which output to choose to compare the different results ? We shall try to answer both questions in the following sections.

## 3. Description of the Three Theories

The theories that we are going to describe with respect to the scheme presented in the above section are denoted by YKM for Yuasa, Knežević and Milani's study, W for Williams's theory and HLM for Henrard, Lemaitre and Morbidelli's. This is in that order that we shall analyze them here, for the clarity of the presentation. This is neither a chronological choice nor an alphabetic order.

### 3.1. THE ANALYTICAL THEORY OF YUASA, KNEŽEVIĆ AND MILANI (YKM)

The first theory which we mention is probably the best known nowadays by the "families builders". It has the enormous advantage to be completely analytical which means that a result can be calculated in a few fractions of seconds for any initial condition. So it is really well appropriate for testing hundreds of test asteroids.

Unfortunately, to keep explicit formulae in the restricted three body problem and to integrate them explicitly, we have to use series expansions in the assumed small parameters of the problem, namely the eccentricities and the inclinations, and to truncate them drastically at some low degree.

Yuasa (1973) started this work; he published a completely analytical method for the calculation of proper elements, using the Hamiltonian formalism and developing Hamiltonian and generating functions in power series of the eccentricities and the inclinations of the asteroid as well as of the planets. He performed by hand the two average processes up to the second order for the first one and up to the first order for the second one. He truncated all of the series at the degree 4 in the eccentricities and inclinations.

Knežević (1988-1989) extended Yuasa's theory by calculating the indirect part of the perturbation mean function and the corresponding first and second orders for the mean elements. He checked also by hand Yuasa's calculations and doing so, corrected a few misprints in Yuasa's original paper.

Milani and Knežević (1990) have rewritten Yuasa's theory and Knežević's additions in a Hamiltonian Lie formalism. They completed the calculation of the proper elements by an iterative process in the second averaging process and they implemented all those results in two FORTRAN programs, available to anybody, the first one for the calculation of the mean elements, the second one for the iterative determination of the proper elements, based on the data file resulting from the first program.

If we want to classify their common work with respect to the six points that we have enumerated in the previous section, we can say:

### 3.1.1. THE INITIAL PROBLEM

- They use a Hamiltonian formalism and consequently canonical differential equations, written in Delaunay or Poincaré variables (which means that they manipulate explicit formulae in eccentricity or inclination)
- They develop the Hamiltonian as a power series of $e$, $\sin i$, $e_j$, $\sin i_j$ up to degree 4, these powers being multiplied by cosinus of linear combinations of all the angles. To give an idea about the size of the problem, the part $\mathcal{H}_1$ takes more than five pages in Yuasa's publication.
- They include the perturbations due to Jupiter, Saturn, Uranus and Neptune and even a contribution due to the inner planets.

### 3.1.2. THE ELIMINATION OF THE SHORT PERIODIC TERMS

The elimination of the short periodic terms is performed up to the second order by hand by means of an explicit generating function again truncated at degree 4 .

### 3.1.3. THE SEPARATION OF $K$ INTO $K_0$ AND $K_1$

- The separation of $K$ into $K_0$ and $K_1$ is performed in a logical way in relation with the previous choices; if the perturbation is developed as a power series of the parameters of both the asteroid and the perturbing planet up to the same degree it is consistent to choose as parameter $\epsilon$ a power of this general small parameter. $\epsilon$ is then chosen as the third power of the small quantity; it means that any product of $e$, $e_j$, $\sin i$ or $\sin i_j$ with three factors or more is relegated into $K_1$ and the terms of degree 2 are kept in $K_0$. However the terms of degree 2 including the pericenter of the asteroid are also included in $K_1$ (typically terms like $e^2 \cos 2\omega$). This is due to the fact that to perform the second averaging process $K_0$ has to be integrable.
- With this choice, $K_0$ regroups the terms in: $\bar{e}^2$, $\sin^2 \bar{i}$, $e_j^2$, $\sin^2 i_j$, $\bar{e}e_j \cos(\bar{\varpi} - \varpi_j)$ and $\sin \bar{i} \sin i_j \cos(\bar{\Omega} - \Omega_j)$. The last two terms are eliminated by a

linear cartesian transformation and the resulting $K_0$ is obviously completely integrable (it is only function of $\bar{e}$ and $\bar{\imath}$). All the orbits are circulators for the induced dynamics. If we forget about the part $K_1$ the resulting values for the eccentricity and the inclination are usually called the **linear proper elements**.

### 3.1.4. THE ELIMINATION OF THE LONG PERIODIC TERMS

- The numerical values for the main planets are taken from Nobili et al (1989).
- The second (first order) average, performed over the long periodic terms, has also been obtained by hand and gives as result an Hamiltonian function independent of the angles. The two basic frequencies $g$ and $s$ are defined by the partial derivatives of $\bar{K}_0$ ($g_0$ and $s_0$) and of $\bar{K}_1$ (adding the corrections $\delta g$ and $\delta s$):

$$g = g_0 + \delta g$$
$$s = s_0 + \delta s \tag{6}$$

- The proper elements are obtained from the mean ones by the generating function of this second average (which takes more than 14 pages in Yuasa's paper !) and the proper semi major axis is the mean semi major axis.

### 3.1.5. THE ITERATIVE PROCESS

A very interesting iterative process is implemented in the theory, based on the following idea: the perturbation is calculated thanks to the $K_0$ integrable part, and depends on $g_0$ and $s_0$ (see equations (6)). Due to this perturbation the frequencies are corrected and become $g$ and $s$; so the idea is to recalculate the correction due to the perturbation but after having replaced $g_0$ and $s_0$ by $g$ and $s$ in $K_0$. This leads to new values of $g$ and $s$ and the process can be iterated leading hopefully to a convergence. Milani and Knežević implemented this iterative procedure in their program and it really improves drastically the results; of course, the iterative process is divergent for some asteroids. The result after one (or more) iteration is then retained but with a very bad quality code. This process is also in some sense empirical since no demonstration is provided for convergence.

### 3.1.6. THE OUTPUTS

For YKM theory the output is obvious; the program ends up with a value of $\bar{\bar{e}}$ and $\bar{\bar{\imath}}$ calculated by an explicit formula with respect to $\bar{e}$ and $\bar{\imath}$ and the mean angles, themselves described by other explicit formulae as functions of the initial osculating elements. So here no ambiguity lies about the choice of a suitable output.

### 3.1.7. CONCLUSIONS

In conclusion we can say that the YKM theory is a very efficient way for calculating proper elements for asteroids with relatively small eccentricities or inclinations (of the same order of those of the main planets); the programs are available and easy to use. Another interesting help that their program offers is the selection of the secular resonant asteroids which are regrouped in a special file. The selection is made according to the size of the denominators playing a role in the generating function of the second averaging process. The results have been checked by comparisons with some numerical integrations (see Knežević (1991))

## 3.2. WILLIAMS'S THEORY

We would like to say here that Williams's theory (see Williams (1969)) was really breakthrough , especially if we recall the context and the ideas known at that time as well as the available computers.

Let us mention first that his theory is usually referred to as a semi analytical method; this means that the ideas are exactly those of a classical perturbation theory but that many calculations are performed numerically; it also means that there are no complete formulae at the end of a paper, allowing to calculate from osculating elements, the corresponding mean and proper elements. In each case there is a numerical part (integration) to perform. This fact is due to the non expansion of the perturbing function in a power series of $e$ and $i$ and from the onset specializes his theory to the calculation of high eccentric or high inclined orbits. Let us describe the different points of our scheme for Williams's theory:

### 3.2.1. THE INITIAL PROBLEM

– Williams uses directly the differential equations in terms of the keplerian variables and not an Hamiltonian formalism.
– He does not develop the perturbation in power series of $e$ or $i$ but he only keeps the first degree terms in $e_j$ or $i_j$.
– Only Jupiter is included as perturbing planet.

### 3.2.2. THE ELIMINATION OF THE SHORT PERIODIC TERMS

– To calculate the mean equations (or the equations for the mean elements) he uses a first order perturbation method i.e. the terms proportional to $m_j^2$ are all omitted.
– He has to perform numerically the double integral (over $\lambda$ and $\lambda_j$); however thanks to the limitation to the first degree in $e_j$ and $\sin_j$, he manages (by a series of algebraic manipulations) to perform the average over $\lambda_j$ analytically, so the first average process with those hypotheses is reduced to calculations of simple integrals.
– The result is obtained without any expansion in $e$ and $\sin i$.

### 3.2.3. THE SEPARATION OF $K$ INTO $K_0$ AND $K_1$

– The decomposition of $K$ into $K_0$ and $K_1$ is done with respect to the small parameter $\epsilon$ chosen as $e_j$ or $\sin i_j$. With this choice we get $K_0$ in close form in terms of the mean eccentricity, mean inclination and mean argument of the pericenter. So we can write:

$$K_0 = K_{(e_j=0,i_j=0)}$$
$$= K_0(\bar{e}, \bar{\imath}, \bar{\omega}) \tag{7}$$

$$K_1 = e_j \frac{\partial K}{\partial e_j}\bigg|_{(e_j=0,i_j=0)} + i_j \frac{\partial K}{\partial i_j}\bigg|_{(e_j=0,i_j=0)}$$
$$= K_1(\bar{e}, \bar{\imath}, \bar{\omega}, \bar{\Omega}, e_j, \sin i_j, \omega_j, \Omega_j) \tag{8}$$

The $K_0$ dynamics has been analyzed by Kozai (1962) and the topology is very interesting; for small values of $\bar{\imath}$ we find a $K_0$ phase space with only circulators (with respect to the angle $\omega$) ; however they are not circles and the corresponding values of $\bar{e}$ can show large oscillations. But for large values of $\bar{\imath}$ (i.e greater than 30°) there is a critical curve separating the whole phase space into two regions of libration (about 90° and 270°) and a region of circulation. We shall refer to this curve as to **Kozai's separatrix**. We present in figures 1 and 2 the two topologies for a choice of variables linked to Delaunay variables $x$ and $y$ for two constant values of $P$:

$$x = \sqrt{2Q} \cos \omega$$
$$y = \sqrt{2Q} \sin \omega$$
$$Q = L - G$$
$$P = L - H \tag{9}$$

where $L$, $G$ and $H$ are the usual Delaunay momenta.

– For the keplerian elements of the main planets and the basic frequencies, Williams uses Brouwer's (1951) values.

### 3.2.4. THE ELIMINATION OF THE LONG PERIODIC TERMS

Williams averages a second time over the long periodic terms by a first order perturbation method; by means of long manual substitutions he succeeds in expressing the proper orbit as function of the simple integrals that he has got for the first average process. The correction is calculated directly on the keplerian mean elements $\bar{e}$, $\bar{\imath}$ and $\bar{\omega}$.

### 3.2.5. THE ITERATIVE PROCESS

Williams calculates together with the differential equations for the elliptic elements variational equations for corrections to these elements; he uses a kind of iterative process to adjust these corrections. We have no indication from the final file whether this process is always convergent or not.

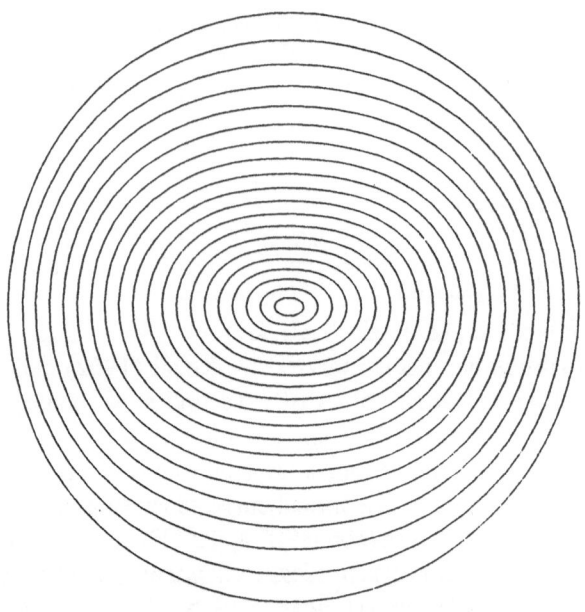

Fig. 1. $K_0$ Phase space for small $i_{max}$ ; $P = 0.044$ and $i_{max} = 20°$

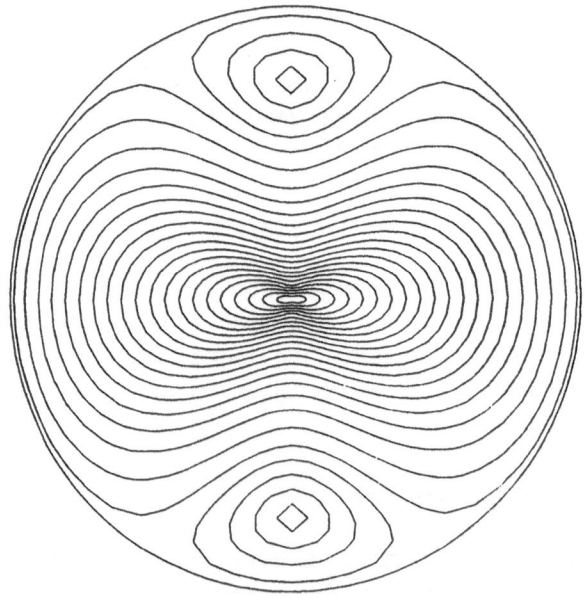

Fig. 2. $K_0$ Phase space for large $i_{max}$ ; $P = 0.17$ and $i_{max} = 40°$

## 3.2.6. THE OUTPUTS

Williams has decided to choose a representative point of his final orbits to be the proper elements; he has chosen the intersection of the orbits with the x-axis (x defined as previously) which corresponds to the point of minimal eccentricity and maximal inclination. This choice also leads to the exclusion of the $\omega$-librators for which Williams's theory fails to give a result.

## 3.2.7. CONCLUSIONS

Williams's theory is very efficient for calculating proper elements for asteroids with high inclinations or eccentricities; two things limit the power of his work: the truncation to the first order in $m_j$ in the first averaging process and, to a less important extend, the limitation to the first degree in the eccentricity and inclination of Jupiter.

### 3.3. HENRARD, LEMAITRE AND MORBIDELLI'S SEMI-ANALYTICAL METHOD (HLM)

The theory for calculating proper elements that we shall present here is quite new, not yet published (Lemaitre and Morbidelli (1992)) and directly based on the semi-analytical method developed by Henrard (1990).

The idea was to combine the positive aspects of both previous theories; of course it is not possible to avoid all their negative aspects and the result appears as a compromise.

## 3.3.1. THE INITIAL PROBLEM

– The HLM theory uses a Hamiltonian formalism and canonical variables; however contrarily to the YKM formulation the chosen canonical variables, after the first average, are non explicit functions of the eccentricity or the inclination. The variables used here are Arnold's action angle variables.
– Concerning the initial Hamiltonian, the hypotheses are the same as Williams: no expansion in series of the eccentricity or the inclination of the asteroid but truncation at the first degree in $e_j$ or $\sin i_j$.
– Only the perturbation due to Jupiter is included; the addition of Saturn is only a problem of CPU time and could be done without any effort but the question arising here is the following: is it logical to include Saturn and to exclude $e_j^2$ for Jupiter ?

## 3.3.2. THE ELIMINATION OF THE SHORT PERIODIC TERMS

– This average is performed numerically by double integrals; in fact, a Fourier series of the long periodic angles is used , the coefficients of which are calculated on a three dimensional grid (for a series of values of $a$, a series of values of $e$ and a series of values of $i$). The Fourier coefficients are then triple integrals, stocked once for ever. Each time that the Hamiltonian or its derivatives are needed, they are calculated by three successive linear interpolations on the grid.

- An important difference with Williams is that this first average is also performed for the second order terms in $m_j^2$ using the principle of the Fourier developments as well. The basic idea comes from Morbidelli (1992).

### 3.3.3. SEPARATION OF K INTO $K_0$ AND $K_1$

- The parameter $\epsilon$ is chosen exactly in the same way as Williams; $(e_j$ or $i_j)$ associated to the same hypothesis of relatively high eccentricity or inclination for the asteroid with respect to the main planets.
- The dynamics of the $K_0$ is illustrated in figures (1) and (2).
- The values for the frequencies and constants for the main planets are taken from Nobili et al (1989).

### 3.3.4. ELIMINATION OF THE LONG PERIODIC TERMS

- Let us recall that action angle variables have been introduced which leads to the following expression for the Hamiltonian $K$:

$$
\begin{aligned}
K_0 &= K_0(\bar{e}, \bar{i}, \bar{\omega}) \\
&= K_0(J, Z) \\
K_1 &= K_1(\bar{e}, \bar{i}, \bar{\omega}, \bar{\Omega}, e_j, \sin i_j, \omega_j, \Omega_j) \\
&= K_1(J, Z, \psi, z, e_j, \sin i_j, \omega_j, \Omega_j)
\end{aligned}
$$

(10)

(11)

$J$ and $Z$ are the momenta , $\psi$ and $z$ the associated angles. An orbit in the $K_0$ integrable problem is described by its action angle variables; the angles $\psi$ and $z$ are linear functions of the time and the momenta $J$ and $Z$ are constant.

- The average over the pericenters and the nodes is performed by the semi numerical method of Henrard (1990); the principles are that the canonical transformation between mean and mean averaged canonical action angle variables is performed by simple integrals along the closed curves of the $K_0$ problem. These integrals are performed numerically.
- This method is a first order averaging process.
- The resulting Hamiltonian, independent of the averaged angles $\bar{\psi}$ and $\bar{z}$ is written as a function of the two averaged momenta $\bar{J}$ and $\bar{Z}$:

$$
\begin{aligned}
\bar{K} &= \bar{K}_0 + \bar{K}_1 + O(\epsilon^2) \\
\bar{K}_0 &= K_0(\bar{J}, \bar{Z}) \\
\bar{K}_1 &= \bar{K}_1(\bar{J}, \bar{Z}, -, -) \equiv 0
\end{aligned}
$$

(12)

The proper elements are $\bar{J}$, $\bar{Z}$ and the basic frequencies $g$ and $s$ are given by $\partial \bar{K}_0 / \partial \bar{J}$ and $\partial \bar{K}_0 / \partial \bar{Z}$. Let us note that the result is here an averaged orbit characterized by its two momenta.

### 3.3.5. THE ITERATIVE PROCESS

Copying the idea of YKM theory an iterative process is also implemented based on the following principle: we start with mean elements and we calculate corresponding proper elements; we obtain the two basic frequencies defining the torus on which the motion takes place. Unfortunately the transformation between mean and proper elements is only a first order one and is not bijective. Calculating the mean elements by the reverse transformation, we do not obtain the initial ones. In other words we find a torus which does not contain the initial point. The idea is then to update the proper elements to find back the initial mean ones by the reverse transformation. This could be modeled up as a Newton Raphson iteration. This process does not converge whenever we cross Kozai's separatrix or a secular resonance.

### 3.3.6. THE OUTPUTS

The final point of HLM theory is a proper orbit; the question is then to decide which point of this orbit has to be chosen to represent the proper elements. It is obvious that the real proper elements considered as quasi invariants of the motion are the mean semi major axis and the two frequencies $g$ and $s$. However up to now the families have been built on proper eccentricities and proper inclinations. Three outputs are then proposed:

—   In order to compare the result with YKM theory a first definition of proper elements is adopted, defining a mean proper eccentricity and a mean proper inclination by considering that the proper area defined by $\bar{J}$ is the area of a circle.

—   A second definition is also used to compare with Williams's results: the point of maximal eccentricity and minimal inclination is chosen which corresponds with the intersection of the proper orbit with the x axis.. However, there is a problem with Williams's definition; Williams does not give any result for $\omega$-libration orbits; YKM's theory gives a result as if these orbits were circulators and so it is non significant. HLM theory is the only one to really take these $\omega$-librators into account.

—   So we would like to adopt a third definition of proper eccentricity and proper inclination which would corresponds to the point of intersection of the orbit with the y axis i.e. the point of the orbit of minimum eccentricity and maximum inclination. This definition does work for circulators and librators.

TABLE I

Features of the three theories

| | YKM | W | HLM |
|---|---|---|---|
| **The initial problem** | | | |
| Hamiltonian formalism | + | - | + |
| Several perturbing planets | + | - | - |
| Degree of expansion in $e$ and $i$ | 4 | $\infty$ | $\infty$ |
| Degree of expansion in $e_j$ and $i_j'$ | 4 | 1 | 1 |
| | | | |
| **Elimination of the short periods** | | | |
| Type of method (An or Num) | A | N | N |
| Order of the theory | 2 | 1 | 2 |
| | | | |
| **Separation of $K$ into $K_0$ and $K_1$** | | | |
| Choice of $\epsilon$ | $e^3, e_j^3, ...$ | $e_j$ | $e_j$ |
| Degree of expansion in $e$ and $i$ | 4 | $\infty$ | $\infty$ |
| $K_0$ dependent on $\omega$ | - | + | + |
| Planets frequencies | Nob. et al | Br | Nob. et al |
| Hamiltonian formalism | + | - | + |
| Action angle variables | - | - | + |
| | | | |
| **Elimination of the long periods** | | | |
| Type of method (An. or Semi Num.) | A | SN | SN |
| Order of the theory | 1 | 1 | 1 |
| | | | |
| **Iterative process** | | | |
| existence | + | + | + |
| convergence study | + | - | + |
| | | | |
| **Outputs** | | | |
| Frequencies $g$ and $s$ | + | + | + |
| Proper eccentricity | Mean | Min | All |
| Proper inclination | Mean | Max | All |

## 3.3.7. CONCLUSIONS

The method keeps the very interesting Wiliams's idea of not using any development in $e$ or $i$; it adds the corrections in $m_j^2$ and a complete iterative process absent in Williams's calculations; however the price to pay is the early truncation in $e_j$ and $i_j$ at the first degree, the maximal precision which could be obtained is of the order of $e_j^2$.

TABLE II
Mean Initial Elements taken from YKM's theory

| n | ω | Ω | i | e | a (AU) |
|---|---|---|---|---|---|
| | Mean Elements | | | | |
| 1 | 76.76967 | 72.72859 | 9.21587 | 0.0777883 | 2.7670769 |
| 4 | 151.16324 | 99.48615 | 5.56213 | 0.0892974 | 2.3615429 |
| 5 | 344.85444 | 150.63399 | 4.14043 | 0.1887523 | 2.5762979 |
| 7 | 138.75558 | 262.47484 | 6.95169 | 0.2301348 | 2.3861404 |
| 9 | 18.13537 | 52.87776 | 4.45547 | 0.1224660 | 2.3864202 |
| 25 | 86.15207 | 214.62271 | 22.09807 | 0.2547551 | 2.4003785 |

## 3.3.8. THE ANGLE CORRECTIONS

In a first version of the theoretical part developed above a complete set of proper elements, namely $\bar{J}$, $\bar{Z}$, $\bar{\psi}$ and $\bar{z}$ was given but actually only the two momenta are used really by the families formers. So it is not necessary to include the proper angles in the program, because they introduce corrections of the second order on the two proper momenta which are not significant in a first order theory and because the calculation of the angles updatings is much more expensive and complicated than the momenta updating and requires a complete set of variational differential equations.

We summarize in Table I the different features of the three theories.

## 4. Comparisons

To compare the results of the HLM and YKM methods we have adopted the following conventions:
1. The YKM's used version is truncated at the first degree in $e_j$ and $\sin i_j$.
2. Only the perturbation of Jupiter is taken into account
3. The $m_j^2$ terms are considered in both methods
4. We adopt the definition of the **mean** proper elements
5. We limit our comparison to small eccentricity and small inclination

We give a table with the results of both methods (proper eccentricity, sinus of the proper inclination, frequency of $\varpi$ and of $\Omega$). The initial conditions (mean elements) are the same and given in the initial table (II). f1 is the frequency of $\varpi$ and f2 of the node $\Omega$.

To compare HLM results with those of Williams:
1. We do not take the $m_j^2$ terms into account
2. We adopt the second definition of proper elements.

As an example we give the result for Ceres (1) and for an inclined asteroid, Phocaea (25)

## TABLE III
### Corresponding Proper Elements

| n | HLM Proper Elements | | | | YKM Proper Elements | | | |
|---|------|-------|----------|-----------|-------|-------|----------|-----------|
|   | $\bar{e}$ | $\sin \bar{i}$ | f1 | f2 | $\bar{e}$ | $\sin \bar{i}$ | f1 | f2 |
| 1 | .1149 | .1699 | 52.19057 | -57.06484 | .1151 | .1692 | 51.81332 | -56.50454 |
| 4 | .0824 | .1175 | 34.67674 | -37.14707 | .0992 | .1160 | 34.47869 | -37.51920 |
| 5 | .2286 | .0807 | 49.53157 | -56.72649 | .2274 | .0813 | 49.13815 | -56.81949 |
| 7 | .2113 | .1083 | 35.98235 | -44.10655 | .2155 | .1070 | 35.64596 | -44.47012 |
| 9 | .1253 | .0826 | 36.53800 | -39.93415 | .1354 | .0845 | 36.21902 | -40.23345 |

## TABLE IV
### Corresponding Proper Elements

| n | HLM Proper Elements | | | | WILLIAMS Proper Elements | | | |
|---|------|-------|----------|-----------|-------|-------|------|-------|
|   | $\bar{e}$ | $\sin \bar{i}$ | f1 | f2 | $\bar{e}$ | $\sin \bar{i}$ | f1 | f2 |
| 1 | .1096 | .1723 | 52.1503 | -56.8997 | .097 | .169 | 50.6 | -58.3 |
| 25 | .1838 | .4153 | 16.7274 | -36.7727 | .183 | .417 | 17.0 | -38.4 |

Let us recall that, even if the calculations were performed exactly in the same way, with the same initial hypotheses, the resulting proper elements would be slightly different because of their definitions; for example, the OUTPUT 1 is not exactly the proper element of the YKM theory, it is the closest possible one.

## 5. Conclusions

As a conclusion we can say that for small eccentricities and small inclinations (it means as far as we know today more or less 13° for the inclination and maximum 0.2 for the eccentricity) the YKM theory is the most precise, the fastest and the easiest to use. For higher values of $e$ and $i$ the HLM method, based on Williams's ideas seems to be the most adequate even if it requires much more CPU.

## Acknowledgements

I would like to thank Claude Froeschlé and Jacques Henrard for their remarks and comments concerning the redaction of this paper.

## References

Bendjoya, Ph., Slezak, E. and Froeschlé, C.: 1991, "The wavelet transform: a new tool for asteroid family determination", *Astron. Astroph.*, in press
Brouwer, D.: 1951, "Secular variations of the orbital elements of the principal planets", *Astron. J.* **56**, 9–32

Henrard, J.: 1990, "A semi numerical perturbation method for separable hamiltonian systems", *Celest. Mech.* **49**, 43–67

Knežević, Z.: 1988, "Asteroid mean orbital elements", *Bull. Astron. Obs. Belgrade* **139**, 1–6

Knežević, Z.: 1989, "Asteroid long periodic perturbations: the second order Hamiltonian", *Celest. Mech.* **46**, 147–158

Knežević, Z.: 1991, "Asteroid long periodic perturbations: derivation of proper elements and assessment of their accuracy", *Astron. Astroph.* **241**, 267–288

Kozai, Y.: 1962, "Secular perturbations of asteroids with high inclinations and high eccentricities", *Astron. J.* **67**, 591–598

Lemaitre A. and Morbidelli, A.: 1992, "A semi numerical method for the calculation of high inclined asteroids", *Celest. Mech.* , In preparation

Milani, A. and Knežević, Z.: 1990, "Secular perturbation theory and computation of asteroid proper elements", *Celest.Mech.* **49**, 247–411

Morbidelli, A.: 1992, "On the successive eliminations of perturbations harmonics", *Celest. Mech.* , In press

Nobili, A.M., Milani, A. and Carpino, M.: 1989, "Fundamental frequencies and small divisors in the orbits of the outer planets", *Astron. Astroph.* **210**, 313–336

Williams, J.G.: 1969, "Secular perturbations in the solar system", *Ph.D Thesis* , Univ. California, Los Angeles

Yuasa, M.: 1973, "Theory of secular perturbations of asteroids including terms of higher orders and higher degrees", *Publ. Astr. Soc. Japan* **25**, 399–445

Zappala, V., Cellino, A., Farinella, P. and Knežević, Z.: 1990, "Asteroids families I: identification by hierarchical clustering and reliability assessment", *Astron. J.* **100**, 2030–2046

# THE HIGH-ECCENTRICITY LIBRATION OF THE HILDAS

## II. SYNTHETIC-THEORY APPROACH

T. MICHTCHENKO and S. FERRAZ-MELLO

*Instituto Astronômico e Geofísico*
*Universidade de São Paulo, São Paulo, Brazil*

**Abstract.** The use of precise numerical integrations and Fourier analysis techniques allowed us an investigation of the regular motions of asteroids near the 3:2 resonance with Jupiter (Hildas). The results are shown and compared to similar results previously obtained with analytical models.

**Key words:** Asteroids – resonance

## 1. Introduction

The analytical modelling of the long-term regular motion of high-eccentricity asteroids near the 3:2 resonance (Ferraz-Mello, 1988) resulted in three approximate laws for the low-amplitude librations: (1) the law of structure, relating the semi-major axis and the eccentricity of libration centers (*i.e.* zero-amplitude librations); (2) the law of periods, giving the proper periods of small-amplitude librations and (3) the law of the second forced mode, giving the amplitude of the long-period oscillation forced by the non-zero eccentricity of Jupiter. The analytical modelling included some important simplifications and was founded on an abridged version of the averaged potential of the disturbing action of Jupiter valid in the neighbourhood of libration centers; it was restricted to low-amplitude librations. The results were confirmed by the few available observational data and some numerical simulations.

In this communication, the same problem is considered once more. But, now, the results are obtained with an accurate numerical technique and are not restricted to the immediate neighbourhood of the libration centers. In order to study these motions, orbits were computed numerically over 50,000 years using Everhart's RA15 integrator (Everhart, 1985) with initial values taken in an array of $40 \times 40$ points in the domain $\{0.75 \leq a \leq 0.79 \, (a_{Jup} = 1), \ 0.005 \leq e \leq 0.4\}$ The initial longitudes of the asteroid and of Jupiter were chosen by taking into account that, in order to have resonance, the initial value of the critical angle

$$\sigma = 3\lambda_J - 2\lambda - \varpi$$

must be kept close to $0°$ (pericentric librations) or $180°$ (apocentric librations). By analogy with the choice done by Wisdom (1983) in the study of the resonance 3:1, we adopted the initial conditions $\sigma = 0$ and $\varpi - \varpi_J = 0$. As the angle $\varpi - \varpi_J$ circulates or librates about $0°$, all possible solutions in the four-dimensional domain under study are considered. In the solutions, the oscillations whose frequencies are above $0.01yrs^{-1}$ were filtered out using the Fourier techniques introduced by Carpino *et al.* (1987). Apocentric librators, with initial values taken in the intervals $0.005 \leq e \leq 0.02, 0.79 \leq a \leq 0.80$ and with initial longitudes such that $\sigma = \pi$,

*Celestial Mechanics and Dynamical Astronomy* **56**: 121–129, 1993.
© 1993 *Kluwer Academic Publishers.*

were also studied; in this case only the oscillations whose frequencies are above $0.013yrs^{-1}$ were filtered out. In both cases, the resulting time series were spectrally analyzed.

The most interesting results thus far obtained concern the pericentric librations and are the following: (1) The spectra show, generally, two independent modes: one free oscillation (libration) and one forced one (the second forced mode due to Jupiter's eccentricity). The peaks are very sharp for small eccentricities but broaden when the libration amplitude increases or when $e$ approaches 0.4. (2) The law of structure is very well reproduced by the low-amplitude librations. (3) The law of the second forced mode is well satisfied while $e$ is not small; when $e$ approaches zero, the results depart strongly from those given by that law and approach the values given by a similar law obtained for low-eccentricity orbits by Greenberg and Franklin (1975). (4) The proper periods show some scattering, but tend to the values given by the law of periods when the libration amplitude tends to zero.

## 2. Description of the Technique

Consider an asteroid moving in the same plane as Jupiter and near the 3:2 resonance with Jupiter. The equations of motion of this dynamical system, in Cartesian coordinates, are:

$$\frac{d^2\vec{r}}{dt^2} = -\frac{k^2\vec{r}}{r^3} + k^2 m_J \left( \frac{\vec{r_J} - \vec{r}}{\Delta^3} - \frac{\vec{r_J}}{r_J^3} \right),$$

$$\frac{d^2\vec{r_J}}{dt^2} = -\frac{k^2\vec{r_J}}{r_J^3}.$$

This system of equations was integrated numerically using Everhart's integrator RA15. The initial values of the asteroid elements (semi-major axis, eccentricity, longitude of perihelion and mean anomaly) and those of Jupiter were varied so as to cover uniformly the domain of the phase space being sampled. The semi-major axis and eccentricity of the asteroid were chosen in the intervals mentioned in the introduction.

The study of the long-term changes in the orbital elements of the asteroid requires an interval of integration large enough so as to allow us a satisfactory determination of the long-period terms and its exhaustive analysis. The numerical integration produces a huge output, which cannot be simply reduced by applying a decimation of the output (that is, by increasing the sampling step). The procedure to be followed has already been described in great detail by Carpino et al. (1987). The result of the numerical integration is a set of values of the orbital elements defined in a discrete set of values of the time variable. This set consists of a discrete time-dependent signal $x_n$ corresponding to values sampled at equal time intervals $\Delta t$, of an ideal, continuous, signal:

$$x_n = x(n\Delta t) \qquad (n \in Z).$$

The discrete time-dependent signal $x_n$ can be represented in the frequency domain by its discrete Fourier transform

$$X(f) = \sum_{n=-\infty}^{+\infty} x_n . e^{-i2\pi n f \Delta t}.$$

The sampling procedure of the continuous signal results in a Fourier Transform containing only a finite interval of frequencies, with an upper bound given by the Nyquist critical frequency:

$$f_c \equiv \frac{1}{2\Delta t}.$$

The discrete Fourier transform is periodic in the frequency domain, with a period equal to twice the Nyquist frequency. This fact is responsible for the phenomenon of aliasing: any frequency component outside the frequency range $(-f_c, +f_c)$ is falsely translated back into this range (aliased), resulting in a pollution of the original spectrum. If we are interested only in the long-period terms, that is, in those Fourier components corresponding to low frequencies, the aliasing effect can be reduced by applying a low-pass digital filter to the sampled signal, followed by the application of output decimation.

We use a finite impulse response linear filter; the output filtered signal is obtained by the discrete convolution of the input signal $x_n$ with a suitable finite sequence $r_n$. The application of the digital filter is founded on the discrete convolution theorem (see Press *et al.*, 1987): If a signal $x_n$ is periodic with period $N$, so that it is completely determined by the $N$ values $x_n$, then, its discrete convolution with a response function $r_n$ of finite duration $N$ is a member of the discrete Fourier transform pair,

$$X_n \cdot R_n,$$

where $X_n$ is the discrete Fourier transform of the values $x_n$ and $R_n$ is the discrete Fourier transform of the values $r_n$. This theorem is of much assistance in the construction of a digital filter: the convolution of any two functions is more easily realized in the frequency domain.

Let $R(f)$ be the Fourier transform of a suitable impulse response function. Given a cut-off frequency $f_{max}$, we must have $R(f) = 0$ for any $f > f_{max}$ and $R(f) = 1$ for $f < f_{max}$. The filter must not modify the phases of the Fourier components of the signal, i.e. $\Im R(f) = 0$. This function can b e approximated by a polynomial of degree $M$ in the variable $z = \cos(2\pi f \Delta t)$ as

$$R(f) = R_0 + 2 \sum_{m=1}^{M} R_m \cos(2\pi m f \Delta t),$$

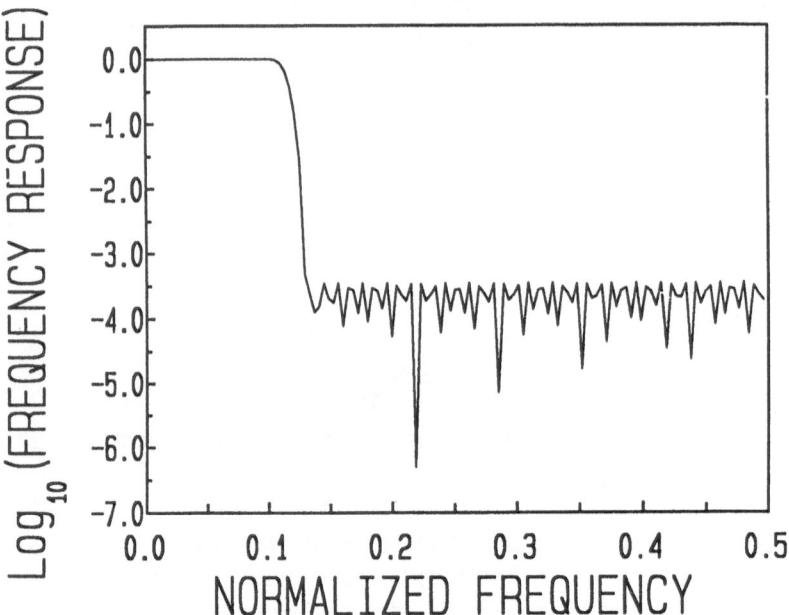

Fig. 1. Frequency response, in logarithmic scale, of the low-pass filter. The normalized frequency is $f/2f_c$.

where $2M+1$ is the filter length, $\Delta t$ is the sampling period and $R_m$ are coefficients which depend on the design parameters of the low-pass filter, obtained from the solution of the minimax problem using the Remez exchange algorithm (Rabiner and Gold, 1975).

Previous studies of the orbital motion of the Hilda asteroids allow us to evaluate the design parameters of the low-pass filter. The design parameters of the low-pass filter, used in the present work, are given in Table I.

Figure 1 shows the frequency response (in logarithmic scale) of the low-pass filter; the logarithm of $R(f)$ is plotted versus the normalized frequency $f/2f_c$.

The filtering procedure was applied during the numerical integration after each 2560 years: 256 points with the time step equal to 10 year; Fast Fourier Transform (FFT), applied for the determination of the Fourier transform, restricts the number of points to a power of two. The filtering to increase the sampling period to 40 years (in this case, $f_c = 1/80yr^{-1}$ is equal to the lower limit of the dark band). The problem of end pollution of each section due to the non-periodicity of the output, required by the discrete convolution theorem was resolved by the overlap-save method (see Press, 1987).

The interval of the numerical integration was determined by the fact that a FFT was also used to study of the resulting time series. If $T$ is the length of the time interval of data, the fundamental frequency of the FFT is $1/T$. This is also the

TABLE I
Design parameter of the low-pass filter

| | |
|---|---|
| Upper limit of the pass band | $1/100 \ yr^{-1}$ |
| Lower limit of the dark band | $1/80 \ yr^{-1}$ |
| Ripple (maximum difference between actual and nominal frequency response in the pass band) | $10^{-5}$ |
| Attenuation (maximum gain in the dark band) | $10^{-4}$ |
| Filter length | 157 |
| Input sampling period | 10 yr |
| Output sampling period | 40 yr |

width of the peak in the power spectrum. The experiments allowed us to choose the time interval as equal to 50,000 years.

## 3. Results

The filtered output of the numerical integration was Fourier analyzed in order to recover the long-period spectral lines and their corresponding amplitudes and phases for the more important quantities expressing the motions such as semi-major axis, eccentricity, critical angle $\sigma$, longitude of perihelion, energy and the rectangular components of the planes $(e, \varpi - \varpi_J)$ and $(e, \sigma)$: $e \cos \sigma$, $e \sin \sigma$, $e \cos(\varpi - \varpi_J)$ and $e \sin(\varpi - \varpi_J)$.

The Keplerian elements of the asteroidal orbit a, e, $e \cos \sigma$, $e \sin \sigma$, $e \cos \varpi$ and $e \sin \varpi$ can be, each, represented as a trigonometric series

$$A_0 + A_1 \cos(2\pi f_1 t + \alpha_1) + A_2 \cos(2\pi f_2 t + \alpha_2) + \cdots$$

where $A_0$ is the mean value of the element in the interval of integration, $f_i$ are the frequency values of power spectrum peaks and $A_i$ and $\alpha_i$ the corresponding amplitudes and phases.

The FFT procedure was applied in order to fit a trigonometric model to given sequences of data and the most relevant lines in the power spectrum were identified. The spectral analysis procedure described in Ferraz-Mello (1981) was then applied in the vicinity of each peak obtained with the FFT in order to obtain an accurate determination of the main frequencies and the corresponding amplitudes. This more accurate procedure uses a basis determined with the functions 1, $\cos 2\pi f t$ and $\sin 2\pi f t$ which is orthonormal with respect to summations over the actual (finite) interval of the time series. All spectra show the two independent modes of oscillation and its combination. In the first mode the longitude of conjunction of the asteroid and Jupiter oscillates about the asteroid's perihelion (libration) with a period in the interval 100-270 years; in the other mode the asteroid's perihelion moves with a period in the interval 200-20,000 years.

The semi-major axis shows a libration. The most relevant frequency in its power spectrum is the libration frequency; the amplitude of the corresponding oscillations of the semi-major axis vary in the range 0.001-0.020. It must be noted that the actual variation of the semi-major axis lies in a larger range (0.001-0.04), the largest observed amplitudes being due to the combination of the amplitude of the semi-major axis libration (that is, the amplitude of the oscillation of the semi-major axis with the proper frequency of the libration) with those of other oscillation modes.

Figure 2 shows the power spectra of the semi-major axis of orbits in a group of solutions with the value of the mean libration amplitude equal to 0.001. For mean eccentricities in the range 0.1-0.2, the libration peak dominates, representing the periodic motion of the asteroid. For the mean values of the eccentricity above this range, the frequency lines corresponding to the linear combination of the two independent modes (libration frequency ± motion of the asteroid perihelion) appear; for increasing $\bar{e}$, they grow and approximate to the libration line resulting, for high $\bar{e}$, in the disappearing of isolated peaks and the apparition of a broad interval of indefinite spectral lines. This phenomenon is characteristic of chaos and is, in this case, associated with the chaotic region emanating from the boundary of usual librations and corotations (see Ferraz-Mello et al., 1992). For mean eccentricities bellow 0.1, the power spectra of the semi-major axis show a line identified with the frequency $f_\varpi$ with an amplitude a few times smaller than the libration amplitude.

For small values of the libration amplitude (shown in figure 2), the spectral indication of chaotic motion occurs for mean eccentricities near 0.35. This limiting value of the mean eccentricity decreases for increasing libration amplitudes and it is impossible to find asteroids in regular motion with large libration amplitude and mean eccentricity above 0.25. Low-eccentricity librations are regular.

Figure 3 gives the distribution of the mean values of the semi-major axes of the pericentric librators corresponding to the estimated value of the eccentricity of the libration center (which, for motions with a librating $\sigma$, is roughly equal to the mean eccentricity). The figure also includes orbits for which the angle $\sigma$ is circulating (these circulations have proper frequencies given by the same law as the librations and are a continuation of these solutions). All solutions were distributed in eleven groups according to the value of the libration amplitude of the semi-major axis. In figure 3, the solutions are represented by small circles; different radius indicate different amplitude in the libration amplitude of $a$ (large radius means large amplitude). This figure clearly shows the existence of limits for the values of the eccentricity of the libration center and the mean semi-major axis. Small-amplitude solutions always exist, with eccentricities from $\sim 0$ to $\sim 0.35$ and mean semi-major axes from $\sim 0.745$ to $\sim 0.763$. It is worthwhile mentioning that this last value is the so-called exact resonance value. The interval of existence of large-amplitude librations is smaller: the upper limit for the eccentricity is $\sim 0.2$ (a limit clearly associated to the approaching of the chaotic zone) and the lower limit for the mean semi-major axis is $\sim 0.76$.

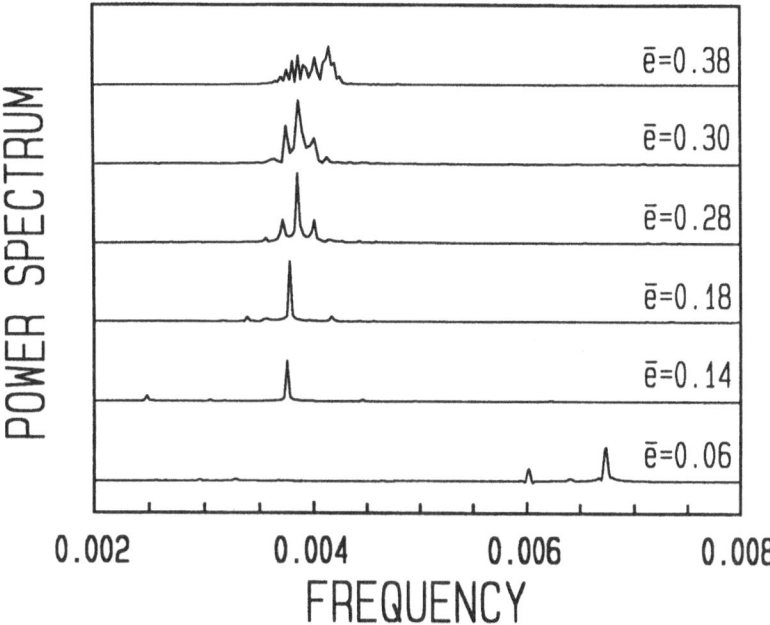

Fig. 2. Power spectra of the semi-major axis of orbits with very small libration amplitude for several values of the mean eccentricity $\bar{e}$

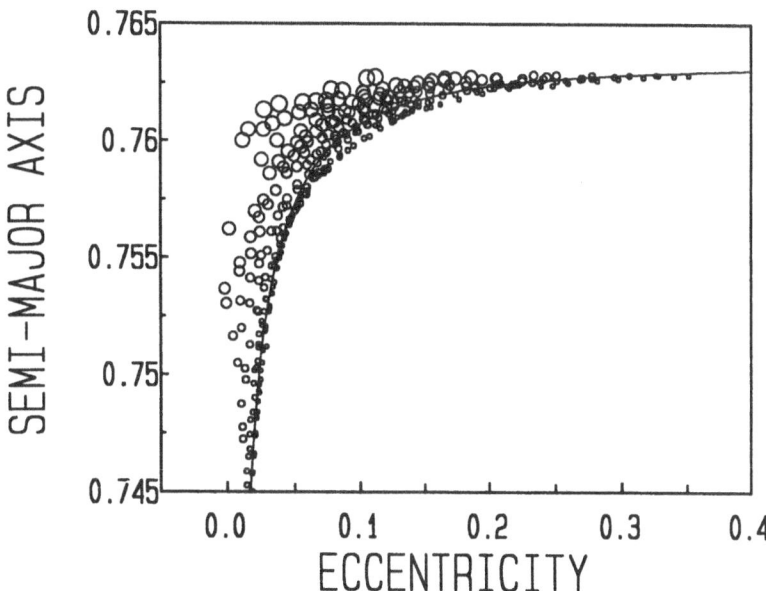

Fig. 3. Distribution of the average semi major-axes. The abscissas are the eccentricities of the libration centers. The size of the circles grows with the libration amplitudes of the solutions (from $\Delta a = 0.001 a_J$ to $\Delta a = 0.02 a_J$). The continuous line is the law of structure taken from (Ferraz-Mello, 1988)

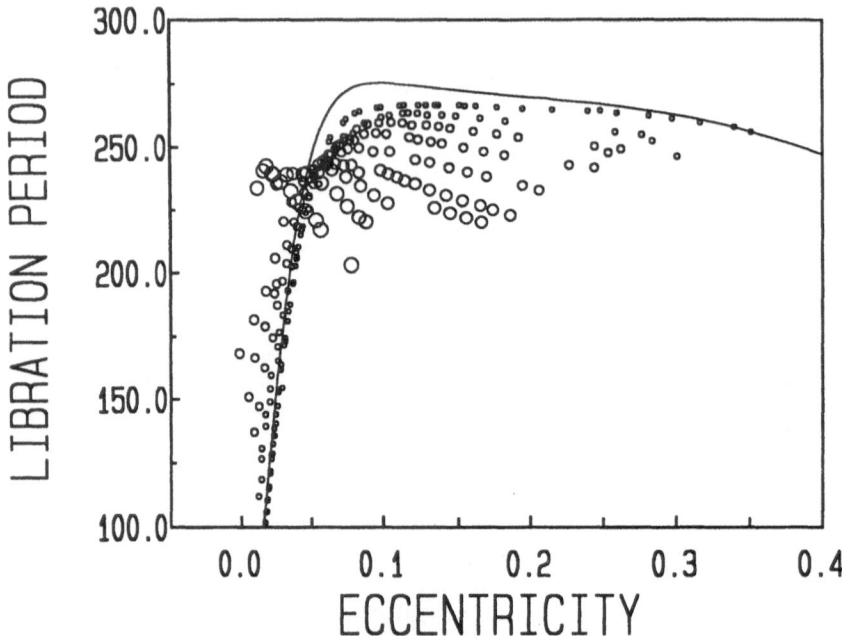

Fig. 4. Distribution of the libration periods. The abscissas and circle sizes are the same of fig. 3. The continuous line is the law of periods taken from (Ferraz-Mello, 1988)

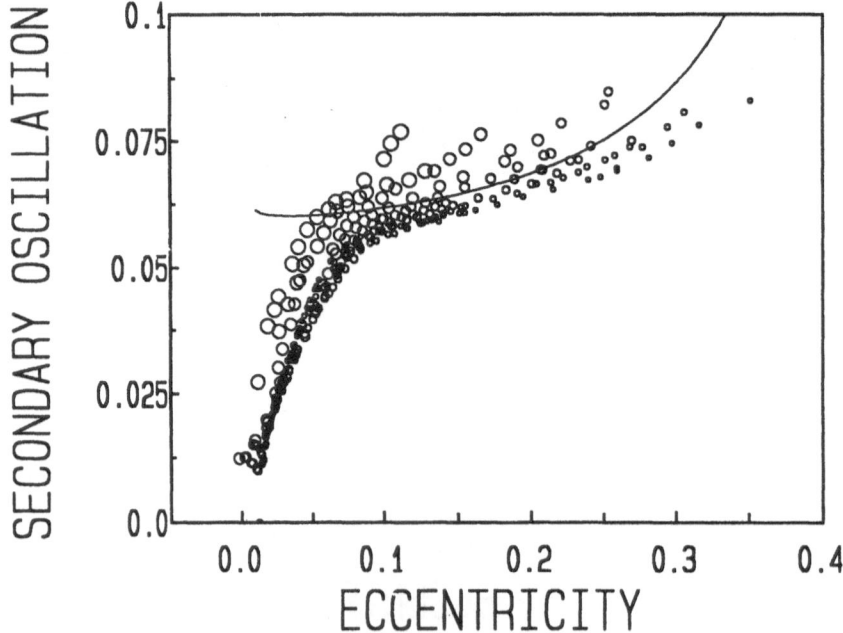

Fig. 5. Distribution of the amplitude of the oscillation of $e$ forced by the eccentricity of Jupiter. The abscissas and circle sizes are the same of fig. 3. The continuous line is the law of the forced mode taken from (Ferraz-Mello, 1988)

For $\bar{e} > 0.1$ the critical angle $\sigma$ librates about zero with an amplitude correlated with the libration amplitude of the semi-major axis; it is of 30° -60° for the
first group of asteroids and above 100° for the last group; also, in these cases the
longitude of the perihelion circulates with periods about ten times the period of
libration. The perihelion circulation period increases for increasing mean eccentricity, reaching 20 thousand years close to the largest mean eccentricities obtained.
For the values of $\bar{e} < 0.1$, the behavior of the critical angle and the longitude of
perihelion corresponds to the analytical model derived by Greenberg and Franklin
(1975). The angles $\sigma$ and $\varpi$ librate and circulate alternately.

Figure 4 shows the period of libration. The abscissas are the eccentricity of
libration centers; as in figure 3, the larger circles correspond to the larger values
of the libration amplitude. The libration period decreases rapidly for small central
eccentricities reaching 100 years.

Figure 5 shows the amplitude of the mode of oscillation in the asteroidal eccentricity forced by the eccentricity of Jupiter. For high eccentricities they follow the
law derived from Ferraz-Mello analytical model; but, as the eccentricity decreases,
the points go away from this law and tend towards the value given by the similar
law derived by Greenberg and Franklin.

## Acknowledgements

This investigation was partly sponsored by the Research Foundation of the State
of São Paulo (FAPESP) and by CAPES. The computations were done with the
Convex C220 of the Computing Center of the University of São Paulo.

## References

Carpino, M., Milani, A., Nobili, A.M.: 1987, "Long-term numerical integrations and synthetic theories
for the motion of the outer planets", *Astron. Astrophys.* **181**, 182-194.
Everhart, E.: 1985, "An efficient integrator that uses Gauss-Radau spacings", in *Dynamics of Comets:
their origin and evolution* (A.Carusi and G.B.Valsecchi, eds.) Reidel, Dordrecht, pp. 185-202.
Ferraz-Mello, S.: 1981, "Estimation of periods from unequally spaced observations" *Astron. J.* **86**,
619-624.
Ferraz-Mello, S.: 1988, "The high-eccentricity libration of the Hildas", *Astron. J.* **96**, 400-408.
Ferraz-Mello, S., Tsuchida, M., Klafke, J.C.: 1992, "On symmetrical planetary corotations", *Celest.
Mech. Dyn. Astron.* (in press).
Greenberg, R., Franklin, F.: 1975, "Coupled librations in the motion of asteroids near the 2:1 reso-
nance", *Mon. Not. R. astr. Soc.* **173**, 1-8.
Ip, W.-H.: 1976, "Dynamical study of the Hilda asteroids. I. Resonant orbital motions of the PLS
objects from Palomar-Leiden survey", *Astrophys. Sp. Sci.* **44**, 337-383.
Press W.H. *et al.*:1987, *Numerical Recipes*, Cambridge University Press, Cambridge, UK.
Rabiner, L.R., Gold, B.: 1975, *Theory and Application of Digital Signal Processing*, Prentice Hall,
Englewood Cliffs, N.J.
Schubart, J.: 1968, "Long-period effects in the motion of Hilda-type planets", *Astron. J.* **73**, 99-103.
Wisdom, J.: 1983, "Chaotic behaviour and the origin of the 3:1 Kirkwood gap", *Icarus* **56**, 51-74.

# A STABILITY STUDY OF ASTEROID FAMILIES NEAR THE 3:1 AND 5:2 RESONANCE WITH JUPITER

G. HAHN and C.-I. LAGERKVIST

*Astronomiska Observatoriet, Box 515, S-751 20 Uppsala, Sweden*

and

B.A. LINDBLAD

*Institutionen för Astronomi, Box 43, S-221 00 Lund, Sweden*

**Abstract.** By using the $D$-criterion Lindblad (1992) has identified 14 asteroid families from a sample of 4100 numbered asteroids with proper elements from Milani and Knežević (1990). Taxonomic types and other physical properties for a significant number of objects in five of the families show strong homogeneity within each family, further strengthening their internal relationship.

To test the hypothesis of a common origin in, e.g., a catastrophic collision event, we have set out to integrate the orbits of the members of the Maria, Dora and Oppavia-Gefion families over some $10^6$ years. The mean distance for the Maria family is close to the 3:1 mean-motion resonance with Jupiter, while the other two families lie close to the 5:2 resonance.

We used a simplified solar system model which included the perturbations by Jupiter and Saturn only and implemented Everhart's variable stepsize integrator RA15. All close encounters between the family members (within 0.1 AU) were recorded as well. Preliminary results from integrations over $\approx 4 \times 10^5$ years are presented here.

The statistics of close encounters show pronounced peaks for several members within each family, while for others no significant levels above the background of random encounters or even very low frequencies were found. This indicates a subclustering within the families. Quite a lot of very close (<0.005 AU) mutual encounters are found, which suggest that, at least for the larger members in a family, the mutual gravitational interactions could be of some importance for the real orbital evolutions.

The encounter statistics between the Dora and Oppavia family members suggest a possible interrelationship between this two groups.

**Key words:** Asteroid – families of asteroids – resonances – numerical integrations

## 1. Introduction

Asteroid families are characterized by almost equal values of the orbital elements $a, e$, and $i$ in proper element space. This was first realized by Hirayama (1918) from a then very limited sample of asteroids. Through the development of more accurate theories for the calculation of proper elements during the last 20 years - Yuasa (1973), Williams (1979), Kozai (1979b), Knežević (1986) and Valsecchi et al. (1989) - and the rapidly growing sample of numbered asteroids, a set of 4100 proper elements is nowadays available (Milani and Knežević, 1990). From this sample, Lindblad (1992) identified 14 dynamical families, each with more than 15 members. The $D$-criterion, which was previously used in many meteor stream studies as well as for studying clustering of asteroid and comet orbits (see e.g. Lindblad and Southworth 1971, Lindblad 1985), can be written in the following form for testing orbital similarity in proper elements $a, e, i$ space

$$D(m, n)^2 = (e_n - e_m)^2 + (q_n - q_m)^2 + (2 \cdot sin\frac{i_n - i_m}{2})^2 \qquad (1)$$

*Celestial Mechanics and Dynamical Astronomy* **56**: 131–141, 1993.
© 1993 *Kluwer Academic Publishers.*

where $m$ and $n$ represent two orbits to be compared, and $q$ is the perihelion distance. The search techniques and selection criteria are described in more detail in Lindblad (1992).

The reality of asteroid families, which have been identified on dynamical grounds, can be checked by studies of the physical properties of the individual members in a family. For most of the larger member in many families such data are available and, indeed, often striking similarities with regard to taxonomic type and albedo are found, which strengthen the hypothesis of a common origin as e.g. emanating from a catastrophic collision event. (For a review on these topics, see e.g. Gradie et al. 1979, Chapman et al. 1989 )

The Maria family contains the following numbered asteroids: 170, 472, 575, 616, 695, 714, 727, 787, 875, 879, 897, 1158, 1160, 1215, 1677, 1996, 2151, 2221, 2429, 2638, 2865, 2903, 2962, 3055, 3066, 3158, 3159, 3167, 3332, 3537, 3594, 3786, 3970, 4099, 4104, 4122.

The Oppavia-Gefion family has the following members: 255, 1272, 1433, 1751, 1839, 2373, 2386, 2493, 2559, 2595, 2631, 2801, 2875, 2977, 3724, 3910, 4020.

To the Dora family belong: 668, 1414, 1734, 1795, 1836, 1970, 2598, 2807, 2940, 3563, 3611, 3630, 3775, 3829, 4135, 4220.

Although it is not the purpose of this paper to compare different methods to define asteroid families, a comparison with the selections found by Zappalá et al. 1990 and Bendjoya et al. 1991 is interesting. The agreement in membership is comparable for all three studies, yielding identical selections for the Dora family and about 75% in common for the Maria family. The Oppavia-Gefion family as defined by Lindblad (1992) is completely represented within the Leto-family in Bendjoya et al. 1991 and most of the members are in common with the Gefion-family in Zappalá et al. 1990.

Data on the physical properties of individual asteroids in each family exist. The Maria family has eight asteroids classified by Tholen (1989) and seven of these are of taxonomic type S and the eight is of type DT. Of the asteroids in this family 13 have albedo determinations (Tedesco, 1989) and all these are characteristic for asteroids of type S.

TABLE I

Mean proper elements $a$, $e$, $i$ and mean values of $D$, taken from Lindblad (1992)

| Family name | N | a | e | i | $\overline{D}$ |
|---|---|---|---|---|---|
| Maria | 36 | 2.555 | 0.091 | 15.0 | 0.021 |
| Oppavia-Gefion | 17 | 2.789 | 0.136 | 9.1 | 0.013 |
| Dora | 16 | 2.787 | 0.196 | 7.8 | 0.007 |

The Dora and Oppavia families have very few asteroids classified into taxonomic types but five asteroids in the Dora family have albedos, all very low, significant for asteroids of type C. Six asteroids in the Oppavia family have albedos determined, all but one are typical for asteroids of types C. This homogeneity hints on a common parent body for both families.

Diameter determinations exist for those asteroids with albedos given by Tedesco (1989). We consider all three families as homogeneous regarding types, as the albedo measurements indicate, in order to get an estimate of the diameter distributions in the three families. The mean visual albedo for the three families were found to be 0.17 (Maria), 0.06 (Oppavia) and 0.04 (Dora). Diameters were then calculated using the following relation (Bowell and Lumme, 1979)

$$\log d(km) = k - 0.5 \cdot p_v - 0.2 \cdot V(0^o) \tag{2}$$

where $k = 3.122$ is a wavelength-dependent constant, $p_v$ is the albedo, and $V(0^o)$ is the absolute magnitude at zero phase angle; the visual magnitude range is assumed.

All asteroids in the three families are small with diameters ranging from 10 to 50 km. This means that the total number of family members should be much larger since the population of numbered asteroids of these sizes is by no means complete.

The aim of the present study is to investigate the stability and homogeneity of some selected families by means of numerical integration techniques. We chose three families from Lindblad's list which are summarized in Table I. The Maria family lies close to the 3:1 mean motion resonance with Jupiter, the Oppavia-Gefion and Dora family are close to the 5:2 resonance. Since the proximity to these low-order resonances could pose strong perturbation on the asteroids, as found in other studies (e.g. Yoshikawa 1989, Hahn et al. 1991), they were selected as being of special interest. Our emphasis here is therefore concentrated on the shorter term stability of these family groupings and to investigate the possible importance of mutual encounters. In particular we wanted to check whether the encounter statistics between family members and non-family members for the two quite similar Oppavia-Gefion and Dora families could be used as an indicator for membership. Preliminary results are presented here.

## 2. Numerical Methods

For the numerical integration of the equations of motion of the asteroids and the planets we used the 15th order Radau (RA15) integrator with variable stepsize, as described in Everhart (1985). The solar system model contained the perturbations by Jupiter and Saturn only and starting values were taken from the DE118 JPL ephemeris. The osculating elements for the asteroids were taken from the MPC orbital database for epoch JD 2448200.5. An initial stepsize of 40 days and an internal accuracy of $\approx 10^{-8}$ was chosen for the integrations. The calculations were run in two batches: one containing the Maria family members plus the planets and the other both the Dora and Oppavia-Gefion asteroids and Jupiter and Saturn. The former were integrated over ≈370,000 years, the latter close to 420,000 years into the future. Coordinates and velocities were stored every 5000 days, corresponding to 13.7 years. In addition, a check for close encounters to the planets (<1.0 AU), and mutually, between the asteroids (<0.1 AU) was performed; the distances and times for such encounters were stored as well.

The analysis of this vast amount of data was performed in two parts: the orbital evolution of each asteroid was derived from the osculating elements, and the close encounters were analysed to study their distribution with time and statistics were evaluated to check their significance against various randoms encounter models.

## 2.1. ORBITAL EVOLUTIONS

Osculating elements were calculated at a rate $20\times$ the original sampling rate, corresponding to a time step of ≈275 years, and plotted versus time. Figures 1 – 3 show these evolutions for the first asteroid in each family. The panels in each plot contain, from bottom to top, semi-major axis $a$, eccentricity $e$, inclination $i$, aphelion distance $Q$, the critical argument $\sigma$(for the 3:1 mean motion resonance for (170) Maria; and 5:2 for (255) Oppavia and (668) Dora), the longitude of the ascending node $\Omega$, and eccentricity vs. ($\tilde{\omega} - \tilde{\omega}_j$), the difference in the longitude of perihelion of the asteroid and Jupiter. As to be expected, these evolutions show a very regular behaviour, characterized by short periodic variations; no secular trends are evident, since the time interval covered is short.

Besides these overall similarities in the evolution of the orbital elements, there are some features which only a few of the members in each family have in common: e.g. the phase of the advance of $\Omega$, or $e$ and $Q$. These evolutions, correlated with the mutual encounter statistics might give us some information on the interrelationship of individual members, the existence of subgroupings, and thereby tell us something about the homogeneity of the family. These peculiarities can in turn be compared with the physical parameters for these particular objects.

From the osculating elements we calculated, for each family member, a value of $D$ according to equation (1) with respect to the mean orbit as taken from Table I. These individual $D$ values, at each sampling time, were used to determine a mean $D$

$$\overline{D} = \sum_{i=1}^{n} D_i/n \tag{3}$$

where $n$ is the number of family members, and their time evolutions are shown in Figure 4 (a – c). A clear periodicity, corresponding to the ≈55,000 yrs $e$ variation cycle of Jupiter and Saturn, is evident, but also secular trends are visible. (Consistent with the evolution of the elements for the giant planets as derived from our integrations.) Integrations over much longer time spans are needed to determine these long-term evolutions; see section 3 below.

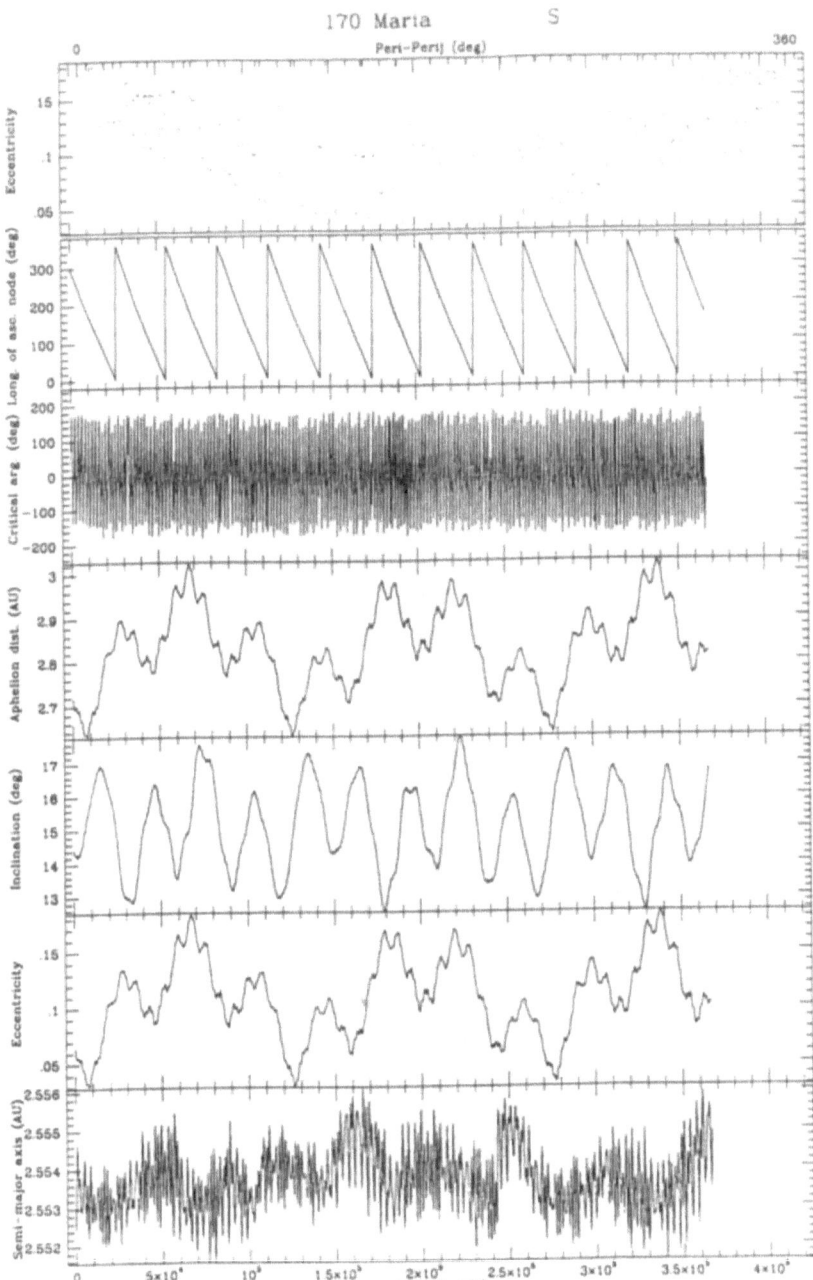

Fig. 1. Orbital evolution for (170) Maria over some 370,000 years into the future. The critical argument (third panel) is defined as $\sigma = 3\lambda_J - 1\lambda - 2\varpi$, where $\lambda$ and $\lambda_J$ are the mean longitudes of the asteroid and of Jupiter, respectively, and $\varpi$ is the longitude of perihelion of the asteroid. The taxonomic type is given at the top of the graph.

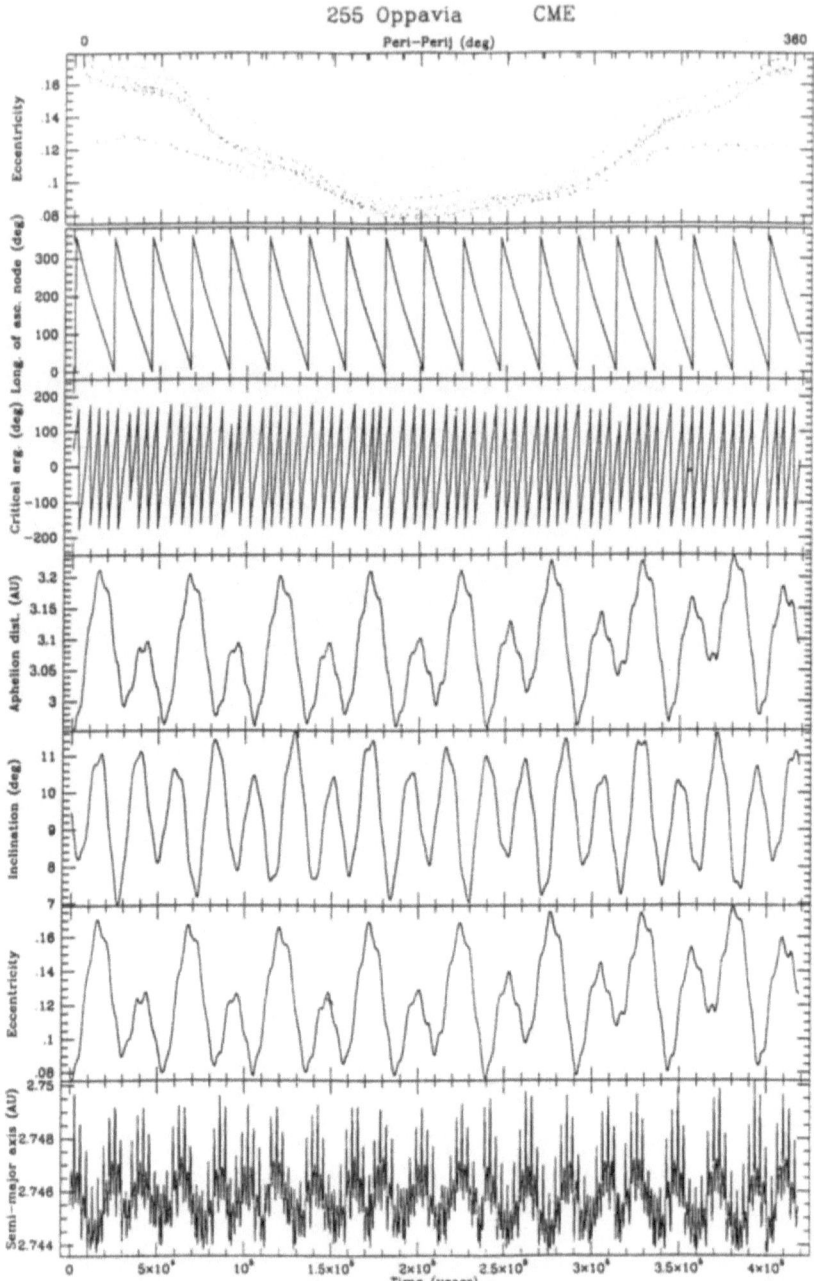

Fig. 2. Orbital evolution for (255) Oppavia over some 420,000 years into the future. The critical argument (third panel) is defined as $\sigma = 5\lambda_J - 2\lambda - 3\varpi$, where $\lambda$ and $\lambda_J$ are the mean longitudes of the asteroid and of Jupiter, respectively, and $\varpi$ is the longitude of perihelion of the asteroid. The taxonomic type is given at the top of the graph.

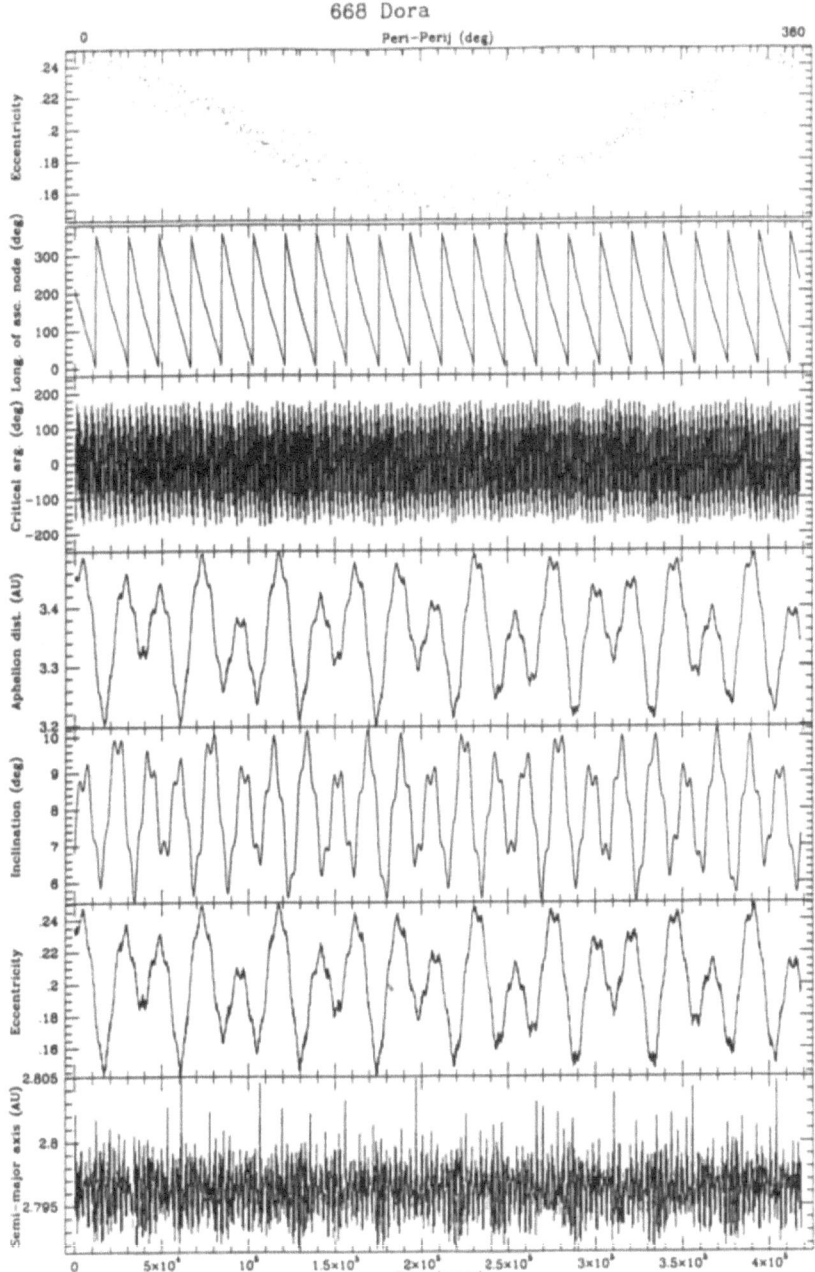

Fig. 3. Orbital evolution for (668) Dora over some 420,000 years into the future. The critical argument (third panel) is defined as $\sigma = 5\lambda_J - 2\lambda - 3\varpi$, where $\lambda$ and $\lambda_J$ are the mean longitudes of the asteroid and of Jupiter, respectively, and $\varpi$ is the longitude of perihelion of the asteroid.

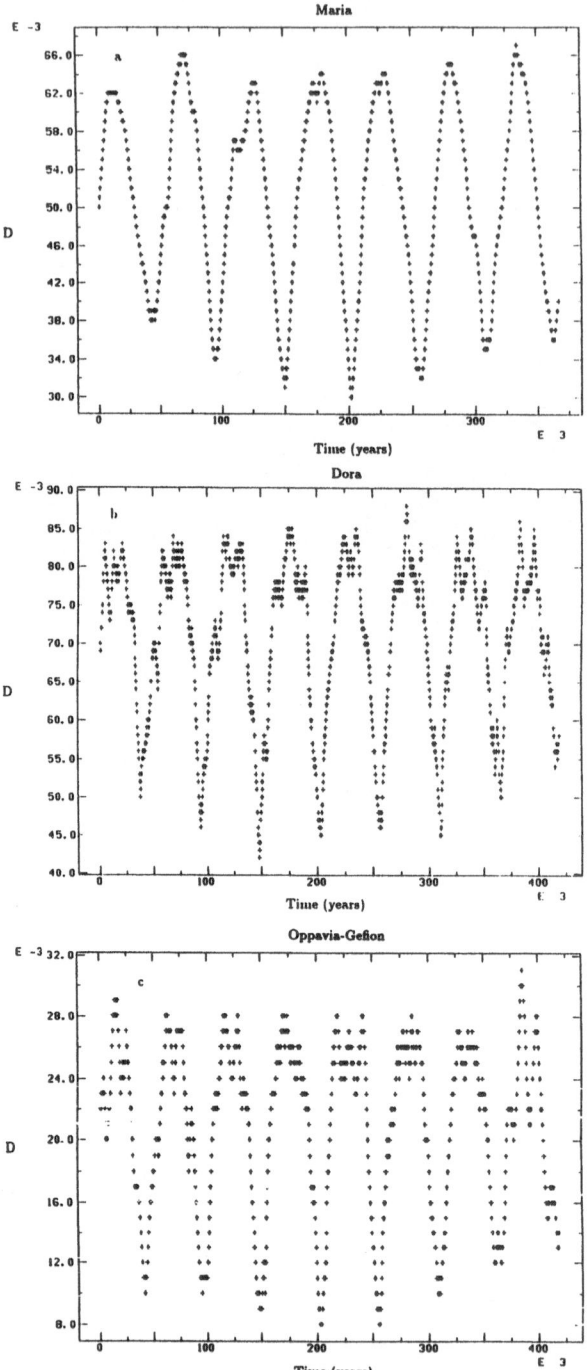

Fig. 4. Time evolution of the mean value of the D-criterion $\overline{D}$ , as defined in (1) and (3) in units of $10^{-3}$, for the Maria (a), Dora (b) and Oppavia-Gefion (c) families. Note the different time scale in panel (a).

## 2.2. CLOSE ENCOUNTERS

The close encounter analysis was done in the following way: first all individual entries were extracted for each pair and written into a separate file, and statistics was produced over the number of all these entries. These pair entries were then analysed to separate individual encounters and their number distribution was determined. As a separate encounter two entries have to be apart by more than one year. Given the approximate stepsize of 40 days, only the long-lasting, deeper encounters, have been registered. Often there is only one entry during each encounter. This means that the closest approach distance is very poorly determined, and some closer encounters might have been missed. On the other hand some asteroids experienced encounters lasting several years.

For the Maria family run, more than one million individual entries have been registered, yielding some 313,000 close encounters. In the Oppavia and Dora data about 700,000 entries, corresponding to approximately 202,000 encounters, were found. In this sample, data from a period of some 50,000 years have been lost due to computer problems, which affects the statistics slightly. A matrix with the numbers registered for each pair was produced; all possible pair combinations yielded encounters.

From these numbers, mean values for each asteroid have been derived, which we considered as the 'background level' or random encounter rate for the family. This background rate was found to be about $500 \pm 50$ per asteroid during the integration period of roughly 400,000 years, which corresponds to one encounter within 0.1 AU every 800 years; roughly equal for all three families. Large deviations (up to $\pm 10\sigma$) from these means have been found, and the time distribution of the close approaches is by no means at random. Periods of many encounters, often long-lasting, up to several years, alternate with low frequency periods. Large encounter rates correspond to the times when the advance of the nodal lines are in phase.

Some 100 mutual encounters within 0.0050 AU have been registered for the Maria family member, with a record close approach of 0.0009 AU. About 70 approaches within 0.0050 AU were found for the Oppavia and Dora families. The Dora pair 4220 - 3829 experienced four episodes of long lasting very close encounters, as illustrated in Figure 5. An extreme period of almost 10 years was found, when these two asteroids were always within 0.1 AU from each other. Further analysis of the data is still in progress.

## 3. Discussion

Our first, rather preliminary results, show that the stability of the orbits of all family members are not affected by the proximity to the 3:1 and 5:2 mean motion resonance with Jupiter. Although much longer time spans have to be covered to reveal secular trends which might evolve some orbits closer into the resonance. Such trends are hinted from the evolutions of the means of the $D$ values, as seen in Figure 4 (a –

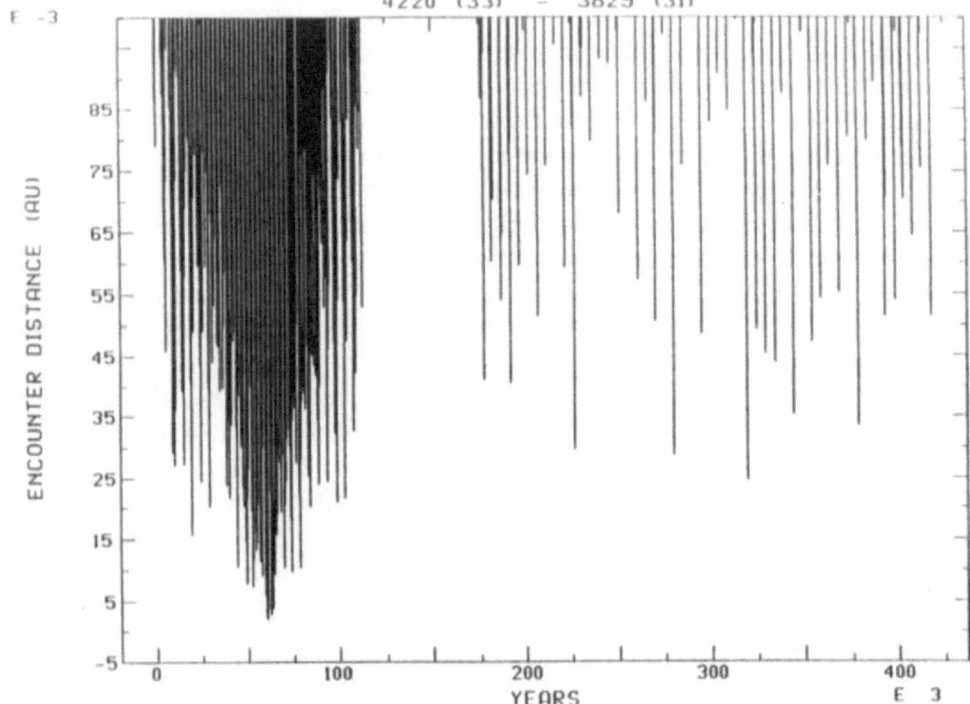

Fig. 5. Distribution of mutual close encounters for the Dora family pair 4220 and 3829 (serial numbers 33 and 31 in our investigation). The encounter distance is plotted, in units of $10^{-3}$ AU, versus time, in units of $10^3$ years. The gap between 110,000 and 160,000 years is due to a loss of data, see text.

c), and should be revealed in our planned integrations over several $10^6$ years. For these integrations a somewhat shorter step size, which can be achieved by a higher internal accuracy, is recommended in order to sample the close approaches more frequently. This should allow us to better determine the closest approach distances and thereby estimate the significance of these encounters for the orbital evolutions. In these future runs we plan to use a solar system model including all four outer planets.

We made a rough comparison between the perturbing effect of Jupiter at 2.5 AU from the asteroid with the pull from another 50 km diameter object passing within 0.001 AU and found the perturbation to be of the order of 1 or 2 % of that of Jupiters. Longer lasting passages should have much more severe effects, which will prompt us to make test runs with assumed masses for some of the asteroids.

From a comparison of the close encounter statistics between the Dora and the Oppavia-Gefion family members, we found that the encounter frequency is only slightly lower for inter-family encounters than for internal approaches, e.g.

between members within each family – suggesting that these two groups might have a common origin. The available physical data-set would support such a conclusion. Further, more detailed investigations of the encounter data are still in progress.

## Acknowledgements

This work has been supported by the Swedish Natural Science Research Council, through contracts FU 9231-301 and FU 204-303.

## References

Bendjoya, Ph., Slézak, E. and Froeschlé, Cl.: 1991, *Astron. Astrophys.* **251**, 312–330
Bowell, E., and Lumme, K.: 1979, in T. Gehrels, ed(s)., *Asteroids*, Univ. of Arizona Press, Tucson, 132–169
Chapman, C.R., Paolicchi, P., Zappalà, V., Binzel, R.P. and Bell, J.F.: 1989, in R.T. Binzel, T. Gehrels and M.S. Matthews, ed(s)., *Asteroids II*, Univ. of Arizona Press, Tucson, 386–415
Everhart, E.: 1985, in A. Carusi and G.B. Valsecchi, ed(s)., *Dynamics of comets: their origin and evolution, IAU Coll. No. 83*, Reidel, Dordrecht, The Netherlands, 185–202
Gradie, J.C., Chapman, C.R. and Williams, J.G.: 1979, in T. Gehrels, ed(s)., *Asteroids*, Univ. of Arizona Press, Tucson, 359–390
Hahn, G., Lagerkvist, C.-I., Lindgren, M. and Dahlgren, M.: 1991, *Astron. Astrophys.* **246**, 603–618
Hirayama, K.: 1918, *Astron. J.* **31**, 185–188
Knežević, Z.: 1986, in C.-I. Lagerkvist, B.A. Lindblad, H. Lundstedt and H. Rickman, ed(s)., *Asteroids Comets Meteors II*, Uppsala Universitet, Uppsala, 129–134
Kozai, Y.: 1979b, in T. Gehrels, ed(s)., *Asteroids*, Univ. of Arizona Press, Tucson, 334–358
Lindblad, B.A.: 1985, in A. Carusi and G.B. Valsecchi, ed(s)., *Dynamics of comets: their origin and evolution, IAU Coll. No. 83*, Reidel, Dordrecht, The Netherlands, 353–363
Lindblad, B.A.: 1992, in A.W. Harris and E.L.G. Bowell, ed(s)., *Asteroids Comets Meteors 1991*, Lunar and Planetary Institute, in press
Lindblad, B.A and Southworth R.: 1971, in T. Gehrels, ed(s)., *Physical studies of minor planets*, NASA SP-267, 337–352
Milani, A. and Knežević, Z.: 1990, *Celest. Mech.* **49**, 347–411
Tedesco, E.F.: 1989, in R.T. Binzel, T. Gehrels and M.S. Matthews, ed(s)., *Asteroids II*, Univ. of Arizona Press, Tucson, 1090–1138
Tholen, D.J.: 1989, in Asteroids II, ed(s)., *R.T. Binzel, T. Gehrels and M.S. Matthews*, 1139–1153, Univ. of Arizona Press, Tucson
Valsecchi, G.B., Carusi, A., Knežević, Z., Kresák, Ĺ. and Williams, J.G.: 1989, in R.T. Gehrels and M.S. Matthews, ed(s)., *Asteroids II*, Univ. of Arizona Press, Tucson, 368–385
Williams, J.G.: 1979, in T. Gehrels, ed(s)., *Asteroids*, Univ. of Arizona Press, Tucson, 1040–1063
Yoshikawa, M.: 1989, *Astron. Astrophys.* **213**, 436–458
Yuasa, M.: 1973, *Publ. Astron. Soc. Japan* **25**, 399–445
Zappalá, V., Cellino, A., Farinella, P. and Knežević, Z.: 1990, *Astron. J.* **100**, 2030–2046

# CHAOTIC BEHAVIOUR OF TRAJECTORIES FOR THE ASTEROIDAL RESONANCES

M. ŠIDLICHOVSKÝ

*Astronomical Institute, Budečská 6,120 23 Praha 2, Czechoslovakia*

**Abstract.** A systematic study of the main asteroidal resonances of the third and fourth order is performed using mapping techniques. For each resonance one-parameter family of surfaces of section is presented together with a simple energy graph which helps to understand and predict the changes in the surfaces of section within the family. As the truncated Hamiltonian for the planar, elliptic, restricted three-body problem is used for the mapping, the method is expected to fail for high eccentricities. We compared, therefore, the surfaces of section with trajectories calculated by symplectic integrators of the fourth and six order employing the full Hamiltonian. We found a good agreement for small eccentricities but differences for the higher eccentricities ($e \sim 0.3$).

**Key words:** Asteroids – Kirkwood gaps –Resonances

## 1. Introduction

In his study of the 3/1 resonance Wisdom (1982,1983) introduced a mapping technique which enabled him to perform very fast calculations of evolution of elements for an asteroid located in this resonance. This leads to finding of a chaotic zone. Chaotic trajectories were later shown (Wisdom, 1985) to cross the so called zone of uncertainty (derived on the basis of the perturbative treatment) where the action for particular oscillations is no longer conserved. Outside the zone of uncertainty the adiabatic approximation yields the guiding trajectories in projection to a two-dimensional subspace of phase space. The projection of the real trajectories keeps very close to the guiding trajectories. The global chaotic region was shown to be formed of trajectories whose guiding trajectories intersect the zone of uncertainty. Chaotic region has a high eccentricity lobe, that is why an irregular switching between high and low eccentricity modes is observed for chaotic trajectories.

These results obtained with the mapping were confirmed by numerical integration. The mapping technique was later applied to the 2/1 and 3/2 resonances (Murray, 1986) and to the 5/2 resonance (Šidlichovský and Melendo, 1986; Šidlichovský 1988).

Šidlichovský (1992) gave the general formulae of mapping for the second, third and fourth order resonances. As the resonant term for the fourth order resonance starts with the fourth degree terms in eccentricities, the mapping was generalized to this degree. The secular evolution between $\delta$-peaks was calculated using Lie-Hori method.

We present here a systematic study of the third and fourth order resonances. Each resonance is described by one dimensional family of the surface of sections. We chose a set of six figures to represent this family. The results are discussed in Section 3. Very good insight into the problem is provided by an energy graph introduced by

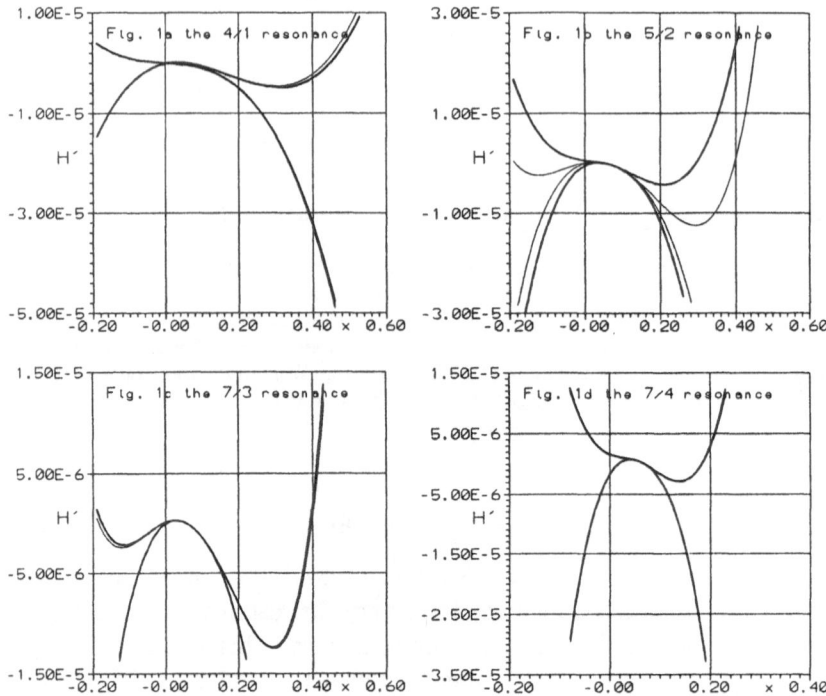

Fig. 1. The energy graphs for the third and fourth order resonances

Šidlichovský (1992). Fig. 1 shows these graphs for the most interesting resonances of the third and fourth order and some basic characteristic of the surface of sections for different energies can be predicted from there.

As the mapping is derived for the Hamiltonian of the planar-elliptic, restricted, three-body problem truncated with respect to degrees of eccentricities, the obtained surfaces of section (Figs 2–4) should give a good representation for small eccentricities but for high eccentricities we cannot expect very high accuracy. To understand how the presented figures will be modified at high eccentricities we used new mapping obtained by application of symplectic mappings (Yoshida, 1990) to our problem. No truncation in eccentricities is then necessary and no fast oscillating terms are neglected. We compared both methods for the 5/2 resonance. The results are shown in Figs 5–7 and discussed in Section 4.

## 2. The Mapping

We choose the system of units in which the gravitational constant G, semimajor axis $a'$ of Jupiter and the sum $\mu_1 + \mu$ of the masses of the Sun and Jupiter are equal to 1. The planar elliptic problem is assumed with $a, e, \lambda, \tilde{\omega}$ denoting the asteroid semimajor axis, eccentricity, mean longitude and longitude of perihelion, respectively. The corresponding values for Jupiter are primed quantities $a' = 1, e' = 0.048, \lambda'$ and $\tilde{\omega}' = 0$. Let us introduce the following set of canonical variables for the $(p + q)/p$ resonance

$$
\begin{aligned}
\Phi &= \sqrt{\mu_1 a}/p, \qquad \phi = p\lambda - (p+q)\lambda', \\
x &= \sqrt{2\sqrt{\mu_1 a}(1 - \sqrt{1 - e^2})} \cos \tilde{\omega}, \\
y &= -\sqrt{2\sqrt{\mu_1 a}(1 - \sqrt{1 - e^2})} \sin \tilde{\omega}.
\end{aligned}
\tag{1}
$$

In (Šidlichovský, 1992) we obtained the Hamiltonian

$$
H' = H_0(\Phi) - H_0(\Phi_r) + H_s(x, y) + A(x, y) \cos(\phi - \phi_0(x, y)),
\tag{2}
$$

where

$$
H_0 = -\left[ \frac{\mu_1^2}{2p^2\Phi^2} + (p + q)\Phi \right].
\tag{3}
$$

$\Phi_r$ is the resonant value of $\Phi$. The terms of the fourth degree in eccentricities were neglected in the secular Hamiltonian $H_s$ as we found that their inclusion and use of the Lie-Hori theory for the corresponding secular part of the mapping (Šidlichovský, 1992) changes the results only very slightly.

$$
H_s = \bar{F}e'x + F(x^2 + y^2),
\tag{4}
$$

The coefficients $\bar{F}, F$ are given in (Šidlichovský, 1992). The last term on r.h.s. of (2) is the resonant Hamiltonian, where

$$
\begin{aligned}
A(x, y) &= \sqrt{C(x, y)^2 + S(x, y)^2}, \tag{5} \\
C(x, y) &= c_0 e'^q + c_1 e'^{q-1} x + c_2 e'^{q-2}\left(x^2 - y^2\right) + c_3 e'^{q-3} x \left(x^2 - 3y^2\right) + \\
&\quad + c_4 e'^{q-4}\left(x^4 - 6x^2 y^2 + y^4\right), \\
S(x, y) &= c_1 e'^{q-1} y + 2c_2 e'^{q-2} xy + c_3 e'^{q-3} y \left(3x^2 - y^2\right) + \\
&\quad + 4c_4 e'^{q-4} xy \left(x^2 - y^2\right),
\end{aligned}
\tag{6}
$$

where the coefficients $c_k$ are again given in (Šidlichovský, 1992). Finally $\phi_0$ is given by

$$
\cos \phi_0(x, y) = -\frac{C(x, y)}{A(x, y)}, \quad \sin \phi_0(x, y) = -\frac{S(x, y)}{A(x, y)}.
\tag{7}
$$

Notice that for the third order resonances $c_4 = 0$. Substituting the resonant Hamiltonian with the effect of calculable kicks as suggested in (31) of (Šidlichovský, 1992) we obtain the mapping consisting of four jumps on $\delta$-functions standing with $\sin\phi$ (sin jumps for briefness), secular evolution and five cos jumps. Actually we do not have to introduce the $\delta$-functions at all, as the mapping obtained can be shown to be equivalent to the first order symplectic mapping derivable immediately as the sequence of evolution on individual integrable parts of the Hamiltonian. Introducing

$$C_k = c_k e^{'q-k},\tag{8}$$

we reach at the following mapping for the $(p+q)/p$ resonance:

$$
\begin{aligned}
y' &= y - 2\pi(C_1 + 2C_2x + 3C_3x^2 + 4C_4x^3)\cos\phi,\\
\Phi' &= \Phi - 2\pi(C_0 + C_1x + C_2x^2 + C_3x^3 + C_4x^4)\sin\phi,
\end{aligned}\tag{9}
$$

$$
\begin{aligned}
x' &= x - 4\pi y(C_2 y - 2C_4 y^2)\cos\phi,\\
\Phi' &= \Phi + 2\pi y^2(C_2 - C_4 y^2)\sin\phi,
\end{aligned}\tag{10}
$$

$$
\begin{aligned}
\Phi' &= \Phi + 6\pi C_3 x y^2 \sin\phi,\\
x' &= x(1 - 6\pi C_3 y \cos\phi)^2,\\
y' &= y/(1 - 6\pi C_3 y \cos\phi),
\end{aligned}\tag{11}
$$

$$
\begin{aligned}
\Phi' &= \Phi + 12\pi C_4 x^2 y^2 \sin\phi,\\
x' &= x\exp(-24\pi C_4 xy \cos\phi),\\
y' &= y\exp(24\pi C_4 xy \cos\phi),
\end{aligned}\tag{12}
$$

$$
\begin{aligned}
\phi' &= \phi + 2\pi\left(\frac{\mu_1^2}{p^2\Phi^3} - (p+q)\right),\\
x' &= \left(x + \frac{\bar{F}e'}{2F}\right)\cos(4\pi F) - y\sin(4\pi F) - \frac{\bar{F}e'}{2F},\\
y' &= \left(x + \frac{\bar{F}e'}{2F}\right)\sin(4\pi F) + y\cos(4\pi F),
\end{aligned}\tag{13}
$$

$$
\begin{aligned}
\Phi' &= \Phi + 2\pi y(C_1 - C_3 y^2)\cos\phi,\\
x' &= x + 2\pi(C_1 - 3C_3 y^2)\sin\phi,
\end{aligned}\tag{14}
$$

$$
\begin{aligned}
\Phi' &= \Phi + 4\pi C_2 xy \cos\phi,\\
x' &= x\exp(4\pi C_2 \sin\phi),\\
y' &= y\exp(-4\pi C_2 \sin\phi),
\end{aligned}\tag{15}
$$

$$\Phi' = \Phi + 6\pi C_3 x^2 y \cos\phi,$$
$$x' = x/(1 - 6\pi C_3 x \sin\phi),$$
$$y' = y(1 - 6\pi C_3 x \sin\phi)^2, \tag{16}$$

$$\Phi' = \Phi + 8\pi C_4 x^3 y \cos\phi,$$
$$x' = x(1 - 16\pi C_4 x^2 \sin\phi)^{-1/2},$$
$$y' = y(1 - 16\pi C_4 x^2 \sin\phi)^{3/2}, \tag{17}$$

$$\Phi' = \Phi - 8\pi C_4 x y^3 \cos\phi,$$
$$x' = x(1 - 16\pi C_4 y^2 \sin\phi)^{3/2},$$
$$y' = y(1 - 16\pi C_4 y^2 \sin\phi)^{-1/2}, \tag{18}$$

Eq. (13) gives the secular evolution for time interval equal to $2\pi$, it is preceded by four cos jumps and followed by five sin jumps. The mapping obtained by successive mappings (9)–(13) yields the change of asteroidal elements for time interval $2\pi$ corresponding to one Jupiter orbital period for the third and fourth order resonances. Of course for the third order resonances mappings (12), (17) and (18) can be omitted as they are identities due to the fact that $C_4 = 0$.

## 3. The Systematic Study of Resonances

Figs 1a–1e show the energy graphs (the dependence of $H_s(x,0) + A(x,0)$ and $H_c = H_s(x,0) - A(x,0)$ on $x$) for the resonances 4/1, 5/2, 7/4, 5/1 and 7/3, respectively. The thin curves in Fig. 1 show the fourth degree approximation in eccentricities for $H_s$. In most cases they are hardly distinguishable from the second degree approximation (bold lines).

From Wisdom's (1985) argumentation we expect $\phi$ to librate at $(x,0)$ for orbits with integral $H'$ between two curves in energy graphs, for $H'$ below both curves $\phi$ circulates. The value $x$ on the $x$-axis are inaccessible for values $H'$ above both curves. A trajectory starting at point $(x,0)$ with $\phi = \phi_0(x,0) + \pi$ and $\Phi = \Phi_r$ is expected to be chaotic. We will describe each of the resonances 5/2, 7/3, 4/1 by surfaces of section in $x, y$ plane. Point $(x,y)$ is recorded when $\phi = \phi_0(x,y)$. It is a good choice as in librating trajectories $\phi$ librates about slowly moving $\phi_0$ so the trajectory can be expected to cross the selected hypersurface quite regularly in both librating and circulating cases. For each $H'$ we get surfaces of sections. After some experience we decided to represent each of the three selected resonances by set of six surfaces of sections. We will take six values of eccentricity $e = 0.05, 0.1$, 0.15, 0.2, 0.25 and 0.3 and we will calculate the corresponding six values of $x$ (for $y = 0$) using (1) with resonant value $a_r$ of the semimajor axis. The surfaces of sections shown in Figs 2–4 correspond to $H_c(e)$ calculated for these six values of $x$. That is why the chaotic region crosses the $x$-axis at values $x$ corresponding to

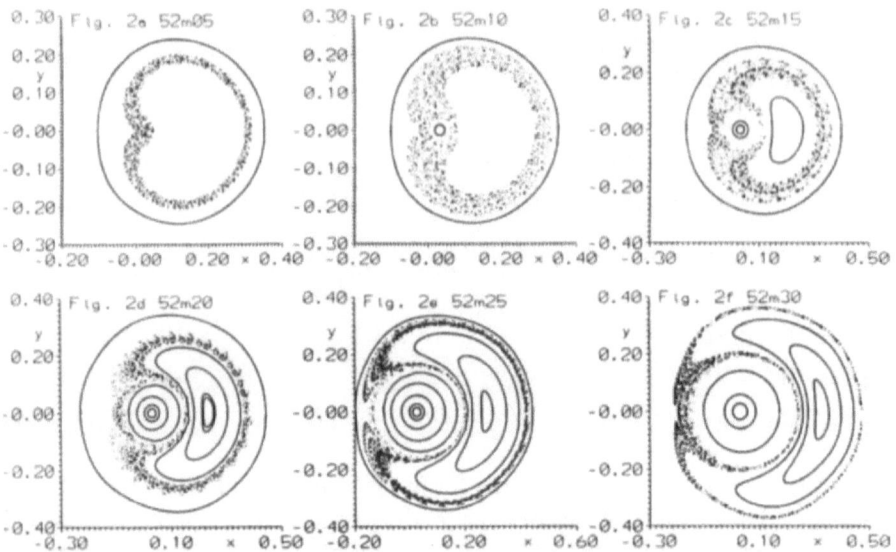

Fig. 2. Surfaces of sections for the 5/2 resonance

$e = 0.05$ in the first of the six figures (Figs 2a, 3a etc.), at values $x$ corresponding to $e = 0.1$ in the second of six figures (Figs 2b, 3b etc.) for each resonance. For each $H_c$ obtained in this way we calculate the surfaces of sections using mapping (9–18) for trajectories starting at $y = 0$ and $x$ corresponding to $e = 0.05, 0.1, 0.15$ if only this eccentricity is accessible for given $H' = H_c$. In practical calculations we use trajectories with initial $\phi = \phi_0(x, 0)$ and initial $a$ is obtained from (2). For inaccessible $e$ it is impossible to find $a$ (as $H_0(\Phi) - H_0(\Phi_r)$ calculated from (2) is positive).

For briefness we will introduce the coded name for each curve in the surfaces of sections. The curve 52m20.05 will denote the curve calculated for the 5/2 resonance (the first two positions in the name), $H' = H_c(e = 0.20)$ so that the information to which of the set of six figures the curve belong is given by positions four and five. Positions after full stop give the initial eccentricity for ($y = 0$). In the above example initial eccentricity was 0.05. The letter m in the third position tells that the mapping (9)–(18) has been used in calculating the curve, which is surface of section. To identify one of the six representative figures we will use the first five positions identifying the participating curves so that Fig. 2a is 52m05, while Fig. 4f is 41m30.

For chaotic trajectory e.g. 52m05.05 in Fig. 2a we can hardly speak about a curve. The important result is that calculations confirmed that all trajectories of the type 52mkn.kn were really found to be chaotic. The same is true for the resonances 7/3 and 4/1. These chaotic trajectories usually reach high eccentricities which correspond to the possibility of switching between low and high eccentricity modes

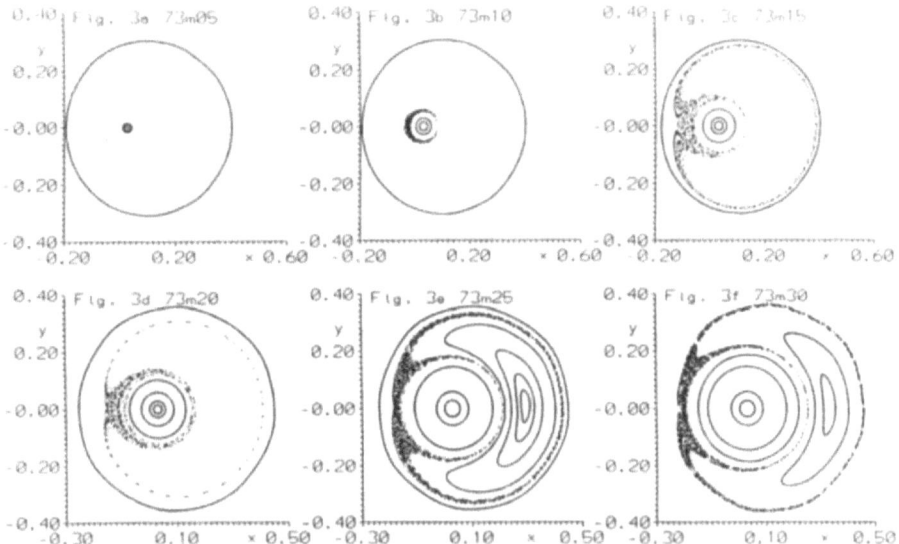

Fig. 3. Surfaces of sections for the 7/3 resonance

found by Wisdom for the 3/1 resonance. The only exceptions were trajectories 41m05.05 and 41m10.10 where we could find no high eccentricity lobe even if we followed these trajectories for more than 8 million years. Chaotic trajectories were usually calculated for 300000 $T_J$ (Jupiter periods) but only each third intersection has been recorded to avoid too big data files. The resulting chaotic regions do not seem to change if the time interval is tripled.

The absence of banana shaped trajectories in Figs 2a,2b,3a–3d,4a–4d reflects the inaccessibility of the region inside chaotic trajectory. This is in agreement with energy graphs Figs 1a, 1b, 1e. In Fig. 2c we added one banana shaped trajectory 52m15.169 as the trajectory 52m15.20 was inaccessible. In Figs 2d–2f there is no more inaccessible region.

The regular curves near the origin corresponding to trajectories with circulating $\phi$ have the periods connected with $x, y$ motion $\sim 10^4$ years. For the 5/2 resonance this period is about 1800 $T_J$, for 7/3 about 1500 $T_J$ and for 4/1 about 4000 $T_J$. The $x, y$ period of the banana shaped curves is still longer (but less than by a factor of 2) depending on position.

## 4. The Symplectic Integration

Recently Wisdom (1991) suggested mapping which can be used for the full Hamiltonian without any truncation. The mapping was again obtained by introducing $N$ $\delta$-functions into the Hamiltonian. For N=2 the method is equivalent to the second order symplectic integrator (Neri, 1988). We decided to use directly symplectic integrator as it can be generalized to the higher orders. Finally we used the fourth and

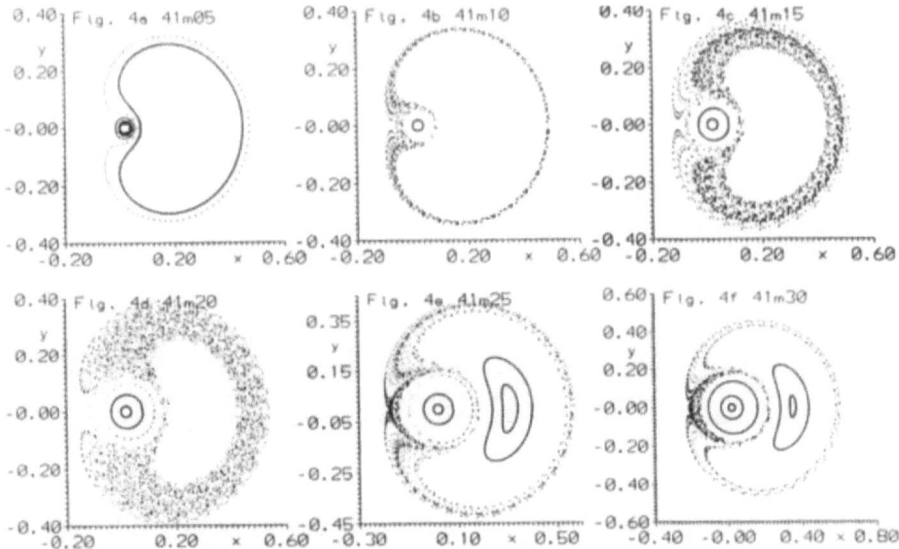

Fig. 4. Surfaces of sections for the 4/1 resonance

six order integrators of Yoshida (1990). It was applied to the following Hamiltonian

$$H = H_0 + H_1, \tag{19}$$

where

$$H_0 = \frac{1}{2}\left(P_X^2 + P_Y^2\right) - \frac{\mu_1}{\sqrt{(X^2 + Y^2)}} + K, \tag{20}$$

and

$$H_1 = -\mu \left( \frac{1}{\sqrt{((X - X'(k))^2 + (Y - Y'(k))^2)}} - \frac{X X'(k) + Y Y'(k)}{\sqrt{(X'(k)^2 + Y'(k)^2)}} \right). \tag{21}$$

Here we introduced a new pair of canonical variables $k, K$ to have time independent Hamiltonian. $X, Y$ and $X', Y'$ are the cartesian coordinates of asteroid and Jupiter. We present here the results of a modified algorithm (Kinoshita et al., 1991) which allows for longer time step. One time step $\tau$ for the fourth order symplectic mapping consists of seven transformations. Four of them advance the orbit for time intervals $c_i\tau$ under the effect of the Hamiltonian $H_0$ only. In between three transformations advancing the orbit for time intervals $d_i\tau$ under the effect of $H_1$ are inserted. Numbers $d_i, c_i$ are given by Yoshida (1990). As both Hamiltonians

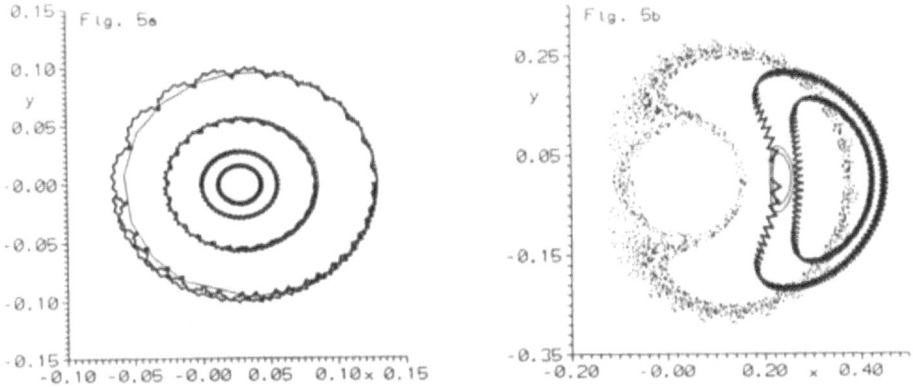

Fig. 5. Comparison of real trajectories with surface of section for small (Fig. 5a) and high eccentricities (Fig. 5b)

$H_0$ and $H_1$ are integrable the solution is straightforward. We compared always one curve from surfaces of sections for the 5/2 resonance with real orbit with the same initial $x, y, \phi, \Phi$. For small eccentricities we obtained very good agreement. Fig. 5a shows a typical situation. Four surfaces of section (smooth curves) 52m20.00, 52m20.05, 52m20.10 and 52m20.15 are compared with corresponding projections of trajectory calculated by symplectic integrator of the fourth order.

On the other hand the numerically integrated trajectories show tendency to reach still much higher eccentricities once they get to the high eccentricity region, even if the topology is very similar. Typical example is shown in Fig. 5b where two banana shaped surfaces of sections 52m20.25, 52m20.30 are compared with corresponding trajectories 52s20.25, 52s20.30, where s in the third position denotes that trajectory was calculated using the symplectic mapping for the full Hamiltonian (without truncation with respect to degrees of eccentricity). The chaotic 52m20.20 is shown there as well.

Fig. 6 shows the trajectory 52s10.10 (for 80000 $T_J$) together with the surface of section for 52m10.10. The typical switching between the low and high eccentricity mode can be seen again. One can see at once that more realistic trajectory is again chaotic but the mode of high eccentricities reaches very soon $e = 0.52$. This is well over the Mars crossing limit. Then the mode with still higher eccentricity ($e \sim 0.78$ with the highest eccentricity for $\tilde{\omega} = \pi$) was found, corresponding to the existence of very-high-eccentricity stable and unstable equilibrium solutions. Note the similarity to results of Ferraz-Mello and Klafke (1992) for the 3/1 resonance. It seems that more realistic calculations work still better towards explanation of Kirkwood gaps than approximation with the truncated Hamiltonian. On the other hand there still is region of regular trajectories with small $e$ and circulating $\phi$ where asteroid could stay.

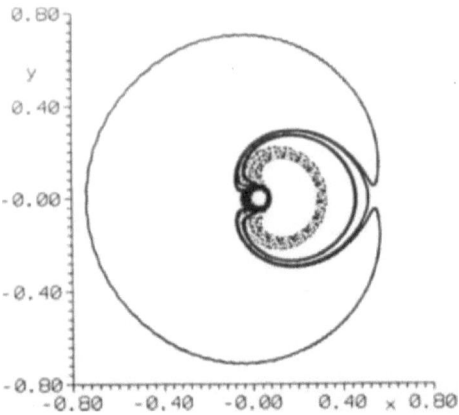

Fig. 6. Chaotic trajectory 52s10.10 calculated with symplectic mapping

# References

Ferraz-Mello, S., and Klafke, J. C.:1992, 'A model for the study of very-high-eccentricity asteroidal motion. The 3:1 resonance', to appear in *Chaos, resonance, and collective dynamical phenomena in the solar system (IAU Symp. 152)*, ed. S. Ferraz-Mello.

Kinoshita, H., Yoshida, H., Nakai, H.: 1991, 'Symplectic integrators and their application to dynamical astronomy', *Celest. Mech.* **50**, 59–71.

Murray, C. D.: 1986, 'Structure of the 2:1 and 3:2 Jovian resonances', *Icarus*, **65**, 70–82.

Neri, F.: 1988, 'Lie algebras and canonical integration', Dept. of Physics, University of Maryland, preprint.

Šidlichovský, M.: 1988, 'The origin of 5/2 Kirkwood gap', *The Few Body Problem (ed. M. Valtonen)* p.117–121, Kluwer Acad. Publishers.

Šidlichovský, M.: 1991, 'Tables of the disturbing function for resonant asteroids', *Bull. Astron. Inst. Czechosl.*, **42**, 116–123.

Šidlichovský, M.: 1992, 'Mapping for asteroidal resonances', accepted in Astron. Astrophys.

Šidlichovský, M., Melendo, B.: 1986, 'Mapping for 5/2 asteroidal commensurability', *Bull. Astron. Inst. Czechosl.*, **37**, 65–80.

Wisdom, J.: 1982, 'The origin of Kirkwood gaps: A mapping for asteroidal motion near the 3/1 commensurability', *Astron. J.*, **87**, 577–593.

Wisdom, J.: 1983, 'Chaotic behavior and the origin of the 3/1 Kirkwood gap', *Icarus*, **56**, 51–74.

Wisdom, J.: 1985, 'A perturbative treatment of motion near the 3/1 commensurability', *Icarus*, **63**, 272–289.

Wisdom, J., Holman, M.: 1991, 'Symplectic maps for the $N$-body problem' *Astron. J*, **102**, 1528–1538.

Yoshida, H.: 1990, 'Construction of higher order symplectic integrators', *Phys. Let. A* **150**, 262–268.

# LOW-ECCENTRICITY MOTION OF ASTEROIDS NEAR THE 2/1
# JOVIAN RESONANCE

JOACHIM SCHUBART

*Astronomisches Rechen-Institut, Heidelberg, Germany*

**Abstract.** (903) Nealley moves on an orbit of low eccentricity with a mean motion that is slightly larger than the 2/1 value of resonance. This orbit and some related fictitious orbits are studied by numerical integrations of the four-body problem Sun-Jupiter-Saturn-asteroid over an interval of 110 000 yr. The author's experience on related cases of resonance allows a study of the variation of suitably defined orbital parameters. The long-term evolution of the orbits is compared with earlier predictions. Some of the librating orbits are temporarily captured in a secondary resonance that refers to three-dimensional motion and is demonstrated by a special example.

**Key words:** Asteroids – evolution of orbits – 2/1 resonance – proper parameter – secondary resonance.

## 1. Introduction

The evolution of asteroidal orbits in the vicinity of the 2/1 Jovian resonance of mean motion is a subject of continuing interest, see the references given in the respective section of a related paper by Yoshikawa (1991). Many studies on this subject depend on the elliptic restricted three-body problem Sun-Jupiter-asteroid treated in two or three dimensions, but Yoshikawa (1991) points to the importance of including the action of Saturn in the model of the forces, especially in work on the 2/1 Kirkwood gap. It is comparatively easy to consider both Jupiter and Saturn in studies that use numerical integration to derive the motion of massive and small bodies in a simultaneous computation. I have done this in work on resonant motion in the outer asteroid belt, especially on the 3/2 case (Schubart 1988, 1991). During this work I have noticed that the special numerical methods of studying single orbits are applicable to some low-eccentricity orbits near the 2/1 resonance as well, in particular to the orbit of (903) Nealley (Schubart 1988). Now I apply the former methods and experience in a detailed study of this orbit that is situated at the sunward border of the Kirkwood gap in the frequency distribution of semi-major axis. This orbit shows a quasi-periodic behaviour in the interval considered. The same turns out for some related fictitious orbits that are closer to the center of the gap, but seem to have no natural counterparts (compare with Wisdom 1987). In a continuation toward the center of the gap, four further orbits clearly show a non-quasiperiodic behaviour. Frequent changes in the ratio of characteristic long periods appear to be a typical feature of these orbits, that show libration of variable amplitude and temporary capture by a secondary resonance.

## 2. Numerical Procedures

In the present study I closely follow the definitions and procedures introduced in an earlier paper (Schubart 1988). As before, the integrations of the problem

*Celestial Mechanics and Dynamical Astronomy* **56**: 153–162, 1993.
© 1993 *Kluwer Academic Publishers.*

Sun-Jupiter-Saturn-asteroid have resulted from the N-body program by Schubart and Stumpff (1966). They have proceeded with a step length of 6.67 days. The accuracy of computation in a step corresponds to about 15 decimal digits. All these integrations cover an interval of 110 000 yr centered close to the present. Since a non-quasiperiodic orbit integrated over 55 000 yr can show comparatively strong effects due to the errors of the method, I have repeated the forward integration of one of the four orbits mentioned at the end of section 1 with considerably increased accuracy in a step, and with a smaller step length. According to this test, the coordinates of Jupiter and Saturn result with an accuracy of several decimal digits from my computations over the interval. However, the considered fictitious orbit starts to show shifts in longitude of more than 1° after about 44 000 yr, and of more than 10° after about 50 000 yr. I have extended the former integration for (903) Nealley (Schubart 1988) by adding a backward computation in the model '$i_S \neq 0$' that considers the mutual inclination of the orbits of the massive planets. However, the recent integrations on fictitious 2/1 orbits depend on the model '$i_S = 0$' which neglects this mutual inclination. Then Jupiter and Saturn move in the plane of reference that is defined to agree with the real orbital plane of Jupiter at a special epoch. In this model, the lack of long periods that are otherwise caused by the moving planes of the major bodies, allows special applications (see section 3).

As before, I use the symbols $a$, $e$, $i$, $\ell$ for semi-major axis [AU], eccentricity, inclination and mean longitude. $\varpi$ and $\Omega$ are the longitudes of perihelion and node. The subscript $J$ refers to an element to Jupiter. The critical argument of the 2/1 resonance is given by $\sigma = 2\,\ell_J - \ell - \varpi$. Again I can use an empirical transformation

$$e_p \cos(\varpi_p - \varpi_J) = e \cos(\varpi - \varpi_J) - \kappa\, e_J;$$
$$e_p \sin(\varpi_p - \varpi_J) = e \sin(\varpi - \varpi_J);$$
$$\bar{\sigma} = \sigma + \varpi - \varpi_p$$

to new variables $e_p$, $\varpi_p$, $\bar{\sigma}$ with a constant $\kappa$ that is suitably chosen for each orbit. $\kappa$ can roughly be fitted in such a way that it approximately removes the asymmetric location with respect to the origin of a corresponding curve plotted in $e \cos(\varpi - \varpi_J)$, $e \sin(\varpi - \varpi_J)$ coordinates, in the analogous new coordinates. A final adjustment can simplify the resulting variations of $\bar{\sigma}$, see section 3. $\kappa$ remains comparatively small in the present applications. In this way I try to remove a part of the influence of the eccentricities of the major bodies from $e$, $\varpi$, and $\sigma$, so that results on the new variables allow a better comparison with analogous results obtained by the circular restricted problem. For example I have found evidence of a permanent libration of $\bar{\sigma}$ for Nealley in this way (Schubart 1988). Proper parameters are given by $\bar{e}_p$, a mean value of $e_p$, $i_p$, here a mean value of $i$, and $\bar{\sigma}_A$, the mean amplitude of the oscillation caused in $\bar{\sigma}$ by the typical period of libration, $T_L$. I have earlier used these parameters to characterize resonant orbits that show libration of $\bar{\sigma}$ and a quasiperiodic behaviour (Schubart 1991). For low-eccentricity orbits, an additional parameter $\mathcal{A}$ is useful to describe the main variations of $e_p$ and $\varpi_p$ (Schubart 1991, section 4). For instance, $e_p$ approximately equals the length of a two-dimensional vector given by the sum of a constant vector of length $\bar{e}_p$

and of a shorter vector of length $\mathcal{A}$ with uniform rotation according to $T_L$. Let $\overline{a}$ be a mean value of $a$ that refers to the interval covered by integration in general. I apply digital filtering in the derivation of all these parameters, see Schubart and Bien (1984) and Schubart (1991).

## 3. Orbital Evolution of Nealley and some Related Examples of Motion

The way of starting the computation for (903) Nealley resembles my earlier procedures (Schubart 1990, 1991). The starting epoch is JDE 2446000.5. The starting elements of Jupiter and Saturn correspond to Table 1 of Schubart (1990), but the orbit of Saturn is rotated to approximate the real inclination. The forward integration brings $\sigma$ of Nealley close to zero at an epoch that is later by 117.64 yr. I use this second epoch and the corresponding result for Nealley to start the integration of the fictitious orbits with similar values of the angular variables that are close to $\varpi - \varpi_J = 337°$, $\Omega - \varpi_J = 148°$, $i = 11°.1$, $\sigma = 0°$. I turn the orbital planes of the major planets into the plane of reference at the second epoch for this integration. Then, I vary the starting value of $a$, $a_0$, away from the orbit of Nealley, N, to obtain a sequence of orbits A, B, $\cdots$, E that approaches the center of the Kirkwood gap. In each case I adjust the starting value of $e$, $e_0$, to obtain a comparatively small effect of libration of $\overline{\sigma}$ with $\overline{\sigma}_A$ near 10°. Table I lists $a_0$ and $e_0$, referred to the second epoch, the derived proper parameters and other values of interest, especially mean periods. Note that the periods $T_P$ and $T_N$ correspond to the mean retrograde revolution of the arguments $\varpi_p - \varpi_J$ and $\Omega$. Fig.1 demonstrates the relation between $\overline{e}_p$ and $\overline{a}$. $a$ oscillates about $\overline{a}$ in case of orbits N and A, B, $\cdots$, E, and all the other considered variations indicate a quasi-periodic evolution of these orbits, although orbit E shows comparatively large shifts of a period related to $T_L$. This behaviour suddenly changes, if I try to continue the sequence of orbits beyond E with values of $a_0$ from 3.263 to 3.267 AU. Four examples started with such values of $a_0$ show irregular variations of $a$ and other effects that clearly indicate a non-quasiperiodic evolution, see section 4.

In the derivation of the proper parameters shown in Table I, I have applied the methods for the main case of Hilda motion to $\overline{\sigma}_A$ and $i_p$, but a special method for small $e_p$ to $\overline{e}_p$ and $\mathcal{A}$ (Schubart 1991). The frequencies resulting in this method allow the derivation of $T_L$ and $T_P$. In an attempt to determine $\overline{\sigma}_A$ by digital filtering, a band of frequencies passes the filter together with the one that corresponds to $T_L$. Therefore, a changing amplitude of $\overline{\sigma}$ appears in a plot versus time, but a final adjustment of $\kappa$ leads to a sufficiently small variation in this amplitude. Along the sequence of orbits A to E, $\kappa$ shows a tendency to approach zero, but this can be a particular feature for values of $i_p$ near 11° and small $\overline{\sigma}_A$. Among other effects, I remove by digital filtering from $\overline{\sigma}$ an oscillation that apparently follows a frequency given by the difference of the absolute values of the frequencies that correspond to $T_L$ and $T_P$. I did not notice such an effect during my former work on Hildas. In case of Nealley, the amplitude of this effect changes from about 6° to 3°, following

TABLE I

Starting values, parameters, and periods

| | N = (903) | A | B | C | D | E |
|---|---|---|---|---|---|---|
| $a_0$ (in AU) | 3.2317 | 3.240 | 3.243 | 3.249 | 3.255 | 3.261 |
| $e_0$ | 0.0701 | 0.042 | 0.043 | 0.046 | 0.050 | 0.056 |
| $\kappa$ | 0.26 | 0.18 | 0.163 | 0.125 | 0.08 | 0.04 |
| $\bar{e}_p$ | 0.034 | 0.034 | 0.036 | 0.042 | 0.049 | 0.057 |
| $\mathcal{A}$ | 0.026 | 0.005 | 0.005 | 0.006 | 0.006 | 0.005 |
| $i_p$ | 11°20 | 11.14 | 11.14 | 11.14 | 11.14 | 11°13 |
| $\bar{\sigma}_A$ | 54° | 10 | 10 | 10 | 10 | 9° |
| $\bar{a}(in AU)$ | 3.2387 | 3.2395 | 3.2420 | 3.2468 | 3.2514 | 3.2556 |
| $T_L$ (in yr) | 308.4 | 313.2 | 330.4 | 366.9 | 403.3 | 435.2 |
| $T_P$ (in yr) | 342.4 | 349.2 | 375.3 | 438.6 | 518.9 | 625.8 |
| $T_N$ (in $10^3$ yr) | 18.0 | 18.3 | 18.5 | 19.2 | 20.0 | 21.4 |

Notes: The starting values $a_0$ and $e_0$ of (903) Nealley and of 5 fictitious orbits refer to a common epoch in the comparatively near future. For the numerical constant $\kappa$ and the following parameters see section 2. $T_L$ is the period of libration. A mean value is given. The mean periods of the retrograde revolution of the arguments $\varpi_p - \varpi_J$ and $\Omega$ are designated by $T_P$ and $T_N$, respectively.

the cycle of the long-period variation of $e_J$. I assume that this amplitude is roughly proportional to both $e_J$ and $\mathcal{A}$, according to the smaller variations of this kind shown by orbits A, B, and C.

In applying my method of derivation of $i_p$, I remove a periodic term that follows the mean period of revolution of the argument $2\Omega - 2\varpi_J$, from the filtered results on $i$, if this is necessary due to the length of this period. In doing so for orbits A to E, I noticed an additional effect depending on the mean period of the argument $2\Omega - \varpi_J - \varpi_S$ with an amplitude of less than 0°05. Here the suffix $S$ refers $\varpi_S$ to Saturn. The use of the model '$i_S = 0$' for these orbits and the corresponding lack of other very long periods has allowed these small effects to show up in my graphs. Effects of this kind are more important for the variations of $i$ of Trojan asteroids, see Fig. 3a of Schubart and Bien (1986).

## 4. Non-Quasiperiodic Types of Motion and Comparison with other Work

In Fig.1 the sequence from B to E corresponds to orbits with small effects of libration. Evidently, the sequence A, B, $\cdots$, E develops in analogy to the pericentric branch of periodic orbits of the circular restricted problem, see Fig.3a of Morbidelli and Giorgilli (1990). In this context it is interesting to follow the evolution of the four examples with non-quasiperiodic motion started close to orbit E, see section

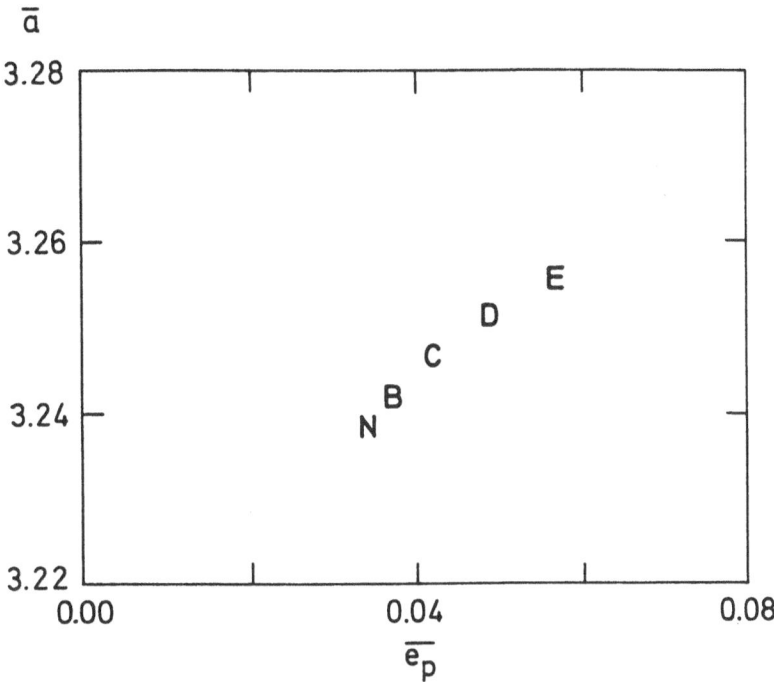

Fig. 1. Mean values of semi-major axis [AU] versus $\bar{e}_p$, a proper parameter related to eccentricity. $a = 3.276$ AU corresponds to the 2/1 ratio of the mean motions of asteroid and Jupiter. N refers to the orbit of (903) Nealley. The letters from B to E represent fictitious orbits that show small effects of libration. A fifth orbit of this kind, A, closely corresponds to the position of the letter N in the figure.

3. I study them with $\kappa = 0$, according to the small value found for orbit E. This means the use of the original elements $e$, $\varpi$, and of $\sigma$. The study shows for the four examples a continuing status of libration of $\sigma$ with $|\sigma| < 100°$, but with strong fluctuations of the amplitude, in the considered interval. In each case the strong irregular variations of $a$ and $e$ are correlated. During limited periods the mean values of $a$ and $e$ of an orbit roughly correspond to members of the pericentric branch mentioned above.

The importance of secondary resonances for the evolution of 2/1 resonant orbits was pointed out by Lemaître and Henrard (1990), see Fig. 3b of Morbidelli and Giorgilli (1990). According to Table 1, the ratio $T_P/T_L$ develops from values near 1.1 to 1.44, the value of orbit E, and passes rational values of interest. Fortunately, the orbit started next to E with $a_0 = 3.263$, $e_0 = 0.0585$ shows comparatively smooth variations during a large part of the backward computation, which includes the starting epoch. $T_P/T_L$ equals about 1.52 in this interval. Maybe the proximity to

a [AU]

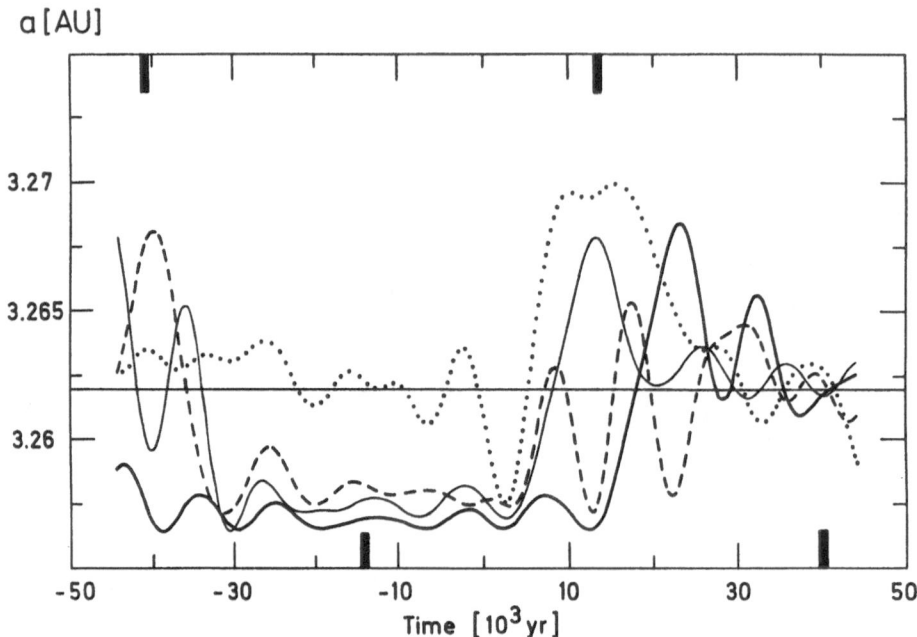

Fig. 2. $a$ [AU] versus time, smoothed results of four orbits. A straight line indicates the level $a = 3.262$. See the analogous Fig.3 for further explanations.

3/2 of the ratio $T_P/T_L$ gives rise to the wild evolution of this orbit. The same can turn out in much more extended computations on orbit E. It is interesting to note that the onset of non-quasiperiodic types of motion found here, qualitatively agrees to the predictions on the structure of the 2/1 resonance by Murray (1986), who has used a simplified model of the three-body problem and a mapping technique in his work. I think that his simplified model approximates the conditions of low-eccentricity motion, but his model is a planar one. Wisdom (1987, p.268) has demonstrated by the numerical integration of an orbit with $e_0 = 0.05$, that the transition from a three-body problem to a model with four major planets can change predictions on the structure of the 2/1 resonance.

I have studied the evolution of $a$, $e$, and $i$ of the four examples with $a_0$ beyond 3.261 AU by means of smoothed curves, eliminating the influence of all periods that are less than about 6000 yr, see Fig.2 for $a$ and Fig.3 for $e$. The following results refer to temporary mean values taken from these curves. An increase and subsequent strong changes of $a$ are apparently triggered by the increase of $e_J$ to a maximum in the forward direction of time. $e$ follows with increase and correlated variations in each case. The librating orbits travel about in an $a$, $e$ domain that contains, for the ratio $T_P/T_L$, the typical 2/1 and 3/1 secondary resonances. The smoothed values of $a$ and $e$ reach 3.270 AU and 0.127 with $T_P/T_L$ temporarily near 4/1 in one of the cases. I note that the respective backward computation even shows proximity to the 5/1 ratio of $T_P/T_L$ at its end. I have observed original

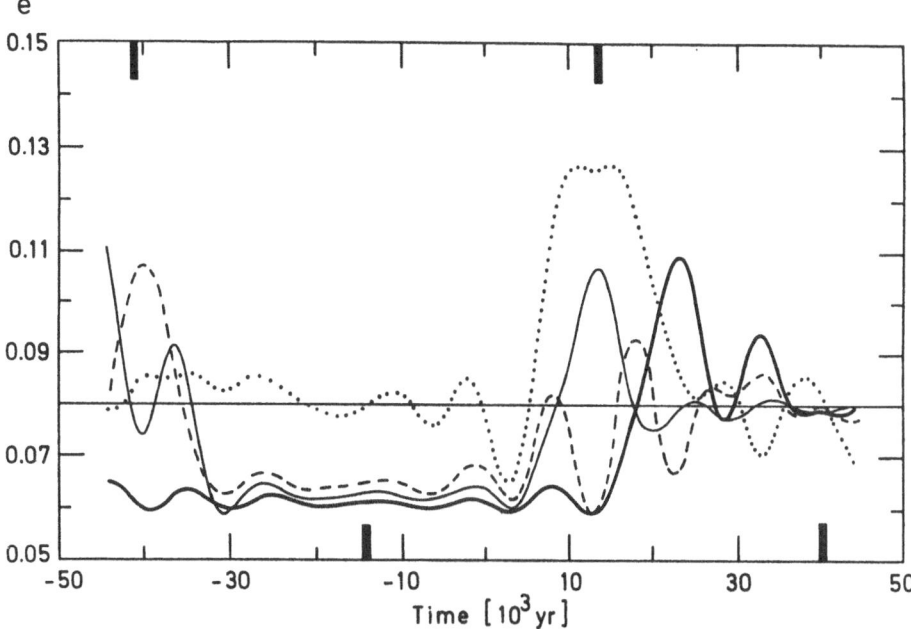

Fig. 3. Four curves smoothed by digital filtering and plotted against time, demonstrate the non-quasiperiodic evolution of the eccentricity of orbits started in an attempt to continue the sequence from A to E. The bars at the upper and lower border indicate maxima and minima of $e_J$, respectively. Note the approach of the curves to the level $e = 0.08$ near the lower right bar. The respective starting values $a_0$ [AU] and $e_0$ of the four orbits are

|  |  |  |  |  |  |
|---|---|---|---|---|---|
| 3.263 | 0.0585 | : heavy line, | 3.264 | 0.0598 | : thin line, |
| 3.265 | 0.0611 | : dashed line, | 3.267 | 0.0640 | : dotted line. |

values of $e$ of up to 0.19.

In the more remote future $e_J$ goes down to reach a minimum. At about this time the smoothed values of $a$ and $e$ of all the four examples show a tendency to approach a domain with $a$ near 3.262 and $e$ close to or a little less than 0.08. In Fig. 1 this corresponds to the upper part of the right border. However, the simultaneous smoothed results on $i$ approach different values in the interval from $10°9$ to $13°1$. This tendency typically appears in case of an orbit started near E with $a_0 = 3.264$, $e_0 = 0.0598$. Since I suspected the influence of a special type of secondary resonance given by the 2/1 ratio of the mean period of revolution of $\omega = \varpi - \Omega$ and $T_L$, I have plotted $e \sin 2\omega$ versus $e \cos 2\omega$ for the respective interval, without smoothing. A polar plot for $e$ and $2\omega$ results in this way. Neglecting short-period variations, I find the subsequent directions from the origin to the maxima of $e$ to librate with a large amplitude about the zero direction of $2\omega$ temporarily. I get the same temporary libration of the directions for all of the four examples near the considered moment of time. I demonstrate it by a related orbit integrated in a simpler model (see below and Fig. 4). This libration corresponds to the evolution of a fictitious Hilda-type orbit studied earlier (see Fig. 2 of Schubart 1990). Here the typical example with

e sin 2 ω

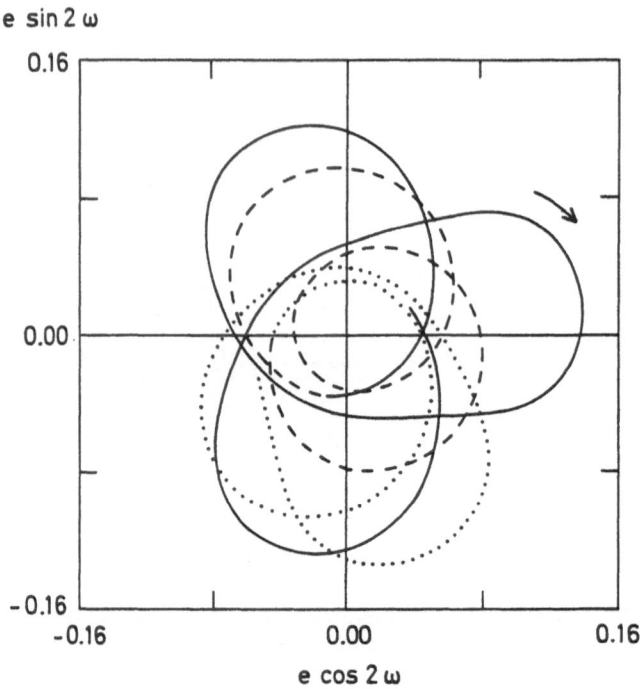

e cos 2 ω

Fig. 4. Polar plot for $e$ and $2\omega$ of a librating orbit of the three-dimensional circular restricted three-body problem Sun-Jupiter-asteroid. The curve starts with a heavy line, continues with dashes and ends as a dotted line. An arrow indicates the sense of motion. The plotted curve corresponds to an interval of about 3360 yr. The period of libration causes extremes of $e$ that change in amount. Note the libration of the direction from the origin to subsequent maxima of $e$ about the zero direction of $2\omega$ (see the text).

$a_0 = 3.264$ shows a period of libration of the direction to the maxima of $e$ of about 3500 yr. $\omega$ decreases with a temporary period that varies about 1000 yr under the influence of the longer period, which simultaneously causes considerable changes in the mean of $e$ of a cycle of $\sigma$ libration. This mean varies between 0.100 and 0.055 according to the period of 3500 yr, and the largest maxima of $e$ reach 0.14. The test integration with increased accuracy mentioned in section 2 refers to this orbit and confirms these results.

Apparently the influence of the considered type of secondary resonance becomes strong during periods of small $e_J$. Perhaps the periodic repetition of such periods can prevent the escape of $e$ to very large values during longer intervals, in case of my four examples with a non-quasiperiodic type of motion, if basic libration according to a period $T_L$ continues to occur. However, forbidden regions and further restrictions used by Froeschlé and Scholl (1976) for an ergodic orbit of a simplified model, are not available for the present results, due to the more complicated model of the forces. Wisdom (1987, p.269) has mentioned results of an extended integration with perturbations of the four major planets for some 2/1 orbits of special interest. One of these orbits reaches a maximum eccentricity of 0.53 and demonstrates the possibility of large long-period changes in inclination. As visible from his Fig.19, this interesting orbit starts with $e_0 = 0.1$ and a comparatively small inclination. According to this and to the initial part of its evolution, that orbit is not closely related to the fictitious orbits considered in the present paper.

To demonstrate the special type of secondary resonance considered above by a comparatively simple example, I use starting values from a forward integration over an interval of 36505 yr of the orbit with the original starting values $a_0 = 3.264$, $e_0 = 0.0598$. I continue the integration with the following simplifications and changes: The mass of Saturn and the eccentricity of Jupiter are neglected, the zero direction of longitude is changed. I start the continuation with $a_J = 5.20282$ AU, $\ell_J = 0°$, and with $\varpi = 305°.9$, $\Omega = 162°.6$, $i = 13°.3$, $a = 3.254$, $e = 0.092$, $\ell = 15°.4$. $\sigma$ librates in the continuation with an amplitude that changes between about 40° and 78°. Fig. 4 refers to a part of this continuation and indicates a 1/1 ratio of the mean period of revolution of $2\omega$ and of the period of libration that causes the sequence of maxima of $e$. A libration of the direction from the origin to subsequent maxima of $e$ about the zero direction of $2\omega$ is visible. During a cycle of this libration of direction large mean values of $e$ occur together with a large amplitude of libration of $\sigma$, but the inclination is comparatively small at the respective phase of this cycle.

## References

Froeschlé, C. and Scholl, H.: 1976, 'On the Dynamical Topology of the Kirkwood Gaps', *Astron. Astrophys.*, **48**, 389–393

Lemaître, A. and Henrard, J.: 1990, 'On the Origin of Chaotic Behaviour in the 2/1 Kirkwood Gap', Icarus **83**, 391–409

Morbidelli, A. and Giorgilli, A.: 1990, 'On the Dynamics in the Asteroids Belt. Part II: Detailed Study of the Main Resonances', *Celest. Mech.*, **47**, 173–204

Murray, C.D.: 1986, 'Structure of the 2:1 and 3:2 Jovian Resonances', *Icarus*, **65**, 70–82

Schubart, J.: 1988, 'Resonant Asteroids between the Main Belt and Jupiter's Orbit', *Celest. Mech.*, **43**, 309–317

Schubart, J.: 1990, 'The Low-Eccentricity Gap at the Hilda Group of Asteroids', *Asteroids, Comets, Meteors III* , eds.: C.-I. Lagerkvist, H. Rickman, B.A. Lindblad, M. Lindgren. Uppsala University, 171–174

Schubart, J.: 1991, 'Additional Results on Orbits of Hilda-Type Asteroids',*Astron. Astrophys.*, **241**, 297–302

Schubart, J. and Bien, R.: 1984, 'An Application of Labrouste's Method to Quasi-Periodic Asteroidal Motion', *Celest. Mech.*, **34**, 443–452

Schubart, J. and Bien, R.: 1986, 'On the Computation of Characteristic Orbital Elements for the Trojan Group of Asteroids', *Asteroids, Comets, Meteors II*, eds.: C.-I. Lagerkvist, B.A. Lindblad, H. Lundstedt, H. Rickman. Uppsala University, 153–156

Schubart, J. and Stumpff, P.: 1966, 'On an N-Body Program of High Accuracy for the Computation of Ephemerides of Minor Planets and Comets', *Veröffentl. Astron. Rechen-Inst. Heidelberg*, No. 18, Verlag G. Braun, Karlsruhe, 1–31

Wisdom, J.: 1987, 'Urey Prize Lecture: Chaotic Dynamics in the Solar System', *Icarus*, **72**, 241–275

JOACHIM SCHUBART

Yoshikawa, M.: 1991, 'Motions of Asteroids at the Kirkwood Gaps II. On the 5:2, 7:3, and 2:1 Resonances with Jupiter', *Icarus*, **92**, 94–117

# NUMERICAL EXPERIMENTS IN THE 3/1 AND $\nu_6$

## OVERLAPPING RESONANCE REGION

CH.FROESCHLÉ and H. SCHOLL

*O.C.A. Laboratoire G.D. Cassini, CNRS URA 1362 B.P. 229 F-06304 Nice Cedex 4,*
*France*

**Abstract.** Numerical experiments of fictitious small bodies with initial eccentricities e=0.1 have been performed in the overlapping region of the 3/1 mean motion resonance and of the $\nu_6$ secular resonance $2.48 \leq a \leq 2.52$AU for different values of the initial inclination $16° \leq i \leq 20°$. An analysis for the $\nu_6$ secular resonance shows that the topology is different from the one found outside the overlapping region: the critical argument for the $\nu_6$ resonance in the overlapping region rotates in opposite direction as compared to the pure $\nu_6$ region. In the 3/1 resonance region the secular resonance $\nu_5$ is dominant, and some secondary secular resonances as $\nu_6 - \nu_{16}$ and $\nu_5 + \nu_6$ are present.

**Key words:** Mean motion and secular resonances, overlapping resonances.

## 1. Introduction

The location of secular resonances close to a mean motion resonance have not been yet determined, except near the 2/1 mean motion resonance, for which Morbidelli (1991 b) has developed a theory based on successive elimination of harmonics. In order to find the deviation of the secular resonance $\nu_6$ near the 3/1 mean commensurability we integrate numerically over 1 Myr orbits of fictitious small bodies in the range $2.48 \leq a \leq 2.52$ AU.

## 2. Numerical Experiments in the Region 2.48 - 2.52 AU

We integrate numerically in the frame of the Sun - Jupiter -Saturn model 25 orbits of alleged fragments.

We used the DVDQ code of Krogh (1970). This integrator is a predictor - corrector scheme of Adam's type, with variable stepsize and variable order. The orbits were integrated over $10^6$ yr. The starting eccentricities of all objects were e = 0.1. All the starting values for the secular argument $\varpi - \varpi_S$ were put equal to 0°, while those of the mean resonant argument $\sigma = (p + q)\lambda_J - p\lambda - q\varpi$ were put equal to 180° ($\lambda_J$ and $\lambda$ are respectively the mean longitude of Jupiter and the alleged body, $\varpi$ is the longitude of perihelion of the alleged body, p=1 ,q=2). Five different starting semimajor axes between a = 2.48 - 2.52 AU ($\Delta a$ =0.01) and five starting inclinations between i= 16° − 20° ($\Delta i = 1°$) were chosen which gives a total of 25 orbits. The results are summarized in tables 1 to 5.

No orbit with starting semimajor axis a=2.48 AU is located in the 3/1 mean motion resonance. Only one body with starting inclination $i = 20°$ is temporarily located in the secular resonance $\nu_6$, the secular resonant argument $\varpi - \varpi_S$ librates about 0° until $6 * 10^5$ years, then circulates (Fig. 1). The circulation is counterclockwise.For all the other 4 orbits the circulation of $\varpi - \varpi_S$ is counterclockwise.

*Celestial Mechanics and Dynamical Astronomy* **56**: 163–176, 1993.

TABLE I
Bodies with initial conditions $e_{\text{starting}} = 0.1$, $a_{\text{starting}} = 2.48 AU$

| Body Number | $i_{\text{starting}}$ ecliptic 1950.0 | $\nu_6$ resonance | 3/1 resonance | $\nu_5$ resonance | s s r | $e_{\text{max}}$ | $i_{\text{max}}$ |
|---|---|---|---|---|---|---|---|
| 3 | 16° | no | no | no | no | 0.26 | 19° |
| 8 | 17° | no | no | no | no | 0.28 | 20° |
| 13 | 18° | no | no | no | no | 0.30 | 22° |
| 18 | 19° | no | no | no | no | 0.34 | 26° |
| 23 | 20° | libr/circ | no | no | no | 0.36 | 28° |

TABLE II
Bodies with initial conditions $e_{\text{starting}} = 0.1$, $a_{\text{starting}} = 2.49 AU$

| Body Number | $i_{\text{starting}}$ ecliptic 1950.0 | $\nu_6$ resonance | 3/1 resonance | $\nu_5$ resonance | s s r | $e_{\text{max}}$ | $i_{\text{max}}$ |
|---|---|---|---|---|---|---|---|
| 4 | 16° | temp. | temp. | temp. | temp. | 0.55 | 28° |
| 9 | 17° | temp. | temp. | temp. | temp. | 0.62 | 32° |
| 14 | 18° | temp. | temp. | temp. | temp. | 0.58 | 28° |
| 19 | 19° | temp. | always | temp. | temp. | 0.62 | 28° |
| 24 | 20° | temp. | temp. | temp. | temp. | 0.48 | 26° |

TABLE III
Bodies with initial conditions $e_{\text{starting}} = 0.1$, $a_{\text{starting}} = 2.50 AU$

| Body Number | $i_{\text{starting}}$ ecliptic 1950.0 | $\nu_6$ resonance | 3/1 resonance | $\nu_5$ resonance | s s r | $e_{\text{max}}$ | $i_{\text{max}}$ |
|---|---|---|---|---|---|---|---|
| 5 | 16° | no | temp. | temp. | temp. | 0.90 | 38° |
| 10 | 17° | no | temp. | temp. | temp. | 0.44 | 26° |
| 15 | 18° | temp. | temp. | temp. | temp. | 0.62 | 26° |
| 20 | 19° | no | temp. | temp. | temp. | 0.55 | 27° |
| 25 | 20° | temp. | temp. | temp. | temp. | 0.46 | 26° |

TABLE IV
Bodies with initial conditions $e_{starting} = 0.1$, $a_{starting} = 2.50 AU$

| Body Number | $i_{starting}$ ecliptic 1950.0 | $\nu_6$ resonance | 3/1 resonance $t$ in $10^5 yrs$ | $\nu_5$ resonance | s s r | $e_{max}$ | $i_{max}$ |
|---|---|---|---|---|---|---|---|
| 6 | 16° | no | temp. | temp. | temp. | 0.46 | 25° |
| 11 | 17° | temp. | temp. | temp. | temp. | 0.52 | 28° |
| 16 | 18° | no | temp. | temp. | temp. | 0.55 | 24° |
| 21 | 19° | no | temp. | temp. | temp. | 0.45 | 27° |
| 26 | 20° | temp. | $t \leq 7$ | temp. | temp. | 0.65 | 31° |

TABLE V
Bodies with initial conditions $e_{starting} = 0.1$, $a_{starting} = 2.50 AU$

| Body Number | $i_{starting}$ ecliptic 1950.0 | $\nu_6$ resonance $t$ in $10^5 y.$ | 3/1 resonance $t$ in $10^5 y.$ | $\nu_5$ resonance $t$ in $10^5 y.$ | s s r $t$ in $10^5 y.$ | $e_{max}$ | $i_{max}$ |
|---|---|---|---|---|---|---|---|
| 7 | 16° | no | no | no | no | 0.26 | 20° |
| 12 | 17° | no | no | no | no | 0.28 | 21° |
| 17 | 18° | no | no | no | no | 0.30 | 24° |
| 22 | 19° | no | no | no | no | 0.34 | 25° |
| 27 | 20° | $8 > t > 7$ | $t > 6$ | $t > 8$ | $t > 8$ | 0.98 | 72° |

Four orbits with $a_{starting} = 2.49$ AU have a resonant argument $\sigma$ which alternatively librates and circulates (Fig. 2a). Body 19 is during 1 Myr in the 3/1 mean motion resonance with $\sigma$ librating around 180°. The behaviour of the secular argument $\varpi - \varpi_S$ is complex (Fig. 2b). After fast clockwise circulations, it librates about 0° (1 libration) then circulates counterclockwise, again librates and circulates. When the bodies are in the 3/1 mean motion resonance the secular resonance $\nu_5$ is always dominant with the resonance argument $\varpi - \varpi_J$ librating about 180°, as in figure 3; the bodies are also temporarily in the secondary secular resonances $\nu_6$ - $\nu_{16}$ and $\nu_5 + \nu_6$. As it is well known the eccentricities reach large values, $0.48 \leq e_{max} \leq 0.62$ while the range of the maximum values of inclinations are between $25° \leq i \leq 32°$. In the following text we refer to secondary secular resonances by ssr.

The 5 orbits with starting a = 2.50 AU are at the center of the 3/1 resonance. The critical argument $\sigma$ circulates and librates about 180° alternatively, or is always librating about 180°. Only two bodies with starting inclination of 18° and 20° respectively are found temporarily in the secular resonance $\nu_6$, the other bodies

CH.FROESCHLÉ AND H. SCHOLL

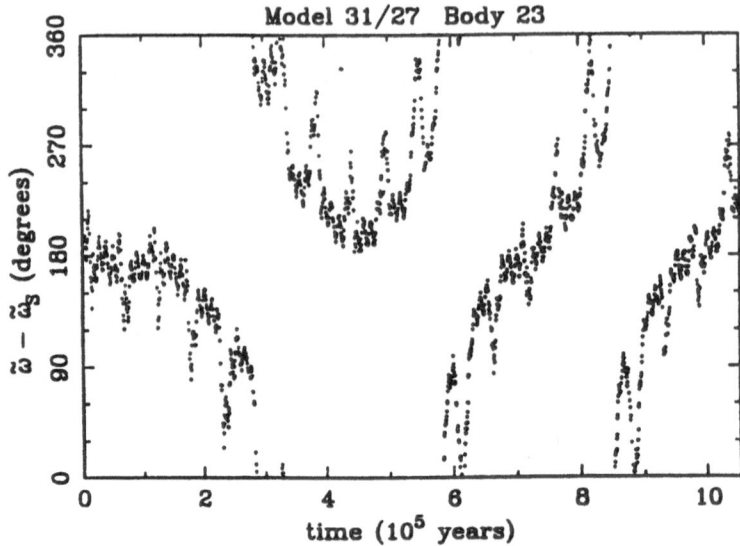

Fig. 1. The $\nu_6$ secular resonance argument $\varpi - \varpi_S$ versus time of Body 23 (starting values $i = 20°$, $e = 0.1$, $a = 2.48\ AU$).

have a secular resonant argument $\varpi - \varpi_S$ which circulates clockwise. Again when the critical argument $\sigma$ is librating about 180° the $\nu_5$ resonant argument librates around 180°, and the ssr are present. Body 5 ($i_{starting} = 16°$) has an eccentricity which reaches 0.9 and a maximum inclination of 38° when it is located in the 3/1 resonance.

We observe the same behaviour of the critical argument $\sigma$ i.e. circulation and libration around 180° alternatively, for the five orbits at starting semimajor axis a=2.51 AU. Two bodies (for instance body 11 with $i_{starting} = 17°$) are located in the secular resonance $\nu_6$. Between $4*10^5 \le t \le 7*10^5$ (Fig. 4) the resonant argument $\varpi - \varpi_S$ librates about 0° otherwise it circulates clockwise. During this period $\sigma$ circulates. Fig.5a shows that until t= $7*10^5$ the body ($i_{starting} = 20°$) is located in the 3/1 mean commensurability, $\sigma$ librates about 180°. It enters the secular resonance $\nu_6$ near $t = 3*10^5$ years (Fig. 5b), the resonant argument $\varpi - \varpi_S$ shows librations with large amplitude alternating with one counterclockwise circulation and two clockwise circulations, then after a last libration, the body leaves the resonance and the secular argument circulates counterclockwise. Notice that the peaks of the eccentricity up to 0.65 (Fig. 6) occur when the secular resonance argument passes from circulation to libration or vice versa. The secular resonance $\nu_5$ is also found when the orbits are in the 3/1 mean motion resonance, and the ssr appear temporarily.

We don't find mean or secular resonances at starting semi major axis a=2.52, and initial inclinations between $16° \le i \le 19°$. The body with $i_{starting} = 20°$ enters the 3/1 resonance near $t_r = 6 \times 10^5$ yrs (Fig. 7a). It is also temporarily in the secular resonance $\nu_6$ (Fig. 7b), (1 libration around 0° then clockwise circulations

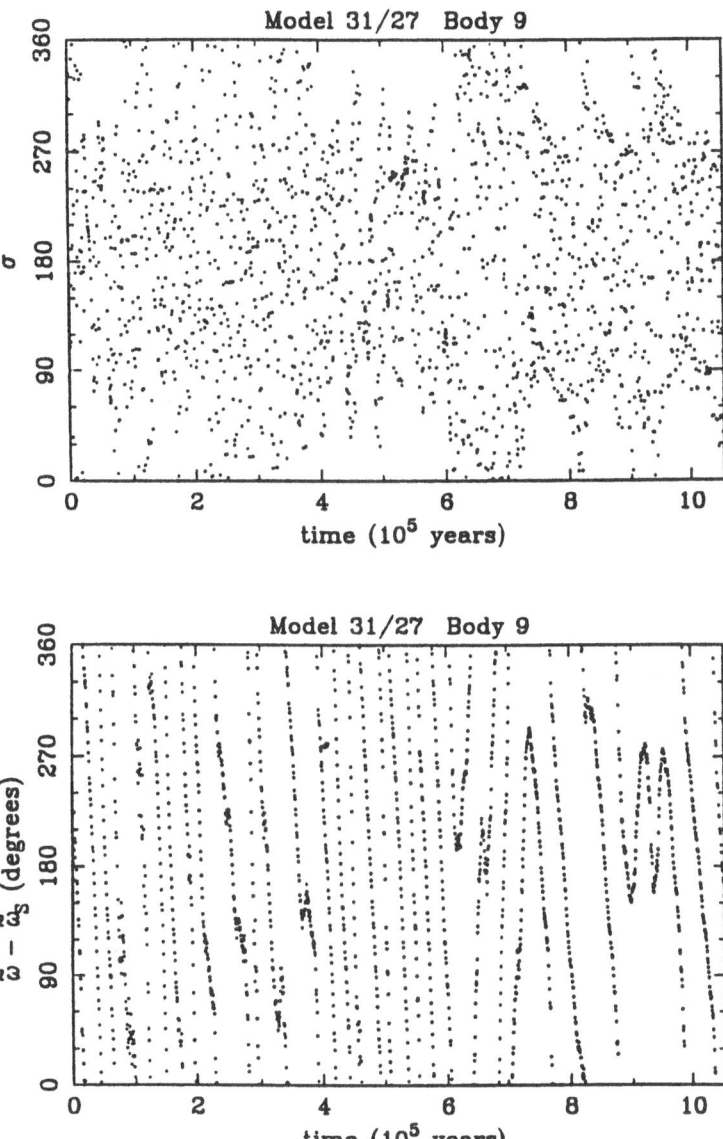

Fig. 2. (a) Time evolution of the critical argument $\sigma$ of Body 9 (starting values $i = 17°$, $e = 0.1$, $a = 2.49 AU$). (b) Time evolution of the $\nu_6$ secular resonance argument $\varpi - \varpi_S$ of the same body.

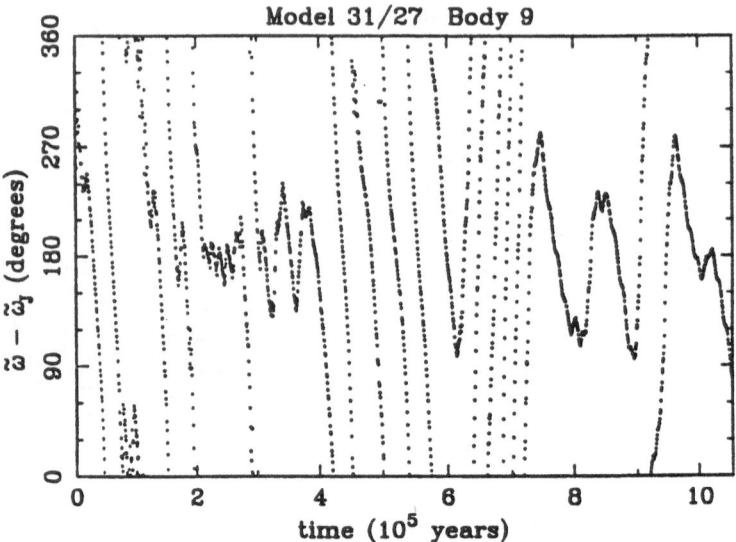

Fig. 3. Time evolution of the $\nu_5$ secular resonance argument $\varpi - \varpi_J$ of Body 9.

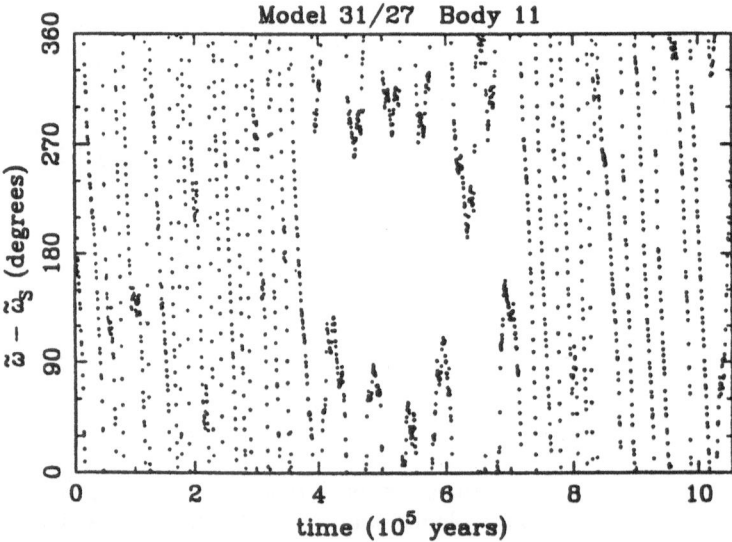

Fig. 4. Time evolution of the $\nu_6$ secular resonance argument $\varpi - \varpi_S$ of Body 11 (starting values $i = 17°$, $e = 0.1$, $a = 2.51 AU$).

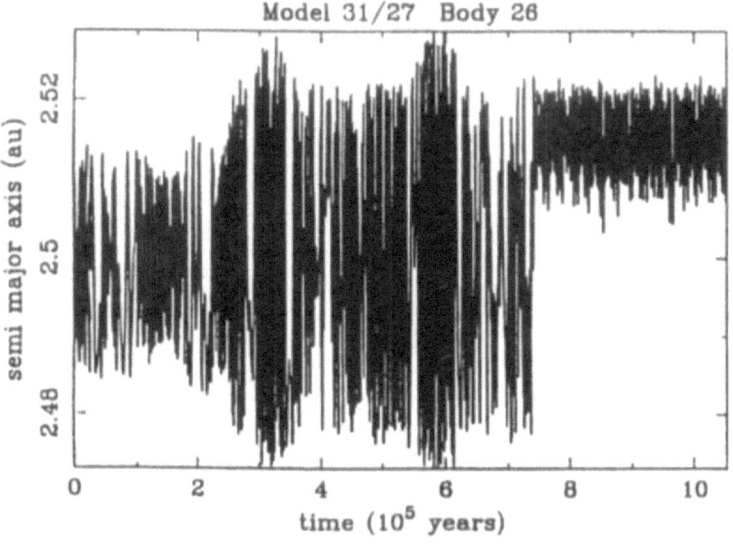

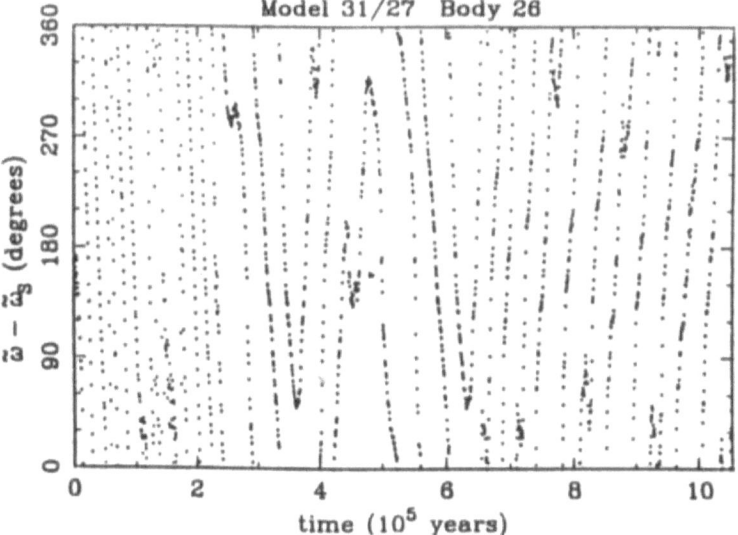

Fig. 5. (a) Time evolution of the semimajor axis of Body 26 (starting values $i = 20°$, $e = 0.1$, $a = 2.51 AU$). (b) Time evolution of the $\nu_6$ secular resonance argument $\varpi - \varpi_S$ of the same body.

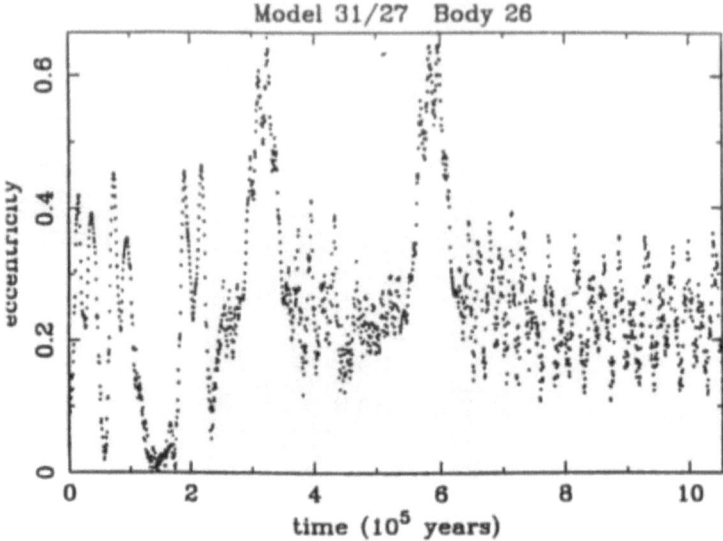

Fig. 6. Time evolution of the eccentricity of Body 26.

while before entering the resonance the circulation was counterclockwise ). At $t \geq 8 \times 10^5$ yrs the resonance argument $\varpi - \varpi_J$ circulates very slowly and begins to librate about 0° (Fig. 8a); the argument of the ssr $\nu_5 + \nu_6$ librates temporarily (Fig. 8b). The eccentricity (Fig. 9a) reaches a maximum of 0.98 and the maximum value of the inclination is about 72° (Fig. 9b).

## 3.  Topological Analysis of the Numerical Integrations

Using the well known "second fundamental model of resonance" (Henrard and Lemaître 1983), Morbidelli (1991 a) provided the topological aspects of a numerical simulation. Following the method of Morbidelli we consider the different characteristics of the integrated orbits, in order to understand the nature of the orbits and the topological structure of the phase space around them.

We have to look at:

1. direction of circulation/libration (clockwise or counterclockwise, as introduced below)

2. maximal and minimal value of the action variable (in our case the eccentricity)

3. ratio between the variation of the resonant angle here $\varpi - \varpi_S$, (in case of libration) and the variation of the action

4. the shape of the orbit

5. modification of the orbit due to the change of one dynamical parameter

6. alternation between libration and circulation due to a separatrix crossing

7. evolution of the critical angle with respect to time.

First of all, analyzing the previous results of Yoshikawa (1987) at a=2.2 AU (bodies A, C, E), the topology of the secular resonance $\nu_6$ is given by model 1 (Fig.

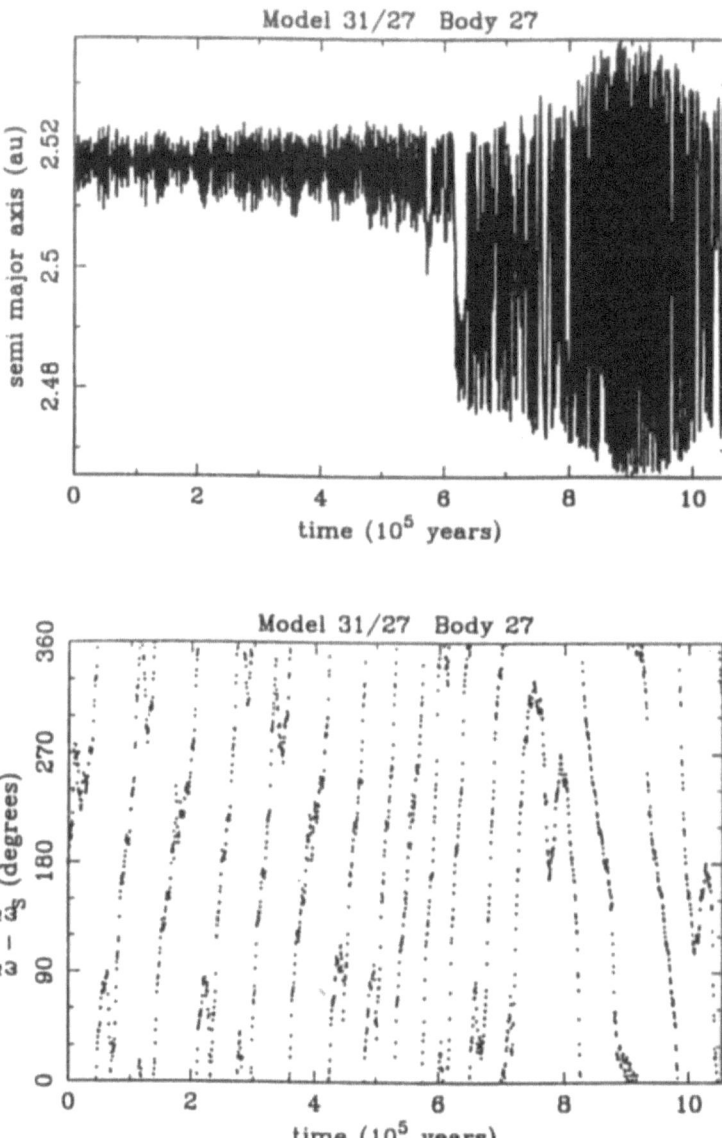

Fig. 7. (a) Time evolution of the semimajor axis of Body 27 (starting values $i = 20°$, $e = 0.1$, $a = 2.52$ AU). (b) Time evolution of the $\nu_6$ secular resonance argument $\varpi - \varpi_S$ of the same body.

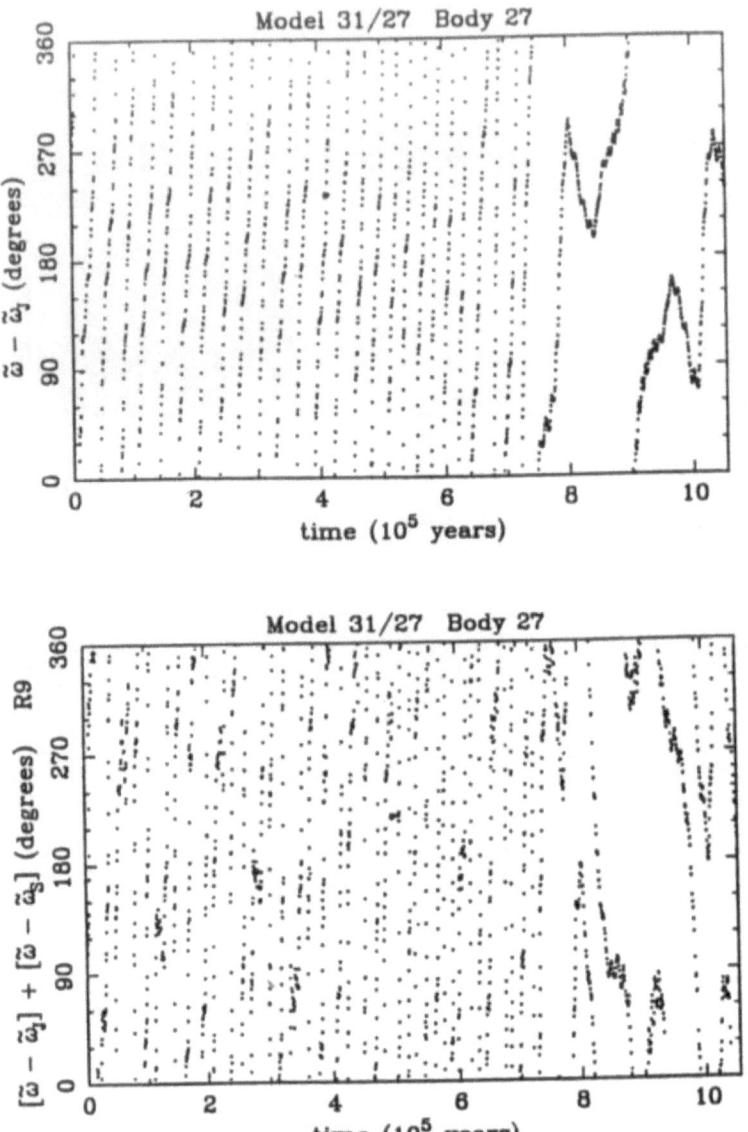

Fig. 8. (a) Time evolution of the $\nu_5$ secular resonance argument $\varpi - \varpi_J$ of Body 27. (b) Time evolution of the secondary secular resonance argument $\nu_5 + \nu_6$ of the same body.

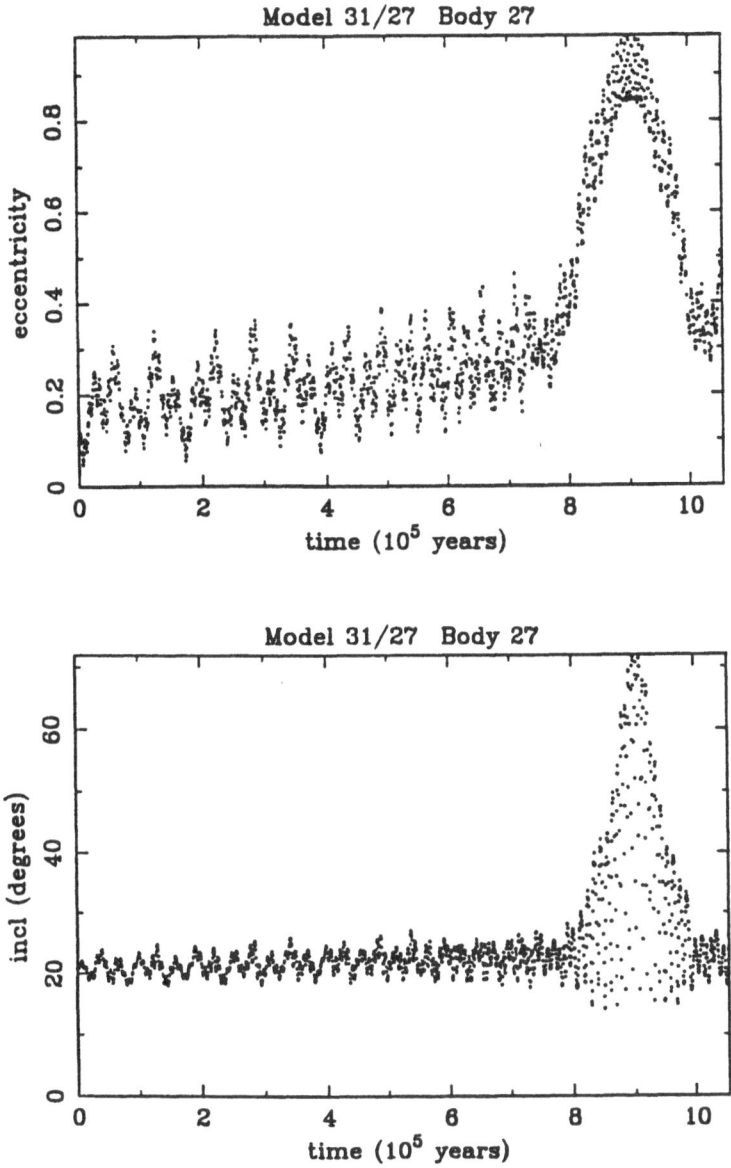

Fig. 9. (a) Time evolution of the eccentricity of Body 27. (b) Time evolution of the inclination of the same body.

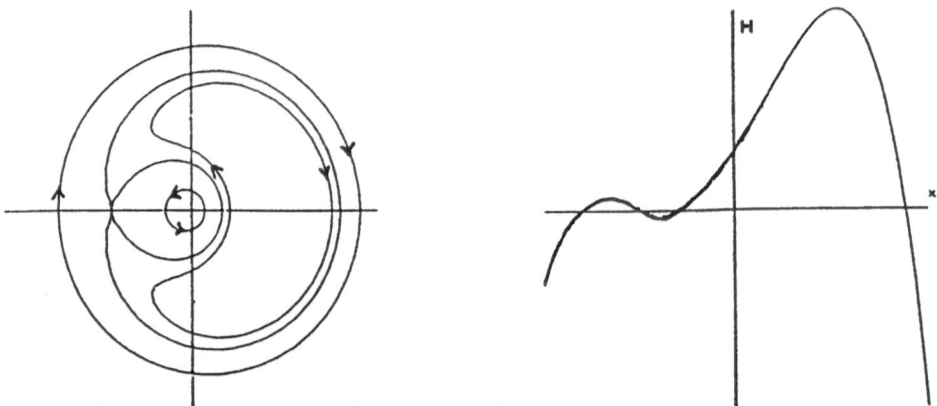

Fig. 10. Topological structure of the dynamics at the borders of the 3/1 mean motion resonance, $(2.2 \leq a \leq 2.46$ AU, and a=2.6 AU), in polar coordinates e, $\varpi - \varpi_S$, and the global shape of the energy H as a function of $x = e \cos(\varpi_S - \varpi)$.

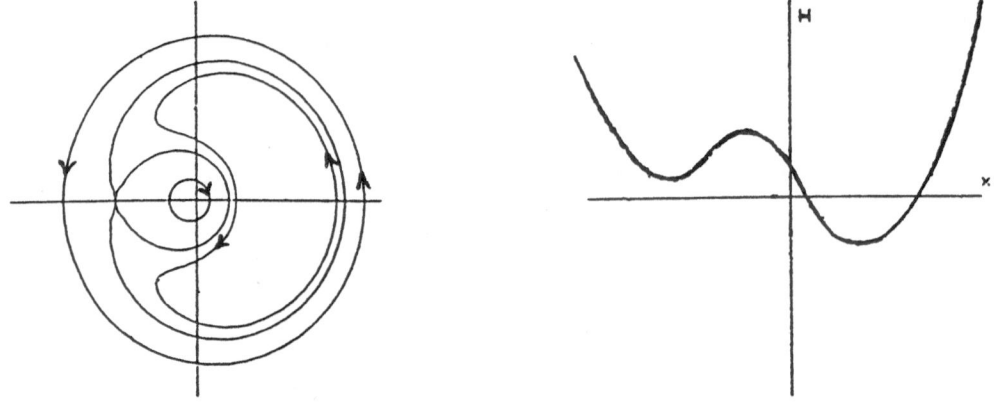

Fig. 11. Same as Fig. 10 for values of semi major axis $2.48 \leq a \leq 2.52$ AU.

10) , where the orbits are plotted in polar diagrams, $e \cos(\varpi - \varpi_S)$, $e \sin(\varpi - \varpi_S)$.

The direction of the inner circulators is counterclockwise, and clockwise for the librators and the outer circulators.

The inner circulators turn around a stable point which is translated on the left of the origin; the eccentricity has a greater value for $\varpi - \varpi_S = 180°$, than at $\varpi - \varpi_S = 0°$.

The same topological model is found by numerical simulations (Gonczi et al. private communication) at a=2.3, 2.42, 2.46 AU.

On the other side of the 3/1 mean commensurability, the numerical integrations of Yoshikawa (1987) at a=2.6AU have also a topology illustrated by model 1. Body H ($i_{starting} = 20.°3$) and Body I ($i_{starting} = 21.°3$) are inner circulators, the direction is counterclockwise and the eccentricity is maximum for $\varpi - \varpi_S = 180°$. Body J ($i_{starting} = 21.°37$) is a temporary librator, then becomes an outer circulator, in both cases the direction is clockwise. The resonant argument $\varpi - \varpi_S$ of Body M also at a=2.6 and starting inclination $i = 22.°3$ circulates clockwise, the eccentricity is maximum for $\varpi - \varpi_S = 0°$ and minimum for $\varpi - \varpi_S = 180°$. Numerical experiments carried out by Morbidelli and Henrard (1991) at a= 2.6AU with initial e=0.35 and with initial i= 26.° and i=25.5°, show that the orbits librate around an equilibrium point placed at $\varpi - \varpi_S = 0°$, the libration is clockwise.

Then we conclude that on the two sides of the mean motion resonance 3/1 up to 2.46 AU on one side and at a $\geq$ 2.6 AU on the other side, the essential features of the dynamics are the same.

We consider now our computational results. At 2.48 AU, (bodies with starting inclinations $i \leq 19.°$) the secular argument $\varpi - \varpi_S$ circulates counterclockwise, the eccentricities are minimum at $\varpi - \varpi_S = 180°$ and maximum at $\varpi - \varpi_S = 0°$. Considering body 23 (a=2.48 AU, i=20.°) it librates about 0°, the libration is counterclockwise, the eccentricity varies from 0.09 to 0.36 then the body circulates counterclockwise with a minimum of the eccentricity e at $\varpi - \varpi_S = 180°$ (Fig. 1).

Between 2.49-2.51 AU the behaviour of the $\nu_6$ secular argument is -as for example (Fig. 5b) - very complex. First the circulation is clockwise, then it librates around 0° counterclockwise, after one circulation counterclockwise, librates again around 0°, then after two circulations clockwise librates and finally leaves the resonance and circulates counterclockwise. During the clockwise circulations the eccentricity is maximum at $\varpi - \varpi_S = 180°$ and minimum at $\varpi - \varpi_S = 0°$, while for the circulations counterclockwise it is the converse i.e. e is maximum at $\varpi - \varpi_S = 0°$ and minimum at $\varpi - \varpi_S = 180°$, (Fig. 6). At 2.52 AU the four first bodies have a secular resonant argument $\varpi - \varpi_S$ which circulates counterclockwise ,with the same variation of the eccentricity described previously. It seems that between 2.48-2.52 AU the dynamical model is illustrated by model 2 see Fig.11.

## 4. Conclusion

Our computations lead to the following conclusions:

1. For a starting eccentricity of 0.1 we found only one body at 2.48 AU and initial inclination $i = 20°$ located in the secular resonance $\nu_6$ during a period of $6 * 10^5$ yrs. The other bodies labeled "temp." exhibit a very complex behaviour of the secular argument $\varpi - \varpi_S$ i.e. few librations alterning with circulations either clockwise or counterclockwise. We conjecture that the secular resonance $\nu_6$ is deviated at larger inclinations when approaching the 3/1 mean motion resonance. The same behaviour was found by Morbidelli (1991 b) near the 2/1 resonance.

2. In most cases when the orbit is located in the 3/1 resonance, the secular

argument $\varpi - \varpi_J$ librates about 180° or 0°.

3. When the bodies are temporarily or always located in the 3/1 resonance, they are also temporary located in the ssr $\nu_5 + \nu_6$ and $\nu_6 - \nu_{16}$.

4. In the range 2.48 – 2.52 AU the dynamics is different from the one found outside this region. The inner circulation of the $\nu_6$ resonance argument is clockwise, while the libration and outer circulation are counterclockwise. These directions are reversed outside the region 2.48 – 2.52 AU region. The dynamics in this region is very complex, more computational and theoretical studies have to be performed.

## References

Henrard, J. and Lemaître, A.: 1983, 'A second fundamental model for resonance', *Celest. Mech.*, **30**, 197–218.

Krogh, F. T.:1970, *JPL Technical Utilization Document*, **No. CP-38**.

Morbidelli, A. (1991a) 'Topological methods for the qualitative analysis of a numerical simulation close to a resonance' in *Interrelations entre la Physique et la Dynamique des Petits Corps du Système Solaire*, Goutelas 1991, D.Benest and Cl.Froeschlé eds. , in press.

Morbidelli, A. (1991b) ' Perturbations methods and asteroids dynamics ', *PhD Thesis University of Namur* ,205–216.

Morbidelli, A. and Henrard, J. (1991) 'The main secular resonances $\nu_6$, $\nu_5$ and $\nu_{16}$ in the asteroid belt', *Celest. Mech.*, **51**, 169–197.

Yoshikawa, M.: 1987, ' A simple analytical model for the $\nu_6$ resonance.',*Celest. Mech.*, **40**, 233–272.

# AN INTRODUCTION TO HAMILTONIAN DYNAMICAL SYSTEMS AND PRACTICAL PERTURBATION METHODS: NEW INSIGHT BY SUCCESSIVE ELIMINATION OF PERTURBATION HARMONICS

ALESSANDRO MORBIDELLI
*Département de mathématique FUNDP*
*8, Rempart de la Vierge, B-5000 Namur, Belgique*

## 1. Introduction

Among the studies on dynamical systems one can generally recognize two different approaches: a theoretical approach, where the general properties of such systems are investigated, along the lines opened by the KAM and the Nekhoroshev theorems, and a more applicative approach, where practical perturbation techniques are developed and applied in order to achieve a satisfactory description of some given specific systems of interest. This last approach is the most common one in Celestial Mechanics.

Evidently with the improvement in the knowledge on dynamical systems, both approaches evolve along parallel lines. Nevertheless, one can still notice a sort of gap between theoretical and computational works.

The purpose of this lecture is to try to fill up partially this gap, by presenting a review on general properties of dynamical systems, practical perturbation methods and their connections. A particular care will be paid to the technique of successive elimination of harmonics, originally conceived by Delaunay (1867) and further developed by the author from both the algorithmic and the theoretical points of view. We will show how, by applying this technique, one follows in practice a piece of the road that leads both to the KAM theorem and to a Nekhoroshev-like result.

This paper is organized in three sections. First the KAM theorem and the Nekhoroshev theorem will be recalled, stressing the different approaches which characterize these two results, also in comparison with the approach that one usually follows in Celestial Mechanics. Furthermore we will analyze two possible implementations of the Lie algorithm by both the use of series expansions and the use of suitable action-angle variables previously introduced in a numerical way; this will lead to present the method of successive elimination of perturbation harmonics as a natural improvement. Finally the theoretical implications of the method of successive elimination of harmonics will be outlined, thus providing a bridge among KAM approach, Nekhoroshev approach and computational algorithm.

## 2. The Dynamical Problem

In the wide class of Hamiltonian dynamical systems, a particularly important role is played by the subset of the so called *integrable* systems. Roughly speaking, these

*Celestial Mechanics and Dynamical Astronomy* **56**: 177–190, 1993.

are the systems such that the solutions of the corresponding differential equations can be found in an analytical way; therefore one can describe the evolution of the orbits for all times, and control the propagation of the error eventually committed in the determination of the initial conditions. More precisely, the integrable systems are those which admit the form $H(p,q) \equiv H_0(p)$ in some suitable canonical variables $p, q$ (Arnold–Liouville theorem; see Arnold, 1963a). In these variables, the actions $p$ are constant of motion (since $\dot{p} = \partial H/\partial q = 0$), while the angles $q$ are linear functions of time. Therefore the phase space is foliated into invariant tori carrying either periodic or quasi periodic motion, depending whether all the frequencies of the angles are resonant or not (see also Giorgilli, 1989).

Historically, it was a general belief that all dynamical systems are integrable, and the problem was thought to be just how to compute complicate solutions. It is only with the work of Poincaré (1892), that the general non–integrability of Hamiltonian systems was proved. The result by Poincaré opened the way to a new rich field in the mathematical research: the investigation of the properties of non–integrable systems.

The general problem of the dynamics can therefore be stated as follows: given an integrable system $H_0(p)$, add a generic perturbation, thus obtaining the Hamiltonian

$$H(p,q) = H_0(p) + \epsilon H_1(p,q)$$

with $\epsilon$ small parameter; what will be the new dynamics?

In particular, let us consider a given invariant torus for the integrable dynamics described by $H_0$. The addition of the perturbation may have one of the two following effects:

1) the torus is deformed by the perturbation, but is still invariant for the dynamics given by $H$. In this case the motion on it is quasi periodic, namely the $p$–variables can be expressed as periodic functions of the $q$–variables, and the latter move with fixed proper frequencies (generally non-resonant ones).

2) the torus is destroyed by the perturbation. This can give origin to irregular motion, where neither any correlation between $p$ variables and $q$ variables is possible, nor the definitions of a full set of $n$ independent local constants of motion ($n$ being the number of degrees of freedom).

Figure 1 shows an example of perturbation of an integrable pendulum, where both the cases outlined above occur, in different regions of the phase space. Therefore the problem becomes to understand which are the conditions which lead to the existence of invariant tori or to irregular motion, and the properties of the latter.

In the following we shall briefly analyze different approaches for the investigation of this dynamical problem.

## 2.1. THE KAM APPROACH

The Kolmogorov–Arnold–Moser (Kolmogorov, 1954; Moser, 1962; Arnold, 1963b) theorem concerns the research of invariant tori. The theorem states that, if the size

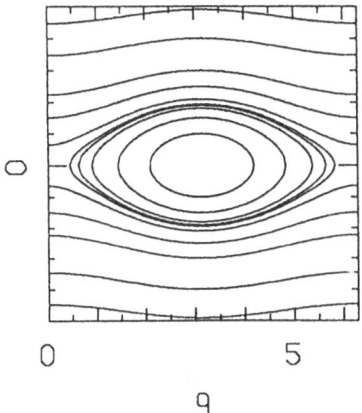 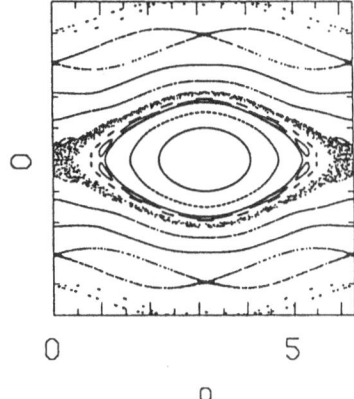

0                    5              0                    5
q                                  q

Fig. 1. On the left side the integrable phase space of a pendulum; on the right side the Poincaré section of the phase space of a perturbed pendulum.

of the perturbation $\epsilon$ is sufficiently small, some invariant tori persist in the phase space of the perturbed Hamiltonian $H$. Moreover, it is always possible to find a local set of new action–angle variables $J, \psi$ such that the $J$–variables are constants of motion on the torus and the angles $\psi$ are linear functions of time. Therefore one can describe exactly the motion on the invariant tori for all times, up to infinity.

The necessary condition to be an invariant torus of the perturbed dynamics is to carry non resonant motion. As a matter of fact, only the strongly non-resonant tori survive in the perturbed phase space, the others being destroyed by the nearby resonances. Since the resonances, which correspond to rational numbers, are dense, the set of the invariant tori is not open (and neither dense), and has the topological structure of a Cantor set. The KAM theorem, nevertheless, states that the volume of this set is large; this means that a lot of invariant tori are present in the phase space. Conversely, the volume taken by a resonance of order $k$ is exponentially small with $k$. This is the reason why, although dense in the phase space, most of the resonances are not visible on an ordinary picture such as Fig. 1b, and entire portions of the phase space look like regular and filled up by invariant tori.

For what concerns the orbits of the perturbed Hamiltonian $H$ which do not lay on an invariant torus, the KAM theorem does not provide any information; these orbits may have a chaotic behaviour. In the case of a two degrees of freedom system, the invariant KAM–tori are insuperable barriers for the motion, so that a chaotic orbit with initial condition in between two KAM–tori will never diffuse in the other regions of the phase space. On the contrary, if the system has more than two degrees of freedom, some orbits which are not on an invariant torus may, in principle, diffuse everywhere. This possibility has been first conjectured by Arnold himself (Arnold, 1964), and is generally called as *Arnold diffusion*.

## 2.2. The Nekhoroshev Approach

The possibility of existence of Arnold diffusion, left open by the KAM theorem, points out the necessity to follow a different approach in order to get results valid all over the phase space, or, at least, on an open subset. This is indeed the work first carried on by Nekhoroshev (1971, 1977, 1979). His revolutionary idea is to look for results which hold over a **finite** time, although a very long one. More precisely, the Nekhoroshev theorem states that, given the Hamiltonian $H(p,q) = H_0(p) + \epsilon H_1(p,q)$ (with a few additional hypotheses on the analyticity of $H$ and the convexity of $H_0$), the variations of the actions $p$, i.e. $|p(t) - p(0)|$, is bounded by $C\epsilon^b$ for all times $t$ such that $|t| < T \equiv A\exp\left(\epsilon_*/\epsilon\right)^\alpha$ , with positive constants $C, b, A, \epsilon_*$ and $\alpha$. In other words, Arnold diffusion may exist, but it is exponentially slow with $-(\epsilon_*/\epsilon)^\alpha$.

The power of Nekhoroshev theorem is the exponential character of its result. The time of diffusion $T$ grows so fastly with $1/\epsilon$ that it may easily exceed the characteristic life time of the system in study (the age of the Solar System in Celestial Mechanics), thus proving the practical stability of the whole system. In these cases the Nekhoroshev theorem can be read as: "the dynamics of the perturbed system is close to that of the unperturbed one for all times of interest".

On the other hand, the Nekhoroshev theorem does not provide a detailed description of the motion. Chaotic layers may well exist, due to resonance overlapping.

## 2.3. The Celestial Mechanics Approach

Neither the KAM theorem, nor the Nekhoroshev theorem are completely satisfactory for the purposes of Celestial Mechanics. Celestial Mechanics was born as the science for the computation of ephemerides, and still looks for a description of the motion of the bodies with the best possible accuracy. Therefore, a result like Nekhoroshev's one, which concerns only the stability of the system within a neighbourhood of the unperturbed one, is not sufficient alone. On the other hand, also the KAM theorem is not very useful in its original formulation. Imagine to have proved, for example, that the Earth–Moon–Sun system lays on an invariant KAM–torus; then, it would be sufficient to add the perturbation given by a small comet to destroy completely this result! The fact that the KAM theorem holds on a non-open set makes the results completely incompatible with the approximations in the definition of the model and of the initial conditions.

The classical approach in Celestial Mechanics is somewhat in between the KAM and the Nekhoroshev approach. Given the system $H(p,q) = H_0(p) + \epsilon H_1(p,q)$, one looks for a canonical transformation $p,q \rightarrow p^{(r)}, q^{(r)}$ in order to transform it into

$$H^{(r)}(p^{(r)}, q^{(r)}) = H_0^{(r)}(p^{(r)}) + \epsilon^r H_1^{(r)}(p^{(r)}, q^{(r)}) .$$

This is a more advantageous form than the original one: indeed the dynamics given by $H_0^{(r)}$ can be described in details, since $H_0^{(r)}$ is integrable and its phase space is

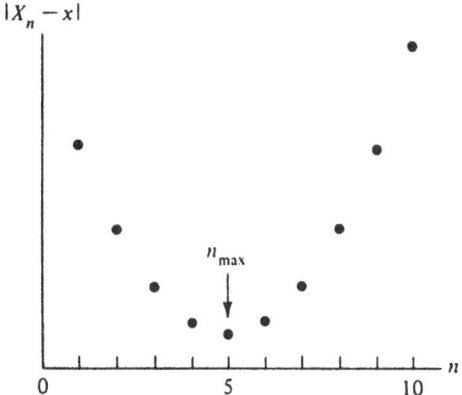

Fig. 2. divergence of asymptotic series for a fixed $\epsilon$. The size of the normalized perturbation first decreases with the order of normalization $n$, then diverges. $n_{max}$ is therefore the optimal order of normalization. From Lichtenberg and Lieberman, 1983

foliated into invariant tori; moreover the new perturbation is much smaller than the original one (being roughly of order $\epsilon^r$) so that one can hope that the new achieved description is much closer to the real behaviour of the system.

One could naively think to carry on this approach up to an arbitrary order $r$, and, therefore, to the limit $r \rightarrow \infty$. If this were generally possible, this result would be incompatible with the Poincaré theorem on the non–integrability of generic systems. As a matter of fact, the size of $H_1^{(r)}$ grows, with the order of normalization $r$, like $r!$. Therefore $\epsilon^r H_1^{(r)}$ diverges with $r \rightarrow \infty$ and has a minimum at $r \sim 1/\epsilon$; the latter can be considered as the optimal order of normalization (see Fig. 2). If one stops at this order, the size of $\epsilon^r H_1^{(r)}$ turns out to be about $\exp(-1/\epsilon)$; this is precisely a Nekhoroshev–like result (the term "like" is due to the fact that the original Nekhoroshev theorem holds globally all over the phase space, while the transformation $p, q \rightarrow p^{(r)}, q^{(r)}$ can be usually defined only in a local way).

Therefore one can conclude that the ideal approach in Celestial Mechanics is to find an open set of quasi–invariant tori, namely tori which are invariant, within a given tolerance neighbourhood, up to a time proportional to $\exp(1/\epsilon)$. (In practice one usually normalizes the Hamiltonian up to a quite low order, thus obtaining shorter diffusion times).

In the following section we will analyze some methods for the practical realization of this approach.

## 3. Some Practical Perturbation Methods

The generally used method for the normalization of the Hamiltonian up to a given order $r$ is the well known Lie algorithm, namely a sequence of canonical transformations introduced via the definition of a sequence of suitable generating functions. We analyze in the following two practical implementations of such an

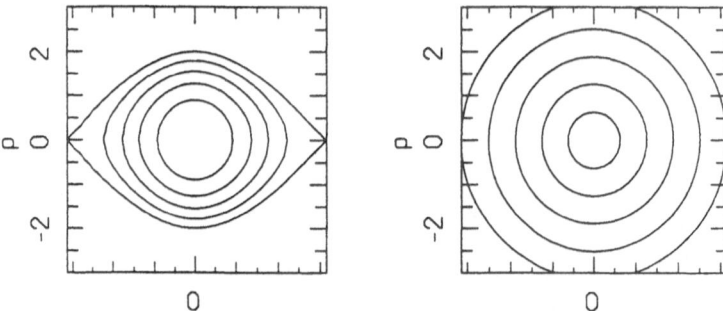

Fig. 3. On the left side, the phase space of the pendulum; on the right side the phase space of the armonic approximation. The two pictures are similar only close to the stable equilibrium point.

algorithm.

## 3.1. IMPLEMENTATION BY SERIES EXPANSIONS

The expansion of the Hamiltonian in power series of the action $p$ and Fourier series of the angles is the most commonly used technique for the implementation of the Lie algorithm, since it allows the use of algebraic manipulators, by which one can achieve the normalization of the Hamiltonian up to a pretty high order.

However this implementation encounters two relevant problems:

1) the results are limited in the neighbourhood of the point of expansion of the original Hamiltonian. The size of this neighbourhood can be difficultly determined a priori, and the problem is generally solved in an experimental way, i.e. by comparing theoretical results and numerical simulations. In Celestial Mechanics the expansion is usually computed in terms of powers of the eccentricity $e$ and of the inclination $i$; the regions at large $e$ or $i$ are still largely unexplored, even in the asteroid belt. Local expansions around a given value $e_*, i_*$ have been computed in some cases, for example by Ferraz–Mello, 1989, but only up to a low order.

2) The unperturbed Hamiltonian $H_0(p)$ must be chosen of a very simple form, i.e. linear in the actions, in order to deal with fixed frequencies for the computation of the generating functions. When the dynamics is strongly non–linear, this fact forces to normalize the Hamiltonian up to a high order, in order to get reasonable results.

For example, imagine to deal with the (integrable) Hamiltonian of a pendulum ($H = 1/2p^2 - \cos q$), and to look for the elimination of the angular variable in the region of libration. Then, one should first expand the Hamiltonian around the stable equilibrium point, i.e.

$$H = \frac{1}{2}(p^2 + q^2) + \sum_{k \geq 2}(-1)^{k+1}\frac{q^{2k}}{(2k)!}$$

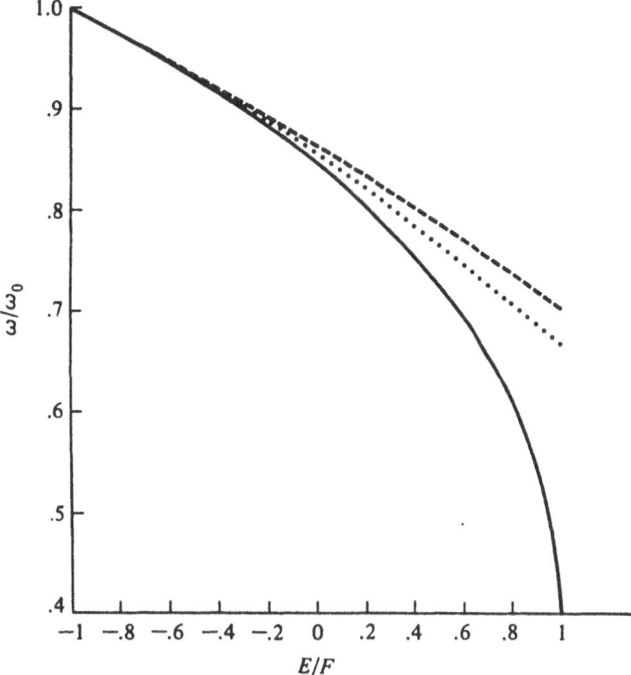

Fig. 4. The frequency–energy relation for the pendulum, in the region of libration (-1 is the energy of the stable equilibrium point, 1 the energy of the separatrix). The solid line is the exact result; the dashed line is obtained using first order perturbation theory. The dotted line is the result from second order perturbation theory. The agreement is good only close to the equilibrium point.

and introduce the new action–angle variables $I, \phi$ defined by $p = \sqrt{2I} \cos \phi$, $q = \sqrt{2I} \sin \phi$, thus obtaining:

$$H = I + \frac{1}{6} I^2 \sin^4 \phi + \frac{1}{90} I^3 \sin^6 \phi + \cdots$$

Furthermore one should apply the Lie algorithm choosing $H_0(I) = I$ as an integrable approximation. Figure 3 shows, on the right side, the phase space of the integrable "approximation" $H_0 = I = (p^2 + q^2)/2$, and, on the left side, the real phase space of the pendulum. As one sees, the two are very similar near the stable equilibrium point, and very different near the separatrix; this suggests that, close to the separatrix, a very high order of normalization will be necessary in order to reconstruct the real phase space starting from a so bad "approximation". Figure 4 makes a similar comparison in the space energy–frequency. According to the integrable "approximation" the frequency is equal to 1 everywhere. The two dashed curves show the energy–frequency relation obtained after normalization up to order 1 and 2 respectively. The bold curve is the real relation for the pendulum. The agreement is good only in a small neighbourhood of the equilibrium point.

Imagine now to have a perturbed pendulum, for which perturbation theory is really necessary. What this simple example shows is that, even if the use of algebraic manipulators allows to go to a quite large order of normalization, most of the work is devoted to the normalization of the integrable part!!

## 3.2. IMPLEMENTATION BY THE USE OF ARNOLD ACTION–ANGLE VARIABLES

In 1963 Arnold proved the already recalled Arnold–Liouville theorem. This theorem states that, given an integrable Hamiltonian $H_0(p, q)$, there exists a set of canonical action–angle variables $J, \psi$ such that the Hamiltonian is independent of the new angles, namely $H_0 \equiv H_0(J)$.

If the phase space of the integrable Hamiltonian is divided into different topological regions (libration, positive and negative circulation, in the case of a pendulum), different Arnold variables must be introduced in each region.

The idea of using Arnold variables for the implementation of the Lie algorithm is due to Henrard (1990). Let us consider, with Henrard, the Hamiltonian

$$H(p, q) = H_0(p) + \epsilon' H'(p, q) + \epsilon_1 H_1(p, q)$$

with $\epsilon_1 \ll \epsilon'$ and $H_0(p) + \epsilon' H'(p, q)$ integrable (the case of the perturbed pendulum is of this form). Then, instead of using $H_0$ as an integrable approximation for the normalization of the Hamiltonian, one first introduces Arnold variables in order to get

$$
\begin{aligned}
H(J, \psi) &= \tilde{H}_0(J) + \epsilon_1 H_1(J, \psi) \quad \text{with} \\
\tilde{H}_0(J) &= H_0(p(J, \psi)) + \epsilon' H'(p(J, \psi), q(J, \psi)) .
\end{aligned}
\tag{1}
$$

Furthermore one can try to normalize by Lie algorithm the real perturbation $\epsilon_1 H_1$, starting from the integrable Hamiltonian $\tilde{H}_0$; the latter is a much better approximation of the real dynamics than $H_0$, since the relevant contribution of $\epsilon' H'$ has been already completely taken into account.

The main problem in this scheme is that the Arnold action–angle variables are not known explicitly, except in some particular cases (like that of the pendulum, and they involve elliptic functions). However they can be computed, for each orbit, in a numerical way. The implementation of the Lie algorithm, on the contrary, requires explicit algebraic expressions. Nevertheless Henrard (1990) has shown that the first order of the Lie algorithm can still be computed in a semi–numerical way.

In Fig. 5 we report an example of application of this method, taken from Moons and Henrard (1992). The Hamiltonian in study is similar to that of a perturbed pendulum and has the form

$$
\begin{aligned}
H &= c_1(N - S)^2 + c_2 N + c_3 S \cos 2\sigma \\
&+ \sqrt{2S}[c_4 \cos(\sigma - \nu) + c_5 \cos(\sigma + \nu)] + 2c_6 \cos 2\nu
\end{aligned}
\tag{2}
$$

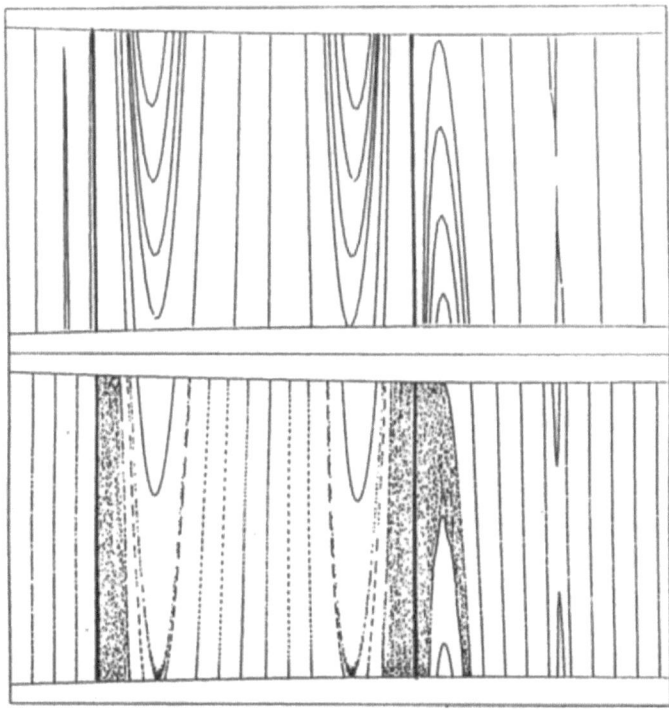

Fig. 5. Poincaré section of (2): above the theoretical result after introduction of suitable action–angle variables for the main pendulum part and first order analysis of the perturbation;
below a numerical simulation. The coordinates are $N - S$ on the $x$–axis and $\sqrt{S}\cos(\sigma + \nu)$
on the $y$–axis. The Poincaré section is made at $\sigma = 0$, so that the three regions of positive
circulation, libration, and negative circulation of $\sigma$ are visible, from left to right. The two
thick vertical lines are the section of the separatrix of the pendulum term. (Moons and
Henrard, 1992.)

(with $S, \sigma$ and $N, \nu$ conjugated pairs and $c_4, c_5, c_6$ much smaller than $c_3$) where
we have isolated in the first line the integrable pendulum–like part. The results are
shown on a Poincaré section at $\sigma = 0$ on a given energy surface. The picture above
is the description of the dynamics obtained by Henrard method after introduction of
suitable action–angle variables for the main pendulum part and first order analysis
of the perturbation. The picture below comes from a numerical integration. The
correspondence between the two pictures is really remarkable and points out the
power of this "first order" method. However one can notice that, at least in this
case, Henrard's method does not provide a good estimation on the size and on the
location of chaotic layers, since regular orbits are expected also where the numerical
simulation reveals stochasticity. This is probably due to the fact that, being a first
order perturbation method, the interaction among secondary resonances and the
primary resonance is not properly taken into account. It would be interesting, then,
to improve this technique, finding a practical way to go further from first order

results, and applicable also to problems where the interaction between the degrees of freedom is strong. This can be done only by looking for an algorithm alternative to the Lie's one. This is the purpose of the new adaptation of Delaunay's idea about the successive elimination of harmonics.

## 3.3. SUCCESSIVE ELIMINATION OF HARMONICS

The basic idea of this approach is simple and connects some known techniques and results. We stress here three basic elements: a) the idea of Delaunay (1867) that one can proceed by successive elimination of harmonics; b) the general result of Arnold (1963a) that an integrable system is characterized by suitable action–angle variables that can be explicitly computed by integrals; c) the suggestion of Henrard (1990) that one can handle the action–angle variables in a practical way by semi–numerical algorithms, thus avoiding the analytical evaluation of complicated integrals.

More precisely, one expands the Hamiltonian $H(I_1, \phi_1, I_2, \phi_2)$ in Fourier series with respect to the angles $\phi_1$ and $\phi_2$, so that one can write

$$H(I_1, \phi_1, I_2, \phi_2) = H_{0,0}(I_1, I_2) + \sum_{n,m} h_{n,m}(I_1, I_2) \cos(n\phi_1 + m\phi_2) ,$$

(here we assume, for simplicity, that only cos components are present in the expansion). Furthermore one chooses one harmonic, $\bar{n}, \bar{m}$ say, and considers the partial Hamiltonian

$$H_{0,0}(I_1, I_2) + h_{\bar{n},\bar{m}}(I_1, I_2) \cos(\bar{n}\phi_1 + \bar{m}\phi_2) .$$

Since this Hamiltonian is integrable, new canonical action–angle variables are introduced in order to take it to be independent of the new angles.

Afterwards, the remainder

$$\sum_{n\neq\bar{n},m\neq\bar{m}} h_{n,m}(I_1, I_2) \cos(n\phi_1 + m\phi_2) ,$$

is transformed in the new variables and expanded in Fourier series with respect to the new angles. Finally the algorithm is iterated in order to eliminate all the principal harmonics. In this way the method proposed here goes beyond the first order approximation of Henrard's method, thus connecting, in some sense, with the possibly infinite expansion used in perturbation theory.

The practical implementation of this algorithm has been deeply discussed in Morbidelli (1992). Moreover, in that paper some criteria are proposed for the determination of the practical convergency of the algorithm, and for the determination of the size of the chaotic layers.

From a practical viewpoint the method of successive elimination of harmonics seems to be particularly indicated for the analysis of systems characterized by a strong coupling among degrees of freedom; indeed, in these cases, the algorithm imposes a fragmentation of the phase space into different "topological regions", according to the location of the main resonances. Moreover, at each step, it compels to leave out of study portions of the phase space (which turn out to be chaotic regions), while elsewhere it allows to define new variables, suitable for the description of the invariant tori in the regular regions.

## 4. Theoretical Implications of the Method of Successive Elimination of Harmonics

The method of successive elimination of harmonics has been analyzed by Morbidelli and Giorgilli (1992) from a theoretical viewpoint. Their work proves that a quantitative perturbation theory can be based on that scheme.

In the following we describe in a very qualitative fashion the guiding lines of this proof, since it provides a connection among KAM result, Nekhoroshev result, and computational algorithm

The original Hamiltonian $H(p,q) = H_0(p) + \epsilon H'(p,q)$ is first expanded in Fourier series, therefore obtaining:

$$H(p,q) = H_0(p) + \epsilon \sum_{k \in \mathbf{Z}^n} h_k(p) e^{ik \cdot q} \ .$$

The size of the coefficients $h_k$ decays exponentially with the order $|k|$ of the harmonic, as results from the Fourier theorem

$$\|h_k(p)\| \le e^{-|k|\sigma} \tag{3}$$

with $\sigma$ radius of analyticity of $H'$ in the complex space.

The elimination of one single harmonic is done by the introduction of Arnold action–angle variables, since the part $H_0(p) + h_k(p) \exp(ik \cdot q)$, with a given $k$, is integrable. In order to give an analytical estimate of this canonical transformation, a strip of suitable size (called resonant strip) is eliminated around the resonant hyperplane correspondent to the considered harmonic. In this way one can prove that, outside of the resonant strip, the transformation to Arnold variables is close to the identity, namely has the form $p = J + f(J, \psi)$ , $q = \psi + g(J, \psi)$ with $\|f\|, \|g\|$ roughly of order $\epsilon$. Therefore, the introduction of the new action–angle variables in the complete Hamiltonian generates new terms the size of which is approximately $\epsilon^2$.

This suggests to proceed in the elimination of a finite number of harmonics only, namely to eliminate all the harmonics such that $\epsilon \|h_k\| > \epsilon^2$ (thanks to 3 this amounts to eliminate about $\ln \epsilon$ harmonics). This is what is called a *quadratic step*, since it reduces the Hamiltonian to

$$H^{(1)}(p,q) = H_0^{(1)}(p) + \epsilon^2 H_1^{(1)}(p,q)$$

(where we have denoted again by $p, q$ the new action–angle variables). Here the demonstration scheme joins the classical one for the proof of the KAM theorem. Indeed it is natural to iterate the quadratic steps to obtain, after $N$ steps

$$H^{(N)}(p, q) = H_0^{(N)}(p) + \epsilon^{2^N} H_1^{(N)}(p, q) .$$

As a matter of fact the size of $H_1^{(N)}$ is not 1, but grows so fast with $N$ that the actual size of the perturbation turns out to be just $\epsilon^N$; anyway this is enough to guarantee the convergence to zero with $N \to \infty$.

The considerations on the size of the perturbation cannot be separated from the considerations on the size and structure of the domain of work. The elimination of several harmonics produces a fragmentation of the phase space, because of the necessary elimination of the correspondent resonant strips. Nevertheless the total amount of volume eliminated with the resonant strips is small; indeed the width of each strip is proportional to the strength of the associated harmonic, and therefore decays also exponentially with the order $k$. Roughly speaking, the total loss of volume during the first quadratic step can be estimated by the series

$$\epsilon \sum_{k \in \mathbf{Z}^n} e^{-|k|}$$

which is convergent and of order $\epsilon$.

Therefore, the loss of volume in the iteration of the quadratic steps can be estimated by the double series

$$\sum_{N \geq 1} \epsilon^N \sum_{k \in \mathbf{Z}^n} e^{-|k|}$$

which is still convergent and of order $\epsilon$.

This allows to conclude that, provided $\epsilon$ is small enough, the domain where the algorithm of successive elimination of harmonics holds is, at each step, a non–empty one.

At this point one can proceed in two different ways:

1)   one can decide to stop after a given number $N$ of quadratic steps. In this way one determines a non–empty set, which is open (since only a finite number of harmonics, $\sim N \ln \epsilon$, has been eliminated) and made up of quasi invariant tori: the diffusion rate of the action variables constructed on these tori is of order $\epsilon^N$. The further condition that such a domain must contain balls of radius $\epsilon$ gives an upper bound to $N$. In this case the diffusion rate turns out to be of order $\exp(-1/\epsilon)$. This can be considered as a Nekhoroshev–like result on the non–resonant part of the phase space.

2)   one can proceed to the limit $N \to \infty$. In this way the open domain of the quasi–invariant tori shrinks to the domain of the invariant tori. Such a domain is neither open (since all the harmonics have been eliminated and the resonances are dense) nor dense (since an open strip is eliminated around each

resonance). This is nothing but the KAM theorem in the global formulation given by Arnold.

The most relevant aspect in the proof qualitatively outlined above is that the KAM result and the Nekhoroshev–like result are both obtained through a unitary scheme: they are just two different answers to two different questions.

On the other hand, the computational algorithm of successive elimination of harmonics, discussed in Morbidelli (1992) and summarized in the previous section, is perfectly consistent with the theoretical scheme which leads to the above result. Therefore one can conclude that Delaunay's method of successive elimination of harmonics allows to point out in a better light, with respect to the Lie algorithm, the interconnections between general properties of dynamical systems, as described by KAM and Nekhoroshev, and computational algorithm for the effective analysis of specific systems.

## References

Arnold, V., I., (1963a): "On a theorem of Liouville concerning integrable problems of dynamics", *Sib. mathem. zh.*, **4**, 2.

Arnold, V., I., (1963b): "Proof of A. N. Kolmogorov's theorem on the conservation of conditionally periodic motions with a small variation in the Hamiltonian.", *Russian Math. Surv.*, **18**, No. 5, 9–36.

Arnold, V., I., (1964): "Instability of dynamical systems with several degrees of freedom", *Sov. Math. Dokl.*, **5**, N1, 581–585.

Delaunay, C., (1867): "Theorie du mouvement de la Lune", *Mem. Acad. Sci.*, Paris, **29**.

Ferraz–Mello, S., (1989): "A semi–numerical expansion of the averaged disturbing function for some very–high–eccentricity orbits.", *Celest. Mech.*, **45**, 65–68.

Giorgilli, A., (1989): "New insights on the stability problem from recent results in classical perturbation theory", in *Modern methods in Celestial Mechanics*, D. Benest and Cl. Froeschlé eds., Editions Frontières.

Henrard, J., (1990): " A semi–numerical perturbation method for separable Hamiltonian systems.", *Celest. Mech.*, **49**, 43–67.

Kolmogorov, A., N., (1954): "Preservation of conditionally periodic movements with small change in the Hamiltonian function", *Dokl. Akad. Nauk SSSR*, **98**, 527–530; English translation in *Lecture notes in Physics*, N.93, 51–56, Casati, G. and Ford, J. eds.

Lichtenberg, A.J., and Lieberman, M.A., (1983): " Regular and stochastic motion.", *Springer Verlag* ed.

Moons, M., and Henrard, J., (1992): "Surfaces of sections in the Miranda–Umbriel 1/3 inclination problem", in preparation.

Morbidelli, A., (1992): "On the successive eliminations of perturbation harmonics.", *Celest. Mech.*, in press.

Morbidelli, A., and Giorgilli, A., (1992): "A quantitative perturbation theory by successive elimination of harmonics.", *Celest. Mech.*, in press.

Moser, J., (1962): "On invariant curves of area–preserving mappings of an annulus", *Nachr. Akad. Wiss. Gottingen, Math. Phys.* 2, 1.

Nekhoroshev, N., N., (1971): "Behaviour of Hamiltonian systems clos to integrable", *Funct. An. and Appl.*, **5**, 338–339.

Nekhoroshev, N., N., (1977): "Exponential estimate of the stability time of near–integrable Hamiltonian systems", *Russ. Math. Survey*, **32**, N.6, 1–65.

Nekhoroshev, N., N., (1979): "Exponential estimate of the stability time of near–integrable Hamiltonian systems, II", *Trudy. Sem. Petrovs.*, **5**, 5–50 (in Russian).
Poincaré, H., (1892): *Les Méthodes Nouvelles de la Mécanique Céleste*, Gauthier–Villars, Paris.

# FREQUENCY ANALYSIS OF A DYNAMICAL SYSTEM

JACQUES LASKAR

*Bureau des Longitudes, Equipe Astronomie et Systèmes Dynamiques,*
*77 Avenue Denfert-Rochereau, F75014 Paris, France*

**Abstract.** Frequency analysis is a new method for analyzing the stability of orbits in a conservative dynamical system. It was first devised in order to study the stability of the solar system (Laskar, Icarus, 88, 1990). It is a powerful method for analyzing weakly chaotic motion in hamiltonian systems or symplectic maps. For regular motions, it yields an analytical representation of the solutions. In cases of 2 degrees of freedom system with monotonous torsion, precise numerical criterions for the destruction of KAM tori can be found. For a 4D symplectic map, plotting the frequency map in the frequency plane provides a clear representation of the global dynamics and describes the actual Arnold web of the system.

**Key words:** Frequency analysis – chaotic motion – symplectic maps

## 1. Frequency Analysis

The method of numerical analysis of the fundamental frequencies was introduced in the study of the stability of the solar system, as modeled by a reduced (but nevertheless complicated) 15 degrees of freedom system (Laskar, 1990). In that case, frequency analysis permitted numerical estimates of the size of chaotic zones in all directions of the 15 degrees of freedom, and revealed that for the inner planets (Mercury to Mars), the chaotic zones were relatively large, while for the outer planets (Jupiter to Neptune), these zones were much smaller.

More generally, the frequency analysis method can be applied to study the stability of the solutions of a conservative dynamical system, and is based on a refined numerical search for a quasiperiodic approximation of its solutions over a finite time span (Laskar, 1990, 1992, Laskar *et al.*, 1992). If $f(t)$ is a function with values in the complex domain, obtained numerically over a finite time span $[-T, T]$ the frequency analysis algorithm will consist in the search for a quasiperiodic approximation for $f(t)$ with a finite number of periodic terms of the form

$$\tilde{f}(t) = \sum_{k=1}^{N} a_n e^{i\sigma_k t} .$$

The frequencies $\sigma_k$ and complex amplitudes $a_k$ are found with an iterative scheme. To determine the first frequency $\sigma_1$, one searches for the maximum of the amplitude of

$$\phi(\sigma) = \langle f(t), e^{i\sigma t} \rangle$$

where the scalar product $\langle f(t), g(t) \rangle$ is defined by

$$\langle f(t), g(t) \rangle = \frac{1}{2T} \int_{-T}^{T} f(t)\bar{g}(t)\chi(t)dt ,$$

*Celestial Mechanics and Dynamical Astronomy* **56**: 191–196, 1993.
© 1993 *Kluwer Academic Publishers.*

and where $\chi(t)$ is a weight function, that is, a positive function with

$$\frac{1}{2T} \int_{-T}^{T} \chi(t)dt = 1 \ .$$

In all computations, the Hanning window filter was used, that is

$$\chi(t) = 1 + \cos(\pi t/T) \ ,$$

although some other weight functions could be used. Once the first periodic term $e^{i\sigma_1 t}$ is found, its complex amplitude $a_1$ is obtained by orthogonal projection, and the process is started again on the remaining part of the function $f_1(t) = f(t) - a_1 e^{i\sigma_1 t}$. As all the different functions $e^{i\sigma_k t}$ are not orthogonal, it is also necessary to orthogonalize the set of functions $(e^{i\sigma_k t})_k$, when projecting $f$ iteratively on these $e^{i\sigma_k t}$. In the case of an hamiltonian system with $n$ degrees of freedom, the frequency analysis of the solutions will give its quasiperiodic expansion and in particular will determine the vector $(\nu_i)_{i=1,n}$ of the fundamental frequencies of the system. In the case of nonintegrable systems, not all the solutions are quasiperiodic, but under certain conditions, for example under the hypotheses of KAM theorems, there still exist many of these. For such solutions, the frequency analysis over a finite time span $[0, T]$ will give the same kind of results as for integrable systems.

Even if an orbit is not regular (quasiperiodic), in case of nearly integrable systems the solution will look very regular on a finite time span. More precisely, this will be the case if the time span is smaller than the characteristic time of divergence of nearby orbits. In this case, the frequency analysis gives a quasiperiodic approximation to the solution which holds only locally in time. In other words, it will give us a frequency vector $(\nu_i(t))_i$ for each value of $t$, obtained by applying the frequency analysis algorithm over the time span $[t, t + T]$. In the case of a quasiperiodic solution, $\nu_i(t)$ does not depend on $t$, while for non-regular solutions, $\nu_i(t)$ will evolve with time, revealing the diffusion of the orbit in phase space. The frequencies are used here instead of the action variables for a more accurate monitoring of the diffusion of the orbit.

## 2. Two Dimensional Twist Map.

Let us first consider the case of a symplectic twist map on $R^2$. As an example, we shall consider the Standard Map,

$$\begin{cases} x' = x - a \sin y & \mod(2\pi) \\ y' = x' + y & \mod(2\pi) \end{cases}$$

As a dynamical system, it is not integrable, and gives rise to the usual features of conservative dynamics, with invariant curves, chaotic regions, and elliptic islands. A simple criterion for the disappearance of irrational curves, based on Birkhoff's theory, was derived from the frequency analysis of such a monotone (increasing) twist map (Laskar et al., 1992).

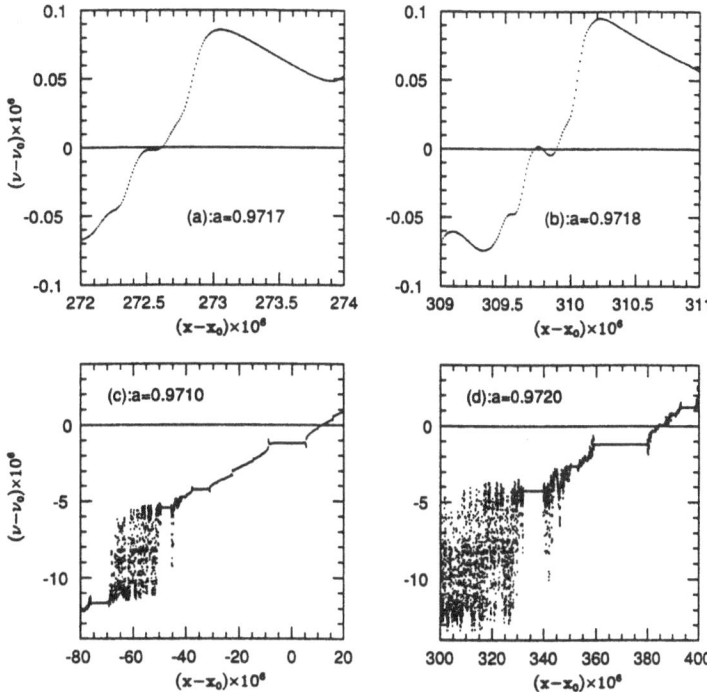

Fig. 1. Variation of the fundamental frequency $\nu$ for the 2D Standard Map for different values of the parameter $a$, in the vicinity of the golden rotation number $\nu_0$ which corresponds to the zero dotted line. The origin in the $x$ scale is arbitrarily taken to be $x_0 = 4.176550$. The origin of frequencies is the golden value $\nu_g = (3 - \sqrt{5})/2$. The unit for $\nu$ and $x$ is $10^{-6}$. If $x_1 < x_2$ and $\nu(x_1) > \nu(x_2)$, we can conclude that there exist no KAM invariant curves of irrational rotation number between $\nu(x_2)$ and $\nu(x_1)$. In Fig. 1b, we can see that that the golden invariant curve does not persist for $a = 0.9718$.

For each initial condition $x$ on the vertical line $y = 0$, we call $\gamma_x$ the orbit obtained by iterating the mapping and $\nu(x)$ the frequency given by the frequency analysis of this orbit during a given time span.

*If there exist two values $x < x'$ on the vertical line $y = 0$ for which $\nu = \nu(x) > \nu' = \nu(x')$, then there are no invariant KAM curves of irrational rotation number $\nu''$ with $\nu' < \nu'' < \nu$.*

This criterion provides a simple way of knowing whether a KAM curve has disappeared by looking at the graph of the frequency map $\nu(x)$, obtained on a given time span $[0, T]$. The figure (1b) was obtained with $T = 12516$, and shows the disappearance of the golden curve for the value $a = 0.9718$ of the parameter, which is very close to and compatible with the value $a_c = 0.971635$ derived by Greene (1979).

## 3. Higher Dimension

We shall consider the case of a symplectic map on $R^{2n}$ written in coordinates $(x, y)$ which are close to angle-action variables.

The angle-like variables $x_0$ are fixed. If we take some initial conditions $y$, we can carry out the frequency analysis for the orbits corresponding to initial conditions $(x_0, y)$ (at $t = 0$) over the time span $[t, t + T]$. We thus define a map

$$F_T : R^n \times R \longrightarrow R^n$$
$$(y, t) \longrightarrow f(y, t)$$

For a given value of $t$, let us denote $F_T^t$ the restriction of $F_T$ to $R^n \times \{y\}$, and let $A$ be the set of $y$-values which correspond to invariant tori of dimension $n$.

a) If $y \in A$ then $F_T(y, .)$ is constant on $R$ (up to the precision of the determination of the frequencies).

b) In the case $n = 1$, for a monotone twist map and for a given value of $t$,

$$F_T^t : A \longrightarrow R$$
$$y \longrightarrow f(y, t)$$

is monotone.

The property a) was already used to study the stability of the solar system (Laskar, 1990); b) was used to study the destruction of golden tori for the two dimensional standard map (Laskar *et al.*, 1992).

The frequency map is exactly defined on the Cantor set of the invariant tori. It can be thought of as a diffeomorphism on this set, which could be extended in some sense to a diffeomorphism on $R^2$ (cf. Pöschel, 1982). Chaotic zones will therefore appear as a loss of regularity for the frequency map. This can be clearly seen around the golden curve for the two dimensional standard map (Fig.1) . As the parameter increases, there are some distorsions in the frequency curves. These distorsions permit statements about the non-existence of KAM tori, but these distorsions also eventually produce complete loss of regularity of the frequency map, which can be taken as an indication of chaotic motion. Moreover, this loss of regularity of the frequency map can be generalized to higher dimensions.

## 4. Application to the 4D Standard Map

We will use the frequency analysis to study the global dynamics of a 4 dimension symplectic map which was first studied by Froeschlé (1972).

$$\begin{cases} x_1' = x_1 + a_1 \sin(x_1 + y_1) + b \sin(\frac{1}{2}(x_1 + y_1 + x_2 + y_2)) & \mod(2\pi) \\ y_1' = x_1 + y_1 & \mod(2\pi) \\ x_2' = x_2 + a_2 \sin(x_2 + y_2) + b \sin(\frac{1}{2}(x_1 + y_1 + x_2 + y_2)) & \mod(2\pi) \\ y_2' = x_2 + y_2 & \mod(2\pi) \end{cases}$$

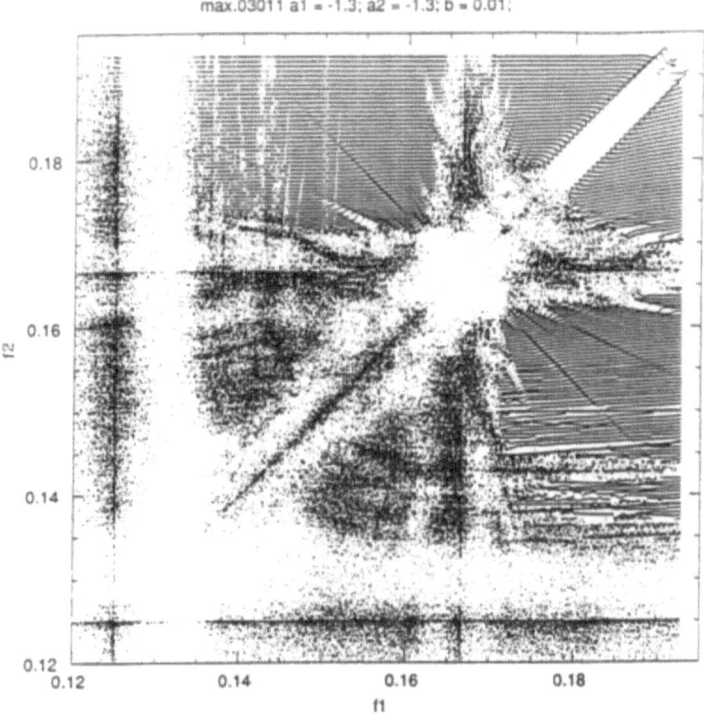

max.03011 a1 = -1.3; a2 = -1.3; b = 0.01;

Fig. 2. Visualisating in the frequency plane $(f_1, f_2)$ of the frequency application $a_1 = a_2 = -1.3, b = 0.01$.

We shall carry out the frequency analysis for the parameter value $a_1 = a_2 = -1.3, b = 0.01$ and we shall consider orbits with initial conditions on the plane $x_1 = x_2 = 0$. We can visualize the complete frequency map $F_T^0$ which is a map from $R^2$ to $R^2$. In the regular regions, the behavior is not very wild, and it will be possible to visualize the map by drawing the images of the lines of initial conditions $R \times y_1$ for various values of $y_1$ (Fig 2). For each initial condition $(y_1, y_2)$, the two main frequencies $f_1$ and $f_2$ of the orbit are determined with the frequency analysis over 516 iterations, and the point of coordinates $(f_1, f_2)$ is indicated on the graph. In a regular region, the image of a line will appear as a smooth curve, which will not be the case in chaotic regions.

In fact, what is pictured here is the Arnold web of the mapping, with the description of the actual strength associated to each of the resonant lines. Resonant lines exist on a dense set of frequencies, but most of them have a negligible effect and are not visible here. Different zones appear on these plots. The first ones are the regular zones, with very smooth, non-distorted frequency curves. The motion will be very regular in these regions. Next are some resonance regions, such as the top of the vertical $f_1 = 1/6$ zone, where the points are regularly spaced in $f_2$, but more erratically in $f_1$. This corresponds to the product of a chaotic motion in $f_1$ with something more regular in $f_2$. In these zones, Arnold diffusion is probably possible. There are also zones of the pictures where the points seem to be erratically distributed in all directions, which ought to correspond to completely chaotic motion. This is the case for example, in the outside zone, for small $f_1$ and $f_2$, which corresponds to the large-scale chaotic motion where most tori are destroyed, even in the uncoupled problem.

The analysis of the regularity of the frequency application presented here allows one to obtain a global picture, in two dimensions, of the dynamics of a 4 dimensional symplectic map, or a 3 degrees of freedom hamiltonian system. This method can also be applied in higher dimensions, and I am convinced that this new method of frequency analysis will become an important tool for the study of many kinds of conservative dynamical systems.

## References

Froeschlé, C.: 1972, 'Numerical study of a four-dimensional mapping',*Astron. Astrophys.*, **16**, 172–189

Greene, J.M.: 1979, 'A method for determining stochastic transitions', *J. Math. Phys.*, **20**, 1183

Laskar, J.: 1990, 'The chaotic behaviour of the solar system: A numerical estimate of the size of the chaotic zones', *Icarus*, **88**, 266–291

Laskar, J.: 1992, 'Frequency analysis of multidimensional systems. Global dynamics and diffusion', *submitted to Physica D*

Laskar, J., Froeschlé, C., Celletti, A.: 1992, 'The measure of chaos by the numerical analysis of the fundamental frequencies. Application to the standard mapping', *Physica*, **D 56** 253–269

Pöschel, J.: 1982, *Commun. Pure. Appl. Math.*, **35** 653

# AN APPLICATION OF KAM THEORY TO THE PLANETARY THREE
# BODY PROBLEM

PHILIPPE ROBUTEL

*Bureau des Longitudes, Equipe Astronomie et Systemes Dynamiques,*
*77 Avenue Denfert-Rochereau, 75014 Paris, France*

In 1963 Arnold proved a theorem of conservation of invariant tori, in degenerate quasi-integrable hamiltonian systems (Arnold, 1963). Then, he applied this theorem to the planetary problem ; he used Leverier's tables and asymptotic expansions of the perturbative function in the semi-major axes. He obtained a result of stability in the planar problem of two planets, in the asymptotic case, when the ratio of the semi-major axes goes to zero. In Arnold's paper, a statement is made about the general case, but the computation does not appear in the paper.

Following the general framework of the proof, I propose, thanks to a new planetary analytical expansion in canonical heliocentric variables, to demonstrate the stability of the two planets spatial problem, for all values of the semi-major axes and for a set of large measure of initial conditions with small inclinations, eccentricities, and planetary mass. In order to obtain this result, we must apply Arnold's theorem, which is briefly summarized as follows :

Let $H$ be an analytical hamiltonian function depending on the canonical variables $(q, p) = (q_0, q_1, p_0, p_1)$ in a domain of $T^{n_0} \times R^{n_1} \times R^{n_0} \times R^{n_1}$ and $\mu$ a small parameter (in our problem the ratio of planet's mass over sun's mass). In these coordinates, $H$ is of the following form :

$$H = H_0(p_0) + \mu H_1(q, p) = H_0(p_0) + \mu \bar{H}_1(q_1, p_0, p_1) + \mu \tilde{H}_1(q, p)$$

where $\tilde{H}_1$ is the short period part, that is $\int \tilde{H}_1(q, p) dq_0 = 0$, and the secular part of order one $\bar{H}_1$ is a normal form of degree six in $(q_1, p_1)$ (that is a polynomial of third degree in $\tau_j = p_{1j}^2 + q_{1j}^2$), plus a remainder.

Many other conditions must be fulfilled, but the only one which is not immediately satisfied is that the application "secular frequency" $\tau \mapsto \partial \bar{H}_1 / \partial \tau$ must be a diffeomorphism in a neighbourhood of zero.

The conclusion of this theorem can be summarized by the existence, in a domain of the phase space which corresponds to a small neighbourhood of zero for secular variables (eccentricites and inclinations in our problem), of a set of large measure consisting of invariant n-dimensional analytic tori filled with quasi-periodic orbits.

To apply this theorem, we must first get an expression of the secular part of the hamiltonian, and then reduce it to a normal form up to order six (in inclinations and eccentricites).

In canonical heliocentric variables (Laskar, 1992), the hamiltonian of the planetary problem takes a simple form, and allows us to make the reduction of the center of mass. With the framework developed by J.Laskar, it is possible to obtain

*Celestial Mechanics and Dynamical Astronomy* **56**: 197–199, 1993.
© 1993 *Kluwer Academic Publishers.*

an expansion, in compact form, of the main part of the perturbative function valid for all values of the semi-major axes (only two Laplace coefficients : $b_{7/2}^{(0)}$ and $b_{7/2}^{(1)}$, are necessary).

Then, we can reduce our problem. Indeed, if we use osculating variables defined with respect to the invariable plane (perpendicular to the angular momentum), it is possible to eliminate both inclinations and longitudes of ascending nodes. Obviously the modulus of the angular momentum $C$ must be kept as a parameter. Rather than $C$, it is better to use the small parameter $D_2$, namely the difference between the square of the angular momentum in a planar circular motion and the square of the actual value of the angular momentum (this quantity tends to zero with the square of eccentricities and inclinations). In this way, we obtain a four degrees of freedom problem, with the small parameter $D_2$ which includes the inclinations.

Expanded in Poincare variables, the secular part of the hamiltonian is the sum of homogeneous polynomials of even degrees which coefficients depending on the semi-major axis ratio and the angular momentum.

The last transformation we must perform, in the process of reducing the secular part to a normal form, is a Birkhoff transformation. But, above everything else, it is essential to diagonalize the quadratic part. The knowledge of the eigenvectors of the linear part of the hamiltonian system, provides us with a symplectic base in which the unperturbed hamiltonian is the sum of two harmonic oscillators of eigen frequencies $\omega_1$ and $\omega_2$ . Because we manage to get explicit formulas for these two frequencies, it is easy to show that, at least for $D_2 \leq 0.01$ (Jupiter-Saturn case), the following inequality is satisfied : $1 < \omega_2(\alpha, D_2)/\omega_1(\alpha, D_2) < 2$ for every value of the ratio of semi-major axes $\alpha$. This relation excludes low order secular resonances, and enables us to perform a Birkhoff transformation up to order six (the first secular resonance occurs at degree eight ).

Up to this point, it is interesting to make a few remarks : If we are far from mean motion's resonances, the secular problem of order one provides us with a good approximation of the average motion. This problem has only two degrees of freedom, thus any invariant torus divides the space phase into separate regions. If we can show the existence of K.A.M. tori, we will obtain stability for an infinite time. Here, the normalization we have done, only provides us with stability of the secular system for a long, but finite, time.

Now, our last task is to check the main condition. After Birkhoff's normalization, the secular part takes the form :

$$\tilde{\tilde{H}} = \omega_1 \tau_1 + \omega_2 \tau_2 + \sum_{i,j=1,2} \lambda_{ij} \tau_i \tau_j + \sum_{i,j,k=1,2} \lambda_{ijk} \tau_i \tau_j \tau_k$$

with $\tau_j = |x_j|^2$ where $(x_i, \bar{x}_j)$ still denote the new canonical variables.

The condition of no degeneracy for the secular frequences becomes :

$$Det(\alpha, D_2) = \lambda_{11}(\alpha, D_2) \cdot \lambda_{22}(\alpha, D_2) - \lambda_{12}(\alpha, D_2) \cdot \lambda_{21}(\alpha, D_2) \neq 0$$

This function, expanded to first degree in $D_2$, is a formal series of 20000 terms and depends of 13 variables. This expression was obtained using the algebraic manipulator TRIP, developed at the Bureau des Longitudes (Laskar, 1989). Because of this great number of terms, the only way to check the condition is a numerical estimation. This computation shows that, for $D_2 = 0$ and for every value of $\alpha$, this function is greater than zero. We also proved that, for a fixed $\alpha$, this quantity is increasing with $D_2$, which finished the proof. This brief proof provides us with a result of stability in the general planetary spatial three body problem. Indeed, for small values of planetary mass, inclinations and eccentricities, almost all initial conditions (or a set of large measure) belongs to a four dimensional invariant torus filled with quasi-periodic orbits which are thus stable over infinite time.

## References

Arnold, V.: 1963, "Small denominators", *Rus. Math. Surv.*, **18**.

Laskar, J.: 1989, "Manipulation des series", *Modern methods in celestial mechanics*, D.Benest, C.Froeschlé (eds), Editions Frontieres.

Laskar, J.: 1992, "Analytical framework in Poincaré variables for the motion of the solar system", *Long-term dynamic behaviour of natural and artificial N-body systems*, A.Roy (ed), Kluwer A.P.

Poincaré, H.: 1893, *Méthodes nouvelles de la mécanique céleste*, volume 1, Gauthier-Villars

# RESONANT MOTION IN THE RESTRICTED THREE BODY PROBLEM

JOHN D. HADJIDEMETRIOU

*Department of Theoretical Mechanics*
*University of Thessaloniki*
*GR-540 06 Thessaloniki, Greece*
*e-mail: CAAZ17@GRTHEUN1*

**Abstract.** The resonant structure of the restricted three body problem for the Sun-Jupiter asteroid system in the plane is studied, both for a circular and an elliptic orbit of Jupiter. Three typical resonances are studied, the 2 : 1, 3 : 1 and 4 : 1 mean motion resonance of the asteroid with Jupiter. The structure of the phase space is topologically different in these cases. These are typical for all other resonances in the asteroid problem. In each case we start with the unperturbed two-body system "Sun-asteroid" and we study the continuation of the periodic orbits when the perturbation due to a circular orbit of Jupiter is introduced. Families of periodic orbits of the first and of the second kind are presented. The structure of the phase space on a surface of section is also given. Next, we study the families of periodic orbits of the asteroid in the elliptic restricted problem with the eccentricity of Jupiter as a parameter. These orbits bifurcate from the families of the circular problem. Finally, we compare the above families of periodic orbits with the corresponding families of fixed points of the averaged problem. Different averaged Hamiltonians are considered in each resonance and the range of validity of each model is discussed.

**Key words:** periodic orbits, resonance, three-body problem.

## 1. Introduction

It is well known that resonances play an important role in the long-term evolution of a dynamical system. Many types of resonances are associated with periodic motion, or equivalently, with fixed points of a Poincaré mapping on a surface of section, and it is clear that the position of the fixed points and their stability determine critically the topology of the phase space on the surface of section.

The study of resonances is very important in a nearly integrable Hamiltonian system, defined by a Hamiltonian $H = H_0(J) + \epsilon H_1(J, \theta)$, where $J$ are the actions, $\theta$ the angles and $\epsilon$ a small parameter. The frequencies in the unperturbed problem are given by $n_i = \partial H_0 / \partial J_i$ and a resonance condition exists when the $n_i$ are in integer dependence. Such resonance conditions are dense, but of particular interest are the low order resonances (the coefficients of the integer dependence are small) because in this case, even for a small value of $\epsilon$ the topology of phase space may change drastically, as we shall see in the following.

There are two important methods to study resonant motion. The first involves the computation of periodic orbits, and the second the method of averaging.

Since a resonance is associated with periodic motion, the computation of families of periodic orbits provides useful information on the resonant structure of the phase space. In the particular case of two degrees of freedom, we can easily find the general topological properties of a Poincaré mapping on a surface of section, from a diagram which gives the characteristic curves of the families of periodic orbits (see for example Hadjidemetriou and Ichtiaroglou, 1984). The knowledge of the characteristic curves of the families of periodic orbits in an autonomous system,

*Celestial Mechanics and Dynamical Astronomy* **56**: 201–219, 1993.
© 1993 *Kluwer Academic Publishers.*

and their stability, is useful because we can find at what energy levels we can take a surface of section for the mapping, so that no main structure of the phase space goes undetected. We also note that it is at the unstable fixed points (periodic orbits) that chaotic motion starts as the perturbation increases.

The method of averaging is a widely used technique in Celestial Mechanics for the study of resonant motion, because near a resonance we can always perform a canonical transformation, so that in the new variables we have a fast and a slow angle (Lichtenberg and Liebermann, 1983). Then, the fast angle is eliminated by the method of averaging and in this way the number of degrees of freedom is reduced. In particular, the averaged Hamiltonian of a system with two degrees of freedom is integrable, because we have only one degree of freedom (see Henrard and Lemaitre, 1983). It can be proved that the fixed points of the averaged Hamiltonian correspond to the periodic orbits of the original dynamical system (Hadjidemetriou, 1991) and the closed level curves of the averaged Hamiltonian correspond to quasiperiodic motion. Moreover, these level curves correspond to the invariant curves of the Poincaré mapping of the original system, provided that we are in an ordered region of phase space.

From the above remarks we see the similarity between the two main methods for the study of resonant motion, the computation of periodic orbits and the method of averaging. Both methods have their advantages and their limitations. The computation of periodic orbits refers to the complete model, i.e. no approximation is involved. It also provides important information on the structure of the phase space on a Poincaré surface of section, and in fact this is the real usefulness of the computation of periodic orbits. Of course, in order to obtain a complete picture of the phase space, we cannot avoid the much more time-consuming computations of mappings, but we can keep the amount of computations to a minimum by properly selecting the surfaces of section. Another difficulty in the computation of periodic orbits, especially in systems with three or more degrees of freedom, is the problem of accuracy in the exact initial conditions and, more important, their stability. This may be a real problem in nearly integrable dynamical systems, because for a zero perturbation there is an infinite number of periodic orbits, but only a finite number survive as soon as the perturbation is switched on (usually only two, one stable and one unstable).

The method of averaging simplifies considerably the problem, because it reduces the number of degrees of freedom. In this way we can do analytic work, in addition to the numerical computations. For example, the computation of periodic orbits is reduced to the computation of the fixed points of the averaged equations and this involves the solution of algebraic equations, rather than the numerical integration of the differential equation of motion. It is also evident that the stability of the fixed points can be computed easily and accurately. However, the averaging method involves series expansions and in most cases only the first few terms are considered. Thus, it is not clear what is the corresponding error of the truncation and consequently, what is the relation of the averaged model with the original system. An important question is: does the averaged model have the same fixed points and the correct stability as the original system?

We will study the resonant structure of the plane elliptic restricted problem, for the Sun-Jupiter mass ratio. This is an important model for the study of the solar system and in particular, for the study of asteroid motion. We will use both methods mentioned above to study the resonant structure of this dynamical system. We believe that both methods are useful and their combined study will provide us with a complete picture of the resonant structure of the phase space. In addition, the comparison of the periodic orbits with the corresponding fixed points of the averaged problem will give useful information on the range of validity of the averaged Hamiltonian. In order to be realistic, the averaged model must have the same fixed points, with the correct stability, as the actual system. If this is true, then the phase space of the averaged Hamiltonian will have the same topological structure as the real system, and consequently it will behave in the same way, in the long-term evolution. If on the other hand, some resonances (fixed points) are missing from the averaged model, the evolution may be quite different, as we will show by an example.

We remark also that if we know the correct resonances from the numerical computations of periodic orbits of the real system, we can correct the averaged Hamiltonian by adding suitable terms which introduce to the averaged model the missing fixed points (Hadjidemetriou, 1992). In this way we obtain a new model which is realistic and is simple, so that we can do analytic work.

In the present work we will study three main resonant cases in the restricted planar three-body problem, namely the 2:1, 3:1 and 4:1 mean motion resonance of an asteroid with Jupiter. These resonances were selected because they are typical cases for all other resonant cases in the restricted problem and because they are important for the study of asteroid motion. Their continuation properties from zero to nonzero mass of Jupiter is topologically different from each other. The existence proof for this type of periodic orbits is given in a review article by Hadjidemetriou (1981,1988).

We start the study by describing the unperturbed Keplerian motion of the asteroid and we show what happens if the gravitational perturbation from Jupiter, revolving around the Sun in a *circular* orbit, is taken into account. Next, we study how the periodic orbits of the circular problem are continued when the perturbation due to a *nonzero* eccentricity of Jupiter is taken into account. These results are compared with similar studies based on the method of averaging.

## 2. The Unperturbed Problem

In order to obtain a clear view of the resonant structure of the restricted problem, for the Sun-Jupiter mass ratio, we start with the unperturbed problem: Jupiter, with zero mass, revolves around the Sun in a circular orbit and we study the motion of a massless body in a uniformly rotating frame $xOy$, whose origin is at the positive

JOHN D. HADJIDEMETRIOU

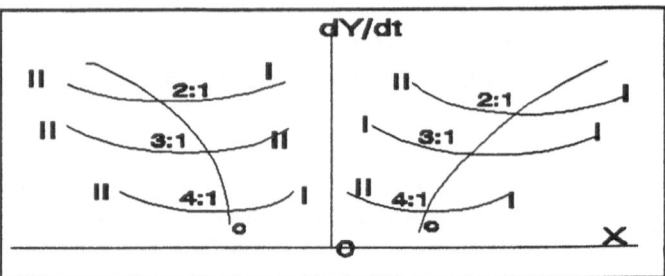

Fig. 1. The characteristic curves of the families of periodic orbits (schematically). The curve $c$ represents the family of circular orbits and along this family the semimajor axis $a$, and the ratio $n : n'$ vary. At the resonant values $n : n' = 2 : 1, 3 : 1, 4 : 1$, bifurcation of families of elliptic periodic orbits exist, along which $a$ and $n : n'$ are almost constant.

$x$-axis in the line Sun-Jupiter. We take the unit of mass to be equal to the total mass of the system (in this case the mass of the Sun), the unit of distance to be equal to the radius of the orbit of Jupiter and the gravitational constant to be equal to unity. In these units, the period of Jupiter is equal to $2\pi$ and the angular velocity of the $xOy$ frame (and of Jupiter) is equal to $n' = 1$.

An asteroid revolving around the Sun in a circular orbit, with radius $a$, is periodic in the rotating frame $xOy$, with period equal to

$$T = 2\pi/(n - n').$$   (1)

This orbit crosses the $x$-axis perpendicularly twice, and the initial conditions of this circular periodic orbit are

$$x_0 = \pm a, \ y_0 = 0, \ \dot{x}_0 = 0, \ \dot{y}_0 = \pm(-a + a^{-1/2}),$$   (2)

where the $+$ sign refers to the intersection with the positive $x$-axis and the $-$ sign with the negative axis. The initial conditions (2) define a family of symmetric periodic orbits, with $a$ as a parameter. This family can be represented in the space $x_0 \dot{y}_0$ by a curve, called *characteristic curve*. In fact we have two such curves, one for $x_0 > 0$ and one for $x_0 < 0$, either of which represents the above family of circular orbits (Figure 1). Along the family the semimajor axis $a$, or the ratio $n : n'$ varies (note that $n = a^{-3/2}$ and $n' = 1$). This means that there exists a dense set of resonant periodic orbits, for each rational value $n : n' = p : q$, where $p, q$ are integers. From each of these resonant periodic orbits there exists a bifurcation of a family of resonant periodic orbits of the asteroid, along which the semimajor axis $a$, and the resonance $n : n'$, remain constant, equal to $p : q$ and the eccentricity increases, starting from zero values. We will restrict ourselves to symmetric periodic orbits only, because it is these orbits that are continued to periodic orbits when the mass $\mu$ of Jupiter increases (its eccentricity remaining equal to zero). The initial conditions of these families are

$$x_0 = \pm a(1 - e), \ y_0 = 0, \ \dot{x}_0 = 0,$$

$$\dot{y}_0 \pm \left[ -x_0 + a^{-1/2}(1+e)^{1/2}/(1-e)^{1/2} \right], \tag{3}$$

where $e > 0$ denotes that the asteroid is at pericenter and $e < 0$ denotes that it is at apocenter, and the $\pm$ sign has the same meaning as in the family of circular orbits, defined by Equation (2). In Equation (3) $a$ is fixed, equal to $a = n^{-2/3} = (p/q)^{2/3}$, for a given resonance, and the eccentricity varies along the family. According to the type of resonance, i.e. the values of $p$ and $q$, we have different phases at the initial condition For the particular resonances we study, a simple geometry gives the following configurations at $t = 0$ and at $t = T/2$:

**2:1 Resonance and 4:1 Resonance**

type I: at $t = 0$, $S - A_p - J$ $(x_0 > 0)$, at $t = T/2$, $A_p - S - J$ $(x_0 < 0)$
type II: at $t = 0$, $S - A_a - J$ $(x_0 > 0)$, at $t = T/2$, $A_a - S - J$ $(x_0 < 0)$

**3:1 Resonance**

type I: at $t = 0$, $S - A_p - J$ $(x_0 > 0)$, at $t = T/2$, $S - A_a - J$ $(x_0 > 0)$
type II: at $t = 0$, $A_p - S - J$ $(x_0 < 0)$, at $t = T/2$, $A_a - S - J$ $(x_0 < 0)$

We note that in each resonance there exist two different branches of the resonant family, differing in phase only, defined as *type I* and *type II*. These are shown (schematically) in Figure 1. We remark that the resonant families at the 2 : 1 and 4 : 1 resonances can be represented by their characteristic curves at the half plane $x_0 > 0$ only of Figure 1, but the 3 : 1 resonant families require the whole plane, $x_0 > 0$ for *type I* and $x_0 < 0$ for *type II*, for their presentation.

## 3. Continuation of the Circular Case

The unperturbed families of periodic orbits described in section 2 can be continued to $\mu > 0$. In all our computations we have used the value

$$\mu = 0.00095387535.$$

For the general theory of the restricted three body problem see Roy (1982) and Szebehely (1967). A review on the existence proofs is given in Hadjidemetriou (1981,1988). The continuation of the circular orbits of the asteroid is possible in all cases except at the resonances 2 : 1, 3 : 2, .... In all other cases these orbits are continued as periodic orbits with a nearly circular orbit of the asteroid, called *periodic orbits of the first kind*. The continuation of the elliptic families is also possible in all cases and we have, for $\mu \neq 0$, symmetric periodic orbits with nearly elliptic orbits of the asteroid, called *periodic orbits of the second kind*.

In what follows, we study each resonance separately and we compare the results with those obtained by the averaging method.

**2:1 Resonance:**The continuation of the family of circular orbits is not possible at this resonance. The unperturbed family of circular orbits breaks down at this point and we have the continuation shown in Figure 2 (for $x_0 > 0$). The unperturbed family is shown with a dotted line (for more information see Colombo *et. al.* 1968, or Hadjidemetriou and Ichtiaroglou, 1984).

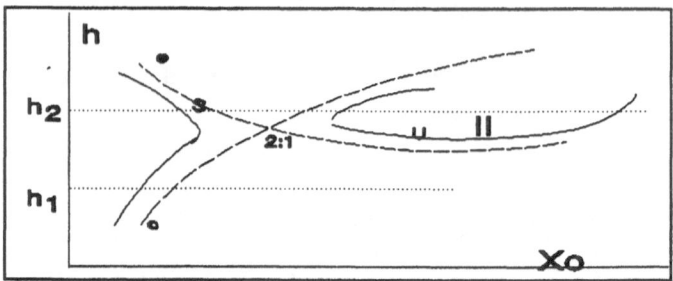

Fig. 2. The families of periodic orbits of the first and second kind near the 2:1 resonance. The dotted curves represent the unperturbed families of circular orbits (c) and elliptic orbits (e).

There exist two families of periodic orbits of the second kind, families I and II, shown in Figure 2, along which the semimajor axis $a$ is almost constant (except for very small values of the eccentricity) and the eccentricity $e$ increases. Family I is stable and family II is unstable, up to a periodic orbit which is a collision orbit. Beyond that point there is a complicated structure with small stable and unstable regions and after that (for $x_0 > 1.0325$) the family II is stable. The details of this structure will be presented in a future publication. The initial configurations of the families I and II are of the type I and II, respectively, as in section 2.

We have used in Figure 2 the energy constant $h$, (Jacobi constant) instead of $\dot{y}$, for the vertical axis. This is more convenient if we wish to see the relation between the families of periodic orbits and the structure of the phase space on a Poincaré surface of section, defined by

$$h = constant, \quad y = 0. \tag{4}$$

For the characteristic curves of Figure 2 we can see that there are only two topologically different types of surfaces of section, for the energy levels $h_1$ and $h_2$ indicated in Figure 2. Some phase diagrams at different energy levels are shown in Figure 3. The fixed points on these diagrams correspond to the periodic orbits. Note the role played by the position of the fixed points and their stability type on the topology of the phase space.

The phase diagrams shown in Figure 3 are topologically similar to the diagrams of the level curves of the averaged Hamiltonian at the 2 : 1 resonance, obtained by Schubart (1964) and recently by Henrard and Lemaitre (1983). The coincidence is good for relatively small values of the eccentricity of the asteroid (this corresponds to low values of the energy constant). For higher values of the eccentricity (or the energy constant), the numerical computations of Hadjidemetriou and Ichtiaroglou (1984) have shown that the invariant curves dissolve, starting from the unstable fixed points. This means that the averaged Hamiltonian is a realistic model for small values of the eccentricity. This should be expected, because the truncation of the series expansions were made at small powers of the eccentricity. On the

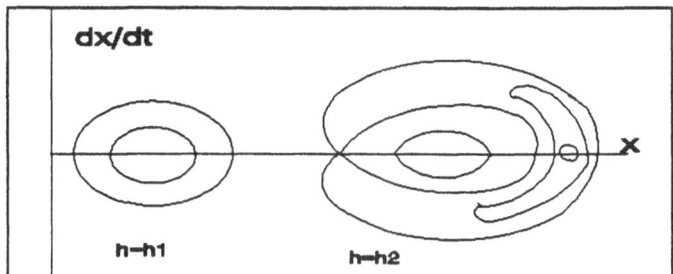

Fig. 3. The structure of phase space at some representative surfaces of section, at the energy levels indicated in Figure 2, near the 2:1 resonance

other hand, we can say, based on the numerical evidence of the appearance of chaotic motion at high values of the eccentricity, that the perturbation series used to obtain the averaged Hamiltonian do not converge for large values of $e$. The main difference between the averaged model and the original system is that the averaged Hamiltonian is integrable (because it is autonomous with one degree of freedom), while the original system (restricted circular three body problem) has two degrees of freedom and is nonintegrable. Consequently, the averaged model cannot describe chaotic motion that may be present in the real system.

### 3:1 Resonance

The continuation of the family of circular orbits for $\mu = 0$, near the $3 : 1$ resonance is possible. However, it can be proved (Hadjidemetriou, 1982, 1985) that a small unstable region AB is generated at the $3 : 1$ resonance (Figure 4). From the critical points A,B that define this unstable region we have a bifurcation of the families II and I, respectively, of periodic orbits of the second kind. Along these latter two families the resonance $n : n'$ is almost constant and the eccentricity $e$ increases, starting from zero values. The phases of the families I, II are the phases I, II, respectively, defined in section 2. The family I is unstable and the family II is stable (Hadjidemetriou, 1992).

As we did for the families at the $2 : 1$ resonance, we can use the energy constant $h$ instead of $\dot{y}$ on the vertical axis. From the characteristic curves of the families of resonant periodic orbits shown in Figure 4, we can easily see that there exist three topologically different surfaces of section, for the energy levels indicated. The corresponding diagrams of invariant curves are presented in Figure 5.

The central fixed point C in the phase diagrams of Figure 5 corresponds to the periodic orbits of the first kind, and the double fixed points $A_1$, $A_2$ and $B_1$, $B_2$, to the stable and unstable periodic orbits of the second kind, respectively.

The phase diagrams of Figure 5 are similar to the diagrams of the level curves of the averaged Hamiltonian at the $3 : 1$ resonance, given by Schubart (1964) and by Lemaitre (1984), valid for small values of the eccentricity of the asteroid. As in the $2 : 1$ resonance, chaotic motion appears for high values of the eccentricity and

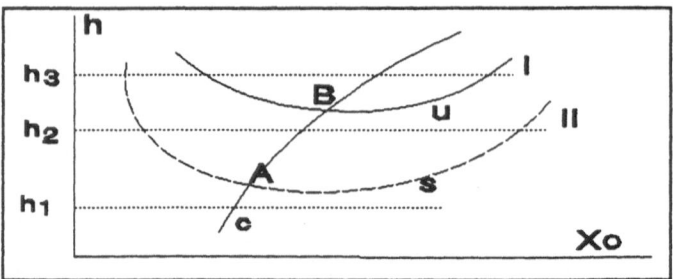

Fig. 4. The family of periodic orbits of the first kind (c) and the families of the second kind (I,II) near the 3:1 resonance (schematically). The family II is shown by a dashed line because it cannot be represented on the present diagram ($x_0 > 0$), because it corresponds to $x_0 < 0$.

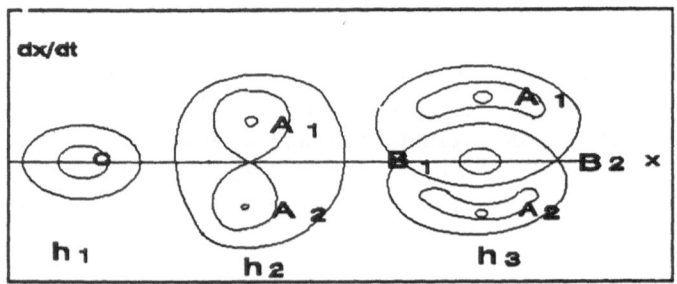

Fig. 5. The structure of the phase space at the energy levels indicated in Fig.4

the same remarks apply also.

### 4:1 Resonance

The continuation of the family of circular orbits near the 4 : 1 resonance is possible, as in the 3 : 1 case, but now no instability is generated. There is only one critical point $A$, as far as the stability is concerned, from which there bifurcate two resonant families of periodic orbits of the second kind, families I and II, along which the resonance $n : n'$ is almost constant and the eccentricity $e$ increases, starting from zero values. The Family I is stable and the family II is unstable. Their initial phases are of the type I and II, respectively, as defined in the previous section.

From the characteristic curves of Figure 6 we can easily see that there are two topologically different surfaces of section, at the energy levels $h_1$ and $h_2$ (Figure 7).

The diagrams of Figure 7 are similar to the level curves of the averaged Hamiltonian near the 4 : 1 resonance, obtained by Schubart (1964) and by Lemaitre (1984). In Figure 7 the fixed point C represents the periodic orbits of the first kind and the triple fixed points $A_1, A_2, A_3$ and $B_1, B_2, B_3$ represent the stable and

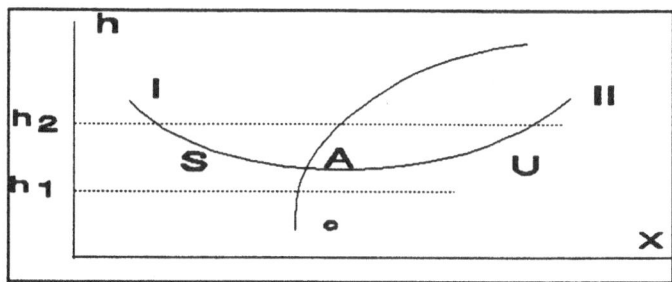

Fig. 6. The family of periodic orbits of the first kind (c) and the families I, II of periodic orbits of the second kind near the 4 : 1 resonance

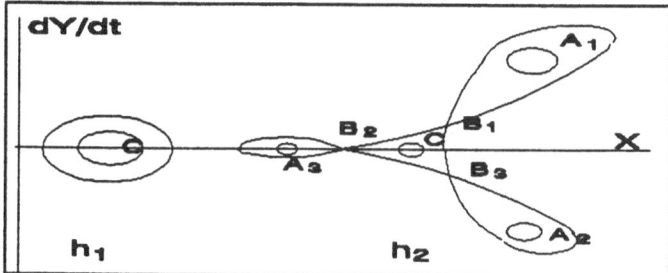

Fig. 7. The structure of phase space at the energy levels indicated in Figure 6

unstable periodic orbits of the second kind.

In the above we presented three typical resonances of the circular three body problem. We restricted ourselves to schematic diagrams only, in order to present clearly the main features of the problem, without being lost in unnecessary details. The exact numerical values of the families presented can be found in the literature cited. These families, especially those for the 2 : 1 and 3 : 1 resonances are well known, but we presented them in order to give a complete picture of the resonant structure of the restricted three body problem and also, because they provide the basic framework for the study of the more realistic model, the elliptic three body problem.

## 4. Families of Periodic Orbits of the Elliptic Restricted Three Body Problem

In sections 2 and 3 we presented the resonant structure of the restricted three body problem and we showed how the topology changes when we go from the unperturbed to the perturbed ,circular, problem. It is the appearance of the stable and unstable fixed points in the case of $\mu \neq 0$ that determines the topology of the phase space. We start from the families of periodic orbits of the circular problem and we study the resonant structure of the elliptic restricted problem. The periodic orbits will be considered in a rotating frame $xOy$ with the origin $O$ at the center of mass of the Sun-Jupiter system and the positive $x$-axis along the line Sun-Jupiter. This is a non-uniformly rotating frame, where the angular velocity of rotation is a known function of the time and Jupiter moves on the $x$-axis according to Keplerian theory. We have a nonautonomous Hamiltonian system with two degrees of freedom, that depends periodically on time with period equal to $2\pi$ (the period of Jupiter's orbit). The energy integral no longer exists, and this will have important consequences on the stability of the periodic orbits, as we will see in the following.

We can study the elliptic restricted three body problem by considering a mapping defined by the intersections of the orbit at the times $t_n = nT$, $n = 1, 2, \dots$. This means that the phase space of the mapping is four-dimensional, in the space $x, y, \dot{x}, \dot{y}$. The periodic orbits are the fixed points of the mapping, but now it is not easy to have a clear geometric picture of the structure of this four-dimensional phase space. The fixed points determine critically the topology of the phase space, as in the two dimensional case of the circular problem, and consequently any model of the elliptic problem, for example an averaged Hamiltonian, must have the correct fixed points (position and stability properties) in order to be realistic.

We study in the following the periodic orbits of the elliptic problem at the $2 : 1$, $3 : 1$ and $4 : 1$ resonances. For a fixed value of the eccentricity of Jupiter the periodic orbits of the elliptic problem are isolated. We can define a family of periodic orbits if we allow the eccentricity $e'$ of Jupiter to vary. (For the theory on the elliptic problem see Broucke, 1968, 1969). The periodic orbits of the elliptic problem that are known are all symmetric with respect to the $x$-axis of the rotating frame. These orbits start perpendicularly from the $x$-axis and at the same time the velocity of Jupiter on the $x$-axis is equal to zero (i.e. Jupiter is either at perihelion or aphelion). The period of a periodic orbit of the elliptic problem is always equal to $2\pi$, or a multiple of it.

We can represent a family of symmetric periodic orbits by a *characteristic curve* in the space $x_0$, $\dot{y}_0$, $e'$, with an indication whether Jupiter is at perihelion or aphelion. All the known families bifurcate from the families of the circular problem that we described in the previous section. The continuation is possible only at those points on the families of the circular problem that their period is equal $(p/q)\pi$, where $p, q$ are integers, for example $\pi$, $2\pi$. This limits very much the possibility of continuation of a periodic orbit to the elliptic case.

### 2:1 Resonance

There exists one periodic orbit only on the family of periodic orbits of the circular problem near the $2 : 1$ resonance, with a period equal to $2\pi$. This is the orbit $A$ with initial conditions $x = 0.1657759$, $\dot{y}_0 = 3.057715$ and is situated on the stable resonant branch of periodic orbits of the second kind. The eccentricity is equal to $e = 0.72$ and two families of periodic orbits bifurcate from this point,

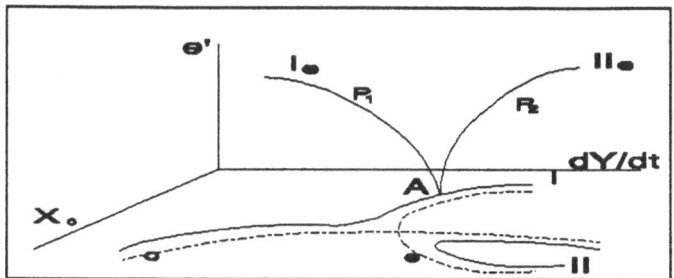

Fig. 8. The bifurcation of the families $I_e$, $II_e$ from the periodic orbit A of the stable branch $I$ of the family of periodic orbits of the circular problem at the 2 : 1 resonance. The dotted curve represent the unperturbed families of circular (c) and elliptic (e) orbits of the asteroid.

families $I_e$ and $II_e$, corresponding to the configuration

$I_e$ :    at $t = 0$, $S - A_p - J_p$, at $t = T/2$, $A_p - S - J_a$

$II_e$ :    at $t = 0$, $S - A_p - J_a$, at $t = T/2$, $A_p - S - J_p$

These two families are shown, schematically, in Figure 8. The initial conditions of some orbits are shown below:

### Family $I_e$: Jupiter at perihelion

| | | | | |
|---|---|---|---|---|
| $e' = 0$ | $x_0 = 0.165776$ | $\dot{y}_0 = 3.057715$ | $e = 0.735$ | $a = 0.6295$ |
| $e' = 0.020$ | $x_0 = 0.172130$ | $\dot{y}_0 = 2.975535$ | $e = 0.725$ | $a = 0.6301$ |
| $e' = 0.048$ | $x_0 = 0.187429$ | $\dot{y}_0 = 2.796163$ | $e = 0.701$ | $a = 0.6308$ |
| $e' = 0.010$ | $x_0 = 0.186246$ | $\dot{y}_0 = 2.785537$ | $e = 0.704$ | $a = 0.6322$ |

### Family $II_e$: Jupiter at aphelion

| | | | | |
|---|---|---|---|---|
| $e' = 0$ | $x_0 = 0.165776$ | $\dot{y}_0 = 3.057715$ | $e = 0.735$ | $a = 0.6295$ |
| $e' = 0.020$ | $x_0 = 0.136000$ | $\dot{y}_0 = 3.473983$ | $e = 0.782$ | $a = 0.6286$ |
| $e' = 0.048$ | $x_0 = 0.109003$ | $\dot{y}_0 = 3.971428$ | $e = 0.824$ | $a = 0.6263$ |
| $e' = 0.075$ | $x_0 = 0.077703$ | $\dot{y}_0 = 4.809806$ | $e = 0.873$ | $a = 0.6202$ |

For the present value of the eccentricity of Jupiter, $e' = 0.048$, we have two isolated periodic orbits, $P_1$ and $P_2$, at the intersection of the characteristic curves of the families $I_e$, $II_e$ with the plane $e' = 0.048$, corresponding to high values of the eccentricity of the asteroid. These orbits should appear as fixed points in an averaged Hamiltonian. However, in the usual way of obtaining the averaged Hamiltonian,(Henrard and Lemaitre, 1983) the series are truncated to low order in the eccentricity and therefore the obtained averaged Hamiltonian cannot describe accurately the resonant structure of the elliptic problem at high values of the eccentricity. Morbidelli and Giorgilli (1990a,b) have obtained an averaged Hamiltonian valid for high values of the eccentricity and they found the above mentioned periodic orbits at $e' = 0.048$, with eccentricities equal to $e = 0.71$ and $e = 0.76$, respectively. The first orbit is stable and the second unstable. The numerical computations we made for the stability of these periodic orbits are not very accurate, because the orbit is very sensitive to the initial conditions, and for this reason we must use the results obtained by the averaging method, as developed by Morbidelli and Giorgilli. Ferraz-Mello *et. al.* (1992) also found two families of fixed points of an averaged Hamiltonian valid for high values of the eccentricity, corresponding to the above mentioned two families of periodic orbits. Their results are in good agreement with the numerical computations.

### 3:1 Resonance

There exist on the families of periodic orbits of the circular problem near the 3:1 resonance, two points of bifurcation of families of periodic orbits of the elliptic problem. One is a periodic orbit on the family of periodic orbits of the first kind, with initial conditions $x_0 = 0.479420$, $\dot{y}_0 = 0.962393$, with a period equal to $\pi$, inside the unstable region AB, and the other is a periodic orbit on the stable family of resonant periodic orbits of the second kind with initial conditions $x_0 = -0.866333$, $\dot{y}_0 = 0.384417$. This latter bifurcation point has an eccentricity equal to $e = 0.798$ and a period equal to $2\pi$. From each of the above periodic orbits of the circular problem there bifurcate two families of periodic orbits of the elliptic problem, namely families $I_c$, $II_c$ from the circular family and families $I_e$, $II_e$ from the elliptic family. Along these families the eccentricity $e'$ of Jupiter varies and the resonance $n : n'$ is almost constant, equal to $3 : 1$. These families are shown schematically in Figure 9. The initial configuration of these families is given below (Hadjidemetriou, 1992).

$$I_c \ : \ at \ t = 0, \ A_p - S - J_p \ at \ t = T/2, \ A_p - S - J_a$$
$$II_c: \ at \ t = 0, \ S - A_p - J_p \ at \ t = T/2, \ S - A_a - J_a$$
$$I_e \ : \ at \ t = 0, \ A_p - S - J_p \ at \ t = T/2, \ A_p - S - J_a$$
$$II_e: \ at \ t = 0, \ A_p - S - J_a \ at \ t = T/2, \ A_a - S - J_p$$

The family $I_e$ is stable and the other three families are unstable. The initial conditions of these families are given in Hadjidemetriou (1992) and will not be given here.

These families of periodic orbits of the elliptic problem have been compared with the corresponding families of fixed points of the averaged Hamiltonian used by Wisdom (1985) and by Henrard and Caranicolas (1989). We present here the main results. For the details see Hadjidemetriou (1992).

The above mentioned averaged Hamiltonian is

$$H = H_0(N, S) + \mu H_1(N, S, \sigma) + \mu e' H_2(S, \sigma, N, \nu), \tag{5}$$

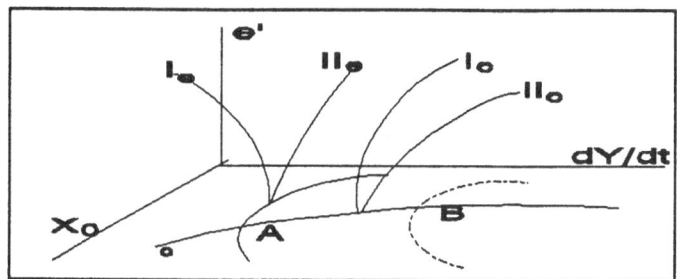

Fig. 9. The bifurcation of the families $I_c, II_c, I_e, II_e$ of periodic orbits of the elliptic problem at the 3 : 1 resonance

where

$$H_0 = -\frac{2(1-\mu)^2}{(N-S)^2},$$

$$H_1 = 2FS - b\frac{S}{N}\cos 2\sigma,$$

$$H_2 = \sqrt{2S}\left[G\cos(\sigma+\nu) + D\cos(\sigma-\nu)\right] + 2\mu e' K\cos 2\nu,$$

and the action-angle variables $S, \sigma, N, \nu$ are defined by

$$S = \sqrt{(1-\mu)a}\left(1 - \sqrt{1-e^2}\right),$$

$$\sigma = \frac{1}{2}(3\lambda' - \lambda) - \omega,$$

$$N = \sqrt{(1-\mu)a}\left(3 - \sqrt{1-e^2}\right),$$

$$\nu = -\frac{1}{2}(3\lambda' - \lambda) + \omega'.$$

The numerical coefficients are
   $b = 2.392398$, $F = -0.205070$, $G = 0.198705$,
   $D = 2.656407$, $K = -0.181477$, $e' = 0.048$.
   It was found that the averaged model defined by the Hamiltonian (5) is a good approximation of the elliptic restricted three body problem near the 3 : 1 resonance, for small values of $e$, as expected. This model does not contain the resonant fixed points corresponding to the families of periodic orbits $I_e$, $II_e$, at high values of the eccentricity. Consequently, this model cannot describe the evolution of an asteroid for high values of $e$. However, a correction term is proved to exist for the averaged

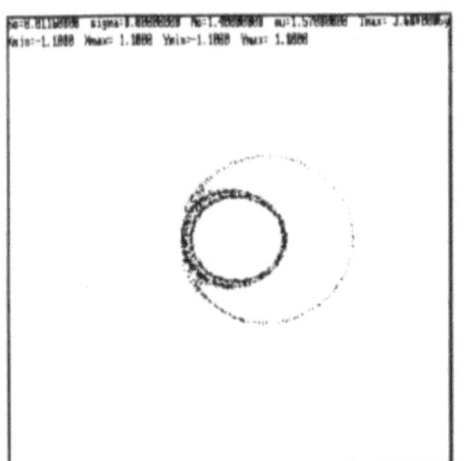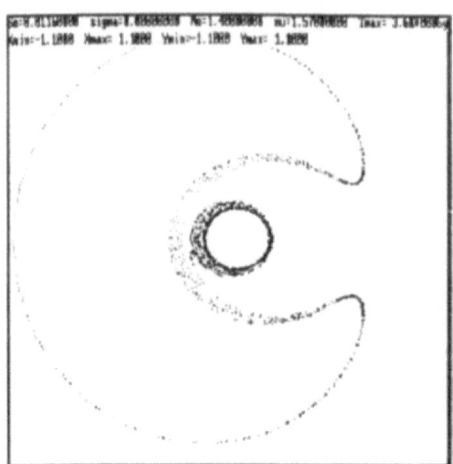

Fig. 10. Evolution of the asteroid in the $X, Y$ plane at the 3 : 1 resonance: (a) the model does not include the high eccentricity resonances, (b) the model includes the high eccentricity resonances

Hamiltonian (5), which introduces the missing fixed points (Hadjidemetriou 1992). This correction term is found to be

$$H_c = -3.4\mu(1 - 6.4e' \cos(\sigma + \nu)) S^3. \tag{6}$$

In this way we obtain a simple model which has all the main features of the real system.

The effect of the resonant fixed points of the families $I_e, II_e$ on the long term evolution of the asteroids is shown in Figures 10a,b. The evolution is given in the axes $X = \sqrt{2S} \cos(\sigma + \nu)$, $Y = \sqrt{2S} \sin(\sigma + \nu)$. In these Figures we show the evolution of an asteroid near the 3 : 1 resonance, obtained by the study of a suitable mapping, based on the averaged Hamiltonian (Hadjidemetriou 1991). In Figure 10a we have used the simple model of Wisdom and Henrard, given by (5), where the resonant fixed points $I_e, II_e$ are missing. In Figure 10b the corrected model $H + H_c$ has been used, so that the above resonant fixed points are also present in the model. The difference is evident. We remark in this respect that if we start with initial conditions such that the eccentricity of the asteroid is small, then the two models behave almost identically, as it was found by numerical computations with the mapping.

The evolution shown in Figure 10b is similar to that obtained by Klafke et. al. (1992) by numerical integrations of the differential equations of an averaged Hamiltonian that they have found, valid for high values of the eccentricity. In their model, the above mentioned resonant fixed points are present.

The fixed points corresponding to the families $I_c, I_e, II_e$ of periodic orbits of the elliptic problem for $e' = 0.048$ are correctly given by Morbidelli and Giorgilli (1990), obtained by an averaged Hamiltonian that they have found, valid for high values of the eccentricity. The fixed points corresponding to the families $I_c, II_c, I_e, II_e$ have been also correctly given by Ferraz-Mello et. al. (1992), obtained by an averaged Hamiltonian that they found, different from that of Morbidelli and Giorgilli, valid also for high values of the eccentricity. These Hamiltonians, valid for high values of $e$, are very useful for the study of resonant motion near the 3 : 1 resonance, but they are not simple and consequently are not convenient for analytic work. On the other hand, the corrected Hamiltonian $H + H_c$ has all the properties of the real system and behaves in the same way as the more accurate model of Ferraz-Mello, as is clear from the comparison between Figure 10a and the corresponding Figure given by Klafke et. al (1992).

### 4:1 Resonance

There exist three periodic orbits of the circular problem near the 4 : 1 resonance, from which there bifurcate families of periodic orbits of the elliptic problem. One orbit is on the family of periodic orbits of the first kind, close to the point A (Figure 6), from which the two families of periodic orbits of the second kind start. This orbit has a period equal to $2\pi/3$ and can be continued to the elliptic problem if we assume that it is described three times, so that its period is equal to $2\pi$. Its initial conditions are $x_0 = 0.395733$, $\dot{y}_0 = 1.190494$. The other two periodic orbits of the circular problem from which we have a bifurcation of families of the elliptic problem are both on the stable resonant family $I$ of periodic orbits of the second kind. Their period is equal to $2\pi$ and their initial conditions are $x_0 = 0.299295$, $\dot{y}_0 = 1.733614$ and $x_0 = 0.077836$, $\dot{y}_0 = 4.700496$. (points $A_1, A_2$ in Figure 11). The first orbit corresponds to an eccentricity of the asteroid equal to $e = 0.243$ and the second to $e = 0.801$. These bifurcation points are denoted in Figure 11 by $A_1$ and $A_2$, respectively.

From each of the above orbits there bifurcate two families of periodic orbits of the elliptic problem, that are shown schematically in Figure 11.The families that bifurcate from the family of periodic orbits of the first kind start with zero eccentricity of the asteroid, and are called families $I_c, II_c$. The families that bifurcate from the family of periodic orbits of the second kind start with nonzero eccentricities of the asteroid and are called families $I_e, II_e$ and $III_e, IV_e$, respectively.

The phase of the above families is given below:

$$I_c \ : \ t = 0, \ S - A_p - J_p, \quad t = T/2, \ A_p - S - J_a$$
$$II_c : \ t = 0, \ S - A_p - J_a, \quad t = T/2, \ A_p - S - J_p$$
$$I_e, III_e : \ \ t = 0, \ S - A_p - J_p, \quad t = T/2, \ A_p - S - J_a$$
$$II_e, IV_e : \ t = 0, \ S - A_p - J_a, \quad t = T/2, \ Ap - S - J_p$$

The families $I_c$ and $II_c$ are unstable and the families $I_e$ and $II_e$ are stable. The stability of the families $III_e, IV_e$ could not be obtained with good accuracy, and we do not give it here.

We give below some initial conditions for the above families:

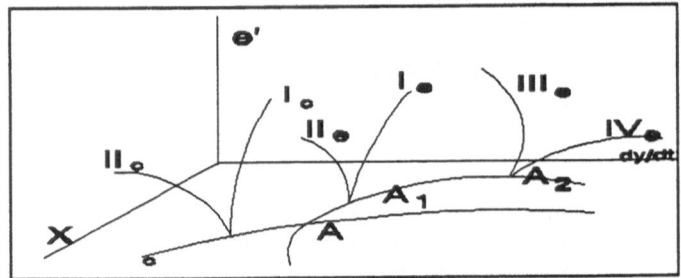

Fig. 11. The families of periodic orbits of the elliptic problem at the 4 : 1 resonance

## Family $I_c$: Jupiter at perihelion

| | | | | |
|---|---|---|---|---|
| $e' = 0$ | $x_0 = 0.395733$ | $\dot{y}_0 = 1.190494$ | $e = 0.0003$ | $a = 0.3972$ |
| $e' = 0.05$ | $x_0 = 0.378593$ | $\dot{y}_0 = 1.237525$ | $e = 0.0447$ | $a = 0.3973$ |
| $e' = 0.10$ | $x_0 = 0.354497$ | $\dot{y}_0 = 1.325581$ | $e = 0.1057$ | $a = 0.3975$ |
| $e' = 0.20$ | $x_0 = 0.300954$ | $\dot{y}_0 = 1.564030$ | $e = 0.2414$ | $a = 0.3980$ |
| $e' = 0.35$ | $x_0 = 0.225103$ | $\dot{y}_0 = 2.016289$ | $e = 0.4346$ | $a = 0.3998$ |

## Family $II_c$: Jupiter at aphelion

| | | | | |
|---|---|---|---|---|
| $e' = 0$ | $x_0 = 0.395733$ | $\dot{y}_0 = 1.190494$ | $e = 0.0003$ | $a = 0.3972$ |
| $e' = 0.01$ | $x_0 = 0.397015$ | $\dot{y}_0 = 1.191898$ | $e = -0.0021$ | $a = 0.3937$ |
| $e' = 0.02$ | $x_0 = 0.396386$ | $\dot{y}_0 = 1.202629$ | $e = -0.0006$ | $a = 0.3941$ |
| $e' = 0.03$ | $x_0 = 0.390295$ | $\dot{y}_0 = 1.240289$ | $e = 0.0147$ | $a = 0.3946$ |
| $e' = 0.05$ | $x_0 = 0.360970$ | $\dot{y}_0 = 1.404728$ | $e = 0.0875$ | $a = 0.3966$ |
| $e' = 0.10$ | $x_0 = 0.292971$ | $\dot{y}_0 = 1.826506$ | $e = 0.2585$ | $a = 0.3964$ |
| $e' = 0.16$ | $x_0 = 0.228092$ | $\dot{y}_0 = 2.321809$ | $e = 0.4213$ | $a = 0.3959$ |

## Family $I_e$: Jupiter at perihelion

| | | | | |
|---|---|---|---|---|
| $e' = 0$ | $x_0 = 0.299295$ | $\dot{y}_0 = 1.733614$ | $e = 0.243$ | $a = 0.3967$ |
| $e' = 0.048$ | $x_0 = 0.217642$ | $\dot{y}_0 = 2.332853$ | $e = 0.450$ | $a = 0.3974$ |
| $e' = 0.080$ | $x_0 = 0.146939$ | $\dot{y}_0 = 3.142349$ | $e = 0.628$ | $a = 0.3979$ |
| $e' = 0.090$ | $x_0 = 0.106069$ | $\dot{y}_0 = 3.892089$ | $e = 0.732$ | $a = 0.3992$ |
| $e' = 0.100$ | $x_0 = 0.052047$ | $\dot{y}_0 = 5.872047$ | $e = 0.870$ | $a = 0.4079$ |

## Family $II_e$: Jupiter at aphelion

| | | | | |
|---|---|---|---|---|
| $e' = 0$ | $x_0 = 0.299295$ | $\dot{y}_0 = 1.733614$ | $e = 0.2430$ | $a = 0.3967$ |
| $e' = 0.01$ | $x_0 = 0.317271$ | $\dot{y}_0 = 1.627262$ | $e = 0.1986$ | $a = 0.3971$ |
| $e' = 0.02$ | $x_0 = 0.336945$ | $\dot{y}_0 = 1.517788$ | $e = 0.1490$ | $a = 0.3971$ |
| $e' = 0.03$ | $x_0 = 0.361906$ | $\dot{y}_0 = 1.386701$ | $e = 0.0452$ | $a = 0.3967$ |
| $e' = 0.04$ | $x_0 = 0.385989$ | $\dot{y}_0 = 1.268952$ | $e = 0.0245$ | $a = 0.3967$ |
| $e' = 0.05$ | $x_0 = 0.410069$ | $\dot{y}_0 = 1.158271$ | $e = -0.0360$ | $a = 0.3966$ |

## Family $III_e$: Jupiter at perihelion

| | | | | |
|---|---|---|---|---|
| $e' = 0$ | $x_0 = 0.077836$ | $\dot{y}_0 = 4.700478$ | $e = 0.801$ | $a = 0.3967$ |
| $e' = 0.05$ | $x_0 = 0.080606$ | $\dot{y}_0 = 4.599587$ | $e = 0.796$ | $a = 0.3990$ |
| $e' = 0.08$ | $x_0 = 0.078027$ | $\dot{y}_0 = 4.682457$ | $e = 0.803$ | $a = 0.4007$ |
| $e' = 0.10$ | $x_0 = 0.071419$ | $\dot{y}_0 = 4.923792$ | $e = 0.820$ | $a = 0.4027$ |
| $e' = 0.12$ | $x_0 = 0.06056432$ | $\dot{y}_0 = 5.40045369$ | $e = 0.849$ | $a = 0.4067$ |
| $e' = 0.15$ | $x_0 = 0.03636179$ | $\dot{y}_0 = 7.10683320$ | $e = 0.914$ | $a = 0.4325$ |

## Family $IV_e$: Jupiter at aphelion

| | | | | |
|---|---|---|---|---|
| $e' = 0$ | $x_0 = 0.077836$ | $\dot{y}_0 = 4.700478$ | $e = 0.801$ | $a = 0.3967$ |
| $e' = 0.010$ | $x_0 = 0.083159$ | $\dot{y}_0 = 4.525573$ | $e = 0.788$ | $a = 0.3963$ |
| $e' = 0.020$ | $x_0 = 0.213797$ | $\dot{y}_0 = 2.398451$ | $e = 0.459$ | $a = 0.3966$ |
| $e' = 0.030$ | $x_0 = 0.239965$ | $\dot{y}_0 = 2.176001$ | $e = 0.392$ | $a = 0.3966$ |
| $e' = 0.035$ | $x_0 = 0.170342$ | $\dot{y}_0 = 2.864093$ | $e = 0.568$ | $a = 0.3964$ |
| $e' = 0.041$ | $x_0 = 0.023039$ | $\dot{y}_0 = 8.956967$ | $e = 0.936$ | $a = 0.3764$ |

These families, that we computed numerically from the full equations of the elliptic restricted three body problem, are in good agreement with the corresponding families of fixed points of the averaged Hamiltonian, obtained by Ferraz-Mello *et. al.* (1992). Some of the above resonance cases were also given by Morbidelli and Giorgilli (1990a,b), namely the fixed points for $e' = 0.048$ corresponding to the families $I_c$, $I_e$ and $II_e$. The values they give coincide with our values.

We make now the following remarks concerning the evolution of the stability of a periodic orbit in the continuation from the circular to the elliptic case. In both cases we have two degrees of freedom and consequently, two pairs of eigenvalues of the monodromy matrix. In the circular problem one of these pairs is the unit pair, due to the existence of the energy integral. The other pair is on the unit circle, in the complex plane (the two eigenvalues are conjugate), if the orbit is stable, or both eigenvalues are real and their product is equal to unity, if the orbit is unstable. When the eccentricity of Jupiter is taken into account, the energy integral no longer exists and the two unit eigenvalues of the circular case are free to move either on the unit circle or on the real axis, one inside and one outside the unit circle. In the latter case instability is generated.

It is interesting to note that in all the resonant cases we considered in this paper, the bifurcation of the two families of periodic orbits of the elliptic problem that start with nonzero value of $e$ takes place from the stable family of periodic orbits of the second kind of the circular problem. Thus, one pair of eigenvalues for $e' = 0$ is of the stable type and the generation or not of instability depends on whether or not the former unit pair moves outside or on the unit circle, respectively.

## 5. Discussion

We have shown in this work the resonant structure of the elliptic restricted problem near some of the main resonances, 2 : 1, 3 : 1, 4 : 1, that are of interest in the study of asteroid motion. We have chosen these particular resonances because they behave in a topologically different way in the continuation from the unperturbed to the circular restricted problem and because the topology of the phase space in their neighbourhood is different. All other low order resonances are similar to these resonances: All resonances of the form $(i+1)/i$ are similar to the 2 : 1 resonance, all resonances of the form $(i+2)/i$ are similar to the 3 : 1 resonance and all other kinds of resonance different from the above are in fact similar to the 4 : 1 resonance (taking into account the differences due to the different order of the resonance, i.e. the different type of multiplicity of the periodic orbit in the rotating frame).

We started with the well known, even trivial, results concerning the families of periodic orbits of the unperturbed motion near the above resonances, in a rotating frame, and we proceeded to the study of their continuation properties when the mass of Jupiter is increased (but its eccentricity remains equal to zero). Most of these results, especially those referring to the 2 : 1 and 3 : 1 resonances are well known, but we included them in order to have a complete picture of the problem. Our main purpose is to show the resonant structure of the elliptic problem, at these resonances, and for this reason the next and final step is to study the continuation of the periodic orbits of the circular problem to the elliptic case. In this hierarchy of perturbations a clear picture of the resonant structure of the restricted three body problem was revealed. The combination of the two different approaches to the problem, namely the numerical computation of periodic orbits of the complete equations of motion and the study of the fixed points of the averaged problem, was necessary in some cases where one of these methods was not accurate enough. Based on these results, it is easy to find out the region of validity of an averaged Hamiltonian. It is even possible, by a combination of the numerical computations and the method of averaging, to find simple correction terms to an averaged Hamiltonian in such a way as to include in the averaged Hamiltonian the missing resonances and thus obtain a new "averaged" Hamiltonian which is realistic for a larger region of phase space.

## Acknowledgements

I thank Dr. S. Ferraz-Mello for useful comments. This research was carried out in the framework of the Community Research Programme with a financial contribution by the Commission of the European Communities.

## References

Broucke, R.: 1968, *Periodic Orbits in the Elliptic Restricted Three Body problem*, JPL Technical Report 32-1360.
Broucke, R. 1969: *Stability of the Periodic Orbits in the Restricted Three Body Problem*, AIAA 7, 1003-1009.
Colombo,G., Franklin, F.A. and Munford, C.M.: 1968, *On a family of Periodic Orbits of the Restricted Three Body Problem and the Question of the Gaps in the Asteroid Belt and in Saturn's Rings*, Astron.J. 73, 111-123.

Ferraz-Mello, S., Tsuchida, M. and Klafke, J.C.: 1992, *On Symmetrical Planetary Corotations*, Celest. Mech. (to appear).

Ferraz-Mello, S., Tsuchida, M. and Klafke, J.C.: 1992, *Corotations in some higher-order Resonances*, Proceedings IAU Symposium 152.

Hadjidemetriou, J.D.: 1981, *The present status of Periodic Orbits*, Celest. Mech. **23**, 277–286.

Hadjidemetriou, J.D.: 1982, *On the Relation between Resonance and instability in Planetary Systems*, Celest. Mech. **27**, 305–322.

Hadjidemetriou, J.D.: 1985, *The Stability of Resonant Orbits in Planetary systems*, in S. Ferraz-Mello and W. Sessin (eds.) *Resonances in the Motion of Planets, Satellites and Asteroids*, Univ. of Sao Paulo.

Hadjidemetriou, J.D.: 1988, *Periodic Orbits*, in M.J. Valtonen (ed.), *The Few Body Problem*, Kluwer Publ., 31–48.

Hadjidemetriou, J.D.: 1991, *Mapping Models for Hamiltonian Systems with applications to Resonant Asteroid motion*, in *Predictability, Stability and Chaos in N-Body Dynamical Systems*, A.E. Roy (ed.), Plenum Press, 157–175.

Hadjidemetriou, J.D.: 1992, *The Elliptic Restricted Problem at the 3:1 Resonance*, Celest. Mech. (to appear).

Hadjidemetriou, J.D. and Ichtiaroglou, S.: 1984, *A qualitative study of the Kirkwood Gaps in the Asteroids*, Astron. Astrophys. **131**, 20–32.

Henrard, J. and Lemaitre, A.: 1983, *A mechanism of Formation of the Kirkwood Gaps*, Icarus **55**, 482–494.

Henrard, J. and Caranicolas, D.: 1990, *Motion near the 3:1 Resonance of the Planar Elliptic Restricted Three-Body Problem*, Celest. Mech. **47**, 99–121.

Klafke, J.C., Ferraz-Mello, S. and Michtchenko, T.: 1992, *Very-High-Eccentricity Librations at some higher order Resonances*, IAU Symposium 152.

Lemaitre, A.: 1984, *Higher Order Resonances in the Restricted Three Body Problem*, Celest. Mech. **32**, 109–126.

Lichtenberg, A.J. and Liebermann, M.A.: 1983, *Regular and Stochastic Motion*, Springer-Verlag.

Morbidelli, A. and Giorgilli, A.: 1990a, *On the Dynamics in the Asteroids Belt, Part I: General Theory*, Celest. Mech. **47**, 145–172.

Morbidelli, A. and Giorgilli, A.: 1990a, *On the Dynamics in the Asteroids Belt, Part II: Detailed study of the main Resonances*, Celest. Mech. **47**, 173–204.

Roy, A.E.: 1982, *Orbital Motion*, Adam Hilger (2nd ed.).

Schubart, J.: 1964, *Long Period effects in nearly Commensurable cases of the Restricted Three Body Problem*, Report No 149, Smithsonian Astrophysical Observatory.

Szebehely, V.: 1967, *Theory of Orbits*, Academic Press.

Wisdom, J.: 1985, *A perturbative treatment of Motion near the 3/1 Commensurability*, Icarus **63**, 272–289.

# A FOURTH-ORDER SOLUTION OF THE IDEAL RESONANCE
# PROBLEM

BÁLINT ÉRDI

*Department of Astronomy, Eötvös University, Budapest, Hungary*

and

JÓZSEF KOVÁCS

*Gothard Observatory of the Eötvös University, Szombathely, Hungary*

**Abstract.** The second-order solution of the Ideal Resonance Problem, obtained by Henrard and Wauthier (1988), is developed further to fourth order applying the same method. The solutions for the critical argument and the momentum are expressed in terms of elementary functions depending on the time variable of the pendulum as independent variable. This variable is related to the original time variable through a 'Kepler-equation'. An explicit solution is given for this equation in terms of elliptic integrals and functions. The fourth-order formal solution is compared with numerical solutions obtained from direct numerical integrations of the equations of motion for two specific Hamiltonians.

**Key words:** Hamiltonian systems – resonance.

## 1. Introduction

The Ideal Resonance Problem (IRP), defined by Garfinkel (1966), can be characterized, as in Henrard and Wauthier (1988), by the one-degree of freedom Hamiltonian system

$$\frac{dx}{dt} = \frac{\partial H}{\partial y}, \quad \frac{dy}{dt} = -\frac{\partial H}{\partial x}, \quad H(x,y) = B(y) - \varepsilon^2 A(y)\cos x, \tag{1}$$

where $\varepsilon$ is the perturbation parameter, $B$ and $A$ are analytical functions of the momentum $y$ and $B' \equiv \partial B/\partial y$ vanishes for some value $y_0$. In the normal form of the IRP it is assumed also that $B''(y_0) \neq 0$ and $A(y_0) \neq 0$. The normal form of the IRP is a model for certain types of resonances and it has been studied thoroughly. In the domain $y - y_0 \approx O(\varepsilon)$, it is close to the pendulum and its solution shows a pendulum-like character with regions of libration and circulation, separated by a separatrix.

There have been two basic approaches to derive a solution for the IRP. Garfinkel (1966), Garfinkel et al. (1971) and Garfinkel and Williams (1974) applied the classical Bohlin perturbation method with mixed-variable canonical transformations to develop a global solution. In contrast, Jupp (1969, 1970, 1972) and Jupp and Abdulla (1984, 1985) used a Lie series perturbation method and obtained a local solution valid in the libration region and in the circulation region within deep resonance. Both theories offer explicit formulas for the solution of the IRP up to second order. These formulas, however, are rather involved due to the presence of the elliptic integrals and functions generated by the pendulum-like nature of the problem.

*Celestial Mechanics and Dynamical Astronomy* **56**: 221–230, 1993.

The complexity of the aforementioned solutions can be substantially simplified as it was recently shown by Howland (1988) and Henrard and Wauthier (1988). Howland found a specific non-canonical transformation which brings the Hamiltonian system of the IRP into a 'quasi-canonical' form. The new Hamiltonian is expressed in terms of trigonometric functions and the solution of the new system, obtained by a traditional perturbation technique, is similarly expressed in trigonometric functions. The elliptic integrals, expected of the system, are introduced only in a final explicit quadrature for a Kepler-type equation in the angular variable. Howland's solution, derived to second order, is restricted to the librational region.

A more general solution, valid for both the librational and circulation region, was obtained by Henrard and Wauthier (1988). They studied a more general problem, representing a large class of perturbations of the pendulum, containing the normal form of the IRP as a special case. They considered a one-degree of freedom Hamiltonian system with a Hamiltonian expressed as a formal power series

$$H(x,p) = \sum_{i \geq 0} \frac{\varepsilon^i}{i!} H_i(x,p), \tag{2}$$

where the coefficients $H_i$ are polynomials in $p$, $\cos x$, and $\sin x$, and $H_0$ is the (properly scaled) pendulum Hamiltonian

$$H_0(x,p) = \frac{1}{2}p^2 - \cos x. \tag{3}$$

Using the non-canonical version of the Lie transform technique, Henrard and Wauthier (1988) showed how one can construct a non-canonical coordinate transformation

$$x = X(\bar{x}, \bar{p}, \varepsilon), \quad p = P(\bar{x}, \bar{p}, \varepsilon) \tag{4}$$

from $(\bar{x}, \bar{p})$ to $(x, p)$ such that $H(x,p)$ is transformed into the function

$$\bar{H}(\bar{x}, \bar{p}) = h(\varepsilon) + C(\varepsilon)(\frac{1}{2}\bar{p}^2 - \cos \bar{x}), \tag{5}$$

where $h(\varepsilon)$ and $C(\varepsilon)$ are formal power series in $\varepsilon$. Up to an additive constant $h(\varepsilon)$ and a scale factor $C(\varepsilon)$, the transformed Hamiltonian is thus the pendulum Hamiltonian itself. (We remark, that the bar in $\bar{x}$, $\bar{p}$ and $\bar{H}$ is just a denotation for the new variables, it does not represent a mean of the old ones.)

The transformation (4) brings the Hamiltonian system

$$\frac{dx}{dt} = \frac{\partial H}{\partial p}, \quad \frac{dp}{dt} = -\frac{\partial H}{\partial x} \tag{6}$$

into the 'quasi-Hamiltonian' equations

$$K\frac{d\bar{x}}{dt} = \frac{\partial \bar{H}}{\partial \bar{p}}, \quad K\frac{d\bar{p}}{dt} = -\frac{\partial \bar{H}}{\partial \bar{x}}, \tag{7}$$

where $K$ is the determinant of the Jacobian matrix

$$K(\bar{x}, \bar{p}) = \det \left\{ \frac{\partial(X, P)}{\partial(\bar{x}, \bar{p})} \right\}. \tag{8}$$

With a change of the time variable from $t$ to $\tau$, defined by

$$\frac{dt}{d\tau} = K(\bar{x}, \bar{p}), \tag{9}$$

the system (7) can be transformed to the Hamiltonian form

$$\frac{d\bar{x}}{d\tau} = \frac{\partial \bar{H}}{\partial \bar{p}}, \quad \frac{d\bar{p}}{d\tau} = -\frac{\partial \bar{H}}{\partial \bar{x}}. \tag{10}$$

Considering (5), the system (10) is essentially the pendulum problem whose solution is known. Equation (9) is termed as the 'Kepler equation'. Its role is that it connects the time variable $t$ with the time $\tau$ of the pendulum.

Henrard and Wauthier (1988) applied this method to the normal form of the IRP and derived a solution up to second order. In this case the coordinate transformation (4) is very simply expressed in terms of elementary functions. Elliptic integrals and functions enter into the solution only through the solution of Equations (10) and (9).

In the second-order solution of the IRP the function $K(\bar{x}, \bar{p})$ has the form

$$K = f_0 + f_1 \bar{p} + f_2 \bar{p}^2, \tag{11}$$

with $f_0$, $f_1$, $f_2$ as constants. Henrard and Wauthier (1988) showed, in principle, how one can determine the coordinate transformation in such a way that $K$ preserve the form (11) to any order.

The method of Henrard and Wauthier (1988) provides an easy way to go above second order in the solution of the normal form of the IRP. In the next section a fourth-order solution is presented; the details of the straightforward calculations are not included. The transformation has been determined according to the principle that K is of the form (11).

## 2. A Fourth-Order Solution of the IRP

First, the system (1) is brought to the form (6) with $H$ given by (2). This is done by developing the Hamiltonian (1) around $y_0$, omitting the redundant constant term $B(y_0)$ and applying a scaling of the time and momentum given by

$$t = \alpha t', \quad p = \beta(y - y_0), \quad \alpha = \varepsilon \sqrt{A_0 B_0''}, \quad \beta = \frac{1}{\varepsilon} \sqrt{B_0''/A_0}. \tag{12}$$

Here $t$ is the new time, $p$ is the new momentum and $A_0 = A(y_0)$, $B_0'' = B''(y_0)$. The new Hamiltonian is

$$H(x, p) = \frac{1}{2}p^2 - \cos x + \varepsilon(b_3 p^3 - a_1 p \cos x) + \frac{\varepsilon^2}{2}(2b_4 p^4 - 2a_2 p^2 \cos x) + \dots, \tag{13}$$

where

$$a_i = \frac{A_0^{(i)}}{i!} \frac{A_0^{\frac{i}{2}-1}}{(B_0'')^{\frac{i}{2}}}, \quad i = 1, 2, \dots, \quad b_i = \frac{B_0^{(i)}}{i!} \frac{A_0^{\frac{i}{2}-1}}{(B_0'')^{\frac{i}{2}}}, \quad i = 3, 4, \dots. \tag{14}$$

Applying the method, described in Henrard and Wauthier (1988), we find for the coordinate transformation up to fourth order

$$x = \bar{x} + \varepsilon^2 x_{21} \sin \bar{x} + \varepsilon^3 x_{31} \bar{p} \sin \bar{x} + \varepsilon^4[(x_{41}\bar{p}^2 + x_{42})\bar{p}^2 \sin \bar{x} + (x_{43}\bar{p}^2 + x_{44}) \sin 2\bar{x}], \tag{15}$$

$$p = \bar{p} + \varepsilon(p_{11}\bar{p}^2 + p_{12} \cos \bar{x}) + \varepsilon^2(p_{21}\bar{p}^2 + p_{22} \cos \bar{x})\bar{p} + \varepsilon^3[(p_{31}\bar{p}^2 + p_{32} \cos \bar{x})\bar{p}^2 + p_{33} + p_{34} \cos 2\bar{x}] + \varepsilon^4[p_{41}\bar{p}^4 + p_{42}\bar{p}^2 + p_{43} + (p_{44}\bar{p}^2 + p_{45}) \cos \bar{x} + (p_{46}\bar{p}^2 + p_{47}) \cos 2\bar{x} + p_{48} \cos 3\bar{x}]\bar{p}, \tag{16}$$

where the coefficients $x_{jk}$, $p_{jk}$ are given in the Appendix as polynomials in $a_i$, $b_i$. The function $K$ is of the form (11) with

$$f_0 = 1 + \varepsilon^2 f_{01} h + \varepsilon^4(f_{02} h^3 + f_{03} h^2 + f_{04} h + f_{05}),$$
$$f_1 = \varepsilon f_{11} + \varepsilon^3 f_{12} h, \tag{17}$$
$$f_2 = \varepsilon^2 f_{21} + \varepsilon^4(f_{22} h^2 + f_{23} h + f_{24}),$$

where

$$h = \frac{1}{2}\bar{p}^2 - \cos \bar{x}, \tag{18}$$

and the coefficients $f_{jk}$ are also given in the Appendix as polynomials in $a_i$, $b_i$. The transformed Hamiltonian $\bar{H}$ is of the form (5) with

$$h(\varepsilon) = -\frac{1}{2}\varepsilon^2 a_1^2, \quad C(\varepsilon) = 1 + \varepsilon^4(a_2 - a_1 b_3)a_1^2. \tag{19}$$

Equation (9) with $K$ given by (11) and (17) can be integrated by means of elliptic integrals. Depending on the value of $h$ ($h$ is constant along the solutions of the pendulum) we distinguish between libration and circulation.

*Case of libration* ($|h| < 1$):

$$t = \{f_0\tau + 2f_1 \arcsin(\sqrt{m}\sin u) + 4f_2[(m-1)\mathbf{F}(u,m) + \mathbf{E}(u,m)]\}C(\varepsilon)^{-1}, \tag{20}$$

with

$$m = \frac{h+1}{2}, \quad u = \operatorname{am}(\tau, m), \tag{21}$$

and $\mathbf{F}(u,m)$, $\mathbf{E}(u,m)$ are the elliptic integrals of first and second kind.

*Case of circulation* ($h > 1$):

$$t = [f_0\tau + 2f_1 u + \frac{4}{\sqrt{m}} f_2 \mathbf{E}(u,m)]C(\varepsilon)^{-1}, \tag{22}$$

with

$$m = \frac{2}{h+1}, \quad u = \operatorname{am}(\frac{\tau}{\sqrt{m}}, m). \tag{23}$$

The algorithm giving the solution of the IRP at the fourth order is as follows:

(i) Given the initial conditions $\bar{p}(0)$, $\bar{x}(0) = 0$, we calculate $h$ from (18), the coefficients $f_i$ from (17) and the modulus $m$ from (21) or (23).

(ii) The position at a later time is obtained by solving the Kepler equation (20) or (22) first, and then using the solution of the pendulum.

*Case of libration:* $\bar{x} = 2\arcsin(\sqrt{m}\sin u)$, $\bar{p} = 2\sqrt{m}\cos u$. $\qquad$ (24)

*Case of circulation:* $\bar{x} = 2u\operatorname{sign}(\bar{p}(0))$, $\bar{p} = \operatorname{sign}(\bar{p}(0))[2(h + \cos\bar{x})]^{1/2}$. $\quad$ (25)

(iii) Then $x$, $p$ are obtained from (15) and (16). Note, that the solutions are singular for $h = 1$ at the separatrix.

Since $f_1 \approx O(\varepsilon)$, $f_2 \approx O(\varepsilon^2)$, the solution of the Kepler equation can be calculated either by means of the Newton-Raphson method or by using the following analytical solution obtained by applying Lagrange's inversion method.

*Case of libration:*

$$w = \frac{tC(\varepsilon)}{f_0}, \quad u = \operatorname{am}(w, m),$$

$$U = \arcsin(\sqrt{m}\sin u), \quad I = (m-1)\mathbf{F}(u,m) + \mathbf{E}(u,m), \tag{26}$$

$$\begin{aligned}
\tau = {}& w + \varepsilon k_{11}U + \varepsilon^2(k_{21}I + m^{\frac{1}{2}}k_{11}^2 U \operatorname{cn} w) + \varepsilon^3[k_{12}U + \\
& k_{11}^3(mU \operatorname{cn}^2 w - \frac{1}{2}m^{\frac{1}{2}}U^2 \operatorname{sn} w \operatorname{dn} w) + k_{11}k_{21}(m^{\frac{1}{2}}I + \\
& mU \operatorname{cn} w)\operatorname{cn} w] + \varepsilon^4\{k_{22}I + 2m^{\frac{1}{2}}k_{11}k_{12}U \operatorname{cn} w + \\
& mk_{21}^2 I \operatorname{cn}^2 w + k_{11}^2 k_{21}[(I + 2m^{\frac{1}{2}}U \operatorname{cn} w)m \operatorname{cn}^2 w - \\
& (m^{\frac{1}{2}}I + mU \operatorname{cn} w)U \operatorname{sn} w \operatorname{dn} w] + k_{11}^4[m \operatorname{cn}^2 w - \\
& \frac{3}{2}m^{\frac{1}{2}}U \operatorname{sn} w \operatorname{dn} w + \frac{1}{6}U^2(m \operatorname{sn}^2 w - \operatorname{dn}^2 w)\operatorname{cn} w]m^{\frac{1}{2}}U\operatorname{cn} w\}. \tag{27}
\end{aligned}$$

*Case of circulation:*

$$w = \frac{tC(\varepsilon)}{f_0\sqrt{m}}, \quad u = \text{am}(w, m), \quad J = \mathbf{E}(u, m), \tag{28}$$

$$
\begin{aligned}
\tau = {} & w + \varepsilon k_{11} u + \varepsilon^2 (k_{21}J + k_{11}^2 u\,\text{dn}\,w) m^{-\frac{1}{2}} + \varepsilon^3 [k_{12}u + \\
& k_{11}^3 (m^{-1} u\,\text{dn}^2 w - \frac{1}{2} u^2\,\text{sn}\,w\,\text{cn}\,w) + m^{-1} k_{11} k_{21} (J + \\
& u\,\text{dn}\,w)\text{dn}\,w] + \varepsilon^4 \{ k_{22}J + 2k_{11}k_{12}u\,\text{dn}\,w + \\
& m^{-1} k_{21}^2 J\,\text{dn}^2 w + k_{11}^2 k_{21}[(J + 2u\,\text{dn}\,w)m^{-1}\,\text{dn}^2 w - \\
& (J + u\,\text{dn}\,w)u\,\text{sn}\,w\,\text{cn}\,w] + k_{11}^4 [m^{-1}\,\text{dn}^2 w - \\
& \frac{3}{2} u\,\text{sn}\,w\,\text{cn}\,w - \frac{1}{6} u^2 (\text{cn}^2 w - \text{sn}^2 w)]u\,\text{dn}\,w\} m^{-\frac{1}{2}}, \tag{29}
\end{aligned}
$$

where $\text{sn}\,w$, $\text{cn}\,w$, $\text{dn}\,w$ are the Jacobi elliptic functions and in both cases the $k_{ij}$ coefficients are

$$k_{11} = -2f_{11}, \quad k_{12} = -2f_{12}h + 2f_{01}f_{12}h,$$

$$k_{21} = -4f_{21}, \quad k_{22} = 4f_{01}f_{21}h - 4(f_{22}h^2 + f_{23}h + f_{24}). \tag{30}$$

## 3. A Numerical Comparison

The fourth-order analytical solution was compared with numerical solutions for two Hamiltonians.

*Example 1:*

$$B(y) = y + \frac{1}{2y^2}, \quad A(y) = y + 1.$$

*Example 2:*

$$B(y) = \frac{y^4}{4} - \frac{y^3}{3} + \frac{y^2}{2} - y + 2, \quad A(y) = y^2 + 3y - 2.$$

This selection of the functions $B(y)$, $A(y)$ corresponds to the examples studied by Jupp and Abdulla (1984, 1985) who compared Garfinkel's and Jupp's first- and second-order theories with numerical solutions. In both examples $B'(1) = 0$, thus $y_0 = 1$. The small parameter $\varepsilon$ was given the respective values $10^{-1}$, $10^{-3/2}$ and $10^{-2}$.

The initial values were $\bar{x}(0) = 0$ and $\bar{p}$ was varied over the libration and circulation regions. The initial values of $x$ and $y$ for the numerical integration were obtained from (15), (16), and (12). The numerical solutions were obtained using the RK7(8) method (Fehlberg, 1968). The constancy of the Hamiltonian was checked and it showed that the integration was very accurate for the comparison with the analytical solutions. The elliptic integrals appearing in the analytical solutions were calculated from algorithms given in Abramowitz and Stegun (1964).

The conclusions of the comparison are very similar to those of Jupp and Abdulla (1984, 1985) except that, as one would expect, the fourth-order solutions show better agreement with the numerical solutions. The general behaviour is the same in both examples and for each value of $\varepsilon$. The agreement is one order of magnitude better for the momentum $y$ than for the coordinate $x$; this is consistent with the finding of Jupp and Abdulla (1984, 1985). Close to the libration centre the agreement is very good, for $\varepsilon = 10^{-2}$ it is of $O(10^{-13})$ and $O(10^{-11})$ for the first example, and $O(10^{-12})$ and $O(10^{-10})$ for the second example. For the same value of $\varepsilon$ and for the second example, half-way between the centre and the separatrix the agreement is of $O(10^{-10})$ and $O(10^{-8})$. Very close to the separatrix but within libration the agreement is of $O(10^{-8})$ and $O(10^{-6})$. Close to the separatrix, but in circulation the agreement is the same. In shallow resonance it is of $O(10^{-7})$, $O(10^{-5})$. For the first example the agreement is somewhat better. At the classical limit, the analytical solution gives unsatisfactory results, but this is to be expected.

## Acknowledgements

We would like to thank Dr. A. H. Jupp for his suggestions on this paper as a referee.

## Appendix

The $x_{jk}$, $p_{jk}$, $f_{jk}$ coefficients of Equations (15), (16) and (17) as polynomials in $a_i$, $b_i$.

$$x_{21} = -\frac{1}{2}a_1^2$$

$$x_{31} = 64b_3^3 - a_1^2 b_3 - 24a_1 b_3^2 + 8a_1 b_4 + 8a_2 b_3 - 48b_3 b_4 - 2a_3 + 8b_5$$

$$p_{11} = -b_3$$

$$p_{12} = a_1$$

$$p_{21} = \frac{5}{2}b_3^2 - b_4$$

$$p_{22} = -3a_1 b_3 + a_2$$

$$p_{31} = -8b_3^3 + 6b_3 b_4 - b_5$$

$$p_{32} = 12a_1 b_3^2 - 4a_1 b_4 - 4a_2 b_3 + a_3$$

## TABLE I
The $x_{jk}$ coefficients at fourth order.

|  | $x_{41}$ | $x_{42}$ | $x_{43}$ | $x_{44}$ |
|---|---|---|---|---|
| $a_1^4$ |  | -1/4 |  | 3/16 |
| $b_3^4$ |  | -129/2 |  |  |
| $a_1^3 b_3$ |  | -5/3 |  | 1/2 |
| $a_1 b_3^3$ | -27/4 | 69/2 | 27/2 |  |
| $a_1^2 b_3^2$ | 1 | -23/4 | -11/8 |  |
| $a_1^2 a_2$ |  | 1 |  | -1/2 |
| $a_1^2 b_4$ | -2/5 | 1/2 | 11/20 |  |
| $a_2 b_3^2$ | 18/5 | -7 | -36/5 |  |
| $b_3^2 b_4$ |  | 114 |  |  |
| $a_1 a_3$ | 1/5 | -3/4 | -11/40 |  |
| $a_1 b_5$ |  | 10 |  |  |
| $a_1 a_2 b_3$ | -8/15 | 13/3 | 11/15 |  |
| $a_1 b_3 b_4$ | 27/10 | -45 | -27/5 |  |
| $a_3 b_3$ | -27/20 | 3/2 | 27/10 |  |
| $b_3 b_5$ |  | -92/3 |  |  |
| $a_2^2$ |  | -1/2 |  |  |
| $b_4^2$ |  | -70/3 |  |  |
| $a_2 b_4$ |  | 10 |  |  |
| $a_4$ |  | -2 |  |  |
| $b_6$ |  | 20/3 |  |  |

## TABLE II
The $p_{jk}$ coefficients at fourth order.

|  | $p_{41}$ | $p_{42}$ | $p_{43}$ | $p_{44}$ | $p_{45}$ | $p_{46}$ | $p_{47}$ | $p_{48}$ |
|---|---|---|---|---|---|---|---|---|
| $a_1^4$ |  |  | 1/8 |  |  |  | -1/8 |  |
| $b_3^4$ | 231/8 |  | 385/4 |  |  |  | -385/4 |  |
| $a_1^3 b_3$ |  |  | 1/12 |  |  |  | -7/12 |  |
| $a_1 b_3^3$ |  | -9/4 | -293/4 | -165/4 | -27/4 | -9 | 293/4 | 27/4 |
| $a_1^2 b_3^2$ |  | -1/2 | 93/4 |  | 7/2 | 1/2 | -9/2 | 1/4 |
| $a_1^2 a_2$ |  |  | 1/4 |  |  |  | 1/4 |  |
| $a_1^2 b_4$ |  | 1/5 | -13/2 |  | -7/5 | -1/5 | 2 | -1/10 |
| $a_2 b_3^2$ |  | 6/5 | 23/2 | 23/2 | 18/5 | 24/5 | -23/2 | -18/5 |
| $b_3^2 b_4$ | -63/2 |  | -105 |  |  |  | 105 |  |
| $a_1 a_3$ |  | -1/10 | 5/2 |  | 7/10 | 1/10 | -1/4 | 1/20 |
| $a_1 b_5$ |  |  | -9 | -5 |  |  | 9 |  |
| $a_1 a_2 b_3$ |  | 4/15 | -38/3 |  | -28/15 | -4/15 | -1/3 | -2/15 |
| $a_1 b_3 b_4$ |  | 9/10 | 109/2 | 61/2 | 27/10 | 18/5 | -109/2 | -27/10 |
| $a_3 b_3$ |  | -9/20 | -11/4 | -11/4 | -27/20 | -9/5 | 11/4 | 27/20 |
| $b_3 b_5$ | 7 |  | 70/3 |  |  |  | -70/3 |  |
| $a_2^2$ |  |  | 1 |  |  |  | 1/2 |  |
| $b_4^2$ | 7/2 |  | 35/3 |  |  |  | -35/3 |  |
| $a_2 b_4$ |  |  | -5 | -5 |  |  | 5 |  |
| $a_4$ |  |  | 1 | 1 |  |  | -1 |  |
| $b_6$ | -1 |  | -10/3 |  |  |  | 10/3 |  |

TABLE III

The $f_{jk}$ coefficients at fourth order.

| | $f_{02}$ | $f_{03}$ | $f_{04}$ | $f_{05}$ | $f_{22}$ | $f_{23}$ | $f_{24}$ |
|---|---|---|---|---|---|---|---|
| $a_1^4$ | | 1/2 | | -3/24 | | -3/4 | |
| $b_3^4$ | | -385/2 | | 385/2 | | 352 | |
| $a_1^3 b_3$ | | 10/3 | | -4/3 | | -5 | |
| $a_1 b_3^3$ | -27 | 165/2 | 27 | -165/2 | -27/2 | -33/4 | -81/4 |
| $a_1^2 b_3^2$ | -1 | 15 | -11/4 | 15/4 | 9/2 | -41/2 | 9/8 |
| $a_1^2 a_2$ | | -2 | | 1 | | 3 | |
| $a_1^2 b_4$ | 2/5 | -4 | 11/10 | -1/2 | -9/5 | 3 | -9/20 |
| $a_2 b_3^2$ | 72/5 | -23 | -72/5 | 23 | 36/5 | 19/2 | 54/5 |
| $b_3^2 b_4$ | | 210 | | -210 | | -264 | |
| $a_1 a_3$ | -1/5 | 3/2 | -11/20 | 3/4 | 9/10 | -9/4 | 9/40 |
| $a_1 b_5$ | | 10 | | -10 | | 15 | |
| $a_1 a_2 b_3$ | 8/15 | -26/3 | 22/15 | -13/3 | -12/5 | 13 | -3/5 |
| $a_1 b_3 b_4$ | 54/5 | -61 | -54/5 | 61 | 27/5 | -95/2 | 81/10 |
| $a_3 b_3$ | -27/5 | 11/2 | 27/5 | -11/2 | -27/10 | -11/4 | -81/20 |
| $b_3 b_5$ | | -140/3 | | 140/3 | | 44 | |
| $a_2^2$ | | 1 | | 1/2 | | -3/2 | |
| $b_4^2$ | | -70/3 | | 70/3 | | | |
| $a_2 b_4$ | | 10 | | -10 | | 15 | |
| $a_4$ | | -2 | | 2 | | -3 | |
| $b_6$ | | 20/3 | | -20/3 | | | |

$$p_{33} = \frac{1}{4}a_1^3 - 32b_3^3 - a_1^2 b_3 + 12a_1 b_3^2 + a_1 a_2 - 4a_1 b_4 -$$

$$4a_2 b_3 + 24b_3 b_4 + a_3 - 4b_5$$

$$p_{34} = -\frac{1}{4}a_1^3 + 32b_3^3 - 2a_1^2 b_3 - 12a_1 b_3^2 + a_1 a_2 + 4a_1 b_4 +$$

$$4a_2 b_3 - 24b_3 b_4 - a_3 + 4b_5$$

$$f_{01} = \frac{1}{2}a_1^2 + 3a_1 b_3 - a_2$$

$$f_{11} = -2b_3$$

$$f_{12} = -64b_3^3 + 48b_3 b_4 - 8b_5$$

$$f_{21} = \frac{15}{2}b_3^2 - \frac{1}{4}a_1^2 - \frac{3}{2}a_1 b_3 + \frac{1}{2}a_2 - 3b_4$$

The $x_{jk}$, $p_{jk}$, $f_{jk}$ coefficients at fourth order are given in Tables I-III. Each coefficient is a polynomial of the terms given in the first column with numerical coefficients in the columns below $x_{jk}$, $p_{jk}$, $f_{jk}$.

# References

Abramowitz, M. and Stegun, I.A.: 1964, *Handbook of Mathematical Functions*, National Bureau of Standards, Applied Mathematics Series 55.

Fehlberg, E.: 1968, *NASA TR* **R-287**, .

Garfinkel, B.: 1966, *Astron. J.* **71**, 657.

Garfinkel, B., Jupp, A. H. and Williams, C. A.: 1971, *Astron. J.* **76**, 157.

Garfinkel, B. and Williams, C. A.: 1974, *Celest. Mech.* **9**, 105.

Henrard, J. and Wauthier, P.: 1988, *Celest. Mech.* **44**, 227.

Howland, R. A.: 1988, *Celest. Mech.* **44**, 209.

Jupp, A. H.: 1969, *Astron. J.* **74**, 35.

Jupp, A. H.: 1970, *Monthly Notices Roy. Astron. Soc.* **148**, 197.

Jupp, A. H.: 1972, *Celest. Mech.* **5**, 8.

Jupp, A. H. and Abdulla, A. Y.: 1984, *Celest. Mech.* **34**, 411.

Jupp, A. H. and Abdulla, A. Y.: 1985, *Celest. Mech.* **37**, 183.

# WAVELET ANALYSIS AND APPLICATIONS TO SOME DYNAMICAL SYSTEMS

PH. BENDJOYA and E. SLEZAK

*Observatoire de la Côte d'Azur BP 229 / F-06304 NICE cedex 4 , France*

**Abstract.** For about five years the wavelet transform technique has given interesting results in a great number of different fields such as mathematics, quantum mechanics, signal analysis and image processing, cluster analysis... The wavelet transform appears as a new time-frequency method which is particularly well-suited to detect and to localize discontinuities and scaling behaviours in signals. The main properties of the wavelet transform and its improvements over classical analyzing methods are summarized. Some results among the first applications to the dynamical systems are presented: solution of partial differential equations, fractal and turbulence characterization, and asteroid family determination from cluster analysis.

**Key words:** time-frequency analysis - wavelet transform - differential equations - fractal - turbulence - asteroid families

## 1. Introduction

Dynamical systems are a more or less intricated set of particles where by definition a temporal and/or spatial evolution occurs. Typical examples are celestial bodies and particles in a fluid. In order to study the behaviour of these systems mathematical models are made which provide signals characterizing their evolution. Different states of the system are possible: periodicity, quasi-periodicity, stochasticity, deterministic chaos or turbulence. Transitions between these different states are observed which can generate hierarchical structures. Moreover, the presence of features at different interacting scales is revealed when considering non linear effects. One main interest in the study of such signals is to point out the different scales present in the model and the location where these features appear or disappear. The goal is to extract frequencies, the time at which these frequencies appear, and their changes as a function of time. Another interest can be the detection of a peculiar scaling law, like a fractal behaviour. Different mathematical tools are available in order to perform such investigations. Among the best suited tools for such purposes, the wavelet transform has been first introduced by Morlet (1982) in order to get time-frequency informations on seismic signals. The specific properties of the wavelet transform led to a great deal of applications, mainly in 1D and 2D signal processing fields (filtering (Starck and Bijaoui 1991), compression (Barlaud *et al.* 1989), partial reconstructions (Bijaoui and Giudicelli 1991) ...). The purpose of this paper is to present briefly this new transform and to review some of its first applications within the framework of dynamical systems. We will mainly focus our interest on the analysis of the corresponding signals and, after a short introduction to the specific problem under examination, we will pay peculiar attention to the improvements due to the use of the wavelet transform with respect to more classical tools.

*Celestial Mechanics and Dynamical Astronomy* **56**: 231–262, 1993.
© 1993 *Kluwer Academic Publishers.*

The remainder of this paper is organized as follow: the first section presents the wavelet transform and a detailed comparison with the Fourier transform and the Gabor transform. The second section is devoted to the wavelet transform as a new method for solving partial differential equations, like the Burgers equation. The third section exposes the results obtained by means of the wavelet transform in the analysis of fractal objects and their impact on the study of the fully developed turbulence. A new cluster analysis method using the wavelet transform is presented in section 4 as well as its application to the determination of asteroid dynamical families. Future perspectives in various fields are summarized in the conclusion.

## 2. The Available Tools.

We consider now a temporal signal $f(t)$ and we will be interested in the analysis of this function both in time and frequency. Spatial signals will be studied by means of a space-scale analysis derived from the temporal formalism.

### 2.1. THE FOURIER TRANSFORM.

A usual way to characterize temporal signals is to compute their Fourier transform. This consists in the decomposition of the signal $f(t)$ onto a basis of planar waves $\{e^{i2\pi\nu t}\}$:

$$\hat{f}(\nu) = < f(t), e^{i2\pi\nu t} > \tag{1}$$

where $< f(t), g(t) > = \int_{-\infty}^{+\infty} f(t)g^*(t)dt$ with $g^*(t)$ standing for the complex conjugate of $g(t)$. The function $f(t)$ can be reconstructed from its Fourier transform using the inversion formula:

$$f(t) = \int_{-\infty}^{+\infty} \hat{f}(\nu)e^{2i\pi\nu t}d\nu \tag{2}$$

The component of $f(t)$ onto the vector $e^{i2\pi\nu_0 t}$ is $\hat{f}(\nu_0)$ and its value indicates the contribution of $e^{i2\pi\nu_0 t}$ to the signal $f(t)$. The Fourier transform $\hat{f}(\nu)$ does not depend on any localization parameter since the functions $\{e^{i2\pi\nu t}\}$ extend from $-\infty$ to $+\infty$. If a given frequency is present in the signal $f(t)$ over a limited time interval, the Fourier transform is unable to accurately detect this frequency and to give any information about its life-time and the moment of its appearance. This transform is therefore well-suited only for stationary signals defined over $\Re$ where all frequencies have an infinite coherence time (lifetime). Let us recall that a stationary signal is defined from a deterministic point of view by the superposition of components with an instantaneous behaviour (amplitude and frequency) which does not depend on time. This stationarity is defined from a stochastic point of view by an expectation and a correlation function which do not depend on time but only on time differences (Flandrin 1987).

Figure 1 presents a sample signal that will be analyzed in order to compare the

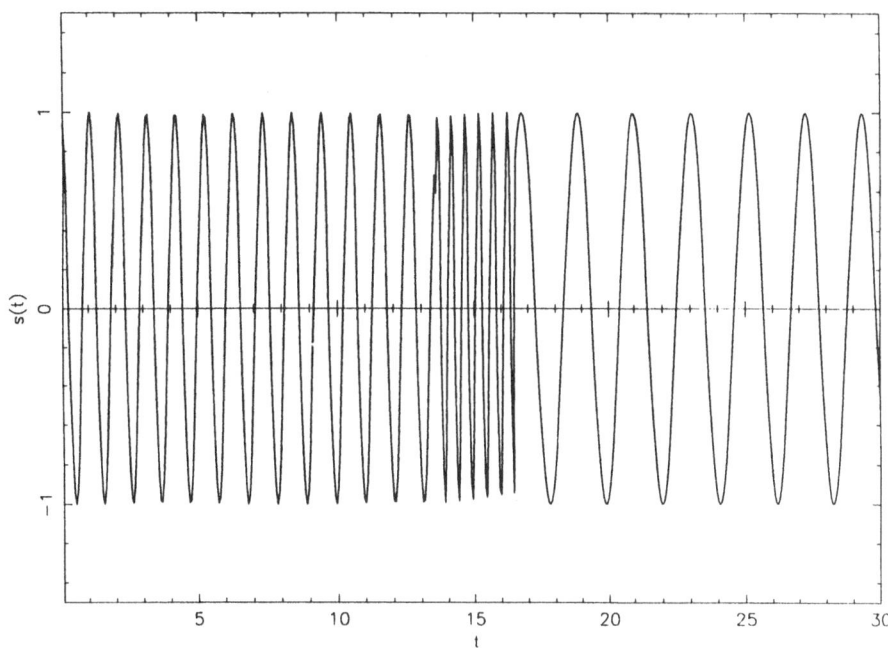

Fig. 1. A sample signal built from sine functions presenting three pulsations: $\omega_1 = 6$, $\omega_2 = 12$, $\omega_3 = 3$ with the same amplitude. The coherence time of $\omega_1$ and $\omega_2$ is identic. Discontinuities are present at the time of change of frequencies. The X-axis is the time, the Y-axis is the value of $f(t)$.

different methods. This function $f(t)$ is built from three successive sine functions with the same lifetime $T_1$ for $\nu_1$ and $\nu_3 < \nu_1$, and a shorter coherence time $T_2$ for $\nu_2 > \nu_1$. This signal presents also two discontinuities when the frequency changes its value. Figure 2 displays the modulus of the Fourier transform, namely the spectrum of $f(t)$. One can notice the good detection of the three frequencies present in the signal, and how the high frequency $\nu_2$ is less precisely detected because of its shorter lifetime. The spurious fluctuations all over the spectrum are due to the discontinuities. Obviously no information can be extracted from the Fourier spectrum about the time at which the different frequencies appear in the signal or about their lifetime: only a frequency analysis is performed, which can be spoiled by discontinuities.

## 2.2. THE WINDOW FOURIER TRANSFORM.

The first improvement in order to build a time-frequency analysis is to introduce a temporal window in the Fourier Transform. Gabor (1946) was the first to introduce such a windowed Fourier transform where the extension of planar waves is limited by a function with a compact support, or at least rapidly vanishing. The window

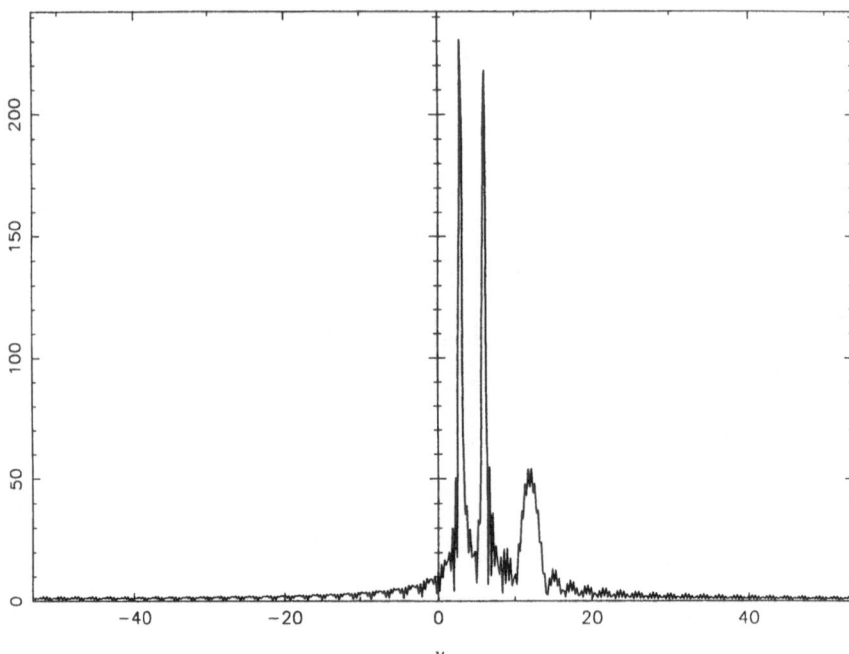

Fig. 2. The Fourier transform of the sample signal. The three frequencies are detected but the shortest one has an obviously lowest weight. The spurious fluctuation all over this spectrum are due to the discontinuities present in the signal

Fourier transform of a function $f(t)$ is a two-dimensional function $G(\nu, b)$ where $\nu$ is the variable associated to the harmonic analysis and $b$ is associated to the location of the window $g(t)$:

$$G(\nu, b) = < f(t), g_{\nu,b}(t) > \qquad \text{with} \qquad g_{\nu,b}(t) = g(t - b)e^{2i\pi\nu t} \qquad (3)$$

$G(\nu, b)$ can be seen as the weight in $f(t)$ associated to the frequency $\nu$ in the neighbourhood of the position $b$. A Gaussian function vanishes rapidly and enables therefore a good localization of the decomposition both in time and frequency since its Fourier transform is still a Gaussian. The so-called Gabor transform is then defined by taking :

$$g(t) = e^{-t^2/2\sigma^2} \qquad (4)$$

The constant $\sigma$, which defines the width of the Gaussian function and the size of the window, can be associated to the coherence time previously defined. >From the set of computed and normalized with respect to $\sigma$ $G(\nu, b)$, a time-frequency $(\nu, b)$ diagram is built where the frequencies present in the signal at a given time can be examined. The original signal can be reconstructed from its Gabor transform using

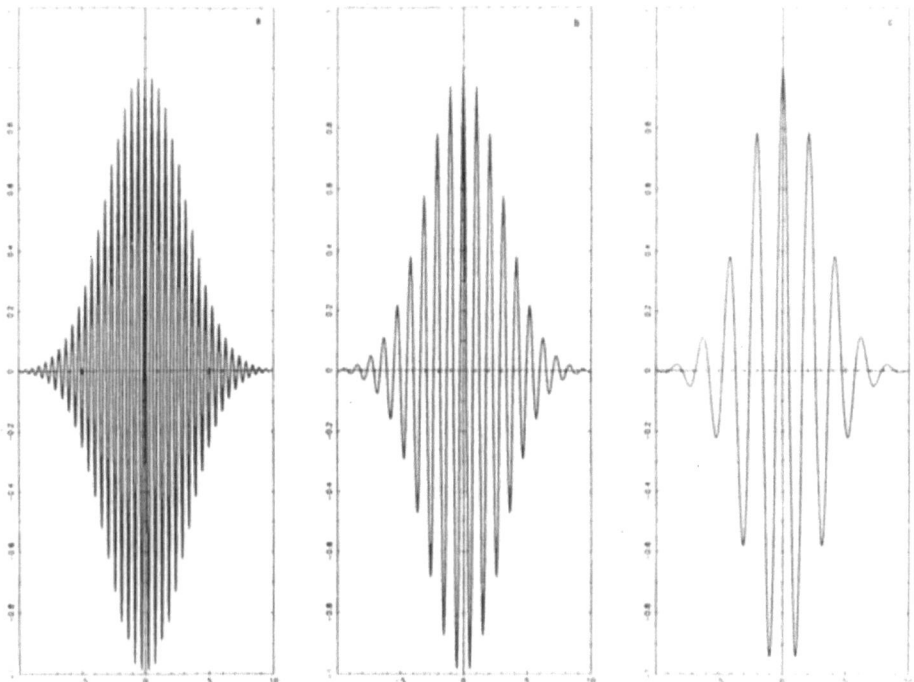

Fig. 3. The filter $g_{\nu_1,0}$ as a function of time (a), the filter $g_{\nu_2,0}$ (b) and the filter $g_{\nu_3,0}$ (c).

the following formula:

$$f(t) = \int_{-\infty}^{+\infty} \int_{-\infty}^{+\infty} G(\nu, b)\, g(t - b)\, e^{2i\pi\nu t}\, d\nu db \tag{5}$$

The signal $f(t)$ can be seen as a sum of localized waves weighted by the profile of the chosen window. Figure 3 show the function $g_{\nu,0}(t)$ for three high (a), middle (b), and low (c) frequencies. It is obvious that the signal $f(t)$ is analyzed by filters which are not the same when the investigated frequency changes. The Gabor transform has been performed on our sample signal for different values of $\nu$ in order to achieve its time-frequency analysis. The value of the parameter $\sigma$ fixing the size of the window has been chosen in order to adapt this size to the coherence time $T_1$. Figure 4 presents the result of this time-frequency analysis using a gray level coding where the highest value of the Gabor coefficients is coded in black and the lowest one in white. The Gabor transform is able to detect the frequencies present in the signal and their temporal location. However, the difficulties to detect frequencies with a lifetime shorter than the size of the window appear clearly. Although all the three frequencies do have the same amplitude in the original signal, this is not the case in the time-frequency diagram where lower coefficients are associated to the highest frequency present in the signal with a very short lifetime. This frequency

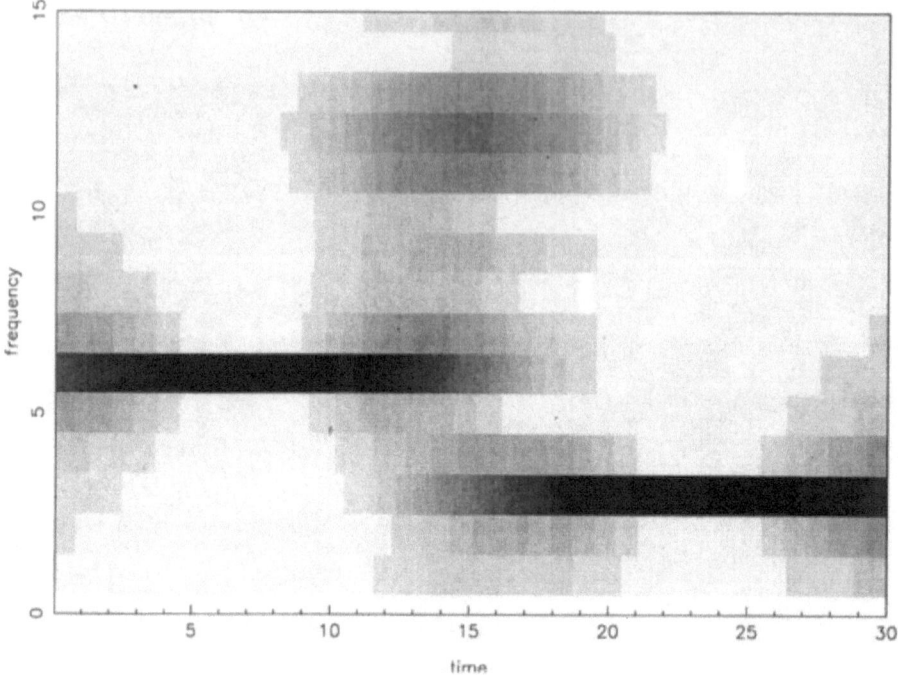

Fig. 4. The time frequency analysis performed by means of the Gabor transform with $\sigma$ adapted to the coherence time of $\nu_1$ and $\nu2$. The Gabor coefficients are coded in gray level: the highest values are black, the lowest white. The X-axis is the time, the Y-axis the pulsations. The pulsation with the shortest lifetime is detected worse than the other for which the size of the window is well suited.

could be interpreted as noise instead of a real signal. The contribution of the short lifetime frequency is obviously under-estimated. Hence, it appears that the windowed Fourier transform is well-suited only for signals with coherence times equal to the temporal size of the window. This is the case for instance for singing signals which do have their coherence time determined by the geometry of the oral cavity. This coherence time is more or less the same for all frequencies and such signals can therefore be studied with great benefit by means of the windowed Fourier transform.

## 2.3. THE WAVELET TRANSFORM.

Obviously a transform resulting into the use of filters with the same shape for each frequency would provide one with a less ambiguous interpretation. Such a transform corresponds to an adaptative window with a size inversely proportional to the frequency under examination. This adaptive window characterizes the wavelet transform introduced by Morlet (1983) in order to get enough resolution whatever the analyzed frequency is. This transform appears suited for analyzing signals where frequencies have a coherence-time proportional to their value. This is necessary for instance in speech analysis, where consonants and vowels have different coherence times, or for studying physical processes which exhibit strong dispersion relationships.

The wavelet transform of a $1D$ function $f(t)$ is a two-dimensional function $C(\nu, b)$ obtained by the decomposition of $f(t)$ onto a basis of functions $\psi_{\nu, b}$ obtained from dilations and translations of a unique function $\psi(t)$ called the analyzing wavelet:

$$C(\nu, b) = \langle f(t), \psi_{\nu, b}(t) \rangle \qquad \text{with} \qquad \psi_{\nu, b} = K(\nu)\psi(\nu(t - b)) \tag{6}$$

where $K(\nu)$ is a normalization factor introduced in order to have the same norm for the whole set of filters $\psi_{\nu, b}(t)$. It is equal to $\nu$ if $f(t)$ belongs to $L^1$ and to $\sqrt{\nu}$ if $f(t)$ has a finite energy $(f(t) \in L^2)$. A scale parameter $a \propto \nu^{-1}$ is very often used in the literature about wavelets. Using this parameter the definition of the wavelet transform becomes:

$$C(a, b) = \langle f(t), \psi_{a, b}(t) \rangle \qquad \text{with} \qquad \psi_{a, b} = K(a)\psi(\frac{t - b}{a}) \tag{7}$$

The function $\psi(t)$ is normalized in the $L^2$ space and it must at least satisfy the so-called admissibility condition in order to be an analyzing wavelet:

$$C_\psi = \int_{-\infty}^{+\infty} \left| \hat{\psi}(\nu) \right|^2 \frac{d\nu}{\nu} < \infty \tag{8}$$

where $\hat{\psi}(\nu)$ is the Fourier transform of $\psi(t)$. Assuming that $\hat{\psi}(\nu)$ is differentiable this implies that $\psi(t)$ has zero average. A wavelet is then a smooth and rapidly vanishing oscillating function, which ensures from a practical point of view a good localization both in time and in frequency. Using the admissibility condition an energy conservation law is obtained

$$\int_{-\infty}^{+\infty} |f(t)|^2 \, dt = \frac{1}{C_\psi} \iint_{\Re \times \Re^+} |C(a, b)|^2 \frac{da\,db}{a^2} \tag{9}$$

which leads to the formula of reconstruction :

$$f(t) = \frac{1}{C_\psi} \iint_{\Re \times \Re^+} C(a, b)\psi_{a, b}(t) \frac{da\,db}{a^2} \tag{10}$$

The first wavelet introduced by Morlet, known as the Morlet wavelet, is:

$$\psi(t) = e^{-t^2/2} e^{i\omega_m t} + e^{-\omega_m^2/2} \tag{11}$$

where the term in $\omega_m^2$ insures the admissibility (it is negligible for $\omega > 5$). Another very popular wavelet is the mexican hat function built from the second derivative of a Gaussian:

$$\psi(t) = (1 - t^2)e^{-t^2/2} \tag{12}$$

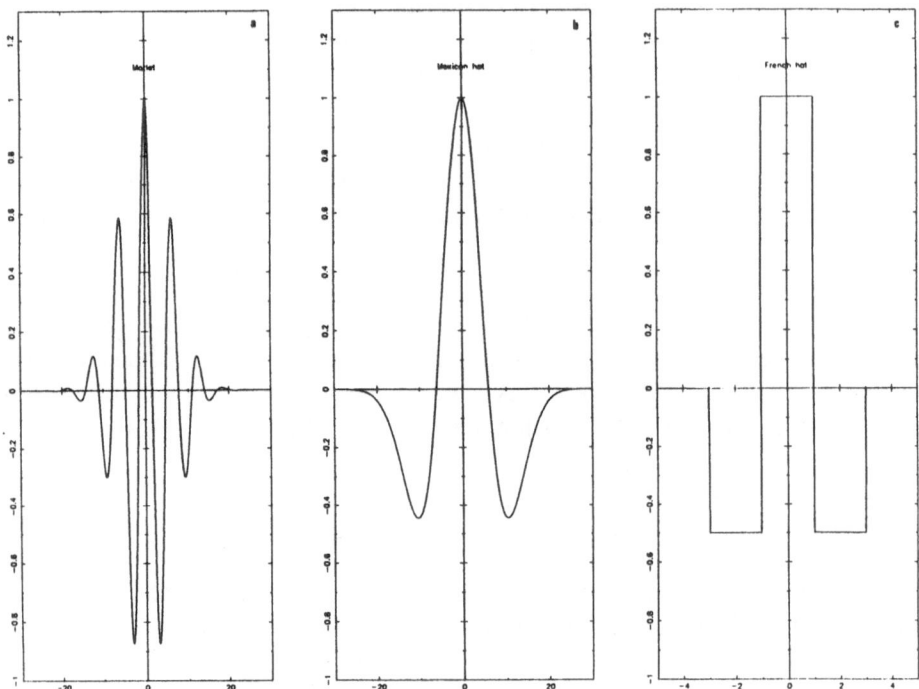

Fig. 5. (a) The Morlet wavelet: $e^{-t^2/2}e^{i\omega_m t}$ with $\omega_m = 6$; (b) The mexican hat function: $(1 - t^2/2)e^{t^2/2}$; (c) The french hat wavelet: $\psi(t) = 1$ if $|t| \leq 1$ and $\psi(t) = -0.5$ if $3 > |t| > 1$ and $\psi(t) = 0$ if $|t| > 3$; (d); (e); (f) are respectively the modulus of the Fourier transform of these functions.

Figures 5 display these two examples of wavelet functions as well as the piecewise constant wavelet, also known as the french hat wavelet.

Figures 6 show the filters implied in the wavelet transform of our sample signal for a high (a), middle (b), and low frequency (c) detection. One can observe how the shape of these filters is invariant under changes of the frequency. This leads to an unambiguous interpretation of the time-frequency analysis performed by means of the wavelet transform. Figure 7 displays the wavelet analysis of our sample signal using the Morlet wavelet with $\omega_m = 6$. The discrepancies with the previous Gabor analysis can be easily pointed out. First of all the three frequencies present in the signal with the same amplitude are detected in the same way. Then the discontinuities are detected by two cones pointing towards the location of these discontinuities at the small scales (see hereafter the localization properties of the wavelet transform).

Let us now present the mathematical properties of the wavelet transform which will be illustrated by examples coming from applications to dynamical systems. Let us denote by $C_f(a, b)$ the wavelet transform of $f(t)$, $g(t) = f(t - t_0)$, and

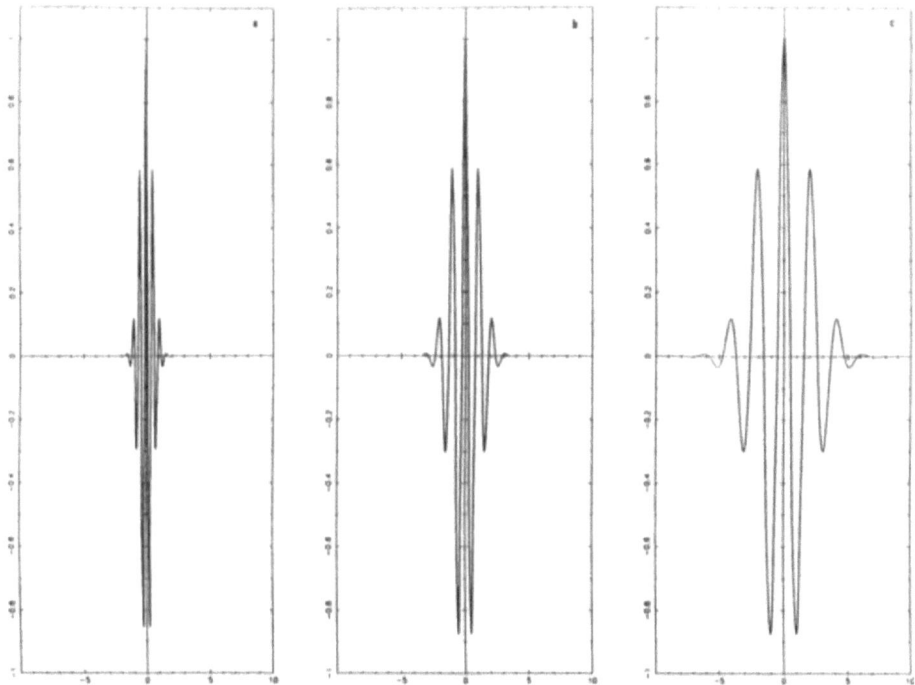

Fig. 6. The same as figure 3 but for the Morlet wavelet with $\omega_m = 6$

$h(t) = f(kt)$. The wavelet transform is
- linear:
$$C_{\alpha f_1 + \beta f_2}(a, b) = \alpha C_{f_1}(a, b) + \beta C_{f_2}(a, b) \tag{13}$$
- invariant under translations:
$$C_g(a, b) = C_f(a, b - t_0) \tag{14}$$
- invariant under dilatations:
$$C_h(a, b) = \frac{1}{k} C_f(ka, kb) \qquad \text{using} \qquad K(a) = a^{-1/2} \tag{15}$$
This property is extensively used for studying fractal behaviours and turbulence.
- localized in time and frequency:

Assuming that $\psi(t)$ vanishes outside some interval $[T_1, T_2]$ the value of $f(t)$ at location $t = t_0$ influences the half-plane $\{a, b\}$ over a limited area. This domain of influence is indeed a cone $t_0 - b$ with vertex at $b = t_0$, since the wavelet scales according to the value of $a$ (Grossmann et al. 1987). Figure 8a shows the wavelet transform of a Dirac $\delta(t - t_0)$ function using the mexican hat wavelet. The cone of influence of the punctual event $\delta(t - t_0)$ can be easily identified. This example shows how the wavelet transform points towards the

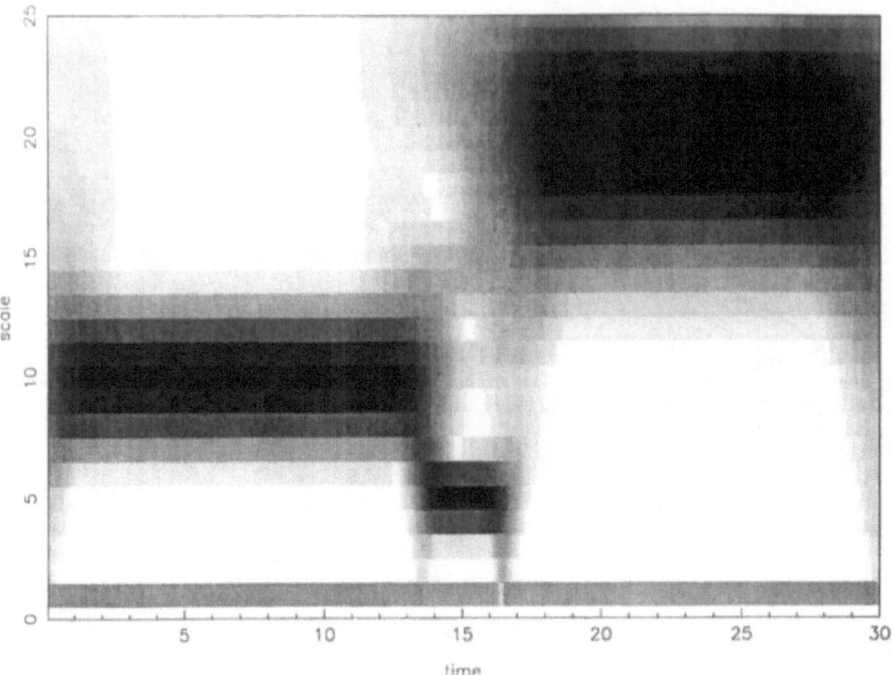

Fig. 7. The time-scale analysis of the sample signal performed by means of the Morlet wavelet for 25 different scales (Y-axis); the time is in abscissae. The gray level coding is the same as in figure 4. The pulsations present in the signal are obtained by: $\omega = \omega_m / scale$. The three detected pulsations have the same weight. Discontinuities are visible at all the scales through the cones pointing toward the location of these discontinuities at the smallest scale. The gray strip at the smallest scale is a sampling effect. (b) The modulus of the wavelet transform described above for 25 different values of the scale.

location of discontinuities at small scales (i.e. high frequencies) and how these discontinuities influence all the scales. More generally the wavelet transform is able to point out events at the scale where they exist and it is often compared to a mathematical microscope (Arneodo *et al.* 1988). This specificity is used in particular for the solution of differential equations.

Now let a monochromatic signal $f(t) = e^{-2i\pi\nu_0 t}$ be analyzed. By definition we get the Fourier transform of $\psi^*(t)$ for $\nu = \nu_0$:

$$C(a,b) = K(a) \int e^{-2i\pi\nu_0 t} \psi^*(\frac{t-b}{a}) dt = K(a) a e^{2i\pi\nu_0 b} \hat{\psi}(a\nu_0) \qquad (16)$$

Assuming that $\hat{\psi}(\nu)$ vanishes outside some interval $[\nu_{min}, \nu_{max}]$, the domain of the half plane $\{a, b\}$ influenced by the frequency $\nu_0$ of the signal is the horizontal strip $\nu_{min}/\nu_0 < a < \nu_{max}/\nu_0$. One can notice that the modulus $|C(a, b)|$ depends only on the scale $a$, and that $C(a, b)$ is in phase with the signal if $\hat{\psi}(\nu)$ is real. This later property is used for determining the temporal

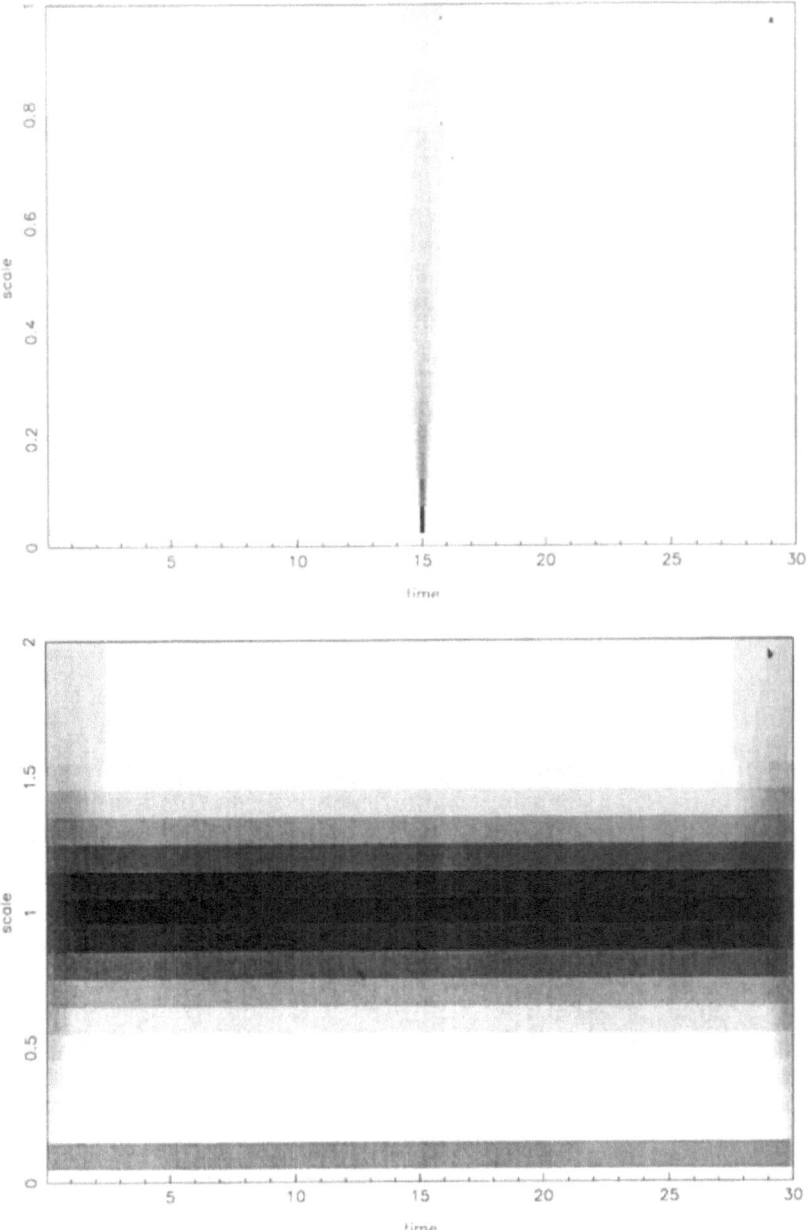

Fig. 8. Wavelet analysis of a Dirac function $\delta(t - t_0)$ with $t_0 = 15$ (a) and of a monochromatic signal $e^{i6t}$ (b). The analyzing wavelet is the Morlet one with $\omega_m = 6$. Values are displayed using a look-up table where the lowest and highest values correspond to white and black respectively. The cone aspect of the time-scale analysis is pointed out in the first case while the strip aspect of this analysis is exhibited in the second case. The gray strip at the smallest scale is the same artifact as the one described in figure 7.

evolution of frequencies (Guillemain *et al.* 1989). Figure 8b displays the wavelet transform of the monochromatic signal $f(t) = e^{i6t}$ using the Morlet wavelet with $\omega_m = 6$. The expected strip is well visible at the right scale. This good localization of the wavelet transform both in space and scale, as well as its ability to detect features at a given scale, has been used to define a new cluster analysis method (Slezak *et al.* 1990) which has been applied to identify asteroid dynamical families (Bendjoya *et al.* 1991a).

A discrete wavelet transform is needed for computer applications and fast algorithm implementation. These aspects will not be considered in this paper, except for the following very brief considerations. It is possible to reconstruct the original signal from a discrete set of wavelet coefficients and to control the error on the norm of the restored function related to a peculiar discretization of parameters $a$ and $b$ (Daubechies 1989). No errors occur when orthogonal wavelet bases are used. Such bases can be very easily obtained within the context of interpolation bases (Bijaoui 1991, Kronland-Martinet *et al.* 1987). The use of an embedded interpolation basis leads to fast algorithms which can make use of the multiresolution concepts (Dutilleux 1989, Mallat 1989).

## 3. Wavelets and the Solution of Differential Equations.

### 3.1. INTERPOLATION BASES

Solutions presenting discontinuities can emerge from the simulations of non linear phenomena in the fluid mechanics and turbulence field. This is in particular the case when modelling the flow of a $2D$ compressible fluid in the vicinity of a flat plate boundary layer; discontinuities in velocity and density appear due to shock waves, generating vortex and small-scale fluctuations along the plate, that is phenomena with different scale lengths (Perrier 1989). The main difficulty is to distinguish intrinsic turbulent fluctuations from numerical fluctuations. Three kinds of methods can be applied in order to solve the partial differential equations which model this class of non linear phenomena: finite differences, finite elements and spectral methods. The first one consists in replacing the differential operator by a difference operator. A set of test functions with small compact support is used in the finite elements method. The equation is integrated against these functions and the final solution appears as a combination of this finite set of functions. In the spectral method the solution is decomposed onto a basis of global support functions, like trigonometric functions, and it is then truncated to a finite number of terms. The robustness of the two former methods in the representation of irregular functions and the accuracy of the spectral method in smooth regions are the two main advantages which are looked for when searching for a mixed method. Although there already exist some methods that mix these two kinds of numerical integration, they are either numerically expensive or the accuracy is geometry dependent (Loisel 1986, Pernaud-Thomas 1988). Discontinuous functions can be approximated using a wavelet basis without spurious fluctuations all over the domain, since wavelets are localized. This characteristic has been used by Perrier (1989), who first built a new method for solving partial differential equations. This method has been improved by Liandrat *et al.* (1990) and it has been applied to the regularized Burgers equation. Let us show how the localization properties of the wavelet transform lead to the advantages of the spectral and finite difference methods.

Let us consider the simple $1D$ transport equation as an example:

$$\frac{\partial f}{\partial t}(x,t) + u(x,t)\frac{\partial f}{\partial x}(x,t) = 0 \qquad \text{with } t > 0 \text{ and } x \in [0,1] \tag{17}$$

where the unknown function is $f$ and $u(x,t)$ stands for the transport velocity assumed to be 1-periodic in space. By choosing $N$ linearly independent functions $\psi_k(x)$ $(k = 1, ..., N)$, an interpolated function $\phi(x)$ can be computed in the vector space generated by $\{\psi_k\}$ which verifies $\phi(x_i) = f(x_i)$ at each collocation point, i.e. points where $f(x)$ is known:

$$\phi = \sum_{k=1}^{N} c_k \psi_k \tag{18}$$

This interpolated function $\phi(x)$ provides an approximation of $f(x)$ and

$$\frac{\partial \phi}{\partial x}(x_i) = \sum_{k=1}^{N} c_k \frac{\partial \psi}{\partial x}(x_i) \tag{19}$$

is an approximation of $\partial f / \partial x(x_i)$. Information can be obtained from the knowledge of $\phi(x)$ on $f(x)$ and on its spatial derivative at the collocation points, and then between them by interpolating. To illustrate this point Perrier (1989) studied how a shock solution $f(x)$ modelled by a 1-periodic Heavyside function (fig. 9a) can be interpolated when using either a Fourier trigonometric basis or an orthonormal wavelet basis. The components $c_k$ are easily computed in the first case using an FFT algorithm and

$$\phi(x) = \sum_{k=-N/2}^{N/2} c_k e^{2i\pi kx} \tag{20}$$

approximates $f(x)$ with $\phi(x_i) = f(x_i)$ at each collocation point. If one takes $\{\psi_k\}$ as an orthonormal wavelet basis by definition one gets:

$$c_k = \int_0^1 f(x)\psi_k(x)dx \simeq \frac{1}{N} \sum_{i=0}^{N-1} f(x_i)\psi_k(x_i) \tag{21}$$

Since $\phi(x_i) = f(x_i)$ is not automatically verified, one needs to solve the following $N^{th}$ order linear system in order to get a satisfactory result:

$$\phi(x) = \sum_{k=0}^{N-1} c_k \psi_k(x) \qquad \text{with} \qquad \phi(x_i) = f(x_i) \qquad \forall x_i \tag{22}$$

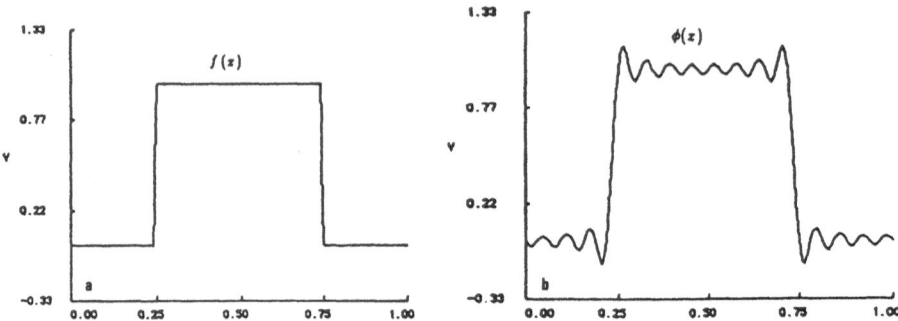

Fig. 9. (a) The 1-periodic Heavyside function $f(t)$ representation a singular solution like a shock. (b) The interpolated function $\phi(x)$ when using a trigonometric function basis. The Gibbs phenomenon (spurious $sinc(t)$ oscillations) is extended all over the domain. The number of collocation points is 32 (from Perrier 1989).

Different wavelet bases have been used and the influence of their regularity has been tested. The interpolated function for the shock solution is plotted in fig. 9b when a trigonometric function basis is used. Figure 10 displays the same function when a wavelet basis with an analyzing wavelet belonging to $C^0$ (10a), $C^2$ (10b) and $C^4$ (10c) is used. It clearly appears that the amplitude of the Gibbs phenomenon strongly correlates with the regularity of the used wavelet. But the fundamental difference between the use of a wavelet basis and the use of a trigonometric function basis is the localization of the Gibbs phenomenon in the first case and its extension to the whole domain in the second case. Hence it is possible to localize the discontinuities when using the wavelet basis of interpolation, as shown by the derivative function $\partial\phi/\partial x$ (fig. 10d). This is impossible when using a trigonometric basis of interpolation, due precisely to the extension of the Gibbs phenomenon.

3.2. APPLICATION TO THE REGULARIZED BURGERS EQUATION.

The wavelets can also be used in order to solve partial differential equations owing to their multiscale property which enables to regularize functions without loosing their discontinuities. Numerical fluctuations can then be filtered without smoothing out the physical ones. Some preliminary work has been performed by Liandrat et al. (1990) in order to develop a scale-adaptive method of numerical integration based on the use of wavelets. This method has been implemented for the Burgers equation with periodic boundary conditions and in the case of small diffusion ($\nu \simeq 10^{-3}$):

$$\frac{\partial u}{\partial t} + u \frac{\partial u}{\partial x} = \nu \frac{\partial^2 u}{\partial x^2} \qquad \text{with } t \geq 0 \text{ and } x \in [0,1] \tag{23}$$

where $u(0, t)$ is given and $u(0, t) = u(1, t)$. This equation can be seen as a $1D$ Navier-Stokes equation with a convective term $u\partial u/\partial x$ and a viscous term $\nu\partial^2 u/\partial x^2$. The multiscale aspect of the solution appears when the non viscid

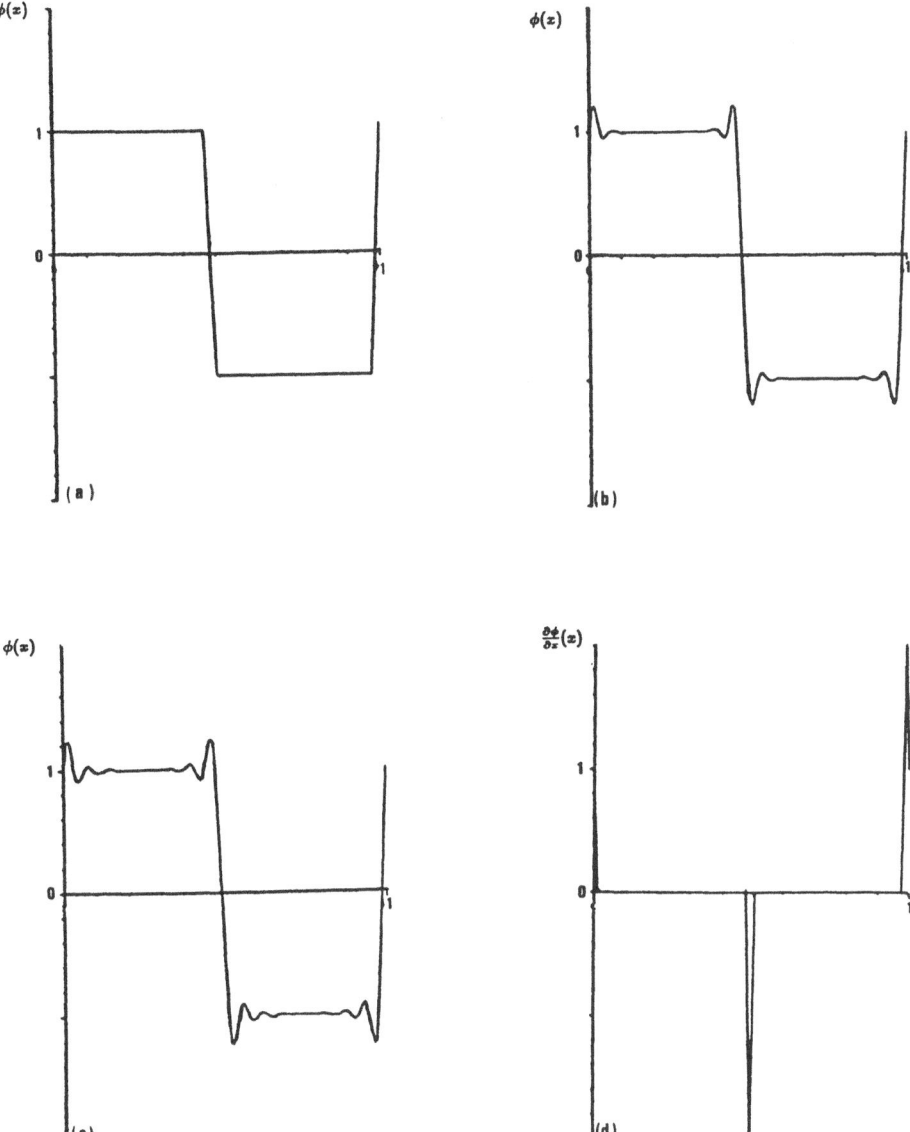

Fig. 10. The same as figure 9b but using a wavelet basis. (a) a $C^0$ wavelet basis. (b) a $C^2$ wavelet basis. (c) a $C^4$ wavelet basis The Gibbs phenomenon ($sinc(t)$ oscillations) is localized near the discontinuities, and its amplitude is related to the regularity of the used wavelet. ($C^n$ is the functional space of the $n$ times derivable functions). (d) The function $\partial\phi/\partial x$ derived from the interpolated function $\phi(x)$ obtained by the use of the $C^0$ wavelet basis. The discontinuities are localized because of the localized Gibbs phenomenon due to the intrinsinc properties of the wavelet. (from Perrier 1989).

Burgers equation is considered. The non linear term generates stronger and stronger gradients leading to the formation of local small-scale structures and to shocks when the solution $u(x, t)$ becomes singular. When the viscosity is taken into account, a competition takes place between the convective term and a growing viscous term. The gradients are damped and discontinuities do not occur after the critical time when the viscous term becomes the strongest: then the solution goes to zero. Due to this multiscale behaviour of the solution (Fournier et al. 1983) an algorithm based on the wavelet decomposition gives a good compromise between accuracy, efficiency and adaptability. This adaptability is to be understood as the ability of the algorithm to automatically determine the smallest scale, and then the corresponding sampling grid, that needs to be take into consideration according to the local gradient. The main idea about the algorithm developed by Liandrat and his coworkers is to decompose the solution onto a wavelet basis $\{\psi_{a,b}\}$. The Burgers equation can be written in a discrete way as follows:

$$(I - \nu \frac{\partial^2}{\partial x^2})u_{n+1} = u_n - \Delta t \, u_n \frac{\partial u_n}{\partial x} \tag{24}$$

$$u_{n+1}(0) = u_{n+1}(1) \tag{25}$$

where $\Delta t$ is a constant time step, $u_n = u(x, n\Delta t)$ and $I$ is the identity operator. Knowing $u_n$ and $\psi_{a,b}$ the component of $u_{n+1}$ on the wavelet 'vector' $\psi_{a,b}$ is the scalar product $< u_{n+1}, \psi_{a,b} >$, which can be written as:

$$< u_{n+1}, \psi_{a,b} > = < u_n, \theta_{a,b} > + \frac{1}{2}\Delta t < u_n^2, \frac{\partial \theta_{a,b}}{\partial x} > \tag{26}$$

where $\theta_{a,b}$ is a precomputed family of functions verifying:

$$(I - \nu\Delta t \frac{\partial^2}{\partial x^2})\theta_{a,b} = \psi_{a,b} \tag{27}$$

This leads to the following expression for $u_{n+1}$:

$$u_{n+1} = \sum_{a,b} < u_n - \frac{1}{2}\Delta t \frac{\partial^2 u_n^2}{\partial x^2}, \theta_{a,b} > \psi_{a,b} \tag{28}$$

Strong gradient regions are pointed out by the wavelet transform. Consequently the component of the solution onto a small-scale wavelet vector indicates whether the analysis must be refined, since a high enough value denotes the existence of a strong gradient region. So, if the energy associated to this small-scale vector component is greater than a predefined threshold, the components of the solution are computed onto smaller scale wavelet vectors within the detected strong gradient region. Adding components corresponding to wavelet vectors at smaller scales is not redundant since $\{\psi_{a,b}\}$ is an orthonormal basis. This criterion gives the expected results since the lattice refinement occurs only for sharp gradients and not for moderate extended gradients. The solution $u(x, t)$ has been computed using (i) a Fourier spectral method,(ii) a non adaptive wavelet decomposition method, or (iii)an adaptive wavelet algorithm. The first solution presents a Gibbs phenomenon which spreads all over its domain. The second method exhibits spurious fluctuations localized in the neighborhood of the discontinuities. These artifacts are due to the lack of resolution in these regions. The solution behaves satisfactorily only with the third method since the strong gradient regions are correctly sampled. It must be noticed that the extrapolation from one grid to another does not affect the rapidity of the algorithm (see Liandrat et al 1990). In the light of these results, wavelets can be used with great benefit to study partial differential equations. They provide indeed better interpolation bases than trigonometric or spline bases since discontinuities are located, numerical fluctuations are smoothed out while shocks are preserved, and a fully adaptive algorithm can be derived. However, a multiscale analysis is time expensive and is often not justified for non hierarchical or regular signals for which classical methods are sufficient.

## 4. Fractals and Turbulence

The study of dissipative dynamical systems consists in defining their asymptotic behaviour from the knowledge of the different kinds of attractors and the transitions between these attractors. Roughly speaking, an attractor is a compact subset of the phase space invariant under the differential flow or the application characterizing the temporal evolution of the dynamical system. The attractor has a null volume in the phase space. It is however connected with the finite volume of the phase space defined by the initial conditions with trajectories attracted by the attractor (Bergé et al. 1984, 1988). All dissipative systems are constrained by a control parameter which determines the temporal behaviour of the system. The system looses its stability and becomes chaotic for a critical value of this parameter. This chaotic behaviour can be charaterized by the presence of a wide band in the power spectrum of the system or by an autocorrelation function which becomes null after some interval of time. Hence, the knowledge of the state of the system during an arbitrarily long time does not permit its later state to be known. These characterizations of chaos are directly applicable to the irregular spatial-temporal behaviour of a hydrodynamic flow, namely to turbulence. A strange attractor is associated to the chaotic behaviour of the system (Ruelle and Takens 1971). The trajectories are sensitive to the initial conditions, and this implies a non integer dimension for these particular attractors. Strange attractors are therefore fractal objects.

The fractal formalism is then best suited to the description of the transition of dynamical systems to chaos through the concept of strange attractors or to the turbulent regime through the repartition of the dissipative eddies at small scales in a turbulent flow. We shall now point out the ability of the wavelet transform to detect and characterize a fractal behaviour. After a brief review of chaos concepts and the fractal formalism, we first present the results obtained by Arneodo et al. (1990)

about the characterization of the fractal behaviour of mathematically interesting cases by means of the wavelet transform. The applications of these preliminary results to real data from fully developed turbulence experiments are then exposed.

## 4.1. FRACTALS

The main property of fractal objects is their self-similarity, since they remain identical to themselves under dilatations. The generalized fractal dimensions $D_q$, known as the Renyi dimensions, have been proposed in order to quantify this property (Jones *et al.* 1988 , Murenzi 1990). Let $\mu$ be a measure defined on a fractal object plunged in $\Re^2$ and let us consider a paving of $\Re^2$ by means of uniform square cells $Q_i$ the size of which is $\epsilon$. Let $\mu_i$ be the ratio of the weight of the cell $Q_i$ over the total measure of the fractal object:

$$\mu_i(\epsilon) = \int_{Q_i(\epsilon)} d\mu(x) \tag{29}$$

Let the partition function $Z_q(\epsilon)$ be:

$$Z_q(\epsilon) = \sum_{i=1}^{N} \mu_i^q(\epsilon) \tag{30}$$

where $N(\sim 1/\epsilon^2)$ is the number of boxes in the grid and $q \in Z$. The generalized fractal dimensions $D_q$ are by definition:

$$D_q = (q - 1)^{-1}\tau(q) \qquad \text{with} \qquad \tau(q) = \lim_{\varepsilon \to 0} \frac{\ln Z_q(\varepsilon)}{\ln \varepsilon} \tag{31}$$

Homogeneous fractals display a linear dependence of $\tau$ on $q$, whereas any departure from linearity implies multifractality. One can recognize in $D_0$ the Haussdorff dimension, first introduced as the fractal dimension; $D_{+\infty}$ or $D_{-\infty}$ gives information on the behaviour of the densest (resp. the least dense) regions of the fractal.

Moreover $\mu_i(\epsilon)$ is related to $\epsilon$ in each cell $Q_i(\epsilon)$ by a power law:

$$\mu_i(\epsilon) \sim \epsilon^{\alpha_i} \tag{32}$$

where the local scale exponent $\alpha_i$ (the Hölder exponent) depends generally on the position of the cell $Q_i$. For a given $\epsilon$ it there are $N_\alpha(\epsilon) \sim \epsilon^{-f(\alpha)}$ cells associated to the same exponent. The spectrum of singularities $f(\alpha)$ is a continuous function, which gives the distribution of dimensionalities that are present in the set. It is connected to the generalized dimensions $D_q$ through a Legendre transformation:

$$f(\alpha) = \alpha q - \tau(q) \tag{33}$$

$$\alpha(q) = \frac{d\tau(q)}{dq} \qquad \text{with} \qquad \tau(q) = (q - 1)D_q \tag{34}$$

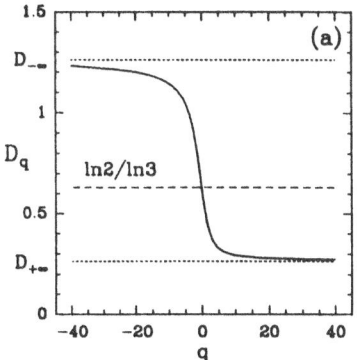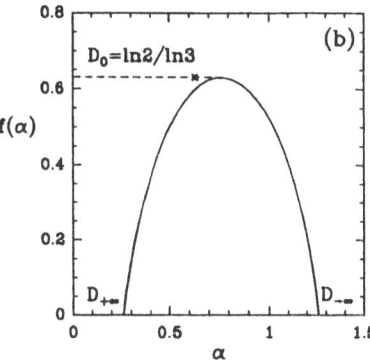

Fig. 11. The generalized fractal dimensions $D_q$ (a) and the spectrum of singularities (b) for a multifractal object (from Arneodo et al. 1989).

All $D_q$ have the same scaling index $\alpha$ for homogeneous fractals, whereas $\alpha$ takes values in a finite range $[\alpha_{min}, \alpha_{max}]$ and $f(\alpha)$ has a maximum equal to $D_0$ for multifractals (see fig. 11a).

The spectrum of singularities (see fig. 11b) has been used in order to describe the fractal or multifractal behaviour of dynamical systems. In spite of interesting results (Halsey et al. 1986, Collet et al. 1987), only the strongest singularities can be detected and any information about their location is lost since the spectrum $f(\alpha)$ gives an average measure over the whole data set (Muzy et al. 1991). The wavelet transform provides a local spectral analysis which enables the singularities to be detected and located. This transform is therefore well suited to locally characterize the scaling invariance of fractals.

Let us assume that the scaling behaviour of the measure $\mu$ of an interval $I$ centered on $x_0$ with a length $\epsilon$ is the following:

$$\mu(I(x_0, \lambda\varepsilon)) \approx \lambda^{\alpha(x_0)}\mu(I(x_0, \varepsilon)) \tag{35}$$

By adapting the definition of the wavelet transform to the measure of a fractal object, one gets:

$$C(a, b) = \frac{1}{a^n} \int \psi^*(\frac{x-b}{a}) \, d\mu(x) \tag{36}$$

where the normalization factor is chosen in order to optimize the visualization of the fractal behaviour. The translation and dilation invariance properties (eqs. 15, 16) of the wavelet transform lead to:

$$C(\lambda a, x_0 + \lambda b) = (\lambda a)^{-n} \int \psi^*(\frac{x - x_0 - \lambda b}{\lambda a}) \, d\mu(x)$$

$$= \lambda^{\alpha(x_0)-n} C(a, x_0 + b) \tag{37}$$

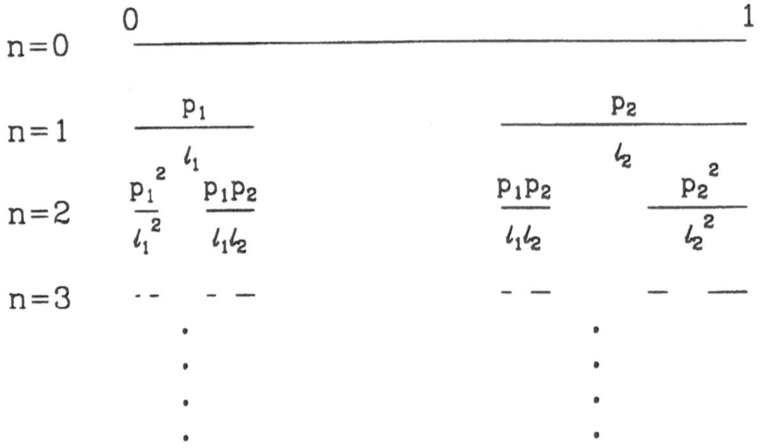

$$
\begin{array}{c}
0 \qquad\qquad\qquad\qquad\qquad\qquad\qquad\qquad\qquad\qquad 1 \\
n=0
\end{array}
$$

n=0

n=1  $p_1$   $p_2$

$l_1$   $l_2$

n=2  $\dfrac{p_1^{\,2}}{l_1^{\,2}}$  $\dfrac{p_1 p_2}{l_1 l_2}$   $\dfrac{p_1 p_2}{l_1 l_2}$  $\dfrac{p_2^{\,2}}{l_2^{\,2}}$

n=3

Fig. 12. The principle of construction of a triadic Cantor set. The $n^{th}$ step consists in removing the central part of each interval of the $(n-1)^{th}$ step and to create two sub-intervals whose length is weighed by $p_1$ and $p_2$. The process is reproduced ad infinitum. The uniform triadic Cantor set is obtained by making $p_1 = p_2 = 1/2$, the non uniform one with $p_1 \neq p_2$, for example $p_1 = 3/4$ and $p_2 = 1/4$ (from Arneodo et al. 1989).

When $\lambda \to 0$, the function $C(a,b)$ has a power law behaviour with an exponent $\hat{\alpha}(x_0) = \alpha(x_0) - n$. This behaviour is verified in most cases only for an infinite set $\lambda_m \approx \beta^m$, $m \in Z$, and not for all $\lambda \in \Re$. The $\alpha$ exponent is then complex and some oscillations of period $\ln \beta$ appear around a straight line with slope $\alpha$ in the graph $\ln C(a,b)$ versus $\ln(a)$, which characterize the lacunarity of such a fractal (Bessis et al. 1987). The local fractal dimension of an object or its multifractal behaviour can therefore be investigated using the wavelet transform. Singularities can be however hidden by some possible polynomial behaviours, and one has by consequence to impose some regularity on the analyzing wavelet (Holschneider 1988).

Let us now present part of the results obtained by Arneodo et al. (1989) about the uniform and non uniform triadic Cantor sets described in fig. 12 (Arneodo et al. 1988, 1989, 1990). The weight of each subinterval in the uniform Cantor set is $1/2$ and this fractal object has a unique characteristic exponent $\alpha = \ln 2/\ln 3$. The wavelet transform is performed using the mexican hat function and taking a normalization exponent $n = 2$. The modulus of the wavelet coefficients is displayed in figure 13a for the mexican hat and in figure 13b for the french hat analyzing wavelet. The construction rules of the Cantor set are clearly exhibited since the $n^{th}$ bifurcation appears at the scale $a \sim 1/3^n$. Figure 14 displays the plot of $\ln(C(a, b^* = 0))$ versus $\ln(a)$. The slope of the resulting straight line is $\ln 2/\ln 3 - 2$, which corresponds to the expected value of $\alpha$. Oscillations of period

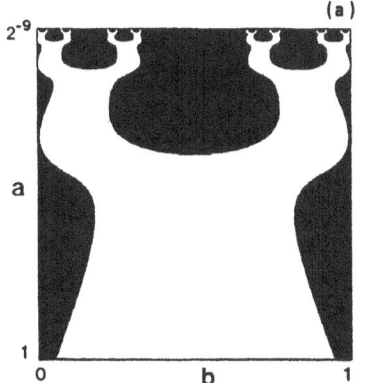
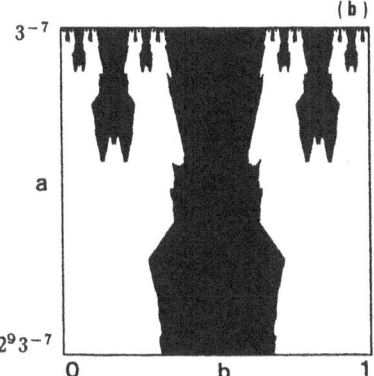

Fig. 13. The wavelet transform of the uniform triadic Cantor set using the mexican hat wavelet (a) and the french hat wavelet (b) with white standing for the highest coefficients and black for the lowest. The rules of construction of the Cantor set are clearly pointed out since the bifurcations appear for the scales $a \sim 3^{-n}$ and are symetric (from Arneodo *et al.* 1989).

$\ln(3)$ are superimposed on the general behaviour of the graph. They reflect the invariancy of this Cantor set under dilations by a factor $\beta = l^{-1} = 3$, since $l$ is the unique scale ratio present in this uniform triadic Cantor set. Another choice on the Cantor set for the point $b^*$ of measurement gives the same results. This indicates the existence of a unique scale exponent, which is the characteristic of a homogeneous fractal.

The non uniform triadic Cantor set is obtained by giving different weights to the two subintervals: $p_1 = p_{Left} = 3/4$ and $p_2 = p_{Right} = 1/4$. Each point of the Cantor set can then be associated to a sequence of L and R. The sequence LLLLL...LL.. is linked with the heaviest singularity whose exponent is $\alpha_{min} = \ln(p_L)/\ln(l) = \ln(3/4)/\ln(1/3)$, and the sequence RRRR...RR.. with the lightest singularity with a scale exponent $\alpha_{max} = \ln(p_R)/\ln(l) = \ln(1/4)/\ln(1/3)$. Other mixed sequences have an exponent $\alpha$ such as $\alpha_{min} < \alpha < \alpha_{max}$. The modulus of the wavelet transform (figs. 15a, 15b) exhibits the non symmetric bifurcations which characterize a multifractal behaviour. The graphs $\ln(C(a, b^* = 0))$ versus $\ln(a)$ and $\ln(C(a, b^* = 1))$ versus $\ln(a)$ (fig. 16a) correspond respectively to the sequences associated to the heaviest singularity and to the lightest one. The slopes of the straight lines are in good agreement with the expected values $\alpha_{min}$ and $\alpha_{max}$. Finally, the same kind of graph is plotted in figure 16b for a value of $b^*$ corresponding to the sequence RRRRRRRRLLL...LL.. where a value $\hat{\alpha} = -1.167$ is expected. One can notice the asymptotic behaviour for the large and small scales as well as the "cross-over" region due to the succession of a sequence of R and a sequence of L.

These results point out the ability of the wavelet transform to locally characterize

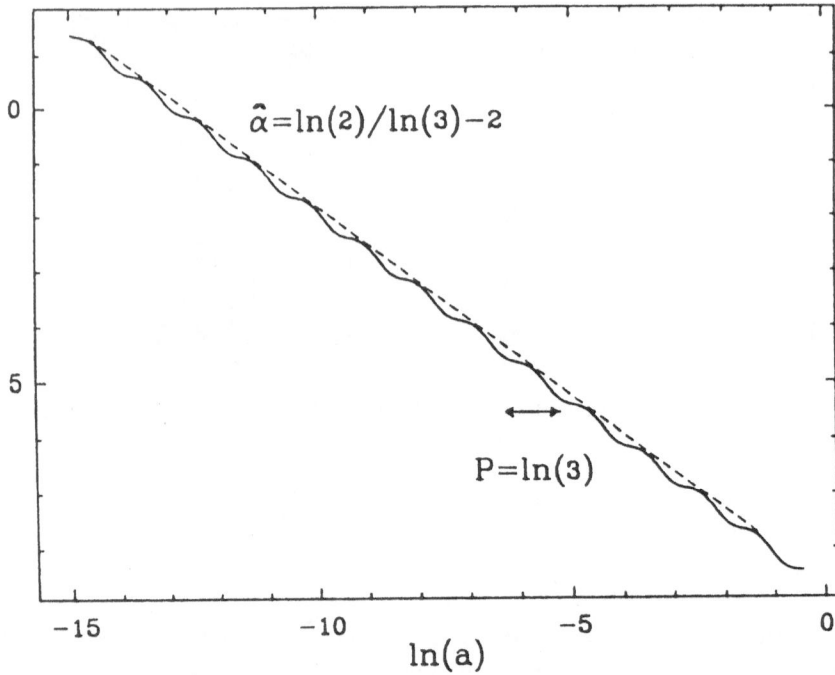

Fig. 14. $\ln(C(a, b^* = 0))$ as a function of $\ln a$ for the uniform Cantor set. The slope of the straight line gives the expected value of $\alpha$. The analyzing wavelet is the mexican hat and $n = 2$.

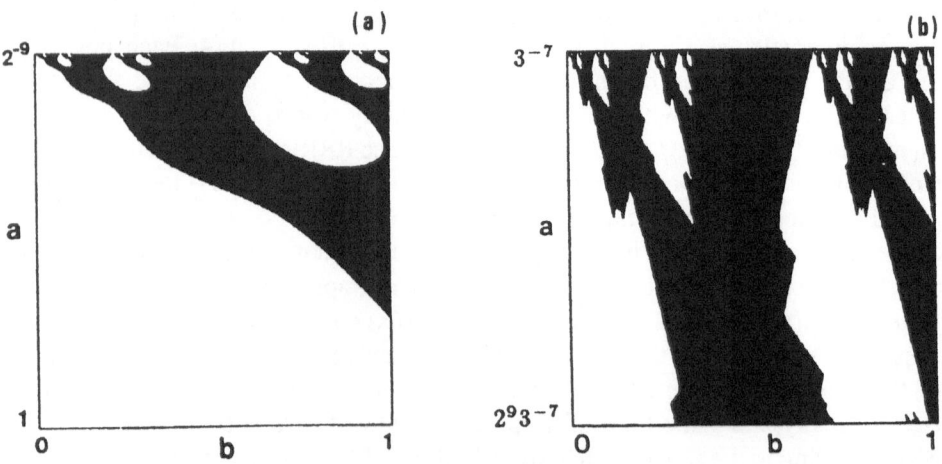

Fig. 15. The same as figure 13 but for the non uniform triadic Cantor set. The bifurcations are no more symetric (a and b), revealing the existence of several characteristic scale exponents (from Arneodo et al. 1989).

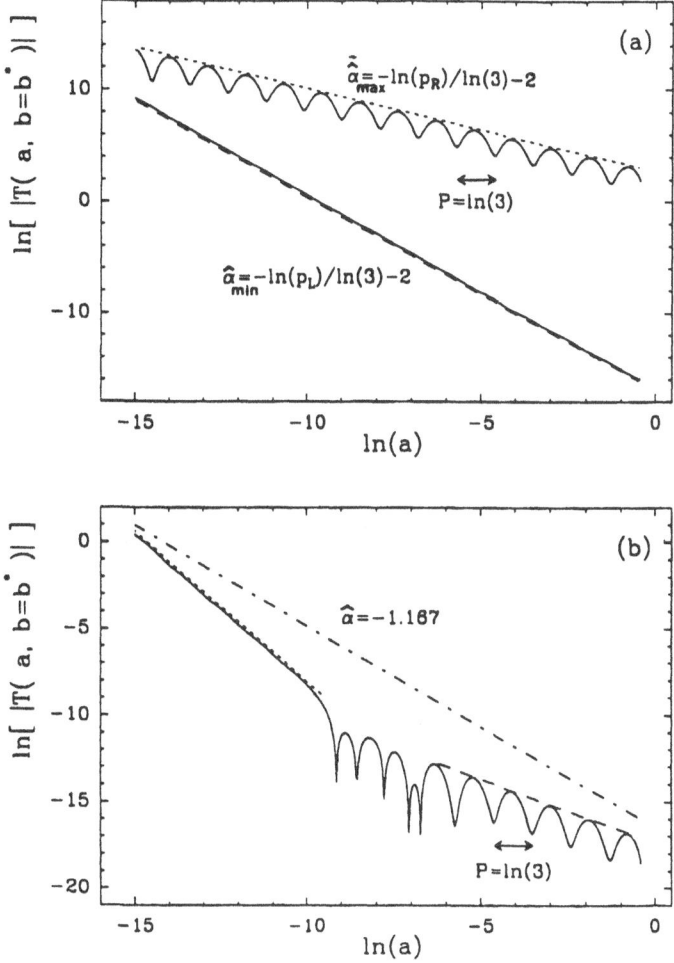

Fig. 16. $\ln(C(a, b^*))$ as a function of $\ln a$ for the non uniform Cantor set. (a) $b^* = 0$, $\hat{\alpha}_{min} = \alpha_{min} - 2$ (LLL..LL..), $b^* = 1$, $\hat{\alpha}_{max} = \alpha_{max} - 2$ (RRR..RR..). (b) $b^* = 1 - 3^{-8}$, $\hat{\alpha} = -1.167$ (RRRRRRRRLL...LL..) (from Arneodo *et al.* 1989).

fractal behaviours. This method of investigation has been therefore used to analyze transitions to chaos, in particular through the period-doubling cascade which yields asymptotically a peculiar Cantor set (Arneodo *et al.* 1989, Argoul *et al.* 1989) . The wavelet transform has been able to reveal the construction rules of this sub-harmonic Cantor set, and to exhibit the universal properties of the transition to chaos trough the period-doubling cascade (Feigenbaum 1978, Coullet *et al.* 1978, 1981, Tresser *et al.* 1978). The wavelet transform allows also to evaluate the critical values of the control parameter for each bifurcation, as they can be identified in graphs where the modulus of $C(a, b)$ is plotted against $a$.

## 4.2. THE FULLY DEVELOPED TURBULENCE

The fully developed turbulence appears in tridimensionnal turbulent incompressible flows with large Reynolds number ($Re > 10^7$), a control parameter defined by:

$$Re = \frac{Lv}{\nu} = \frac{\text{typical scale} \times \text{typical velocity}}{\text{kinematic viscosity}} \qquad (38)$$

This spatial-temporal chaos with self-similar properties presents a hierarchy of flow structures (eddies) whose length scales extend from $l_0$, namely the scale of the boundary and/or of the initial conditions, to the dissipative scales $l_{diss} \ll l_0$, where viscous and inertial forces are comparable. The chaotic aspect of the fully developed turbulence can be characterized in two ways. First, in a given flow realization, the trajectory of each fluid particle can be extremely intricated, leading to a strong mixing of the flow and to enhanced transport properties compared to those existing in a laminar state. Another characterization can be found in instability properties. A small amount of noise in initial conditions is amplified and grows a significant level independently of its initial value.

The transfer of energy between the scales of energy production (typically $l_0$) and the scales of energy dissipation ($l_{diss}$) has been first described by Richardson (1922) in terms of a cascade where eddies have the tendency to generate smaller eddies of comparable size at the same scale. It appeared that this cascade had to be modelled by intermittency (Frisch *et al.* 1978, Frisch 1985a). The assumption is that the eddies at scale $l_{n+1}$ are concentrated in a fraction $\beta$ of the volume occupied by the eddies at scale $l_n$, leading to a spottiness of the smallest scales. This $\beta$ model (Frisch *et al.* 1978) has been improved by Frisch (1985b) through the stochastic $\beta$ model. The latter assumes that the ratio between the volumes occupied by the eddies at two successive scales is no more a constant equal to $\beta$ but varies randomly as a function of the scale.

By applying the wavelet transform to the data provided by some experiments in the wind tunnel of ONERA (Gagne 1987), it has been possible to visualize for the first time the multifractal aspect of the Richardson cascade. The analyzed signal was the spatial behaviour of the flow velocity. Figure 17 shows the signal and its wavelet transform for the largest scales. The french hat wavelet has been used in order to improve the visualization of the conic structures and to detect the characteristic scales of the eddies. It first appears that the structures are randomly distributed at the largest scales. The wavelet transform for smaller scales shows a behaviour comparable to the one observed for the non uniform Cantor set. The asymmetry in the bifurcations suggests a non unique scale ratio, and it is tempting to conclude positively about a multifractal repartition of the singularities in the signal. The ratio between the scales at which two successive bifurcations appear seems to take

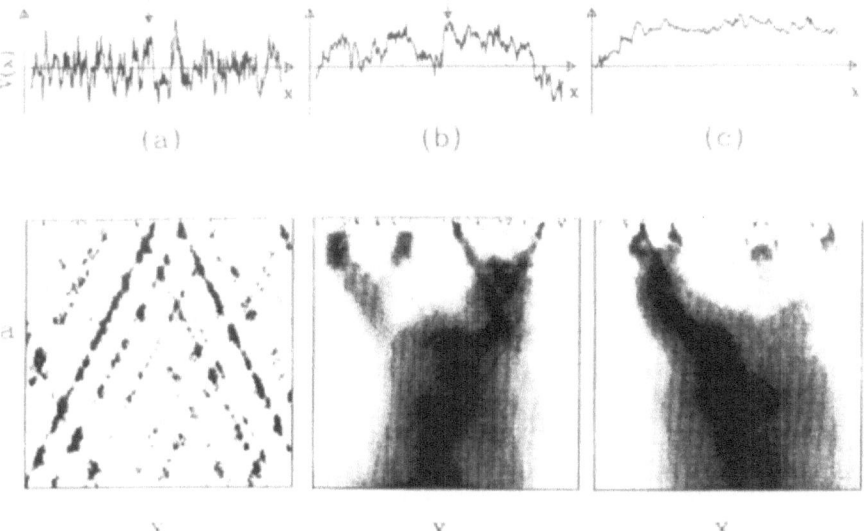

Fig. 17. Analysis of a signal whose velocity is turbulent obtained from wind tunnel of
ONERA. The experimental conditions were $Re = 2720$, $l_0 = 15$ $m$, $l_{diss} = 0.35$ $mm$
[46]. The wavelet transforms are represented under the corresponding signals: (a) the global
signal (total length 852 m) $a \in [l_0/10, 28l_0]$. (b) analysis of a portion of the signal centered
on the location pointed by the vertical arrow. The scale of length has been dilated 20 times.
(c) A zoom of the signal (b) with the same length scale dilatation. The wavelet analysis has
been performed by means of the french hat wavelet. (from Arneodo et al. 1989).

different values, which could be another indication for multifractality. If the scaling
behaviour of the signal is studied by plotting the logarithm of the absolute value
of the wavelet coefficient versus the logarithm of the scale, great care has to be
taken before concluding to lacunarity. Such plots present oscillations for statistical
self-similar processes (Vergassola and Frisch 1991), and further tests are therefore
necessary to discriminate these fluctuations from those originating from lacunarity.
Moreover the measurement of the local scaling exponent from a finite set of scales
is difficult to perform unambiguously.

## 5. The Wavelet Cluster Analysis.

When the signal can be modelled by a set of point the wavelet transform can be
used with great benefit to exhibit significant clusters versus chance and to point out
the intrication of these structures. An application has been made by Gilbert et al.
(1993) to the standard map which is discussed in this issue. We shall rather expose
in this paper how the wavelet transform has been used as a new cluster analysis
method in order to define dynamical asteroid families (Bendjoya et al. 1991a).
The concept of asteroid family is due to Hirayama (1918) who first noticed the

presence of clusters in the asteroid distribution. The precise determination of these families can shed some light on the collisional evolution of the asteroids, which is in particular related to the abundance of solid material in the primordial asteroid belt (Zappala *et al.* 1984, Davis *et al.* 1989). The common hypothesis retained in order to model the creation of an asteroid family is the break-up of a parent body due to a collisional event, but there was until a few years ago a poor agreement among investigators about family memberships (Brouwer 1951, Arnold 1969, Lindblad *et al.* 1971, Williams 1971, Kozai 1979, Carusi and Valsecchi 1982, Milani *et al.* 1991).

The made variables to be analyzed in order to find families are the (approximate) first integrals of the motion, since these integrals keep track of ancient violent events. These so-called proper elements can be accurately computed by the secular perturbation theory of Milani and Knezevic (1990). An asteroid is then represented by a Dirac function in the $3D$ space defined by the proper semi-major axis $(a')$, proper eccentricity $(e')$, and sine of the proper inclination $(\sin i')$. Bendjoya *et al.* (1991a) used the wavelet transform in order to detect hierarchical structures in the proper element space with a confidence level attached to the detection. Following the method first introduced by Slezak *et al.* (1990) to analyze galaxy counts, they modelled the sample of $N (\simeq 4100)$ asteroids by a sum of $N$ spatial Dirac functions, which leads to:

$$C(a, \mathbf{b}) = K(a) \sum_{j=1}^{N} \psi_{a, \mathbf{b}}(\mathbf{x}_j) \qquad \text{with} \qquad \psi_{a, \mathbf{b}}(\mathbf{x}) = \psi(\frac{\| \mathbf{x} - \mathbf{b} \|}{a}) \qquad (39)$$

where $\| \mathbf{x} - \mathbf{b} \|$ denotes the distance between $\mathbf{x}$ and $\mathbf{b}$ expressed in a metric suited to the problem. This metric is derived from the Gauss perturbation equations, where the osculating elements have been replaced by the proper elements. It has the dimension of a velocity and the identification of clusters of points in the $(a', e', \sin i')$ space leads to what is called asteroid dynamical families. A straightforward algorithm has been used to get the set of wavelet coefficients: a network is superimposed to the set of data and a wavelet coefficient $C(a, \mathbf{b}_j)$ is computed at each node $j$ of the network by means of equation (40). The wavelet used for this analysis is an isotropic mexican hat the multidimensionnal expression of which is:

$$\psi_{a, \mathbf{b}} = (n - \frac{\|\mathbf{x} - \mathbf{b}\|^2}{a^2}) \exp(-\frac{\|\mathbf{x} - \mathbf{b}\|^2}{2a^2}) \qquad (40)$$

where $n$ is the dimension of the space where the signal is defined. The wavelet transform identifies at a given scale heterogeneities in the cloud of data points by locating density contrast variations at this scale. A coefficient $C(a_0, \mathbf{b}_{j_0})$ with a high enough positive value indicates a structure located in the neighbourhood of node $j_0$ with a typical size $a_0$, whereas a value close to zero denotes homogeneous distribution of the data around this position for scale $a_0$. The statistical significance level of the detection is estimated from the wavelet coefficients computed for a set of pseudo-random distributions with the same local density as the real data. A threshold is computed from these simulations and it is applied to the wavelet coefficients map of the real data distribution in order to select the coefficients higher than this threshold. If the detection threshold is chosen close enough to the positive end of the random coefficients histogram, the selected coefficients define structures which have a very small quantified probability to be chance fluctuations. The catalogue of all the existing structures is then obtained by scanning the scales within reach and the clumps of asteroids are modelled by ellipses. The hierarchy of these embedded structures is cut by applying a criterion designed to select structures isolated enough. In the first attempt to identify asteroid families, the wavelet transform has been performed in each of the three projected planes $(a', e')$, $(a', \sin i')$ and $(e', \sin i')$, and $3D$ families have been defined by intersecting the subsets of asteroids inside the ellipses which modelled the significant structures in each plane.

This method has been applied to a set of 4100 proper elements computed by the theory of Milani and Knezevic (version 4.2) (Milani and Knezevic 1990). The results are in good agreement with those obtained by Zappala *et al.* (1990) on the same set of data but using a single linkage hierarchical clustering algorithm. This latter method (named hereafter HCM) is based on the agglomeration of the nearest neighbours in a unique object and on the updating of the distance between this new object and the rest of the catalogue in order to restart the agglomeration procedure. The same number of statistically significant families is found with both the wavelet analysis method (WAM) and the HCM method (Bendjoya and Cellino 1991b). Among the 21 defined families not only the big well known families such as Eos, Koronis and Themis have a high percentage of identical members, but also the smallest ones. The main discrepancies arise in the Flora region where the background is very dense and WAM method seemed to be best suited for such a peculiar case. Such an agreement is encouraging for confirming the physical reality of these families. Figure 18a shows the whole set of proper elements (version 4.2) projected on the plane $(a', e')$ while the WAM determined families are displayed in the same plane in fig. 18b.

A direct $3D$ analysis has been subsequently made in the proper elements space using a parallel computer which enables the use of a $128^3$ network with a short computation time and without space memory problems (Bendjoya *et al.* 1991c). The same method has been applied in order to select the wavelet coefficients. The detected structures are now modelled directly from the selected coefficients using what is called the "skeleton". A skeleton is a set of significant wavelet coefficients at a given scale. Two significant coefficients belong to the same skeleton if they are two nearest neighbours on the network. From these skeletons one is able to model the associated structure of points, at the considered scale. A sphere is centered on each node of a given skeleton, its radius is proportional to the value of the normalized coefficient at this node and to the studied scale ( the normalization of the coefficients is made within the considered skeleton ). All the points inside

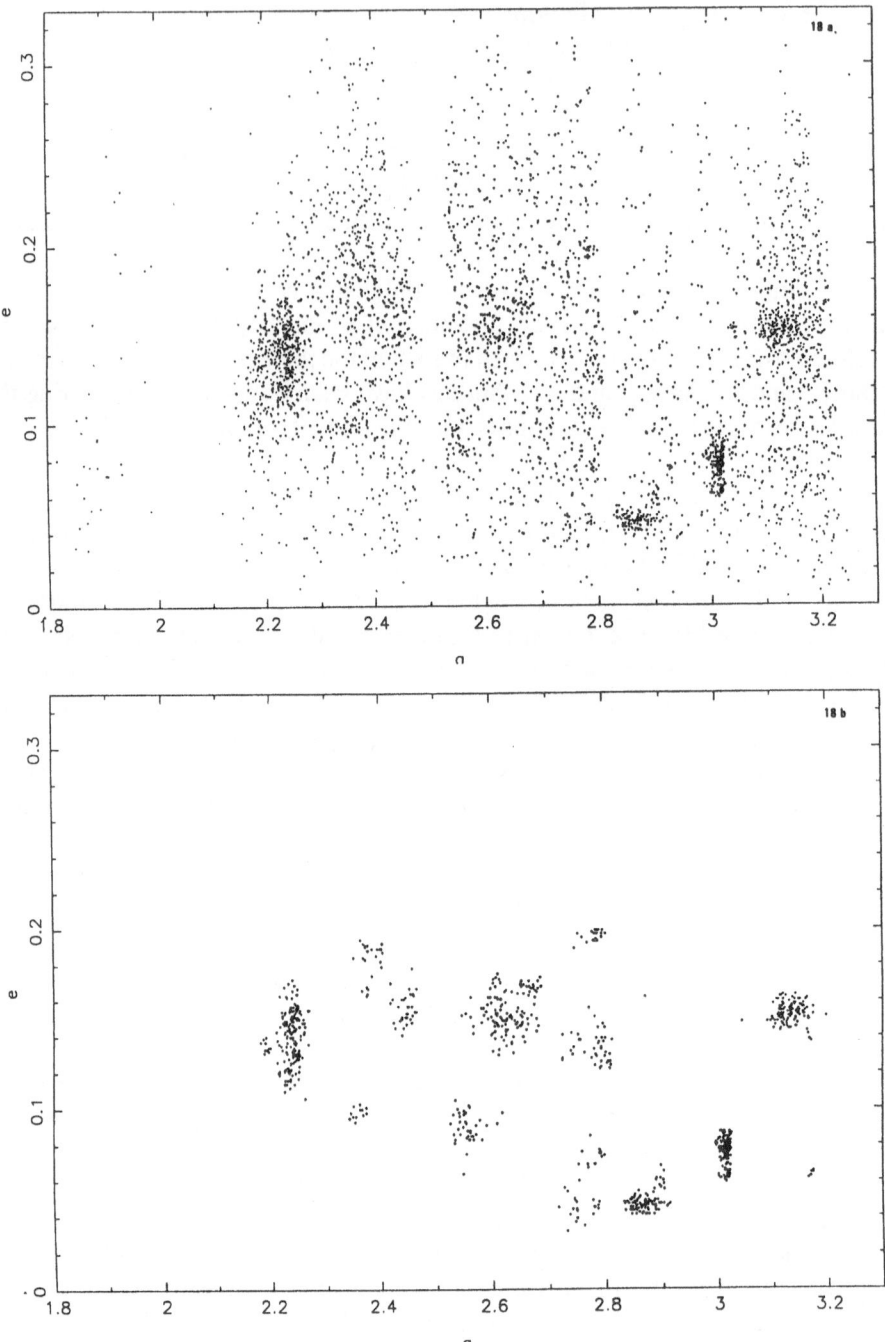

Fig. 18. (a) The set of proper element obtained by the theory of Milani and Knezevic (version 4.2), projected on the plane $(a', e')$. (b) The projection, on the same plane of the families, determined by means of the wavelet cluster analysis method.

each sphere define the structure of points associated to this skeleton. A criterion similar to the one defined previously is then applied to build families from the set of embedded structures points. This new wavelet cluster analysis method has been applied to several simulated asteroid families, and its robustness has been tested from *a priori* known situations against the background density, the spread of the family, the presence of anisotropic features, and its ability to separate families close to each other (Bendjoya 1992). The method passed these tests in a positive way, and this allows to be quite confident about the families found previously in the real asteroid belt.

## 5.1. CONCLUSION

A non exhaustive presentation of the wavelet transform techniques has been made within the framework of applications to the dynamical systems Its property of invariance under dilatations of the signal yields an adaptive multiscale analysis which enables to locate singularities. The improvements and the limitations of the wavelet transform have been discussed for the considered topics when compared to more classical methods of investigation. It appears that wavelets can offer a new basis of interpolating functions in order to solve partial differential equations leading to singular solutions. Their properties allow to develop algorithms which automatically adapt the resolution of their analysis according to the local strength of the singularities in the computed solution. Another result is the ability of the wavelet transform to quantify scaling behaviours and to characterize fractal objects. Even if the interpretations to the fully developed turbulence field, require a careful analysis one disposes nevertheless of a performing tool leading to scenic visualization of the hierarchical organization of the structures. Finally a new cluster analysis method based on the wavelet transform has been developed leading to the detection of asteroid dynamical families in the main belt with a high degree of significance against chance.

The wavelet transform is peculiarly well suited to study signals exhibiting hierarchical features or discontinuities, but this tool must not be considered as a panacea. A preliminary study of the signal has to be performed in order to choose the best technique to analyze it. The Fourier transform for instance is the right tool for stationary functions. Another difficulty lies in the overabundant amount of informations due to the time-frequency analysis. This results from the unfolding of the signal in a space of greater dimension. Some improvements are then developed in order to define criteria and algorithms allowing a selection of the relevant informations. For instance new tools such as wavelet packets (Wickerhauser 1991) or spectral 'skeleton' estimation (Guillemain *et al.* 1989) are actually developed in order to adapt locally the scale of the analyzing wavelet as a function of the local properties of the signal. This last method appears to be well suited to the determination of the temporal evolution of the frequencies in a signal. A study of the evolution of the fundamental frequencies of the motion of celestial bodies in order to detect chaos is for instance in progress (Bendjoya private com.). For what concerns the solution of partial differential equations, the adaptative algorithm based on the wavelets is currently improved in order to find criteria which flag not only the strongest gradients but also the weaker ones. Applications of the wavelet transform in order to analyze systems far from equilibrium are also developed in different domains like biology, economy, physics and mathematics. Taking benefit of the results obtained from the $3D$ wavelet cluster analysis, 7000 sets of asteroid proper elements are presently analyzed This new set includes the multiopposition asteroids (asteroids not yet numbered) and contained improved proper elements.

# References

Argoul F., Arneodo A., Grasseau G., Gagne Y., Hopfinger E.J. Frisch U.: 1989, "Wavelet analysis of turbulence reveals the multifractal nature of the Richardson cascade", *Nature* **338**, 51–53

Arneodo A., Grasseau G., Holschneider M.: 1988, "Wavelet transform of multifractals", *Phys. Rev. Lett.* **61**, 2281–2284

Arneodo A., Grasseau G., Holschneider M.: 1989, "Wavelet transform analysis of invariant measures of some dynamical systems", *Wavelets, Time-Frequency Methods in Phase Space*, (Combes J.M., Grossmann A. and Tchamitchian Ph. Eds.) Springer-Verlag, 182–196

Arnéodo A., Argoul F., Grasseau G.: 1990, "Transformation en ondelettes et renormalisation", *Les ondelettes en 1989*, (Lemarié P.G. Ed.),125–187

Arnold: 1969, *Astron J.*, **74**, 1235–1242

Barlaud M., Mathieu P., Antonini M.: 1989), *Proc. of the 6$^{th}$ IEEE Multidimensional Signal processing Workshop*, New York, 103

Bendjoya Ph., Slézak E., Froeschlé Cl.: 1991, " The wavelet transform: a new tool for asteroid family determination", *Astron. Astroph.* **251**, 312-330

Bendjoya Ph., Cellino A.: 1991b, " Asteroid families identified by two different methods", *Compte rendus de l 'école de Goutelas* Froeschlé Cl., Benest D. (Eds) Edition Frontiere, (in press)

Bendjoya Ph., Cellino A., Zappala V., Froeschlé Cl.: 1991c, "Simulated families: a test for different methods of family identification", *Asteroid Comets Meteors 91*, (Harris A., Bowell T. Eds), in press

Bendjoya Ph., Cellino A., Froeschlé Cl., Zappala V.: (1992) *in preparation*

Bergé P., Pommeau Y., Vidal Ch.: 1984, " L'ordre dans le chaos", *Hermann Ed.*, Paris

Bergé P., Dubois M.: 1988, "Etude experimentale de transition vers le chaos en convection de Rayleigh-Benard", *Le Chaos* (Eyrolles Ed.),1–83

Bessis D., Fournier J.D., Servizi G., Tuschetti G., Vaienti S.: 1987, *Phys. Rev. A*, **36**, 920

Bijaoui A., Giudicelli M.: 1991, *Experimental Astronomy*, **1**, 347–363

Bijaoui A.: 1991, "Algorithme de la transformée ondelette: application à l'imagerie astronomique", *Ondelettes et paquets d'ondes. Cours CEA, EDF, INRIA*, 125–140

Brouwer D.: 1951, *Astron J.*, **56**, 9

Coullet P., Tresser C.: 1978), *J. Physique, Colloq.*, **39**, C5

Carusi and Valsecchi G.: 1982, *Astron. Astroph.*, **115**, 327

Coullet P., Treeser C., Tirapegui : 1981, *Field Theory, Quantization and Statistical Physic*, (Reidel D. Publishing Company, Dordrecht), 249

Collet P., Lebowitz J., Porzio A.: 1987, *J. Stat. Phys.*, **47**, 609

Daubechies I.: 1989, "Wavelets: a tool for time-frequency analysis", *Proc. of the 6$^{th}$ IEEE Multidimensional Signal processing Worshop*, New York, 98

Dutilleux P.: 1989, "An implementation of the 'algorithme a trou' to compute the wavelet transform", *Wavelets, Time-Frequency Methods in Phase Space*, (Combes J.M., Grossmann A. and Tchamitchian Ph. Eds.), Springer-Verlag, 298–304

Feigenbaum M.J.: 1978), *J. Stat. Phys.*, **19**, 25

Flandrin P.: 1987, "Some aspect of non-stationary signal processing with emphasis on time frequency and time-scale methods", *Wavelets Time-Frequency Methods in Phase Space* (Combes J.M., Grossmann A. and Tchamitchian Ph. Eds.), Springer-Verlag, 68–99

Fournier J. D., Frisch U.: 1983, "L'équation de Burgers déterministe et statistique", *Journal de Mécanique Théorique et Appliquée*, **2**, 699–750

Frisch U., Sulem P.L., Nelkin M.:1978, *J. Fluid. Mech.*, **87**, 719

Frisch U.: 1985, "Turbulence and predictability in geophysical fluid dynamics and climate dynamics", *Compte-Rendus Ecole Intern. "Enrico Fermi"*, Ghil M., Benzi R., Parisi G. Eds (North Holland, Amsterdam)

Frisch U.: 1985, *Physica Scripta*, **T9**, 137

Gilbert A.D., Froeschlé Cl. and Frisch U.: 1993, *in this volume*

Gabor D.: 1946, *J. IEEE*, **93**, (III) 429–457

Gagne Y.:1987, *Thesis*, Univ. de Grenoble

Grossmann A., Kronland-Martinet R., and Morlet J.: 1987, "Reading and understanding continuous wavelet transform", *Wavelets, Time-Frequency Methods in Phase Space*, (Combes J.M., Grossmann A. and Tchamitchian Ph. Eds.), Springer-Verlag, 2–20

Guillemain P., Kronland-Martinet R., Martens B.: 1989, "Application de la transformée en ondelette en spectroscopie RNM", *Note-LMA*, 112.

Halsey T.C., Jensen M.H, Kadanoff L.P., Procaccia I., Shraiman B.I.: 1986), *Phys. Rev. A*, **33**, 1141

Hirayama K.: 1918, "Group of asteroids probably of common origin", *Proc. Phys.-Math. Soc. Japan*, **11**, 354-361

Holschneider M.: 1988, *J. Stat. Phys.*, **50**, 963

Kozai Y.: 1979, *Asteroid*, (Gehrels T. Ed. University of Arizona, Tucson), 334–358

Kronland-Martinet R., Morlet J., Grossmann A.: 1987, "Analysis of sound patterns through wavelet transform", *International Journal of Pattern Recognition and Artificial Intelligence*, 273-302

Liandrat J., Perrier V. Tchamitchian Ph.: 1990, "Numerical resolution of the 1D regularized Burgers equation using a spatial wavelet approximation algorithm and numerical results", *ICASE Report, NASA, Hampton, Virginia*

Linblad B.A., Southworth: 1971), *Physical Studies of Minor Planets*, (Gehrels T. Ed. NASA SP-267, Washington DC: US Government Printing Office), 337–352

Loisel P.: 1986, "Resolution des equations de Navier-Stokes compressibles instationaires par methodes spectrales de Tchebycheff", *Thesis*, Paris VI

Jones B.J.T., Martinez V.J., Saar E., Einasto J.: 1988, "Multifractal description of large scale structure of the universe", *Astroph. J.*, **352**, L1–L5

Mallat S. G.: 1989, " A theory for multiresolution signal decomposition: the wavelet representation", *IEEE Trans. on Pattern Anal. and Machine Intell.*, **11**, 674–693

Morlet J., Arens G., Fourgeau I., Giard D.: 1982,"Wave propagation and sampling theory", *Geophysics*, **47**, 203–236

Milani A., Knezevic Z.: 1990, *Celestial Mechanics*, **4**, 347–412

Milani A., Farinella P., Knezevic Z.: 1991, "On the search for asteroid families", *Compte rendus de l 'école de Goutelas*, Froeschlé Cl., Benest D. (Eds) Edition Frontiere, (in press)

Murenzi R: 1990, "Ondelettes multidimensionnelles et applications à l'analyse d'images", *Thesis*, Université catholique de Louvain

Muzy J.F., Barcy E., Arneodo A.: 1991, "Wavelets and multifractal formalism for singular signals: application to turbulence data", *Phys. Rev. Let.*, **67**, 3515-3518

Pernaud-Thomas B.: 1988, "Méthodes numériques d'ordre élevé appliquées aux calculs d' écoulements compressibles", *Thesis*, Paris VI

Perrier V.: 1989, " Towards a method for solving partial differential equations using wavelet basis", *Wavelets, Time-Frequency Methods in Phase Space*, (Combes J.M., Grossmann A. and Tchamitchian Ph. Eds.), Springer-Verlag, 269–283

Ruelle D., Takens F.: 1971, "On the nature of turbulence", *Mathematical Physics*, **20**,167

Richardson L.F.: 1922, *Weather Prediction by Numerical Process*, Cambridge Univ. Press

Starck J.L., Bijaoui A.: 1991, "Filtering and restoration with the wavelet transform", $3^{rd}$ *ESO data analysis workshop*

Slézak E., Bijaoui A., Mars G. : 1990, "Identification of structures from galaxy count: use of the wavelet transform", *Astron. Astroph.*, **227**, 301–316

Tresser C., Coullet P.: 1978, *C. R. Acad. Sci*, **287**, 577

Vergassola M., Frisch U.: 1991, "Wavelet transforms of self -similar processes", *Physica D*, **54**, 58–64

Wickerhauser M. V.: 1991, "INRIA Lectures on wavelet packets algorithm", YALE University preprint

Williams J.G.: 1971, *Physical studies of minor planets* (Gehrels T. Ed. NASA SP-267, Washington DC: US Government printing office), 177–181

Zappala V., Farinella P., Knezevic Z., Paolicchi P.: 1984, "Collisional origin of the asteroid families: Mass and velocity distributions", *Icarus*, **59** 261–285

Zappala V., Cellino A., Farinella P., Knezevic Z.: 1990, "Asteroid families: identification by hierarchical clustering and reliability assessment", *Astron J.*, **100**, 2030–2046

# WAVELET ANALYSIS OF THE STANDARD MAP:

## STRUCTURE AND SCALING

A.D. GILBERT

*Observatoire de Nice, B.P. 229 - 06304 Nice Cedex 4, France,*
*and D.A.M.T.P., Silver St., Cambridge CB3 9EW, U.K.*

and

C. FROESCHLÉ and U. FRISCH

*Observatoire de Nice, B.P. 229 - 06304 Nice Cedex 4, France.*

**Abstract.** Wavelet analysis is applied to distributions of points generated by iterating the standard map. The initial condition is chosen so that the points fill the largest chaotic region. When the standard map parameter $k = 1.3$, the distribution of points contains many voids corresponding to islands in the chaotic region. The wavelet transform is dominated by contributions from these islands. For $k = 10$ the chaos fills phase space and no structure is apparent; the wavelet transform reveals statistical fluctuations in the distribution of points.

**Key words:** Wavelet analysis – standard mapping – chaotic motion

## 1. Introduction

Wavelet transform is a method of analyzing data and picking up scaling behaviour and singularities (see review by Bendjoya and Slezak in this volume). Among other applications it has been applied to elucidate the fractal structure of galaxies (see Slezak *et al.* 1991) and for clustering analysis of asteroid families (see Bendjoya *et al.* 1991). It is of interest to apply the wavelet distribution to a model in which large numbers of points can easily be generated. One method is by iterating a chaotic area-preserving map. For certain parameter ranges the map has a large region of chaos with a hierarchy of islands embedded, and for others the chaotic region fills all of space. In this paper we apply the wavelet transform to distributions of (varying numbers of) points in the chaotic region of the standard map. We see how the structure of the phase space is reflected in the wavelet transform and its scaling properties.

The wavelet transform is here applied to distributions of points generated by iterating the standard map:

$$x' = x + k \sin y \qquad \mod 2\pi \tag{1}$$
$$y' = y + x' \qquad \mod 2\pi . \tag{2}$$

The properties of the standard map have been explored in detail and are reviewed in Lichtenberg *et al.* (1983). The value of the parameter $k$ strongly influences the structure of the phase space. When $k = 0$ the map is integrable and the phase space is filled with invariant curves. As $k$ is increased the resonant invariant curves break up, leaving chaotic regions and families of islands. By $k = 1.3$ there is a large region of chaos (as well as smaller regions) with hierarchies of islands intermingled (figure (1)), and for $k = 10$ the phase space appears to be totally chaotic.

*Celestial Mechanics and Dynamical Astronomy* **56**: 263–272, 1993.

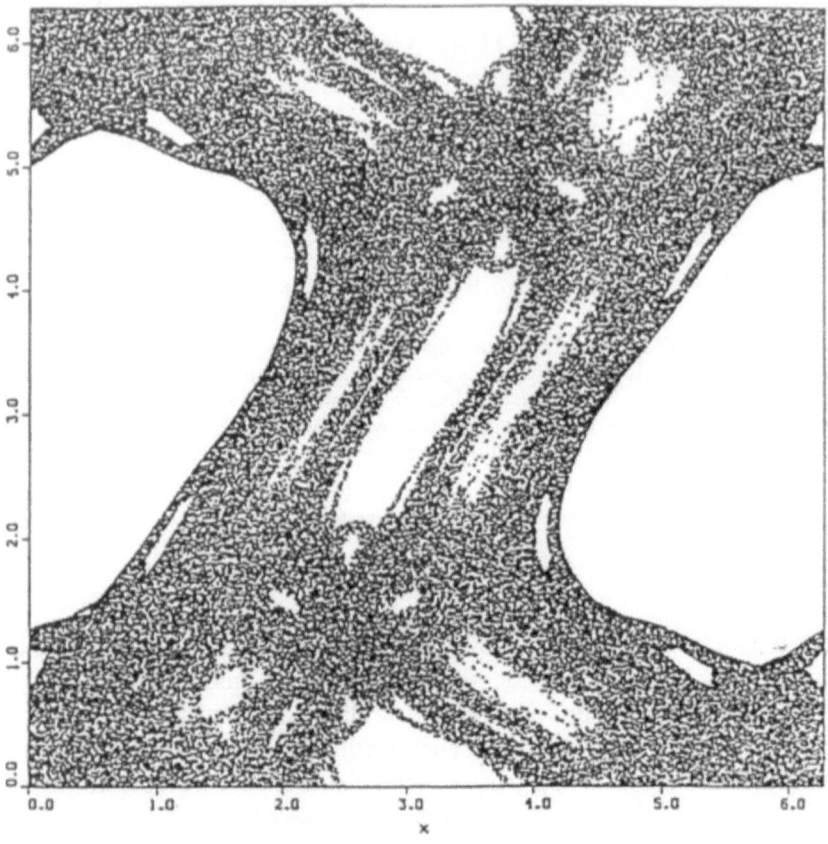

Fig. 1. $10^5$ iterates of the standard map with $k = 1.3$ from the initial condition $(x, y) = (\frac{1}{2}, \frac{1}{2})$.

We generate a distribution of points by iterating the standard map on a single initial condition placed in the largest chaotic region for $k = 1.3$ and $k = 10$. We apply wavelet analysis to these distributions of points. Our aim is to explore to what extent the wavelet transform reveals the hierarchy of islands present in the phase space when $k = 1.3$ for different number of points and whether it reveals any hidden structure in the apparently totally chaotic phase space when $k = 10$.

Finally we recall that the standard map is the Poincaré map of a perturbed pendulum, i.e. of a Hamiltonian system. Since celestial mechanics deals with Hamiltonian dynamics the same complicated structures is present and was emphasised by Poincaré, therefore such mappings have been used already in this field (see Lichtenberg and Lieberman 1983)

## 2. Wavelet Analysis

We apply a wavelet transform to the distribution of points given by iterating the standard map (1) on an initial condition in the largest chaotic region. It is inexpensive numerically to generate a large number $N$ of points and, rather than using the precise positions of these points in calculating the wavelet transform, we adopted the following procedure. We divide the two-dimensional $2\pi$-periodic phase space $T^2$ into $128^2$ boxes of width $2\pi/128$ and only record the number of points falling into each box. This gives a discretized density $D(\mathbf{r})$ (here $\mathbf{r} = (x, y)$) of points in the chaotic region, normalized so that

$$\int_{T^2} D(\mathbf{r})\, d^2\mathbf{r} = 1.$$

This discretization will not affect the result of the wavelet transform provided the scale of the wavelets is restricted to be larger than $2\pi/128$. The advantages of this approach are that one can take very large numbers of points and use fast Fourier transforms to evaluate the wavelet transform.

The wavelet transform $d(\mathbf{r}, a)$ of the density $D(\mathbf{r})$ is given by:

$$d(\mathbf{r}, a) = a^{-2} \int_{T^2} D(\mathbf{r}')g(\mathbf{r} - \mathbf{r}', a)\, d^2\mathbf{r}'.$$

Here $g(\mathbf{r}, a)$ is the family of wavelets parametrized by $a$, the scale of the wavelet. If we define the Fourier transform $\hat{f}(\mathbf{k})$ of a function $f(\mathbf{r})$ on $T^2$ by:

$$f(\mathbf{r}) = \sum_{\mathbf{k}} \hat{f}(\mathbf{k}) \exp(i\mathbf{k} \cdot \mathbf{r}),$$

(where $\mathbf{k}$ has integer components), then we have:

$$\hat{d}(\mathbf{k}, a) = (2\pi)^2 \hat{D}(\mathbf{k})\, \hat{g}(\mathbf{k}, a).$$

We shall use the family $g(\mathbf{r}, a)$ of periodic wavelets defined by the Fourier transform:

$$\hat{g}(\mathbf{k}, a) = (a^4 k^2/2\pi) \exp(-a^2 k^2/2). \tag{3}$$

These are the periodic versions of the well-known isotropic mexican hat wavelets:

$$g_{\text{mex}}(\mathbf{r}, a) = -a^2 \nabla^2 \exp(\mathbf{r}^2/2a^2);$$

the latter has the Fourier transform (3), but with $\mathbf{k}$ ranging over all *real* values. Note that, strictly speaking, wavelets $g(\mathbf{r}, a)$ with different values of $a$ are not related by dilations, as is the case with $g_{\text{mex}}$, this being impossible to realize exactly on $T^2$.

In Bendjoya *et al.* (see this volume) references are given showing how wavelet analysis of data can be used to detect and characterize "singularities", that is, points near which the data are self-similar under rescaling. If the density $D(\mathbf{r})$ is self-similar under rescaling about some point $\mathbf{r}_0$ then, locally:

$$D(\lambda(\mathbf{r} - \mathbf{r}_0)) \simeq \lambda^p D(\mathbf{r} - \mathbf{r}_0), \tag{4}$$

for some scaling factor $\lambda$ and scaling exponent $p$. This is reflected in the wavelet transform at the point $\mathbf{r}_0$ which scales as:

$$d(\mathbf{r}_0, a) \sim a^p F(\log a) \qquad \text{as} \qquad a \to 0. \tag{5}$$

Here $F$ is a periodic function with period $\log \lambda$; such oscillations are typical in the scaling behaviour lacunar of fractals (see Arneodo *et al.* 1988), and arise because a distribution is generally self-similar under a *discrete* set of rescalings, in this case by $\lambda^n$ for any integer $n$. Since the position of a singularity is not known *a priori* this definition is not easy to use. We shall only look for the strongest singularities, and we shall define:

$$d_{\max}(a) = \max_{\mathbf{r}} d(\mathbf{r}, a),$$
$$d_{\min}(a) = \min_{\mathbf{r}} d(\mathbf{r}, a)$$

and see how these scale with the size $a$ of the wavelet.

## 3. Standard Map for $k = 1.3$: Chaos and Islands.

In figure (1) we show the largest chaotic region of the standard map for $k = 1.3$ with $10^5$ points plotted from the initial condition $(x, y) = (\frac{1}{2}, \frac{1}{2})$. For this parameter value there are families of islands embedded in the chaotic "sea". In figure (2) we show the wavelet transform $d(\mathbf{r}, a)$ using varying numbers of points; in fig. 2a-d $N = 10^8$, in fig. 2 e-h $N = 10^5$ while for fig. 2 i-l $N = 10^3$; reading down each column, $d$ is shown for $a = \pi/4, \pi/8, \pi/16$ and $\pi/32$. White is plotted where $d$ is negative or zero, while positive values are shown by grey-scale shading, with black corresponding to the maximum value in each plot. For $N = 10^3$ and $10^5$ we in fact iterated the standard map $10^3$ times for each point used to generate the density $D$ i.e. we have kept one point per thousand iterates over $10^6$ iterations . This was to minimize the effect of the orbit remaining trapped close to island chains for many iterations and thus covering the chaotic region poorly.

Consider first the case when $N = 10^8$ (figure (2a-d)); we see how the wavelet transform reveals the island structure apparent in figure (1). For a given wavelet width $a$ the wavelet transform shows the islands in the map at scales of $a$ and larger; the strongest contribution is coming from structure at the scale $a$. In fact the islands give a strong negative contribution to the wavelet transform, and the positive contributions come largely from the edges of the islands. Note that the the plots (2a-d) are approximately invariant under the symmetry of the standard map:

$$x \leftarrow 2\pi - x, \qquad y \leftarrow 2\pi - y \tag{6}$$

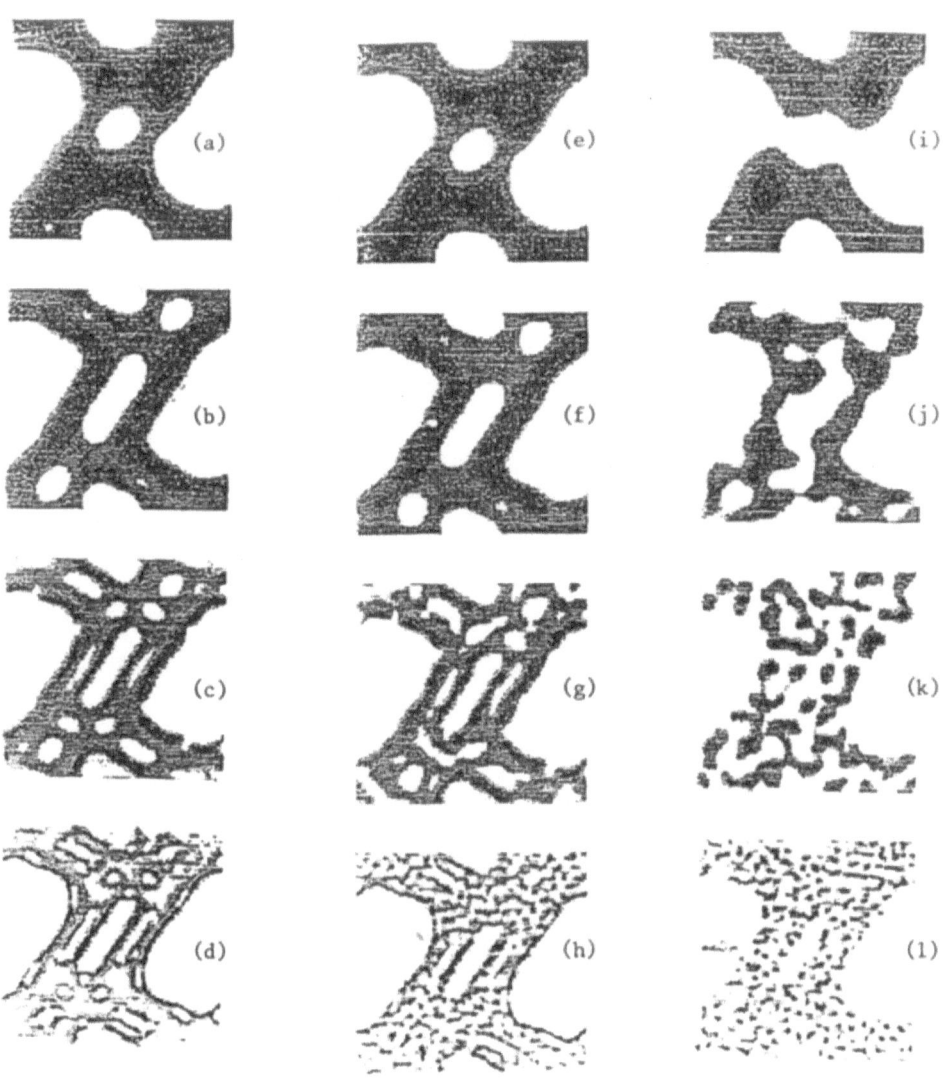

Fig. 2. Grey-scale plots of wavelet transforms $d(\mathbf{r}, a)$ for $k = 1.3$. The number of iterates used in (a-d) is $N = 10^8$, (e-h), $N = 10^5$ and (i-l), $N = 10^3$. The width of the wavelets in (a,e,i) is $a = \pi/4$, (b,f,j), $a = \pi/8$, (c,g,k), $a = \pi/16$ and (d,h,l) $a = \pi/32$. Negative or zero values are shown in white; positive values are shown by grey-scale shading.

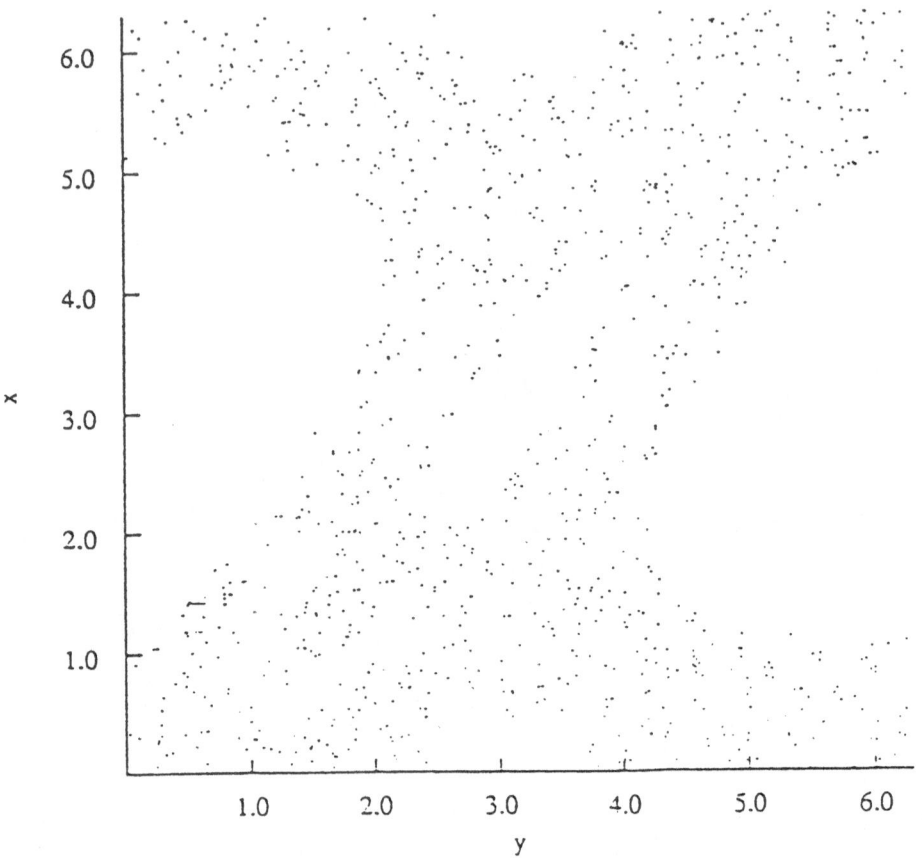

Fig. 3. $10^3$ points (one per thousand iterates) of the standard map with $k = 1.3$ from the initial condition $(x, y) = (\frac{1}{2}, \frac{1}{2})$.

confirming that the islands are indeed a feature of the chaotic region, rather than of the particular orbit used.

Now consider the effect of reducing the number of points to $N = 10^5$ (figures (2e-h)) and $N = 10^3$ (figures (2i-l)). We see that this leads to changes in the wavelet transform for increasing values of the wavelet width $a$. For example, the regular pattern of figure (2d) dissolves into a mess (figures (2h,l)) which bears no apparent relation to the islands in figure (1). In fact the small-scale structure seen in figures (2,h,k,l) is not invariant under the symmetry (6) and so results from the particular distribution of points in the orbit of the standard map (see fig. 3), rather than from the geometry of the chaotic region. Thus, if too few points are used in the transform, spurious structure may be generated. $10^8$ points are sufficient to reveal all genuine structure down to the scale $a = \pi/32$.

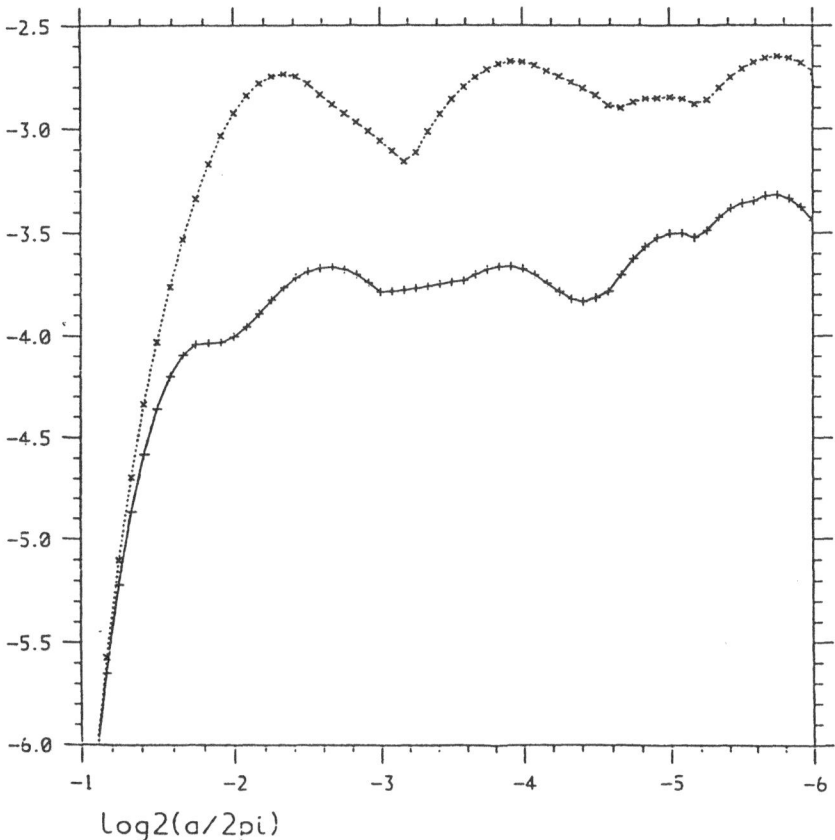

Fig. 4. Maximum $d_{max}(a)$ (solid, $+$) and minimum $-d_{min}(a)$ (dashed, $\times$) are plotted against $a/2\pi$ for the standard map with $k = 1.3$ and $N = 10^8$ points. The scales are logarithmic base two.

Using the data-set of $10^8$ points we calculated the scaling factors $d_{max}(a)$ and $-d_{min}(a)$, which are plotted against $a$ on $\log_2$-$\log_2$ scales in Figure (4). The minimum of the wavelet transform is much larger in absolute value than the maximum; both show strong oscillations, the peaks in $d_{min}$ occurring when $a$ is same scale of a family of islands. The oscillations are perhaps superimposed on a weak increase, however this is unclear.

## 4. Standard Map for $k = 10$: Completely Chaotic

For $k = 10$ the whole phase space of the standard map appears to be chaotic; points appear to be spread randomly and evenly in space. We calculated the wavelet transform using $N = 10^7$ iterates, beginning from $(x, y) = (\frac{1}{2}, \frac{1}{2})$. Plots of the wavelet transform are given in figure (5) and show interesting structure. However

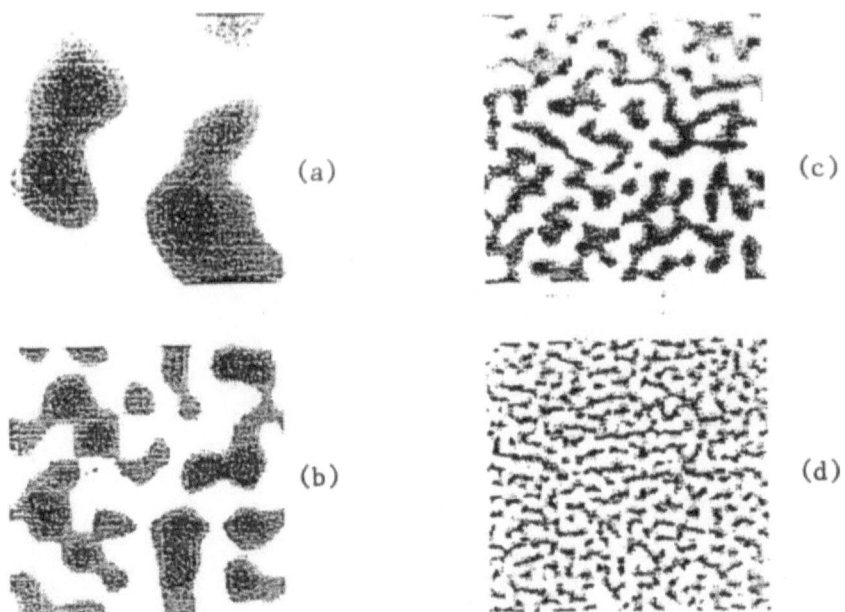

Fig. 5. Grey-scale plots of wavelet transforms $d(\mathbf{r}, a)$ for $k = 10$ using 107 iterates. The width of the wavelets is given by $a = \pi/4$ (a), $a = \pi/8$ (b), $a = \pi/16$ (c) and $a = \pi/32$ (d). Negative or zero values are shown in white; positive values are shown by grey-scale shading.

this structure is, spurious, since the plots do not satisfy the symmetry (6), even approximately. This suggests that the results of the transform depend on $N$ and the orbit chosen, rather than revealing properties of the phase space. We confirmed this by varying the number of iterates used, taking $N = 10^6$ and $10^8$; we found that plots of the transform change completely, even on large scales. Thus any apparent structure seems to be the result of statistical fluctuations in an effectively random distribution of points. Rather surprisingly, however, the wavelet transform shows clear scaling behaviour for each value of $N$. In figure (6) we plot $d_{\max}(a)$ and $-d_{\min}(a)$ against $a$ in $\log_2$-$\log_2$ coordinates, calculated using $N = 10^6$, $10^7$ and $10^8$ iterates of the map. For each number of points, we see a clear exponential growth of the maximum and minimum of the transform as $a$ decreases. However as the number of points is increased for a given value of $a$, the magnitude of the transform decreases. This reflects the fact that the wavelet transform is only showing statistical fluctuations in the distribution of points (which appears to be tending to a uniform density as $N \to \infty$), rather than any structure intrinsic to the phase space. Note also that the maximum and minimum curves in figure (6) are approximately the same, unlike the case $k = 1.3$.

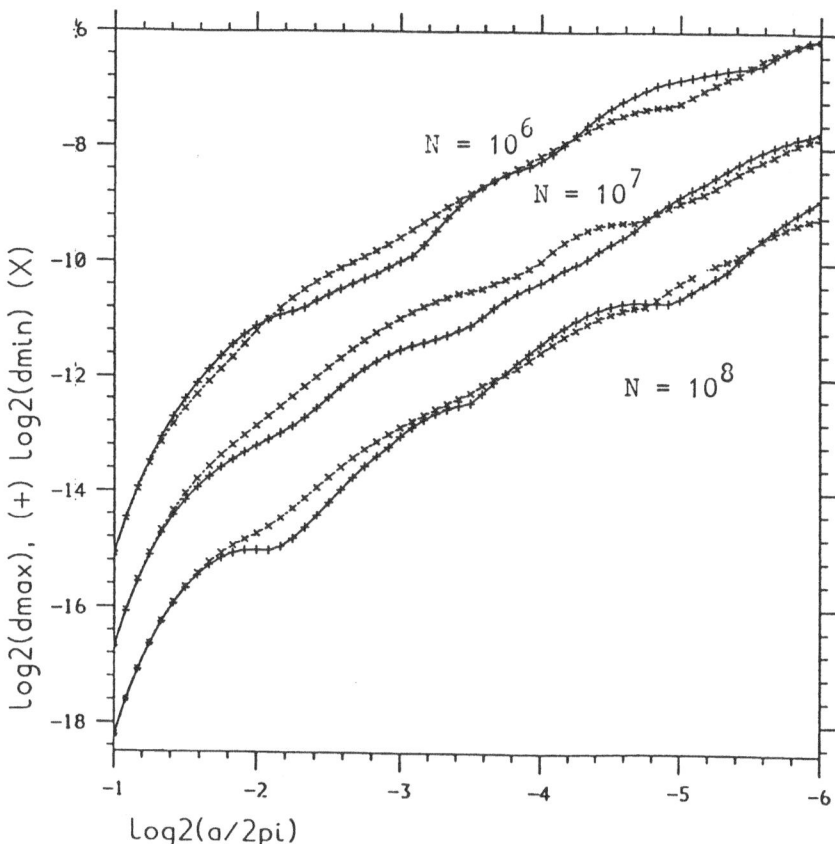

Fig. 6. Maximum $d_{\max}(a)$ (solid, +) and minimum $-d_{\min}(a)$ (dashed, ×) are plotted against $a/2\pi$ for the standard map with $k = 10$ for $N = 10^6$, $10^7$ and $10^8$ points. The scales are logarithmic base two.

## 5. Conclusions

The standard map appears to be a good model to test wavelet transform analysis. When the phase space of the map shows clear structure, this is reflected in the wavelet transform if sufficiently many points are used. For the standard map with $k = 1.3$, approximately $10^8$ points were required to reveal structure varying in scale over a range of about 50.

When the map appears completely random the wavelet transform reveals structure, but it is not real, being caused by statistical fluctuations. The wavelet transform, however, shows clear scaling behaviour, a result of statistical effects. To see this it is necessary to vary the number of points used in the analysis.

Thus the moral for galaxy applications could be to vary the number of points used, to see whether the structures perceived by the transform are real, and whether any scaling behaviour observed is simply the effect of statistical fluctuations.

## Acknowledgements

One of us (A.D.G.) gratefully acknowledges support by the E.E.C. under contract ST-2J-0029-1-F and by Gonville & Caius College, Cambridge.

## References

Arnéodo, A., Grasseau, G. & Holschneider, M.: 1988, "Wavelet transform of multifractals," *Phys. Rev. Lett.*, **61**, 2281-2284.

Bendjoya, Ph., Slezak, E. and Froeschlé, Cl.: 1991, "The wavelet transform : a new tool for asteroid family determination", *Astron. Astrophys*, **251**, 312-330.

Lichtenberg, A.J. & Lieberman, M.A.: 1983, *"Regular and Stochastic Motion,"* Applied Mathematical Sciences Vol. 38, Springer-Verlag.

Slezak, E., Bijaoui, A. & Mars, G.: 1990, " Identification of structures from galaxy counts: use of the wavelet transform," *Astron. Astrophys.*, **227**, 301-316.

# ASTEROIDS: 2/1 RESONANCE AND HIGH ECCENTRICITY

MICHÈLE MOONS AND ALESSANDRO MORBIDELLI
*Département de mathématique FUNDP*
*8, Rempart de la Vierge, B-5000 Namur, Belgique*

## 1. Introduction

The motion of asteroids in a mean motion commensurability with Jupiter has been studied many times these last years and extensive references to that subject can be found in Yoshikawa (1989) and in Henrard (1988).

The present work consists in a systematic exploration of the dynamics in mean motion resonances and is original in at least three aspects:

1. in developing a general theory, based on the use of action-angle variables, which can be applied to any isolated mean motion resonance in the main asteroid belt.
2. in taking into account also the inclination of the asteroid (3 dimensional model) and avoiding expansions in the eccentricity.
3. in taking into account the secular perturbations introduced by the Jupiter-Saturn system.

The work is actually in progress and we present here only briefly the procedure we have adopted and the very preliminary results we have obtained in the particular case of the 2:1 resonance in the frame of the planar restricted three-body problem Sun-Jupiter-Asteroid. The underlying structure is the semi-numerical perturbation method of Henrard and the reader should consult (Henrard, 1990) for more informations on that subject.

## 2. The Hamiltonian

To study the 2:1 mean motion resonance between asteroids and Jupiter, we take the well-known Hamiltonian of the restricted three-body problem (see, for instance, Szebehely, 1967) and introduce canonical variables appropriate to the 2:1 resonance.

With the usual notations and the choice of units $G = m_S + m_J = a' = 1$, this Hamiltonian takes the form

$$\mathcal{H} = \Lambda' - \frac{1-\mu}{2a} - 2\sqrt{(1-\mu)a} - \mu\left(\frac{1}{|\mathbf{r}-\mathbf{r}'|} - \frac{\mathbf{r}|\mathbf{r}'}{r'^3}\right) \tag{1}$$

and the canonical variables are (Poincaré, 1902)

$$
\begin{aligned}
\sigma &= 2\lambda' - \lambda - \varpi & , \quad S &= \sqrt{(1-\mu)a}\left(1 - \sqrt{1-e^2}\right) \\
-\nu &= 2\lambda' - \lambda - \varpi' & , \quad N &= \sqrt{(1-\mu)a}\left(2 - \sqrt{1-e^2}\right) \\
\lambda' & & , \quad \Lambda' &
\end{aligned}
\tag{2}
$$

*Celestial Mechanics and Dynamical Astronomy* 56: 273–276, 1993.
© 1993 *Kluwer Academic Publishers.*

Having in mind a study of the dynamics at high eccentricity, we do not perform the classical developments in power series of the eccentricity nor in Fourier series of the angular variables. Following Ferraz-Mello and Sato (1989), we compute the value of the averaged Hamiltonian

$$H = H_0(\sigma, S, N) + e'H_1(\sigma, S, \nu, N) \tag{3}$$

using Poincaré-like variables which are non-singular for small eccentricities. The explicit form of this Hamiltonian, however, is not known, but its value can be computed with accuracy for any given value of the phase space variables $(\sigma, S, \nu, N)$.

## 3. The unperturbed problem

The unperturbed part, i.e. the circular part $H_0(\sigma, S, N)$ of the Hamiltonian (3), can be considered as a two-degree of freedom separable Hamiltonian the phase space portrait of which is well-known.

Following Henrard (1990), we reduce it to a function $K_0(J_1, J_2)$ of the momenta by the introduction of the action-angle variables

$$\psi_1 = \frac{2\pi}{T_1}t \qquad , \qquad J_1 = \frac{1}{2\pi}\oint S \, d\sigma$$

$$\psi_2 = \nu - \rho(\psi_1, J_1, J_2) \quad , \quad J_2 = N \tag{4}$$

and compute the unperturbed frequencies

$$\omega_1 = \dot{\psi}_1 = \frac{\partial K_0}{\partial J_1} = \frac{2\pi}{T_1} \quad , \quad \omega_2 = \dot{\psi}_2 = \frac{\partial K_0}{\partial J_2} = \frac{1}{T_1}\int_0^{T_1} \frac{\partial H_0}{\partial N} \, dt \tag{5}$$

by numerically integrating along the periodic trajectories of the integrable Hamiltonian $H_0(\sigma, S, N)$; $T_1$ is the period of the trajectory and $\rho(\psi_1, J_1, J_2)$ is a periodic function.

Let us recall that, as the topology of the phase space associated to $H_0$ is not uniform, there is a discontinuity in the definition of the action-angle variables when crossing the separatrix of the first degree of freedom. The canonical transformation (4), however, can be considered as a global transformation of the phase space as it conserves its topological structure.

## 4. The perturbation

When the action-angle variables (4) are introduced into the Hamiltonian (3), this latest takes the form

$$K = K_0(J_1, J_2) + e'K_1(\psi_1, J_1, \psi_2, J_2) \tag{6}$$

In view of the application of a perturbation method in order to study the dynamics associated to such an Hamiltonian, we locate the secondary resonances that can

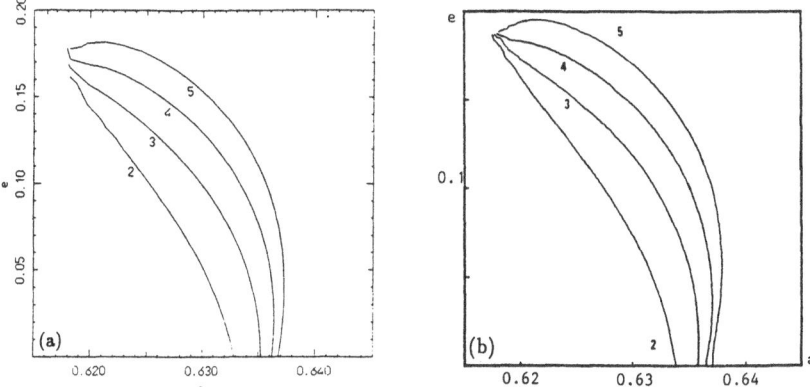

Fig. 1. (a) Secondary resonances $\omega_1/\omega_2 = 2, 3, 4, 5$ in our model. (b) The same in the model of Lemaître and Henrard (1990).

be excited by the perturbation and split the phase space into zones, each of which containing only one secondary resonance $j_1\omega_1 + j_2\omega_2 = 0$. In each zone we introduce then new action-angle variables, appropriate to the resonance involved, by means of the canonical unimodular transformation

$$
\begin{aligned}
\theta_1 &= j_1\psi_1 + j_2\psi_2 &,& \quad I_1 = k_2 J_1 - k_1 J_2 \\
\theta_2 &= k_1\psi_1 + k_2\psi_2 &,& \quad I_2 = -j_2 J_1 + j_1 J_2
\end{aligned}
\tag{7}
$$

Finally, we perform an averaging with respect to the "fast" angular variable $\theta_2$ and the Hamiltonian becomes

$$
\bar{K} = \bar{K}_0(\bar{I}_1, \bar{I}_2) + e'\bar{K}_1(\bar{\theta}_1, \bar{I}_1, \bar{I}_2)
\tag{8}
$$

We get thus a "quasi-integral" of motion

$$
\bar{I}_2 = -j_2 J_1 + j_1 J_2 + e'\frac{\partial W}{\partial \theta_2}
\tag{9}
$$

the level curves of which can be plotted on given surfaces of section in order to show some important features of the dynamics.

## 5. Preliminary results

In Figure 1a, we have plotted the secondary resonances, computed from the un-perturbed Hamiltonian $H_0$, as functions of the initial conditions of integration $(a, e, \sigma = 0)$. For the set of initial conditions we have chosen, there are no col-liding trajectories with Jupiter. In Figure 1b, we show the secondary resonances found by Lemaître and Henrard (1990) using an analytic model truncated at degree 8 in eccentricity; as one can see, the agreement is very good at low eccentricity.

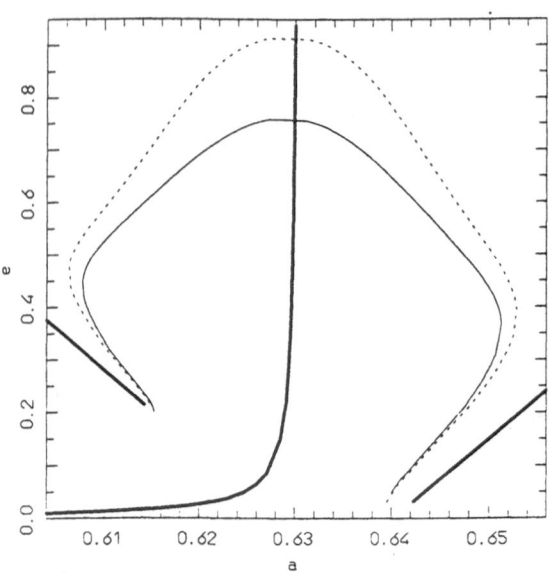

Fig. 2. Secular resonances $\nu_5$ (solid line) and $\nu_6$ (dashed line). The thick lines represent the separatrices and the stable family of the first degree of freedom.

In Figure 2, we show the location of $\nu_5$ and $\nu_6$ secular resonances; these correspond to 1/1 commensurabilities of the frequency of the longitude of perihelion $\varpi$ with the fundamental frequencies of the Jupiter-Saturn system $g_5$ and $g_6$ respectively.

## References

Ferraz-Mello, S. and Sato, M.: 1989, "The very-high-eccentricity asymmetric expansion of the disturbing function near resonances at any order", *Astron. Astrophys.*, **225**, 541-547.
Henrard, J.: 1988, "Resonances in the planar elliptic restricted problem", in *Long-Term Dynamical Behaviour of Natural and Artificial N-Body Systems* (A. E. Roy ed.), Kluwer Academic Publishers, 405-425.
Henrard, J.: 1990, "A semi-numerical perturbation method for separable Hamiltonian systems", *Celest. Mech.*, **49**, 43-68.
Lemaître, A. and Henrard, J.: 1990, "On the Origin of Chaotic Behavior in the 2/1 Kirkwood Gap", *Icarus*, **83**, 391-409.
Poincaré, H.: 1902, "Sur les planètes du type d'Hécube", *Bull. Astron.*, **19**, 289-310.
Szebehely, V.: 1967, "Theory of orbits", *Academic Press*.
Yoshikawa, M.: 1989, "A survey of the motions of asteroids in the commensurabilities with Jupiter", *Astron. Astrophys.*, **213**, 436-458.

# ON THE SECOND ORDER LONG-PERIOD MOTION OF HYPERION

P.J. MESSAGE

*Department of Applied Mathematics and Theoretical Physics,*
*University of Liverpool, United Kingdom*

**Abstract.** A brief description is given of the present state of some aspects of work on the theory of the long-period features of the orbit of Saturn's satellite Hyperion. Expressions resulting from the use of a Lie series transformation, which give a theory of the long-period motion to the second order in the mass of Titan, and also a differential correction process, provide a procedure for the construction of a dynamically consistent set of parameters of the long-period motion, including co-efficients of the long-period terms, using a fit to observationally determined values of such parameters. This is carried out with two sets of data from the reduction of the two main sequences of observations of the satellite, those between 1875 and 1922, and also those between 1967 and 1983.

**Key words:** Saturn's satellites – Hyperion – ephemeris

## 1. The Development of the Disturbing Function and its Derivatives

An earlier paper (Message 1989) gave a brief outline of the method which has been used to develop theoretical expressions for the long-period features of the perturbations of the orbit of Hyperion, using a Lie series transformation to enable the long-period terms to be treated separately from those of short period. The disturbing function for the action of Titan and its derivatives are developed as Fourier series in the two angular arguments $\omega$ and $\sigma$, and Taylor series in $a - a_0$ and $e - e_0$. Here $\sigma$ is the difference between the apse longitudes of Hyperion and Titan, and $\omega$ is related to the critical argument

$$\vartheta = 4\lambda - 3\lambda' - \varpi ,\tag{1}$$

by

$$\vartheta = \pi + q \sin \omega ,\tag{2}$$

for a fixed value of $q$ (54° was chosen so that only real values of $\omega$ occur in the actual motion of Hyperion). Also $a_0$ and $e_0$ are chosen close to the mean values taken in the motion by the major semi-axis, $a$, and the eccentricity, $e$. (Here $\lambda$, $\varpi$, $a$, and $e$ refer to Hyperion's orbit, and $\lambda'$ and $\varpi'$ to Titan's.) In the equations of motion in the long-period system, after separation from the short-period terms by the Lie series transformation, the disturbing function $R$ is replaced by, to complete the transformation to second order in the mass of Titan,

$$R^* = \bar{R} + \frac{1}{2}\overline{\{R, S_1\}} + \text{terms of third \& higher orders} ,\tag{3}$$

where $\bar{R}$ is the mean of $R$ over the short-period argument $\phi = \lambda - \lambda'$, and $\overline{\{R, S_1\}}$ is the mean over $\phi$ of the Poisson bracket of $R$ and $S_1$, where $S_1$ is the first-order part of the generating function of the Lie series transformation.

*Celestial Mechanics and Dynamical Astronomy* **56**: 277–284, 1993.

So $R^*$ takes the form

$$R^* = \sum_{m=0}^{\infty} \sum_{k=0}^{\infty} \{L_{m,k} \cos 2m\omega \; \cos k\sigma + M_{m,k} \sin(2m+1)\omega \; \sin k\sigma\} \; , \qquad (4)$$

where $L_{m,k}$ and $M_{m,k}$ are expanded as Taylor series in $a - a_0$ and $e - e_0$ (For some further details see the earlier paper (Message 1989); a more complete account of the Lie series transformation will be given in a later paper, together with a discussion of the motion of the orbit plane and its relation to the motion in the orbit plane.)

## 2. The Librational Solution: Using Observationally-Derived Values of the Parameters

It has been known since the investigations of Hill (1888) and Newcomb (1891) that the critical argument $\vartheta$ does not circulate, but librates about 180°, the amplitude being approximately 36°, and the period about 21 months. The form of the solution of the equations of motion for the long-period part of the motion in the orbit plane corresponding to this libration is

$$\vartheta = 180° + \sum \vartheta_{ij} \sin(i\tau + j\zeta) \; ,$$
$$\sigma = \zeta + \sum \varpi_{ij} \sin(i\tau + j\zeta) \; ,$$
$$a = a_{00} + \sum a_{ij} \cos(i\tau + j\zeta) \; ,$$
$$e = e_{00} + \sum e_{ij} \cos(i\tau + j\zeta) \; , \qquad (5)$$

where $\tau = \nu t + \tau_0$ is the argument of the libration, and $\zeta = \chi t + \zeta_0$ is the linear part of the difference between the apse longitudes ($\sigma = \varpi - \varpi'$), which has period approximately 18 3/4 years. The summations are over all integer pairs $(i, j)$ with $i \geq 0$ (There are of course additional small periodic terms resulting from the influence of sources other than Titan, but these are added later). It is indicated in the earlier paper (Message 1989) how a set of such expressions is used to put the long-period part of the disturbing function, and its derivatives with respect to the orbital elements, into the form of a double Fourier series in the two arguments $\tau$ and $\zeta$. Then the Lagrange equations for the long-period part of the action of Titan on the orbital elements describing motion in the orbital plane are put into the form

$$\dot{\lambda} = n + m' \sum \Lambda_{ij} \cos(i\tau + j\zeta) + \Lambda' \; ,$$
$$\dot{\varpi} = m' \sum P_{ij} \cos(i\tau + j\zeta) + P' \; ,$$
$$\dot{a} = m' \sum A_{ij} \cos(i\tau + j\zeta) \; , \qquad (6)$$
$$\dot{e} = m' \sum E_{ij} \cos(i\tau + j\zeta) \; ,$$

where $n$ is $\sqrt{\mu/a^3}$, $m'$ is the ratio of the mass of Titan to that of Saturn, $\Lambda'$ is the contribution to $\dot{\lambda}$ from sources other than Titan (solar terms, the effect of the figure of Saturn, and the attractions of the satellites other than Titan), and $P'$ is the contribution to $\tilde{\varpi}$ from these sources other than Titan (The constant parts of $\Lambda'$ and $P'$ are included here: the effect of the periodic parts, which are much smaller, are added later). We replace the first of these equations by

$$\ddot{\vartheta} = m' \sum T_{ij} \sin(i\tau + j\zeta) ,$$

to ease the handling of the longer-period terms in $\vartheta$. Equating co-efficients of $\cos(i\tau + j\zeta)$ or $\sin(i\tau + j\zeta)$ in each of these equations then leads to

$$
\begin{aligned}
-(i\nu + j\chi)^2 \vartheta_{ij} &= m' T_{ij} , \\
(i\nu + j\chi)\, \varpi_{ij} &= m' P_{ij} , \\
-(i\nu + j\chi)\, a_{ij} &= m' A_{ij} , \\
-(i\nu + j\chi)\, e_{ij} &= m' E_{ij} ,
\end{aligned}
\tag{7}
$$

The constant term in the first equation of (6) gives

$$\hat{n} = \bar{n} + m'\Lambda_{00} + \bar{\Lambda}' \tag{8}$$

where $\hat{n}$ is the mean value of $\dot{\lambda}$, $\bar{n}$ is the mean value of $n$, and $\bar{\Lambda}'$ is the mean part of $\Lambda'$. The constant term in the second equation of (6) gives

$$\chi = \bar{\tilde{\varpi}} - \bar{\tilde{\varpi}}' = m' P_{00} + \bar{P}' - \bar{\tilde{\varpi}}' \tag{9}$$

where $\bar{\tilde{\varpi}}$ is the mean value of $\dot{\varpi}$, $\bar{\tilde{\varpi}}'$ is the mean value of $\dot{\varpi}'$, and $\bar{P}'$ is the mean part of $P'$. Also we have the commensurability condition

$$4\hat{n} - 3n' - \bar{\tilde{\varpi}} = 0 . \tag{10}$$

These equations leave just two independent quantities amongst the set $\vartheta_{ij}$, $\varpi_{ij}$, $a_{ij}$, $e_{ij}$, (for all pairs $i,j$), $\nu$, $\chi$, and $\hat{n}$. The phase constants $\tau_0$ and $\zeta_0$ complete the set of four constants of integration in motion in the orbit plane.

The procedure indicated above was first carried out using values of the co-efficients and other parameters obtained from a new analysis of the opposition mean values of the orbital elements given by Woltjer (1928) from his reduction of the main sequences of observations made between 1875 to 1922. This gave the estimates:

$$
\begin{aligned}
e_{00} &= +0.10419 \pm 0.00027 , & e_{01} &= +0.02414 \pm 0.00044 , \\
e_{02} &= -0.00183 \pm 0.00040 , & e_{10} &= -0.00401 \pm 0.00034 , \\
\varpi_{01} &= -13°.905 \pm 0°.273 , & \varpi_{02} &= +0°.754 \pm 0°.249 , \\
\varpi_{10} &= -0°.314 \pm 0°.262 , & \varpi_{20} &= -0°.795 \pm 0°.304 , \\
\lambda_{01} &= -0°.054 \pm 0°.019 , & \lambda_{02} &= +0°.007 \pm 0°.018 , \\
\lambda_{10} &= +9°.112 \pm 0°.018 , & \lambda_{20} &= +0°.039 \pm 0°.018 ,
\end{aligned}
$$

$$\hat{n} = 16°.9199890 \pm 0°.0000027 \quad \text{per day} ,$$

and

$$\nu = 0°.562025 \pm 0°.000025 \quad \text{per day} .$$

Also used were the values

$$n' = 22°.5770122 \quad \text{per day} , \quad \bar{\bar{\omega}}' = 0°.528 \quad \text{per year} ,$$
$$\bar{\Lambda}' = 0°.001601 \quad \text{per day} , \quad \bar{P}' = 0°.271 \quad \text{per year} ,$$
$$\text{and} \quad e' = 0.0291 .$$

The consistency of a set of values for the co-efficients can be tested by comparing the estimates of $m'$ given by the equations (7). From the mean motion of the apse (equation (9)), we obtain, in units of $10^{-4}$ of the mass of Saturn,

$$
\begin{array}{lll}
& 2.350 \pm 0.012 & \text{(i.e.} \quad 1/m' = 4256 \pm 21) , \\
\text{from the value of } \vartheta_{10} , & 2.361 \pm 0.026 & \text{(i.e.} \quad 1/m' = 4235 \pm 47) , \\
\text{from the value of } e_{01} , & 2.290 \pm 0.053 & \text{(i.e.} \quad 1/m' = 4367 \pm 96) , \\
\text{from the value of } \varpi_{01} , & 2.442 \pm 0.099 & \text{(i.e.} \quad 1/m' = 4095 \pm 176) , \\
\text{from the value of } e_{10} , & 2.442 \pm 0.208 & \text{(i.e.} \quad 1/m' = 4095 \pm 372) ,
\end{array}
$$

giving results satisfactorily consistent within the uncertainties, which follow from those of the observationally derived data.

## 3. The Differential Correction Scheme

To outline the principle underlying the construction of the differential correction scheme used to derive a dynamically consistent set of co-efficients and parameters, let us consider a vector, $x$, whose $N$ components, $x_k$, are those co-efficients, $\vartheta_{ij}$, $\varpi_{ij}$, $a_{ij}$, $e_{ij}$, which are to be included in the solution. Then the equations (7) are of the form

$$\pm(i\nu + j\chi)^p x_k = m' X_k , \tag{11}$$

(where $p$ is 2 if $x$ is a $\vartheta_{ij}$, and 1 otherwise, and the sign is "+" if $x_k$ is a $\varpi_{ij}$, and "−" otherwise.) Then, to first order in the changes, $\delta x_k$, in the $x_k$, from the values used in the theoretical developments, and in the change $\delta m'$ in the value $m'$ of the mass of Titan used provisionally,

$$\pm(i\nu + j\chi)^p \delta x_k \pm p(i\nu + j\chi)^{p-1}(i\delta\nu + j\delta\chi)x_k = m' \sum_{\ell=1}^{N} \frac{\partial X_k}{\partial x_\ell} \delta x_\ell + X_k \delta m' \tag{12}$$

Also (8) gives

$$\delta\hat{n} = -\frac{3\bar{n}}{2a_{00}} \delta a_{00} + m' \sum_{\ell=1}^{N} \frac{\partial \Lambda_{00}}{\partial x_\ell} \delta x_\ell + \Lambda_{00} \delta m' , \tag{13}$$

and (9) gives

$$\delta\chi = m' \sum_{\ell=1}^{N} \frac{\partial P_{00}}{\partial x_\ell} \delta x_\ell + P_{00}\delta m' , \tag{14}$$

while (10) gives

$$4\delta\hat{n} = \delta\chi , \tag{15}$$

(It was of course necessary to calculate the partial derivatives $\partial X_k/\partial x_\ell$, $\partial \Lambda_{00}/\partial x_\ell$ and $\partial P_{00}/\partial x_\ell$ for the calculation in section 3 of the standard errors of the estimates of the mass of Titan from the observationally determined estimates of individual co-efficients, so they are in any case available for the present calculation). The motion in the orbit plane has four free parameters. Two of these are the phases, $\tau_0$ and $\zeta_0$, of the long-period arguments, which do not enter into the equations relating the co-efficients. The latter of course is related to the values at epoch of the mean longitudes and apses, through

$$4\lambda_0 - 3\lambda_0' - \varpi = 0 \quad \text{and} \quad \zeta_0 = \varpi_0 - \varpi_0' .$$

There therefore remain two free parameters amongst $\nu$, $\chi$, $\hat{n}$, and the co-efficients $x_k$. The choice made was of $\hat{n}$, the mean part of $\dot{\lambda}$, and of $\vartheta_{10}$, the co-efficient of $\sin\tau$ in the critical argument $\vartheta$, i.e. of the 21 month libration. The equations (12), (13), (14) and (15) are solved to give $\delta\nu$ and each of the $\delta x_k$ (apart from $\delta\vartheta_{10}$) as linear expressions in $\delta\hat{n}$, $\delta\vartheta_{10}$ and $\delta m'$:

$$\delta x_k = A_k\delta\vartheta_{10} + B_k\delta\hat{n} + C_k\delta m' , \tag{16}$$

and

$$\delta\nu = A_\nu\delta\vartheta_{10} + B_\nu\delta\hat{n} + C_\nu\delta m' . \tag{17}$$

Those equations (16), for those $x_k$ for which there is an observationally determined value, and equation (17), as well as the observationally determined values of $\hat{n}$ and $\vartheta_{10}$, provide the equations of condition.

## 4. The Solution of the Equations to Give a Dynamically Consistent Set of Co-efficients, Using the 1875 to 1922 Data

The least-squares solution was carried out, using these equations of condition, with the estimates of co-efficients and of $\hat{n}$ and of $\nu$ given in section 3 above, derived from the 1875 to 1922 observations. This gave:

$$\hat{n} = 16°.9199888 \pm 0°.0000067 \quad \text{per day} ,$$
$$\vartheta_{10} = 36°.877 \pm 0°.182 ,$$
$$m' = 0.000236122 \pm 0.000000008 \quad (\text{i.e.} \quad 1/m' = 4235.10 \pm 0.14) .$$

(The standard errors given here are those internal to this solution, and so reflect the degree of consistency of the values of the co-efficients, which were used as the data for this solution, and not the consistency of the individually observed positions.) (For comparison, the value for the mass of Titan derived from the Pioneer and Voyager mission data is (Campbell and Anderson, 1989)

0.000236693 ± 0.000000026 (i.e.    $1/m' = 4224.87 \pm 0.47$)

The relations of the type indicated by the equations (16) and (17), which embody the theoretical relations between the co-efficients and other parameters of the long-period motion, are then used, by substitution into them of these three estimates, to give a set of values of the parameters of the long-period motion, dynamically consistent within the second-order theory here developed. This gave

$$\nu = 0°.562024 \quad \text{per day} , \tag{18}$$

and the co-efficients in the following set of expressions:

$$e = 0.10473 \begin{array}{l} +0.02453\cos\zeta - 0.00143\cos 2\zeta + 0.00017\cos 3\zeta \\ -0.00389\cos\tau - 0.00005\cos 2\tau \\ +0.00018\cos(\tau - \zeta) - 0.00014\cos(\tau + \zeta) , \end{array}$$

$$a = a_{00}\{1 \begin{array}{l} -0.003227\cos\tau \\ +0.0000084\cos(\tau - \zeta) - 0.0000058\cos(\tau + \zeta)\} , \end{array}$$

$$\varpi = \zeta + \varpi' \begin{array}{l} -13°.579\sin\zeta + 1°.626\sin 2\zeta - 0°.257\sin 3\zeta \\ -0°.430\sin\tau - 0°.017\sin 2\tau \\ +0°.360\sin(\tau - \zeta) - 0°.272\sin(\tau + \zeta) , \end{array}$$

$$\lambda = \hat{n}t + \epsilon \begin{array}{l} -0°.070\sin\zeta - 0°.002\sin 2\zeta + 0°.001\sin 3\zeta \\ +9°.113\sin\tau + 0°.004\sin 2\tau - 0°.017\sin 3\tau \\ -0°.220\sin(\tau - \zeta) + 0°.190\sin(\tau + \zeta) . \end{array}$$

## 5. A Solution of the Equations Using 1967 to 1983 Data

A second least-squares solution was carried out, using equations of condition of the same form, with observational data provided by the analysis of the main observational sequences between 1967 to 1983 (Taylor et al., 1987, improved in Taylor 1993). This analysis, which had been carried out by a quite different method from that which had been used for the earlier series, gave the following estimates for the parameters of the long-period theory:

$$e_{00} = +0.1044817 \pm 0.0000045 , \quad e_{01} = 0.0242121 \pm 0.0000070 ,$$
$$e_{02} = -0.0013759 \pm 0.0000068 , \quad e_{03} = 0.0001849 \pm 0.0000068 ,$$
$$e_{10} = -0.0040718 \pm 0.0000068 , \quad e_{1-1} = 0.0002326 \pm 0.0000068 ,$$
$$e_{11} = -0.0001691 \pm 0.0000068 ,$$

$$a_{10} = -0.003518 \pm 0.000011 , \quad a_{1-1} = 0.000084 \pm 0.000011$$
$$a_{11} = -0.000069 \pm 0.000011 ,$$

$$\varpi_{01} = -13°.4701 \pm 0°.0037 , \quad \varpi_{02} = -1°.5629 \pm 0°.0.0038 ,$$
$$\varpi_{01} = -0°.2604 \pm 0°.0038 , \quad \varpi_{10} = -0°.4461 \pm 0°.0.0038 ,$$
$$\varpi_{1-1} = +0°.3543 \pm 0°.0038 , \quad \varpi_{10} = -0°.2635 \pm 0°.0.0038 ,$$

$$\lambda_{01} = -0°.09011 \pm 0°.00061 , \quad \lambda_{02} = -0°.00963 \pm 0°.00065 ,$$
$$\lambda_{03} = -0°.00782 \pm 0°.00062 , \quad \lambda_{10} = +9°.12785 \pm 0°.00062 ,$$
$$\lambda_{1-1} = -0°.21032 \pm 0°.00062 , \quad \lambda_{11} = +0°.22779 \pm 0°.00062 ,$$

(the $a_{ij}$ are expressed as factors of $a_{00}$) and

$$\nu = 0°.5622095 \pm 0°.00000074 \quad \text{per day} .$$

To these was added the estimate of the mean motion from the earlier data set, which covers a longer time span:

$$\hat{n} = 16°.9199890 \pm 0°.0000027 \quad \text{per day} ,$$

and the estimate $e' = 0.02887$ was used. The result of this solution is:

$$\hat{n} = 16°.919939 \pm 0°.000040 \quad \text{per day} ,$$
$$\vartheta_{10} = 36°.955 \pm 0°.096 ,$$
$$m' = 0.000236292 \pm 0.000000072 \quad (\text{i.e. } 1/m' = 4232.05 \pm 0.13) .$$

Use of these results in equations (16) and (17) then gives

$$\nu = 0°.5622094 \quad \text{per day} .$$

and the co-efficients in the following set of expressions, (corresponding of course to a dynamically consistent motion different from the numerical integration from which the co-efficients used as data were derived, because of the use here of the better determined value of the mean motion):

$$e = 0.104496 \quad +0.024321\cos\zeta - 0.001408\cos2\zeta + 0.000165\cos3\zeta$$
$$-0.003906\cos\tau - 0.000050\cos2\tau + 0.000019\cos3\tau$$
$$+0000183\cos(\tau - \zeta) - 0.000147\cos(\tau + \zeta) \,,$$

$$a = a_{00}\{1 \quad -0000038\cos\zeta - 0.003234\cos\tau$$
$$+0.000084\cos(\tau - \zeta) - 0.000060\cos(\tau + \zeta)\} \,,$$

$$\varpi = \zeta + \varpi' \quad -13°.4657\sin\zeta + 1°.6029\sin2\zeta - 0°.2519\sin3\zeta$$
$$-0°.4336\sin\tau - 0°.0178\sin2\tau - 0°.0004\sin3\tau$$
$$+0°.3597\sin(\tau - \zeta) - 0°.2726\sin(\tau + \zeta) \,,$$

$$\lambda = \hat{n}t + \epsilon \quad -0°.0642\sin\zeta - 0°.0030\sin2\zeta + 0°.0004\sin3\zeta$$
$$+9°.1311\sin\tau + 0°.0035\sin2\tau - 0°.0172\sin3\tau$$
$$-0°.2196\sin(\tau - \zeta) + 0°.1941\sin(\tau + \zeta) \,.$$

The agreement between the two sets of estimates of the co-efficients of the long period terms from the two series of observations is encouraging; the indication being that the later series of observations gives the more accurate values for these co-efficients, though not for the mean motion, due to the shorter time covered. Work is in progress to extend the procedures to encompass higher multiples of $\zeta$ (as is clearly necessary in $\varpi$), and on means to deal with both series of observations in a consistent way, so as to enable the entire span of observations to be encompassed in a single solution.

## References

Campbell, J.K. and Anderson, J.D.: 1989, *Astron. J.*, **97**,1485–1495.
Hill, G.W.: 1888, *Astron. J.*,**8**, 57–62.
Message, P.J.: 1989, *Celest. Mech.*, **45**, 45–53.
Newcomb, S.: 1891, *Astron. Papers of the American Ephemeris*, **III**.
Taylor, D.B., Sinclair, A.T., and Message, P.J.: 1987, *Astron. & Astrophys.*, **181**, 383–390.
Taylor, D.B.: 1993, *Astron. & Astrophys.*, in press.
Woltjer, J.: 1928, *Annalen van de Sterrewacht te Leiden*, **XVI**, Pt.3.

# THE PROBLEM OF CRITICAL INCLINATION COMBINED
# WITH A RESONANCE IN MEAN MOTION
# IN ARTIFICIAL SATELLITE THEORY

FABIENNE DELHAISE* and JACQUES HENRARD
*Department of Mathematics*
*Facultés N.D. de la Paix*
*Rempart de la Vierge,8 B-5000 NAMUR ,Belgium*

In this paper, the problem of a geosynchronous artificial satellite orbiting near the critical inclination is investigated. At the critical inclination the secular $J_2$ effect in the argument of perigee is zero. This allows the apogee to be stabilized above the desired (e.g. northern European) coverage area by adjusting other orbital elements, thus providing a greatly enhanced coverage of the target regions. Apart from the improved visibility from high-latitude-regions, highly inclined orbits are becoming attractive due to the increasing collision hazard of geostationary satellites with debris and de-activated spacecrafts which "pollute" the relatively confined geostationary ring.

The aim of this paper is to give a qualitative study of the global dynamics of such orbits. More specifically the 12-hr Molniya-type and the 24-hr Tundra-type orbits, both located at the critical inclination, are analyzed. The Hamiltonian takes into account the gravitational perturbations arising from an aspherical Earth. The expansion of the geopotential includes the main zonal harmonic $J_2$ together with all main critical tesseral harmonics coefficients. It is worth noting that luni-solar perturbation effects are non-negligible for such high elevation orbits but modeling these effects is beyond the scope of this paper. The luni-solar effects on such type of orbits will be examined in a future paper.

The study of this type of orbits deals with a double resonance problem: a mean motion commensurability combined with the critical inclination. To tackle this two-degrees-of-freedom problem, a local study is performed first. The Hamiltonian is expanded up to the quadratic terms around a linearly stable equilibrium point. The Hessian matrix is then diagonalized by means of a canonical transformation. The values of the proper frequencies of the two critical resonant arguments are thus deduced together with the set of variables in which the Hamiltonian is transformed into a two-decoupled-harmonic-oscillators form.

The semi-numerical method developed by Henrard (1990) is then applied in order to explore in a global way the secular dynamics of the problem. This theory can describe the perturbations of non-trivial separable systems and has the distinct advantage of not being confined to the analysis of a small neighborhood of their periodic orbits. Approximate surfaces of section are constructed in the plane of the inclination and argument of perigee. The values of the "mean" eccentricity

* Research Assistant for the Belgian National Fund for Scientific Research

are also denoted on the right hand side of each picture. >From these figures the qualitative and some of the quantitative aspects of the perturbed dynamical system can easily be read. The chaotic regions of the phase space are identified around the critical curve associated to the resonance in mean motion. Another interesting feature highlighted by this analysis is the stable and unstable equilibria switch positions in the argument of perigee for a given energy level which designates the change of sign of the coefficients of the main harmonics of the critical angle.

These results have been confirmed by the construction of a mapping obtained through numerical integration of the full Hamiltonian system. The main difference between numerical and semi-numerical results is in the equilibrium value of the inclination which can be either slightly smaller or larger than the so-called critical inclination, depending on the value of the considered energy surface.

A complete paper on this subject has been submitted to Celestial Mechanics.

## Acknowledgements

The authors wish to thank Mr. Alessandro Morbidelli for very useful discussions during the course of this study.

## References

Brouwer, D.: 1959, 'Solution of the Problem of Artificial Satellite Theory without Drag', *Astron. J*, **64**, 378–397.

Delhaise, F.: 1989, 'Geopotential Perturbations of the Tundra and Molniya Orbits', *ESA, MAS Working Paper*, **288**.

Giacaglia,G. E.: 1976, 'A Note on the Inclination Functions of the Satellite Theory', *Celest. Mech.*, **13**, 503–509.

Garfinkel, B.: 1959, 'The orbit of a Satellite of an Oblate Planet', *Astron. J*, **64**, 353.

Jupp, A.H.: 1988, 'The Critical Inclination Problem - 30 Years of Progress', *Celest. Mech.*, **43**, 127–138.

Henrard, J.: 1990, 'A Semi-Numerical Perturbation Method for Separable Hamiltonian systems', *Celest. Mech.*, **49**, 43–67.

Kozai, Y.: 1959, 'The Motion of a Close Earth Satellite', *Astron. J.*, **64**, 367–377.

Lecohier, G., Guermonprez, V. and Delhaise, F.: 1989, 'European Molniya and Tundra Orbit Control', *CNES, Mécanique Spatiale, Symposium International en Mécanique Spatiale*, Toulouse (France), 165–191.

Moser, J.: 1958, 'New Aspects in the Theory of Stability of Hamiltonian Systems', *Comm. Pure App. Math.*, **XI**, 81–114.

Sochilina, A.S.: 1982, 'On the Motion of a Satellite in Resonance with its Rotating Planet', *Celest. Mech.*, **26**, 337–352.

# METEORITES FROM THE ASTEROID 6 HEBE

PAOLO FARINELLA

*Dipartimento di Matematica, Università di Pisa,*
*Via Buonarroti 2, 56127 Pisa, Italy*

and

CHRISTIANE FROESCHLÉ and ROBERT GONCZI

*OCA – Observatoire de Nice, B.P. 229, 06304 Nice Cedex 04, France*

**Abstract.** We have numerically integrated the orbits of 18 fictitious fragments ejected from the asteroid 6 Hebe, an S–type object about 200 $km$ across which is located very close to the $g = g_6$ (or $\nu_6$) secular resonance at a semimajor axis of 2.425 $AU$ and a (proper) inclination of $15°.0$. A realistic ejection velocity distribution, with most fragments escaping at relative speeds of a few hundreds $m/s$, has been assumed. In four cases we have found that the resonance pumps up the orbital eccentricity of the fragments to values $> 0.6$, which result into Earth–crossing, within a time span of $\approx 1\ Myr$; subsequent close encounters with the Earth cause strongly chaotic orbital evolution. The closest Earth and Mars encounters recorded in our integration occur at miss distances of a few thousandths of $AU$, implying collision lifetimes $< 10^9\ yr$. Some other fragments affected by the secular resonance become Mars–crossers but not Earth–crossers over the integration time span. Two bodies are injected into the 3 : 1 mean motion resonance with Jupiter, and also display macroscopically chaotic behaviour leading to Earth–crossing. 6 Hebe is the first asteroid for which a realistic collisional/dynamical evolution *route* to generate meteorites has been fully demonstrated. It may be the parent body of one of the ordinary chondrite classes.

**Key words:** Meteorites – asteroids

## 1. Introduction

A very popular theory on the origin of (most) meteorites and Near–Earth Asteroids (NEAs) is that they are asteroidal fragments, ejected from their parent bodies as a consequence of impacts and channeled into chaotic dynamical routes, associated with mean motion and secular resonances (see e.g. Greenberg and Chapman, 1983; Wetherill, 1985, 1987; Wetherill and Chapman, 1988; Greenberg and Nolan, 1989). Only in the last decade the availability of fast computers has led to test this theory by suitable numerical experiments, which have shown that sizeable chaotic zones are present in the real asteroid belt, and that in some cases the corresponding orbits can reach within $\approx 1\ Myr$ eccentricities of $\approx 0.6$, sufficient to cross the orbit of both Mars and the Earth.

Two specific source locations have been identified for such planet–crossing objects: the 3 : 1 mean motion resonance with Jupiter near 2.5 $AU$ (Wisdom, 1983, 1985) and the inner edge of the main belt near 2.1 $AU$, where the dynamics is dominated by the $g = g_6$ (or $\nu_6$) secular resonance (Wetherill and Williams, 1979; Scholl and Froeschlé, 1991). Recently, Knežević *et al.* (1991) and Farinella *et al.* (1991, 1992a) have argued that the $g = g_6$ resonance is probably effective in collecting asteroid fragments in two other regions of the main belt, i.e. at semimajor axes of about 2.4 and 2.7 $AU$ and moderate inclinations (15 to 20°).

*Celestial Mechanics and Dynamical Astronomy* **56**: 287–305, 1993.
© *1993 Kluwer Academic Publishers.*

TABLE I

Proper elements and secular apsidal frequency $g$ of Hebe and its fictitious fragments whose orbits have been integrated, as predicted by the secular perturbation theory of Milani and Knežević ($g$ cannot be derived for bodies in the 3:1 mean motion resonance). The fragment label numbers are conventional.

| $n^o$ | $a$ (AU) | $e$ | $i$ (deg.) | $g$ (arcsec/yr) |
|---|---|---|---|---|
| Hebe | 2.4253 | 0.1686 | 15.052 | 29.048 |
| 137 | 2.5002 | 0.1127 | 13.875 | — |
| 171 | 2.3320 | 0.1426 | 14.554 | 28.427 |
| 221 | 2.3991 | 0.1894 | 14.839 | 29.289 |
| 228 | 2.3690 | 0.1585 | 14.750 | 28.892 |
| 262 | 2.4258 | 0.1592 | 15.118 | 28.918 |
| 319 | 2.4948 | 0.3899 | 16.469 | — |
| 372 | 2.2910 | 0.1173 | 14.554 | 27.560 |
| 401 | 2.3322 | 0.1360 | 14.513 | 28.478 |
| 404 | 2.3051 | 0.1331 | 14.418 | 28.034 |
| 415 | 2.3488 | 0.1446 | 14.904 | 28.301 |
| 418 | 2.3483 | 0.1455 | 14.673 | 28.599 |
| 560 | 2.4292 | 0.1850 | 15.646 | 28.178 |
| 620 | 2.4660 | 0.1445 | 14.371 | 28.578 |
| 626 | 2.4079 | 0.1594 | 15.005 | 29.039 |
| 877 | 2.4757 | 0.1550 | 13.710 | 28.047 |
| 953 | 2.3934 | 0.1591 | 14.862 | 29.095 |
| 964 | 2.4158 | 0.2027 | 19.050 | 22.518 |
| 987 | 2.3432 | 0.1351 | 14.460 | 28.758 |

The latter authors also noted that a small number of sizeable asteroids lie very close to the resonance surfaces in these regions. As these asteroids are probably abundant sources of fragments ejected at relative speeds of a few hundreds $m/s$ as a consequence of impacts, and a significant fraction of these ejecta is likely to "fall" inside the chaotic zones associated with the resonance, it is possible that many meteorites/NEAs come from a relatively small number of parent asteroids, whose composition is not necessarily representative of the general asteroid population.

In particular, the asteroid 6 Hebe — a 200–$km$ sized S–type object whose proper elements, as computed by version 5.3 of the theory of Milani and Knežević (1990, 1992) are: semimajor axis $a = 2.425\ AU$, eccentricity $e = 0.169$, inclination $i = 15°.05$ — lies so close to the secular resonance as located by the same theory that a large fraction of its fragments were considered as plausible candidates for achieving chaotic orbits, eventually leading to Earth encounters. In the present paper, we have tested this hypothesis by carrying out a suitable set of numerical experiments on the long–term behaviour of the orbits of fictitious Hebe fragments.

TABLE II

Osculating elements (semimajor axis in $AU$, angles in degrees) and ejection velocities (in $m/s$) of Hebe and its fictitious fragments whose orbits have been integrated. The labels in the last column refer to the dynamical behaviour of the body as inferred from the integrations: EC = Earth crosser; 3 : 1 = Earth–crosser in 3 : 1 mean motion resonance; MC = Mars crosser; NC = non–crosser.

| $n^o$ | $a$ | $e$ | $i$ | $\omega$ | $\Omega$ | $M$ | $V_{ej}$ | Dyn. |
|---|---|---|---|---|---|---|---|---|
| Hebe | 2.4251 | 0.2019 | 14.787 | 238.40 | 138.347 | 358.201 | — | — |
| 137 | 2.4951 | 0.2097 | 15.088 | 231.84 | 137.908 | 309.443 | 281 | 3:1 |
| 171 | 2.3318 | 0.1869 | 14.978 | 229.28 | 137.758 | 71.128 | 397 | EC |
| 221 | 2.3991 | 0.1946 | 15.369 | 237.74 | 136.797 | 66.234 | 283 | NC |
| 228 | 2.3689 | 0.2000 | 14.384 | 230.83 | 139.511 | 69.687 | 282 | NC |
| 262 | 2.4250 | 0.1854 | 14.788 | 241.01 | 138.337 | 239.162 | 359 | NC |
| 319 | 2.5015 | 0.2347 | 14.352 | 239.52 | 138.107 | 327.360 | 568 | 3:1 |
| 372 | 2.2907 | 0.1597 | 14.727 | 229.55 | 139.639 | 37.171 | 526 | EC |
| 401 | 2.3322 | 0.1716 | 14.696 | 242.01 | 138.089 | 341.744 | 320 | EC |
| 404 | 2.3051 | 0.1645 | 14.474 | 245.77 | 137.438 | 339.170 | 455 | EC |
| 415 | 2.3488 | 0.1802 | 15.105 | 242.81 | 138.750 | 328.350 | 295 | MC |
| 418 | 2.3483 | 0.1808 | 14.873 | 243.49 | 138.458 | 328.087 | 270 | MC |
| 560 | 2.4309 | 0.2221 | 15.974 | 240.77 | 139.663 | 114.687 | 595 | NC |
| 620 | 2.4653 | 0.1768 | 14.819 | 238.59 | 139.200 | 211.639 | 348 | NC |
| 626 | 2.4077 | 0.1957 | 14.750 | 236.05 | 138.148 | 358.439 | 191 | NC |
| 877 | 2.4771 | 0.1783 | 14.108 | 244.13 | 138.413 | 94.866 | 445 | NC |
| 953 | 2.3932 | 0.1947 | 14.580 | 234.73 | 139.433 | 51.500 | 174 | NC |
| 964 | 2.4173 | 0.2230 | 15.500 | 238.58 | 138.500 | 101.459 | 446 | NC |
| 987 | 2.3428 | 0.1754 | 14.792 | 234.04 | 138.302 | 42.827 | 355 | MC |

## 2. The Collisional and Dynamical Model

As explained in detail by Farinella *et al.* (1991, 1992a), we have developed a program which simulates the isotropic ejection, from the surface of any given "target" asteroid, of a large number of fragments, as a result of either impact break–up or a sequence of cratering events. Isotropic ejection means that the direction of the fragment ejection velocity vector was taken at random; while this is of course not the case for any single collisional event, it is a reasonable assumption when one averages over a large number of independent events. The magnitude of the ejection velocity $V_{ej}$ was assumed to match a truncated power–law distribution, in agreement with the data from laboratory experiments on hypervelocity impacts (see Gault and Heitowit, 1963, and Stöffler *et al.*, 1975, for craters; Davis and Ryan, 1990, and Nakamura and Fujiwara, 1991, for break–up events). Only the fragments for which $V_{ej} > V_{esc}$ (the surface escape velocity of the parent asteroid) were assumed to escape and reach "infinity" (i.e., an independent heliocentric orbit) with a relative velocity of magnitude $V = (V_{ej}^2 - V_{esc}^2)^{1/2}$. The exponent of the number vs. ejection velocity distribution of the fragments and the lower cutoff of

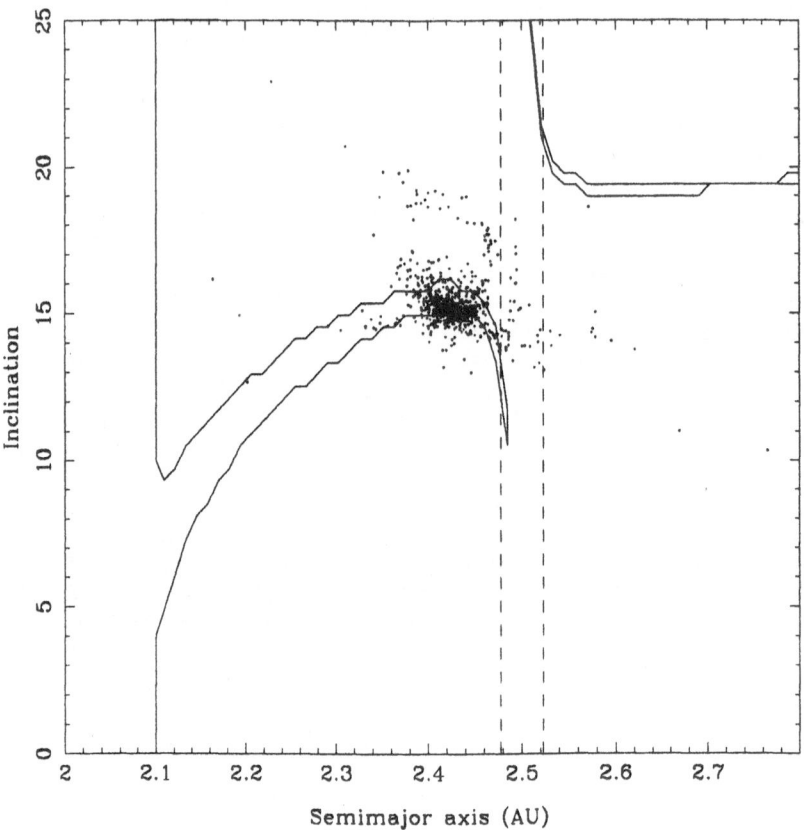

Fig. 1. The "fragment cloud" ejected from 6 Hebe (according to the algorithm described in the text) in the proper $i$ vs. $a$ plane. The $g = g_6$ resonant strip has been drawn for $\delta = 1\ arcsec/yr$ and proper $e = 0.15$; the 3 : 1 chaotic zone boundaries near $a = 2.5\ AU$ have also been plotted (dashed lines).

the distribution (i.e., the minimum ejection velocity) were assumed to be $-3.25$ and $100\ m/s$, respectively. The former value has been derived from the experimental results quoted above, while the latter one is consistent with the observed dispersion of proper elements in asteroid families (Zappalà *et al.*, 1984, 1990), which are the likely outcomes of large–scale asteroidal collisions; according to Farinella *et al.* (1983), a similar lower cutoff occurred in the collisional break–up of the Saturnian satellite Hyperion.

Since the secular resonance surfaces are defined in the space of proper elements and not in that of osculating elements, in order to test whether a fragment does actually "fall" into the secular resonance we had to compute its proper elements. We adopted the proper elements derivation code of Milani and Knežević (1990,

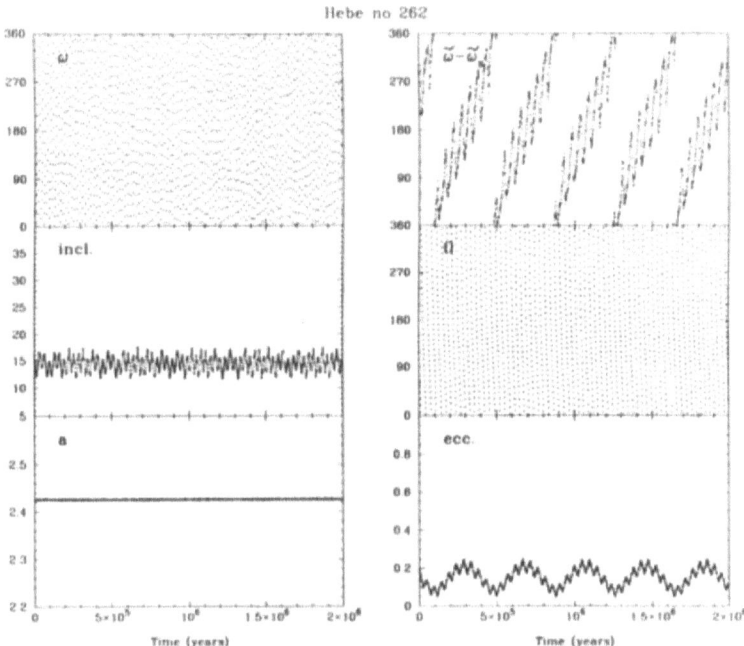

Fig. 2. Orbital evolution of fragment 262. The figure shows the perihelion argument $\omega$, the inclination $i$ (degrees), the semimajor axis $a$ $(AU)$, the $g = g_6$ critical argument $(\varpi - \varpi_6)$, the nodal longitude $\Omega$ and the eccentricity $e$ vs. time over the integration time span of $2\ Myr$.

1992; version 5.3), kindly provided to us by the authors. According to an extensive set of numerical integration tests carried out over time spans of $\approx 10^6\ yr$, the accuracy — i.e., stability in time — of these proper elements is of the order of $10^{-3}$ (corresponding to relative velocities of $\approx 10\ m/s$) in most of the main asteroid belt, up to values of the eccentricity and the (sine of) inclination of about 0.3 (Milani and Knežević, 1992).

The proper elements of every fragment were then tested to assess whether the fragment's apsidal secular frequency $g$ was such that $|g - g_6| < \delta$, where $\delta$ is the assumed half-width of the chaotic zone associated to the resonance (an input parameter, see later) and for the secular frequency $g_6$ we adopted the value 28.2455 $arcsec/yr$, derived from the *LONGSTOP 1B* numerical integration of the outer solar system (Nobili *et al.*, 1989). The test was made by interpolating $g$ within a numerical grid in the three–dimensional proper elements space, up to $e = 0.5$ and $i = 40°$; in the grid, $g$ was computed at a large number of equally spaced points by an algorithm fully consistent with the proper elements derivation procedure of Milani and Knežević (see Knežević *et al.*, 1991, for further details about this

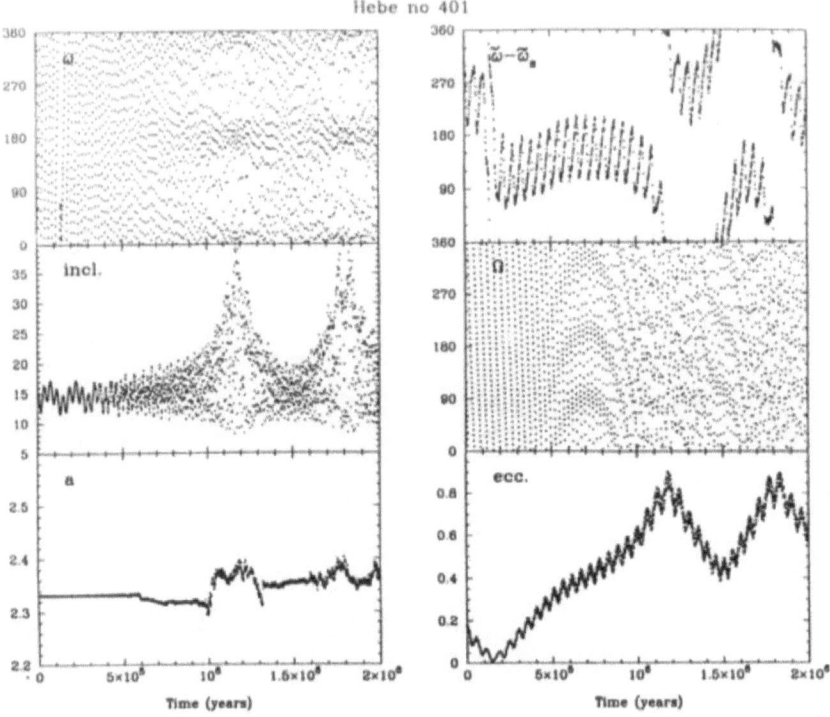

Fig. 3. The same as in Fig. 2, but for fragment 401.

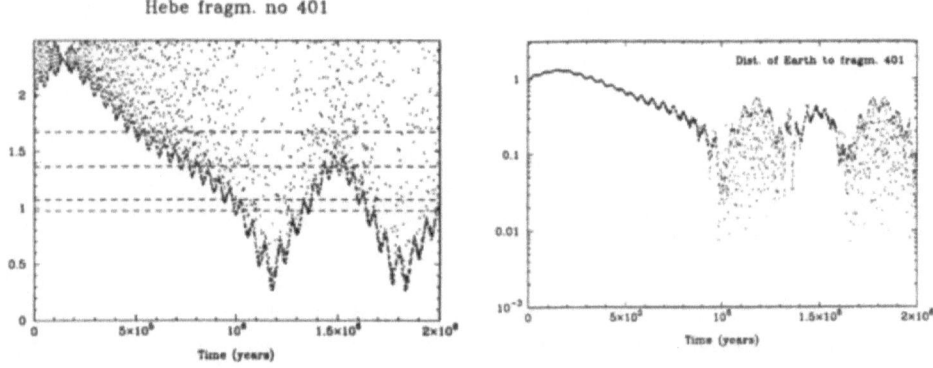

Fig. 4. The left part of the figure shows, as a function of time, the heliocentric distances (in $AU$) at which fragment 401 crosses the ecliptic plane (dashed lines correspond to the perihelion and aphelion distances of Mars and the Earth). The right part of the figure shows the minimum approach distances the Earth for the same fragment, recorded over successive time intervals of $10^3$ $yr$ (notice the logarithmic vertical scale; units are $AU$).

method). If the outcome of the test was positive, the fragment was assumed to lie inside the $g = g_6$ secular resonance.

We also made a similar test for the 3 : 1 mean motion resonance with Jupiter. Here the condition for lying inside the chaotic zone was the following: $(2.497 - e/8.85)$ $AU < a < (2.510 + e/9.615)$ $AU$, in agreement with the results of numerical experiments on the dynamics of this resonance and the observed width of the corresponding Kirkwood gap (Wisdom, 1983, 1985; Yoshikawa, 1990, Fig. 2).

Fig. 1 shows the locations in the proper $i$ vs. $a$ plane of 1000 fictitious fragments escaped from 6 Hebe (100 were assumed to be ejected at each of 10 randomly chosen mean anomalies of Hebe's current orbit). With the ejection velocity distribution described above and assuming for Hebe an escape velocity $V_{esc} = 115$ $m/s$ (corresponding to a mean density of about 2.5 $g/cm^3$), the average value of $V_{ej}$ for the escaped fragments was 157 $m/s$. The $g = g_6$ resonant strip has been drawn for $\delta = 1$ $arcsec/yr$ and proper $e = 0.15$, and the 3 : 1 chaotic zone boundaries near $a = 2.5$ $AU$ were derived from the relationship given above. As this figure shows, a large fraction of the fragments (702 over 1000) lie in the strip bordering the $g = g_6$ surface, and therefore are possibly subject to chaotic dynamical evolution. 40 fragments (4% of the total sample) were also found to lie inside the boundaries of the 3 : 1 resonance.

Numerical experiments by Yoshikawa (1987, Fig. 9f) and Froeschlé and Scholl (1991) had already shown that fictitious bodies starting inside the $g = g_6$ resonant strip near $a = 2.4$ $AU$ and with moderate eccentricities and inclinations (e.g., $e = 0.14$ and $i = 16°$) do become Earth–crossers within a time scale of the order of $10^6$ $yr$. To test whether this really happens also for some of our fictitious Hebe fragments, we have integrated numerically the orbits of 18 of them. They were chosen either to span the typical range of orbital elements over which the "fragment cloud" of Fig. 1 extends, or because *a priori* they appeared to us as potentially interesting (this applies in particular to two bodies lying in the 3 : 1 resonance). Of course, owing to the limited number of numerical experiments which were feasible, this choice was somewhat arbitrary, and we cannot in any sense claim to have extracted a statistically "representative" fragment sample. However, the results of the numerical experiments have provided us some general understanding of the behaviour of Hebe fragments.

The dynamical model included the Sun and four planets (Jupiter, Saturn, Mars and the Earth) in addition to the (assumedly) massless fictitious fragments, whose initial conditions were derived from their respective ejection velocity vectors by applying standard two–body formulae to the osculating elements of 6 Hebe. We used a variable stepsize Burlisch–Stoer integration method (Stoer and Burlisch, 1980), allowing an accurate treatment of close planetary encounters. Note that the inclusion of Mars and the Earth — essential to model the effect of possible encounters with the inner planets — is likely to shift the position of the secular resonance by a few tenths of $arcsec$ with respect to that predicted by a secular

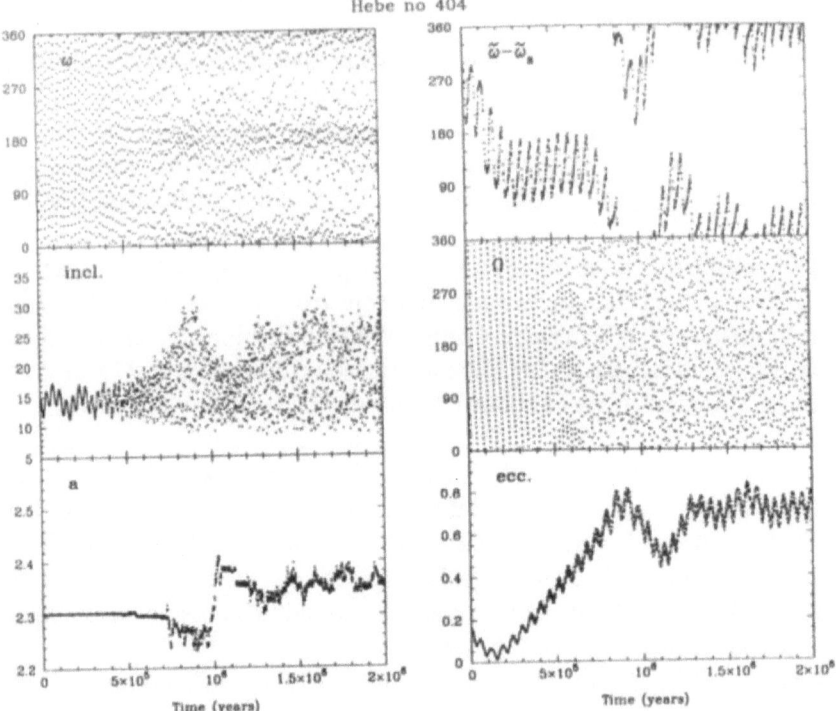

Fig. 5. The same as in Fig. 2, but for fragment 404.

perturbation theory without inner planets, such as that used to find the resonance boundaries as described above (Knežević *et al.*, 1991); a similar small effect is caused by the absence of Uranus and Neptune. The integration time span was 2 *Myr*, enough for resonant secular perturbations to build up large orbital changes and possibly result into chaotic behaviour. Table 1 lists the proper elements of the fragments chosen for the integration (numbered in a conventional way by the fragment generation code), together with the corresponding values of the secular frequency $g$. Table 2 gives their osculating elements (epoch JD 2448600.5, ecliptic 1950.0), the magnitude of the assumed ejection velocity from Hebe and a label referring to their dynamical behaviour as observed from the integrations.

## 3. Results

Table 3 provides some quantitative parameters on the results of the numerical experiments described above, including the minimum distances from the Earth, Mars and Jupiter reached by the fictitious fragments during the integration time span. Figs. 2 to 12 illustrate the orbital evolution of some interesting fictitious fragments.

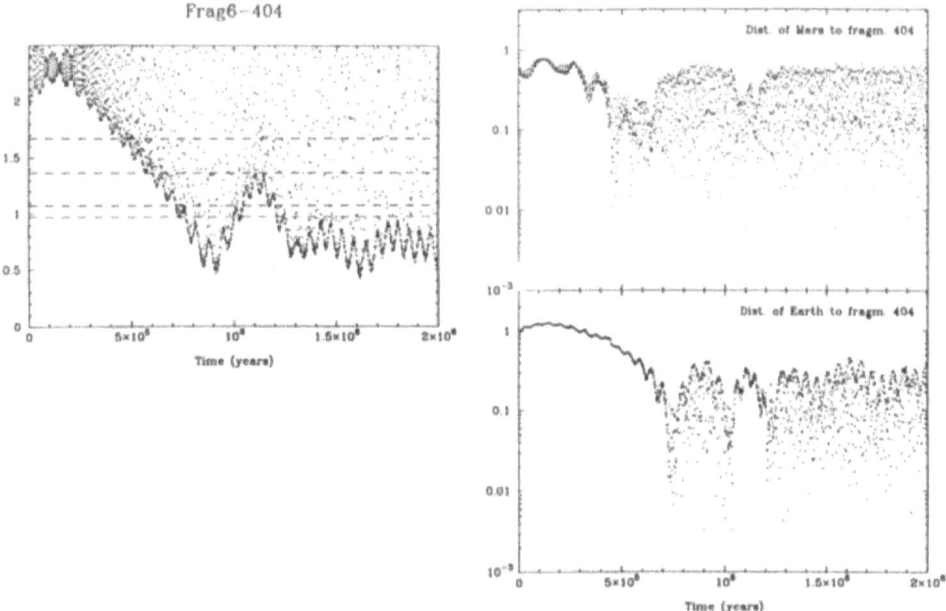

Fig. 6. The same as in Fig. 4, but for fragment 404 (the minimum approach distances to both Mars and the Earth are shown in the right part of this figure).

Two extreme cases are fragments 262 and 401. The former one, which starts at a semimajor axis very close to that of Hebe, is about 3 $arcsec/yr$ away from the resonance (this can be inferred from the fact that the critical argument $(\varpi - \varpi_6)$ completes a circulation in about $4 \times 10^5$ $yr$, see Fig.2), and as a consequence both the eccentricity and the inclination, albeit undergoing sizeable secular perturbations, display a regular, quasi–periodic behaviour and never exceed moderate values (about 0.25 and 17°, respectively; see Fig. 2 and Table 3). As for the reason why the analytically derived value of $g$ underestimates the real one for this body, we conjecture that this is due to the proximity of the 3 : 1 mean motion resonance with Jupiter, whose secular effects are not fully accounted for by the theory. This conjecture is supported by the observation that, closer to the mean motion resonance the fragments are, larger is the discrepancy between the numerically and analytically derived values of $g$ — for instance, fragment 620, another non–planet–crosser starting at a semimajor axis of 2.465 $AU$, completes a circulation of the critical argument in only $250,000$ $yr$.

Fragment 401, on the other hand, starts at a semimajor axis of 2.332 $AU$. As shown by Fig. 3, the secular resonance pumps up its eccentricity until, after about 1 $Myr$ since the beginning of the integration, it becomes Earth crosser. Subsequently, close encounters with the Earth cause a strongly chaotic behaviour,

Hebe no 415

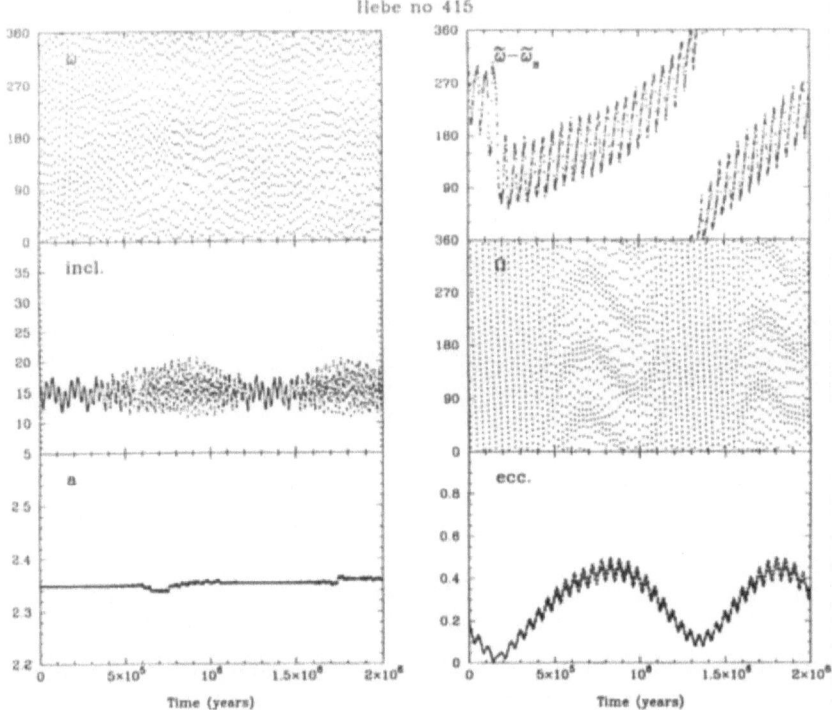

Fig. 7. The same as in Fig. 2, but for fragment 415.

with the eccentricity staying at very large values for all the remaining integration time span and the inclination oscillating wildly between $\approx 10°$ and $40°$. The critical secular argument $\varpi - \varpi_6$ displays short–period ($\approx 5 \times 10^4 \; yr$) oscillations due to the fact that the longitude of perihelion of Saturn is not a linear function of time, $g_6$ being just its average rate. When these short–periodic terms are filtered out, the critical argument shows two distinct behaviours: after a quick clockwise circulation starting from $\approx 240°$, first it stays in the neighbourhood of $180°$ (but on the other side with respect to the starting value) for about 1 $Myr$, and subsequently it librates clockwise with a large amplitude around $0°$. These behaviours are due to the fact that the orbit gets close to a homoclinic point separating the circulation and libration regions of the phase space — indeed, for the $g = g_6$ resonance Morbidelli and Henrard (1991) have shown that the unstable equilibrium point is at $\varpi - \varpi_6 = 180°$. After the eccentricity has been pumped up to $\approx 0.6$ by the secular resonance, the signature of the Earth encounters becomes apparent in the evolution of the semimajor axis, which at $t \approx 1.0 \times 10^6 \; yr$ starts a quasi–random walk pattern leading to changes of $\approx 0.1 \; AU$ during the integration time span. The most effective encounters in changing the semimajor axis are the quasi–tangent ones, occurring when the fragment perihelion is at about 1 $AU$ from the Sun.

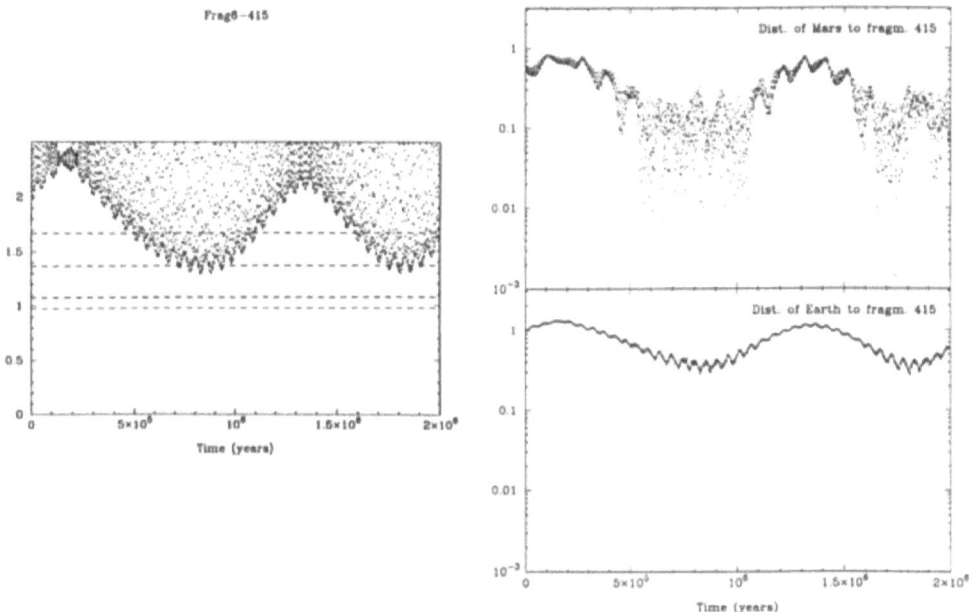

Fig. 8. The same as in Fig. 6, but for fragment 415.

The occurrence of Earth encounters can be monitored in two ways, illustrated by Fig. 4: the left part of this Figure shows, as a function of time, the variable $a(1 - e^2)/(1 + e\cos\omega)$, that is the heliocentric distance at which the body crosses the ecliptic plane (dashed lines correspond to the perihelion and aphelion distances of Mars and the Earth); the right part of the Figure directly shows the minimum Earth approach distances for the same body, recorded by the integration program over successive time intervals of $10^3$ $yr$. As shown by this Figure (see also Table 3), the closest approaches to the Earth occur at a distance smaller than 0.01 $AU$. The behavior of this fictitious Earth–approaching body (like other similar ones found in our sample) recalls that of the real Earth–crossing asteroids classified by Milani *et al.* (1989) in the Oljiato class. Note, however, that these authors could not monitor the occurrence of secular resonances during their integrations, extending over $2 \times 10^5$ $yr$ only.

A similar case is that of fragment 404 (see Figs. 5 and 6), which starts at a semimajor axis of 2.305 $AU$. Here again the critical argument stays not far from 180° in the first $\approx 8 \times 10^5$ $yr$, and later on it librates clockwise about 0°; but in the last $6 \times 10^5$ $yr$ of the integration time span it stays very close to the equilibrium point, with a nearly constant eccentricity of $\approx 0.7$. For this body the closest approach to the Earth occurs at about 0.003 $AU$, i.e., as an order of magnitude, the orbital distance of the Moon; but it gets even closer to Mars, i.e., less than

Hebe no 137

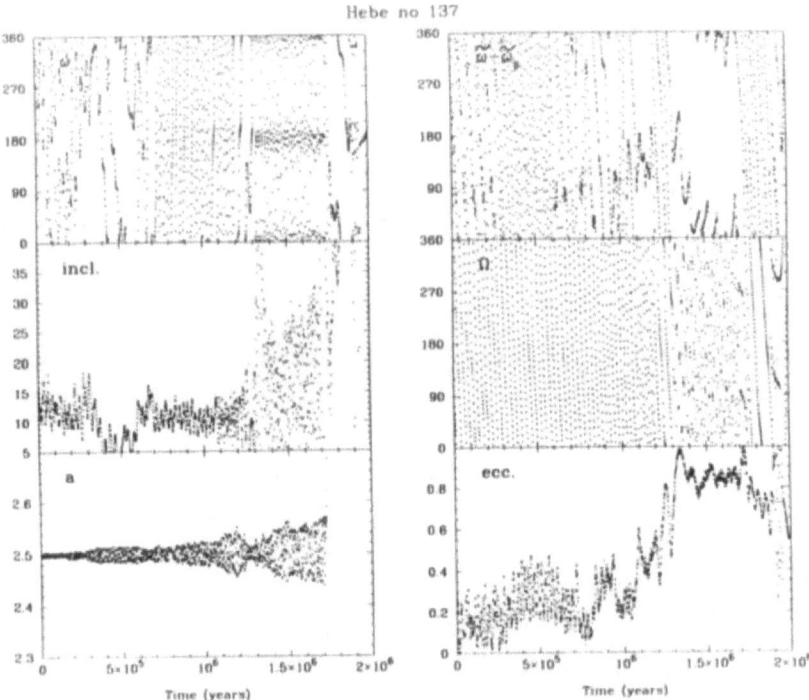

Fig. 9. The same as in Fig. 2, but for fragment 137.

0.001 $AU$ (out of scale in Fig. 6). The closest approach to the Earth in our entire sample of fictitious Hebe fragments has been recorded for body 372 (miss distance 0.0013 $AU$, see Table 3). A simple cross-section calculation shows that, over time spans a few hundred times longer than that of our integration, a physical impact with the terrestrial planets would be expected to occur for such bodies.

An intermediate case is that of fragments, such as 415 (see Figs. 7 and 8), for which the secular resonance, within a time span of 2 $Myr$, increases the eccentricity just to about 0.5, resulting into close Mars encounters but no close Earth encounter. Mars encounters are not very effective in changing the semimajor axis, due to the comparatively small mass of the planet — but they also cause chaotic evolution and it is reasonable to expect that, over longer time spans, most of these bodies will also switch to a strongly chaotic behaviour including Earth approaches. Longer integrations are needed to directly observe this evolution.

Finally, we have integrated a couple of fragments (nos. 137 and 319) injected into the 3 : 1 mean motion resonance with Jupiter, near 2.5 $AU$. As shown by Figs. 9 to 12, their behaviour shows a complex interplay of resonant effects. These temporarily include also $g = g_6$ and $g = g_5$. The eccentricity reaches maximum values (near unity) when the body is located in both the 3 : 1 and the $g = g_6$

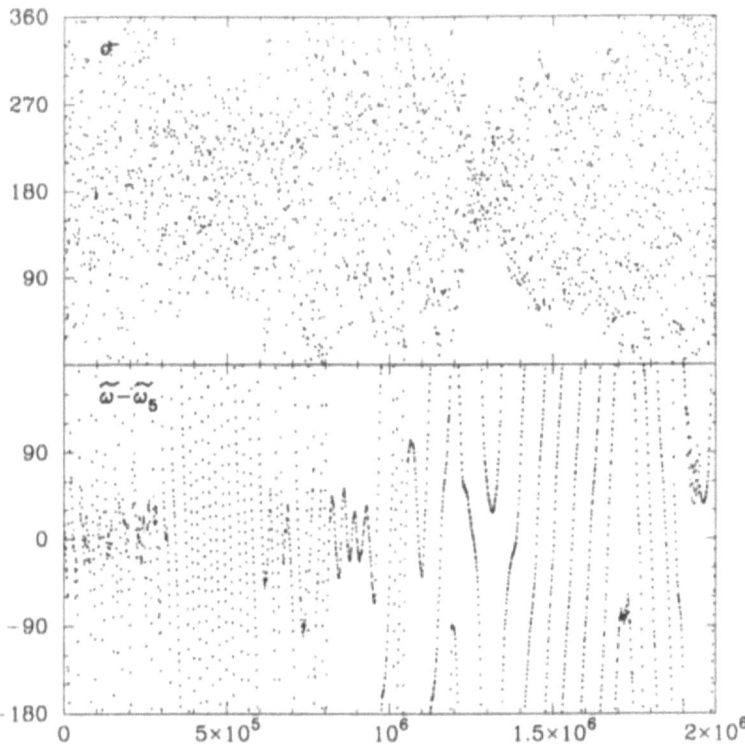

Fig. 10. The 3 : 1 resonance critical argument $(3\lambda_5 - \lambda - 2\varpi)$ and the $g = g_5$ critical argument $(\varpi - \varpi_5)$ vs. time for fragment 137.

resonances. As noted by Yoshikawa (1990) and recently investigated in more detail by Froeschlé and Scholl (1992), this resonance alternation/overlapping is very effective in causing large growths of the eccentricity and inclination — and then, of course, close planetary approaches also enter into play. These objects, which at the end of the integration display comet–like orbits, could be classified in Milani's *et al.* (1989) Alinda class of Earth-approaching objects. It is interesting to note that their strongly chaotic behaviour shows up in the evolution of the orbital elements from the very beginning of the integration (contrasting with bodies affected by the secular resonance only, whose initial behaviour is fairly regular).

How many Hebe fragments can be expected to become planet–crossers, namely candidate meteorites? As we explained in Sec. 2, our sample is too limited to draw any reliable statistical inference. Actually, we found that 4/16 of the fragments predicted by the analytical theory to lie in the secular resonance did rapidly become Earth–approachers. However, as a general rule, we have observed that the Earth–crossing fragments were always located at starting semimajor axes $a < 2.34\ AU$ — or in the 3 : 1 resonance, where both the fragments we have integrated became

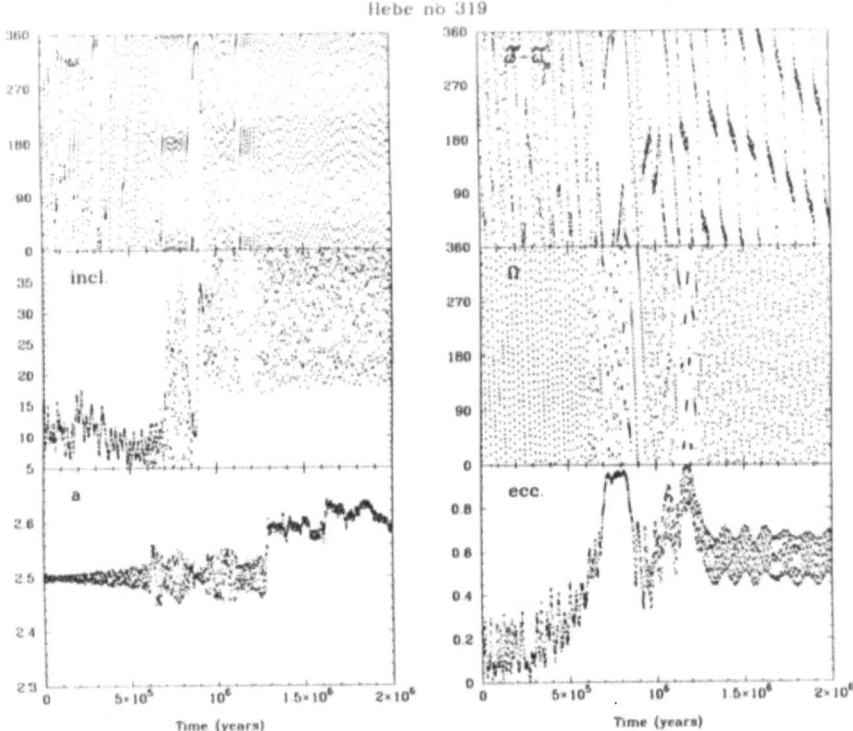

Hebe no 319

Fig. 11. The same as in Fig. 2, but for fragment 319.

Earth–crossers. We had 12 and 40 of these bodies in our fragment set, respectively; 7 more bodies were initially located between 2.34 and 2.35 $AU$, where we found the Mars–crossers. This would yield a total of about 6% of planet–crossers in our overall fragment set. However, while we could not identify planet–crossers near 2.4 $AU$, as we quoted earlier such bodies were found in previous investigations (e.g., Froeschlé and Scholl, 1991); so we believe that probably a more comprehensive search would lead to find planet–crossing fragments in other regions of the proper elements space. A likely possibility is that the current angular elements of Hebe's orbit, which were used in the fragment generation code and yield for all fragments initial values of the critical argument $\varpi - \varpi_6 \approx 270°$, are not the most favourable ones to yield fragments located in the chaotic zone associated with $g = g_6$ (see Morbidelli and Henrard, 1991).

## 4. Conclusions

In this work we have shown for the first time that fragments ejected from an existing large asteroid according to plausible collisional physics can approach the Earth — i.e., become candidate meteorites — within a time span of the order of 1 $Myr$.

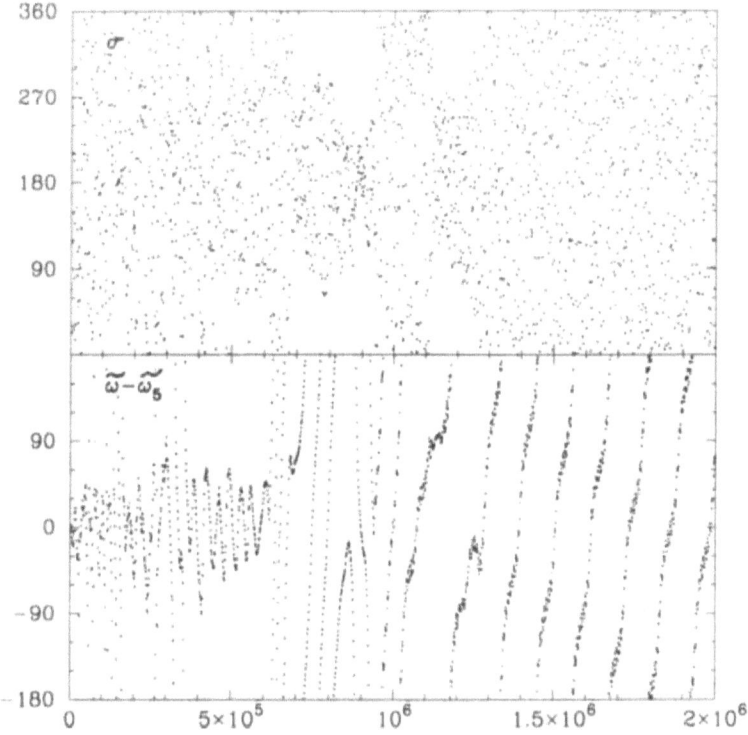

Fig. 12. The same as in Fig. 10, but for fragment 319.

While further work remains to be done to improve the statistics and study the orbital evolution of fragments from other possible parent asteroids, we can point out some interesting conclusions on the asteroid/meteorite connection which can be drawn from the results described above:

(1) Wetherill and Chapman (1988) have discussed in detail the so–called "spectrophotometric paradox", namely the fact that ordinary chondrites (the absolute majority class among meteorite falls) do not find well–matching analogues in the typical reflectance spectra of any common asteroid taxonomic type (and in particular of S–types, most abundant in the inner asteroid belt). As Wetherill and Chapman noted, however, there is the possibility that most ordinary chondrites may come from relatively few S–type asteroids, such as Hebe, which could be efficient fragment deliverers and have chondritic compositions, unlike those of most other S–types. This solution of the paradox — discussed in more detail by Farinella *et al.* (1992a) — would be consistent with the meteoritic evidence, which suggests that almost all ordinary chondrites probably come from a small number of parent bodies of diameter between 100 and 300 $km$ (Sears and Dodd, 1988; Lipschutz *et*

## TABLE III

Some quantitative parameters describing the outcomes of our integrations of fictitious Hebe fragments over a time span of 2 $Myr$. For every fragment, the table gives: the minimum and maximum values of the osculating semimajor axis ($AU$), eccentricity and inclination (degrees), the minimum perihelion and the maximum aphelion distances and the minimum distances from the Earth, Mars and Jupiter ($AU$). For the planet–crossing fragments, the corresponding epochs (in $Myr$, numbers in parentheses) are also given.

| $n^o$ | $a_{min}$ | $a_{max}$ | $e_{min}$ | $e_{max}$ | $i_{min}$ | $i_{max}$ | $q_{min}$ | $Q_{max}$ | $d_{Earth}$ | $d_{Mars}$ | $d_{Jup}$ |
|---|---|---|---|---|---|---|---|---|---|---|---|
| 137 | 2.438 (1.704) | 10.092 (1.819) | 0.006 (0.016) | 0.999 (1.726) | 2.74 (0.482) | 141.08 (1.989) | 0.002 (1.726) | 17.787 (1.815) | 0.0049 (1.629) | 0.0023 (1.430) | 0.052 (1.734) |
| 171 | 2.287 (1.575) | 2.397 (1.688) | 0.032 (0.208) | 0.920 (1.776) | 8.60 (1.067) | 48.13 (1.781) | 0.190 (1.997) | 4.575 (1.773) | 0.0032 (1.816) | 0.0027 (0.580) | 0.82 (1.158) |
| 221 | 2.397 | 2.403 | 0.022 | 0.292 | 11.91 | 18.10 | 1.698 | 3.100 | 0.71 | 0.16 | 2.2 |
| 228 | 2.367 | 2.372 | 0.033 | 0.277 | 11.01 | 17.19 | 1.713 | 3.026 | 0.78 | 0.14 | 2.2 |
| 262 | 2.424 | 2.429 | 0.049 | 0.251 | 11.62 | 17.48 | 1.815 | 3.035 | 0.88 | 0.27 | 2.2 |
| 319 | 2.449 (0.675) | 2.641 (1.633) | 0.010 (0.272) | 0.993 (1.176) | 3.76 (0.797) | 84.84 (1.183) | 0.017 (1.176) | 5.013 (1.194) | 0.0040 (0.938) | 0.0052 (0.607) | 0.77 (0.779) |
| 372 | 2.096 (0.807) | 2.351 (1.000) | 0.055 (0.097) | 0.908 (1.999) | 8.80 (1.779) | 42.16 (1.998) | 0.211 (1.999) | 4.403 (1.775) | 0.0013 (0.936) | 0.0012 (1.128) | 0.95 (1.585) |
| 401 | 2.294 (1.001) | 2.402 (1.792) | 0.004 (0.148) | 0.905 (1.178) | 8.19 (1.166) | 42.31 (1.184) | 0.226 (1.179) | 4.547 (1.182) | 0.0073 (1.656) | 0.0014 (0.599) | 0.78 (1.800) |
| 404 | 2.233 (0.900) | 2.410 (1.033) | 0.014 (0.151) | 0.841 (1.614) | 8.27 (1.660) | 32.87 (1.623) | 0.374 (1.614) | 4.342 (1.610) | 0.0033 (1.034) | 0.0009 (1.127) | 0.94 (1.586) |
| 415 | 2.337 (0.752) | 2.367 (1.770) | 0.002 (0.143) | 0.498 (1.803) | 10.96 (1.817) | 20.60 (0.926) | 1.183 (0.829) | 3.534 (1.803) | 0.27 (1.811) | 0.0014 (1.743) | 1.8 (1.834) |
| 418 | 2.317 (1.999) | 2.352 (0.655) | 0.003 (1.166) | 0.407 (0.666) | 11.21 (0.682) | 18.95 (0.719) | 1.393 (0.666) | 3.303 (0.666) | 0.47 (1.751) | 0.0024 (1.921) | 2.0 (0.636) |
| 560 | 2.427 | 2.433 | 0.041 | 0.320 | 12.49 | 18.95 | 1.652 | 3.205 | 0.73 | 0.13 | 2.1 |
| 620 | 2.463 | 2.470 | 0.062 | 0.223 | 11.59 | 17.21 | 1.917 | 3.016 | 0.99 | 0.34 | 2.1 |
| 626 | 2.406 | 2.411 | 0.049 | 0.263 | 11.40 | 17.36 | 1.773 | 3.040 | 0.84 | 0.25 | 2.2 |
| 877 | 2.472 | 2.480 | 0.076 | 0.225 | 10.99 | 16.61 | 1.920 | 3.034 | 0.97 | 0.37 | 2.2 |
| 953 | 2.392 | 2.397 | 0.044 | 0.260 | 11.16 | 17.18 | 1.756 | 3.030 | 0.82 | 0.23 | 2.2 |
| 964 | 2.414 | 2.425 | 0.055 | 0.309 | 11.89 | 18.64 | 1.674 | 3.167 | 0.76 | 0.14 | 2.1 |
| 987 | 2.341 (0.566) | 2.372 (1.928) | 0.007 (0.144) | 0.364 (1.316) | 11.02 (1.869) | 18.19 (1.278) | 1.502 (1.316) | 3.221 (1.910) | 0.56 (0.606) | 0.0017 (0.568) | 2.0 (1.937) |

*al.*, 1989; Pellas and Storzer, 1981).

(2) The Příbram meteorite, one of the three photographically recorded meteorite falls, had the following preterrestrial orbital elements: $a = 2.4008 \pm 0.0023\ AU$, $e = 0.6712 \pm 0.0003$ and $i = 10°.481 \pm 0°.004$ (Ceplecha, 1977). Since this orbit is undistinguishable from those of our Earth–approaching fictitious Hebe fragments channeled via the $g = g_6$ resonance, we suggest that this meteorite (an H5 ordinary chondrite) is a fragment ejected from Hebe. Note that according to Pellas and Fieni (1988) and Lipschutz *et al.* (1988, p. 749) the parent body of the H chondrites was originally an onion–shell layered asteroid 80 to 100 $km$ in radius. Several Apollo and Amor asteroids have also osculating orbital elements close to those of our Earth–approaching, $g = g_6$ resonant Hebe fragments: 2329 Orthos, 1982 TA, 1991 TB2 (Apollos), 1983 LB, 1989 RC, 1989 RS1, 1990 SB, 1990 VB, 1991 FB (Amors).

(3) Can we estimate the total meteorite flux to the Earth coming from Hebe? Consider for instance large cratering events, such as caused by projectile asteroids 1 $km$ in diameter. The average frequency of these events for a target as big as Hebe can be estimated at one over $2 \times 10^7\ yr$ (assuming an intrinsic collision probability of $2.8 \times 10^{-18}\ km^{-2}\ yr^{-1}$ and a projectile population of about $2 \times 10^6$; see Farinella and Davis, 1992, and Farinella *al.*, 1992b). This is comparable with the typical cosmic–ray exposure ages of ordinary chondrites — H chondrites actually display an exposure age peak at about 8 $Myr$ (Anders, 1964; Crabb and Schultz, 1981), while the exposure age of the Příbram meteorite has been determined between 20 and 40 $Myr$ (Stauffer and Urey, 1962; Fireman and DeFelice, 1963; Lavrukhina *et al.*, 1974). We further assume that the mass ejected from the crater is $\approx 500$ times that of the projectile, a plausible guess for a mean impact velocity of 6.6 $km/s$, as found for Hebe by Farinella and Davis (1992), and a (presumably) regolith–covered surface. We thus get an ejecta production of $\approx 3 \times 10^7\ kg/yr$. Let us also assume that: (i) some 10% of this material is injected into the resonances and can achieve Earth–crossing orbits (according to the results discussed in Sec. 3); (ii) 10% of the latter has meteorite–like sizes (smaller than $\approx 1\ m$; plausible for crater ejecta); and (iii) another factor 10 is lost *en route* to the Earth due to collisional disruption, impact with other planets, or ejection from the solar system (this is consistent with the typical exposure ages of ordinary chondrites, which are about a factor 10 shorter than the Earth–impact lifetimes of Earth–crossing fragments). Thus we get an Earth influx of some $3 \times 10^4\ kg/yr$, which is about 10% of the overall meteorite influx as estimated (with an uncertainty of perhaps a factor 10) by Halliday *et al.* (1984) and Wetherill (1985). Although it is apparent that the estimate obtained above for the meteorite flux from Hebe, being the product of several uncertain factors, must be considered as highly uncertain, it shows that the hypothesis that Hebe is a significant deliverer of ordinary chondrites cannot be excluded by simple order–of–magnitude arguments.

## Acknowledgements

We are grateful to the reviewer G. Hahn and to D.R. Davis, C. Froeschlé, R. Greenberg, Z. Knežević, A. Milani, G.W. Wetherill and V. Zappalà for useful discussions and comments. We are also indebted to A. Milani and Z. Knežević for putting at our disposal their code for the derivation of proper elements and secular frequencies. This project was partially supported by the EEC research contract no. SC1–0011–C(GDF). The work of P.F. was also supported by the Italian Ministry of University and Scientific Research (MURST).

## References

Anders, J.B.: 1964, 'Origin, age, and composition of meteorites.', Space Sci. Rev., 3, 583–714.

Ceplecha, Z.: 1977, 'Fireballs photographed in central Europe.', Bull. Astron. Inst. Czech., 28, 328–340.

Crabb, J., and Schultz, L.: 1981, 'Cosmic–ray exposure ages of the ordinary chondrites and their significance for parent–body stratigraphy.', Geochim. Cosmochim. Acta, 45, 2151–2160.

Davis, D.R. and Ryan, E.: 1990, 'On collisional disruption: Experimental results and scaling laws.', Icarus, 83, 156–182.

Farinella, P., Milani, A., Nobili, A.M., Paolicchi, P. and Zappalà, V.: 1983, ' Hyperion: Collisional disruption of a resonant satellite.', Icarus, 54, 353–360.

Farinella, P., Gonczi, R., Froeschlé, Ch., and Froeschlé, C. 1991. Injecting asteroid fragments into resonances. Proceedings International Conference Asteroids Comets Meteors 1991 (Flagstaff, Arizona, June 1991), in press.

Farinella, P., and Davis, D.R.: 1992, 'Collision probabilities and impact velocities in the main asteroid belt',. Icarus, 97, 111–123.

Farinella, P., Gonczi, R., Froeschlé, Ch., and Froeschlé, C.: 1992a, 'The injection of asteroid fragments into resonances.', Icarus, submitted.

Farinella, P., Davis, D.R., Cellino, A., and Zappalà, V.: 1992b, 'The collisional lifetime of the asteroid 951 Gaspra.', Astron. Astrophys., 257, 329–330.

Fireman, E.L., and Defelice, J.: 1964, 'Multiple fall of Příbram meteorites photographed 7. The tritium and argon–39 in the Příbram meteorite.', Bull. Astron. Inst. Czech., 15, 113.

Froeschlé, Ch., and Scholl, H.: 1987, 'Orbital evolution of asteroids near the secular resonance $\nu_6$.', Astron. Astrophys., 179, 294–303.

Froeschlé, Ch., and Scholl, H.: 1991, 'The effect of secular resonances in the asteroid region between 2.1 and 2.4 AU.' Proceedings of the International Conference Asteroids Comets Meteors 1991 (Flagstaff, Arizona, June 1991), in press.

Froeschlé, Ch., and Scholl, H.: 1992, 'Numerical experiments in the 3 : 1 and $\nu_6$ resonant region.' Celest. Mech., this volume.

Gault, D.E., and Heitowit, E.D.: 1963, In Proceedings of the 6th Hypervelocity Impact Symposium, Vol. 2, pp. 419–456. Firestone Rubber Co., Cleveland, OH.

Greenberg, R., and Chapman, C.R.: 1983, 'Asteroids and meteorites: Parent bodies and delivered samples.', Icarus, 55, 455–481.

Greenberg, R., and Nolan, M.C.:1989, 'Delivery of asteroids and meteorites to the inner solar system.' In Asteroids II (R.P. Binzel, T. Gehrels, and M.S. Matthews, eds.), pp. 778–804, Univ. of Arizona Press, Tucson.

Halliday, I., Blackwell, A.T. and Griffin A.A.: 1984, 'The frequency of meteorite falls on the Earth.', Science, 223, 1405–1407.

Knežević, Z., Milani, A., Farinella, P., Froeschlé, Ch., and Froeschlé, C.: 1991, 'Secular resonances from 2 to 50 AU.', Icarus, 93, 316–330.

Lavrukhina, A.K., Fisenko, A.V., and Kolesnikov, E.M.: 1974, 'Multiple fall of Příbram meteorites photographed 11. Preatmospheric size and radiation age of the Příbram chondrite.', Bull. Astron. Inst. Czech., 25, 122–126.

Lipchutz, M.E., Gaffey, M.J., and Pellas, P.: 1989, 'Meteoritic parent bodies: Nature, number, size and relation to present–day asteroids.' In *Asteroids II* (R.P. Binzel, T. Gehrels, and M.S. Matthews, eds.), pp. 740–777, Univ. of Arizona Press, Tucson.

Milani, A., Carpino, M., Hahn, G., and Nobili, A.M.: 1989, 'Dynamics of planet–crossing asteroids: Classes of orbital behavior.', *Icarus*, **78**, 212–269.

Milani, A., and Knežević, Z.: 1990, 'Secular perturbation theory and computation of asteroid proper elements.', *Cel. Mech.*, **49**, 347–411.

Milani, A., and Knežević, Z.: 1992, 'Asteroid proper elements and secular resonances.', *Icarus*, in press.

Morbidelli, A. and Henrard, J.: 1991, 'The main secular resonances $\nu_6$, $\nu_5$ and $\nu_{16}$ in the asteroid belt.', *Celest. Mech.*, **51**, 169–198.

Nakamura, A.and Fujiwara, A.: 1991, 'Velocity distribution of fragments formed in a simulated collisional disruption.', *Icarus*, **92**, 132–146.

Nobili, A.M., Milani, A. and Carpino, M.: 1989, 'Fundamental frequencies and small divisors in the orbits of the outer planets.', *Astron. Astrophys.*, **210**, 313–336.

Pellas, P.., and Storzer, D.: 1981, $^{244}$Pu fission track thermometry and its application to stony meteorites.', *Proc. R. Soc. Lond. A*, **374**, 253–270.

Pellas, P.., and Fieni, C.: 1988, 'Thermal histories of ordinary chondrite parent asteroids.', *Lunar Planet. Sci.*, **XIX**, 915–916.

Scholl, H., and Froeschlé, Ch.: 1991, 'The $\nu_6$ secular resonance region near 2 AU: A possible source of meteorites.', *Astron. Astrophys.*, **245**, 316–321.

Sears, D.W.G., and Dodd, R.T.: 1988, 'Overview and classification of meteorites.' In *Meteorites and the Early Solar System* (J.F. Kerridge and M.S. Matthews, eds.), pp. 3–31, Univ. of Arizona Press, Tucson.

Stauffer , H., and Urey, H.C.: 1962, ' Multiple fall of Pŕíbram meteorites photographed III. Rare gas isotopes in the Velká stone meteorite.', *Bull. Astron. Inst. Czech.*, **13**, 106–108.

Stoer, J., and Burlisch, R.: 1980, *Introduction to Numerical Analysis*, Springer Verlag, New York.

Stöffler, D., Gault, D.E., Wedekind, J. and Polkowski, G.: 1975, 'Experimental hypervelocity impact into quartz sand: Distribution and shock metamorphism of ejecta.', *J. Geophys. Res.*, **80**, 4062–4077.

Wetherill, G.W.: 1985, 'Asteroidal source of ordinary chondrites.', *Meteoritics*, **20**, 1–21.

Wetherill, G.W.: 1987, 'Dynamical relations between asteroids, meteorites and Apollo–Amor objects.', *Phil. Trans. R. Soc. Lond. A*, **323**, 323–337.

Wetherill, G.W., and Williams, J.G.: 1979, 'Origin of differentiated meteorites.' In *Origin and Distribution of the Elements* (L.H. Ahrens, ed.), pp. 19–31, Pergamon Press, Oxford.

Wetherill, G.W. and Chapman, C.R.: 1988, 'Asteroids and meteorites.' In *Meteorites and the Early Solar System* (J.F. Kerridge and M.S. Matthews, eds.), pp. 35–67, Univ. of Arizona Press, Tucson.

Wisdom, J.: 1983, 'Chaotic behavior and the origin of the 3/1 Kirkwood gap.', *Icarus*, **56**, 51–74.

Wisdom, J.: 1985, 'Meteorites may follow a chaotic route to earth.', *Nature*, **315**, 731–733.

Yoshikawa, M.: 1987, 'A simple analytical model for the secular resonance $\nu_6$ in the asteroidal belt.', *Celest. Mech.*, **40**, 233–272.

Yoshikawa, M.: 1990, ' Motions of asteroids at the Kirkwood gaps, I. On the 3:1 resonance with Jupiter.', *Icarus*, **87**, 78–102.

Zappalà, V., Farinella, P., Knežević, Z. and Paolicchi, P.:1984, 'Collisional origin of the asteroid families: Mass and velocity distributions.', *Icarus*, **59**, 261–285.

Zappalà, V., Cellino, A., Farinella, P. and Knežević, Z.: 1990, 'Asteroid families. I. Identification by hierarchical clustering and reliability assessment.', *Astron. J.*, **100**, 2030–2046.

# GENERALIZED LYAPUNOV CHARACTERISTIC INDICATORS AND CORRESPONDING KOLMOGOROV LIKE ENTROPY OF THE STANDARD MAPPING

C. FROESCHLÉ and CH. FROESCHLÉ

*Observatoire de la Côte d'Azur - B.P. 229, 06304 Nice Cedex 4, France*

and

E. LOHINGER*

*Institute of Astronomy, University of Vienna, Türkenschanzstr. 17 - 1180 Wien, Austria*

**Abstract.** Lyapunov characteristic indicators are currently defined as the mean, i.e. the first moment, of the distribution of the local variations of the tangent vector to the flow. Higher moments of the distribution give further informations about the fluctuations around the average.

**Key words:** Lyapunov characteristic numbers – Kolmogorov entropy – standard mapping

## 1. Introduction

It is well known that chaoticity of a dynamical system (in the sense of exponential divergence of nearby trajectory) is usually estimated by the largest Lyapunov indicator (see for a review Froeschlé 1984). However, this quantity, even for $t$ very large, does not give a full description of the chaotic flow since it is an asymptotic quantity. Actually the Lyapunov characteristic indicators are nothing but the first moment of the distribution of the local variations of the tangent vectors to the flow (see Benettin 1980, Froeschlé 1984). Since we have no prior knowledge about the mathematical forms of the distribution, we will try to characterize this distribution using not only the two first moments (the mean and the $r.m.s.$) but also the Fisher coefficients $\gamma_1$ and $\gamma_2$, which measure respectively the asymmetry and the flatness with respect to the normal distribution. Actually generalized Lyapunov exponents of order 2 have been already introduced (Crisantin et al. 1988, Fujisaka 1983) in the context of theoretical physics mainly for dissipative flows.

In Section 2, we describe briefly the definition of local Lyapunov indicators and the statistical tools used to study their distribution. In Section 3, we give in the first subsection the variations with time of the generalized Lyapunov indicators (GLI) for typical orbits of the standard map. Using the GLI we make an exploration of a section of the standard map in the second subsection, and, in the third one, we perform a global study of the standard map by means of Kolmogorov like entropy varying the parameter.

## 2. Generalized Lyapunov Indicators (GLI)

For a given dynamical system it is well known that two orbits initially close diverge either linearly or exponentially depending on whether the two starting points lie in

* Present address : Observatoire de la Côte d'Azur, Nice

an integrable region or in a stochastic one.

The Lyapunov characteristic exponents (LCE) provide a more precise quantitative definition of stochasticity. Let us recall the essential features of the theory (see Benettin et al. 1980, Froeschlé 1984). Let $M$ be an $N$-dimensional compact differentiable manifold, $\mu$ a normalized measure on it and $\Phi^t$ a measure-preserving flow. The set $(M, \mu, \Phi^t)$ is called a dynamical system. Let $P \in M$; $T_P(M)$ denotes the tangent space of $M$ at the point $P$ and $D\Phi_P^t$, the tangent mapping of $\Phi^t$ from $T_P(M)$ onto $T_{\Phi^t(P)}(M)$. Given a nonzero vector $w \in T_P(M)$, one defines the quantity

$$\gamma_w^t(P) = \| w \|^{-1} \, ln \, \| \, D\Phi_P^t(w) \, \|$$

and, if the limit exists,

$$\chi(P, w) = \lim_{t \to \infty} \frac{1}{t} \gamma_w^t(P)$$

(here $\| \, . \, \|$ denotes the norm associated with the metric $\mu$). The limit is shown to exist for almost all initial points $P$ and all vectors $w$ and it is called the LCE of the flow $\Phi^t$ relative to $P$ and $w$. Of course the compactness of the phase space is rarely realized in celestial mechanics and therefore we are dealing with what could be called heuristic Lyapunov characteristic indicators. Moreover, as the vectors $w$ scan the whole tangent space, $\chi(P, w)$ takes at most $N$ distinct values $\chi_i(P)$, $i = 1, ..., N$, and there exists at least one basis $(e_1, ..., e_N)$ of $T_P(M)$ such that

$$\chi_i(P) = \lim_{t \to \infty} \frac{1}{t} \gamma_{e_i}^t(P)$$

For the sake of simplicity we shall denote $\gamma_{e_i}^t$ by $\gamma_i^t$ and choose the indices such that $\chi_1(P) \geq \chi_2(P) \geq ... \geq \chi_N(P)$. Furthermore, for a Hamiltonian system or for a symplectic mapping, $\chi_i(P) = -\chi_{N+1-i}(P)$. This set of $\chi_i(P)$ is a sensitive indicator of stochasticity in the sense that, if there exist $p$ isolating integrals i.e. uniform integrals functionally independent and in involution. Then there are $2p$ vanishing $\chi_i(P)$. As a consequence, in an integrable situation, all the $\chi_i(P)$ vanish. However for the standard map we are interested in the computation of the largest and unique positive LCE. Let us consider a mapping $F$ and the corresponding tangential mapping defined as follow

$$X_{n+1} = F(X_n) \quad , \quad Y_{n+1} = (\frac{\partial F}{\partial X_n}) Y_n \, .$$

We iterate simultaneously these two mappings, taking the norm of the initial vector $Y_0$ equal to 1, the evolved vector $Y_n$ are renormalized at arbitrary times $j\tau$ (here $\tau = 1$ is the period of the mapping and $j = 1, ...n$). Then we get :

$$\chi(X_0, n) = \frac{1}{n\,\tau} \sum_{j=1}^{n} ln(\alpha_j) \quad , \quad \chi_1(X_0) = \lim_{n \to \infty} \chi(X_0, n) ,$$

where $\chi_1(X_0)$ is the largest LCE and is of course approximated with a finite $n$. The Lyapunov characteristic indicator $\chi(X_0, n)$ is no more than the first moment, i.e the mean value of the distribution of local variations $ln(\alpha_j)$. See Froeschlé 1984 for the mathematical definition of $\alpha_j$, which result from the Gram-Schmidt orthonormalization of the vectors $Y_n$.

Actually this distribution summarizes a lot of informations concerning the dynamics of an orbit. For instance the presence of cantori around invariant zones will be reflected in the distribution, since for a while the motion is roughly quasi periodic. Besides the mean $m$ and the $r.m.s.$ $\sigma$ given by the first and second moment of the distribution :

$$m = E(X) \quad , \quad \sigma = \sqrt{E(X-m)^2} \,,$$

we compute the fisher coefficients

$$\gamma_1 = \mu_3/\sigma^3 \quad , \quad \gamma_2 = \mu_4/\sigma^4 - 3 \,,$$

with $\mu_p = E(X-m)^p$.

In the case of a normal distribution $\gamma_1$ and $\gamma_2$ vanish; when the shape of the distribution is symmetric $\gamma_1$ is equal to zero, in the unimodal case while $\gamma_2$ reflects the flatness of the distribution with respect to a normal one. These quantities could be considered as generalized Lyapunov exponents (GLE). However like for the definition of Lyapunov characteristic exponents we necessarily consider the results of a finite number of iterations (or integrations for a continuous system). Therefore we will rather consider the concepts of generalized Lyapunov indicators, which will be the values of $m$, $\sigma$, $\gamma_1$, $\gamma_2$ estimated from a finite number $n$ of iterations, and will be called in the following GLI.

## 3. Numerical Results

### 3.1. CONVERGENCE OF THE GENERALIZED LYAPUNOV INDICATORS (GLI)

We study orbits of the standard map

$$x_1 = x_0 + a\sin(x_0 + y_0) \quad , \quad y_1 = x_0 + y_0 \quad (mod2\pi) \,,$$

for different initial conditions and different values of the parameter $a$.

Fig. 1 a, b, c, d show respectively the variations with the number of iterations of $m$, $\sigma$, $\gamma_1$, and $\gamma_2$ previously defined. In the well studied case (see Lichtenberg et al. 1983) $a = -1.3$ for a regular orbit, i.e an invariant curve. We notice as expected the convergence of $m$ to the value zero, the fast convergence of $\sigma$ to a constant value. While the quantity $\gamma_1$ and $\gamma_2$ converge more slowly to a non-zero value. This is not surprising since the distribution for $10^5$ iterations (Fig. 5 a) is bimodal, then consequently neither symmetric unimodal nor normal.

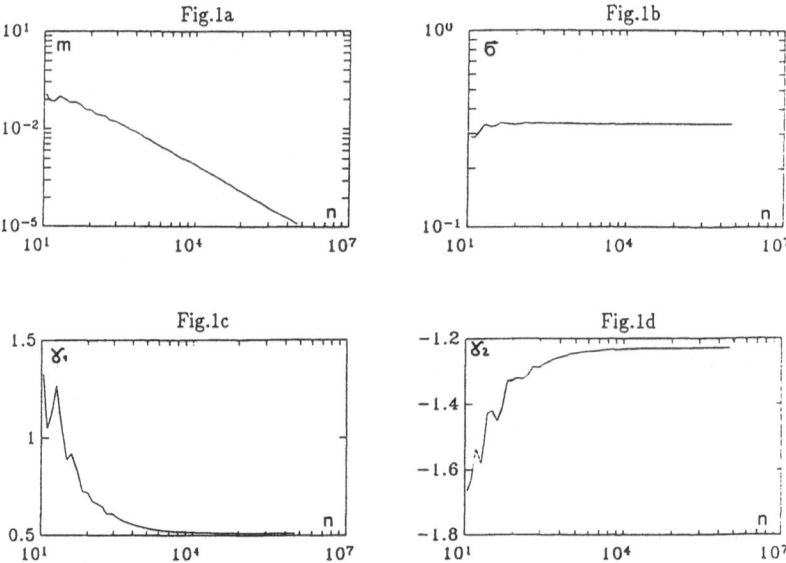

Fig. 1. (a), (b), (c), (d) show respectively the variation of the generalized indicators $m$, $\sigma$, $\gamma_1$, $\gamma_2$, for initial conditions corresponding to an invariant curve ($a = -1.3$, $x_0 = 1.$, $y_0 = 0.$)

The variations of the GLI $m$, $\sigma$, $\gamma_1$ and $\gamma_2$ for a chaotic orbit ($a = -1.3$, $x_0 = 2$, $y_0 = 0$) are respectively shown on Fig. 2 a, b, c and d. As it is well known, the mean does not go to zero, and a good approximation to a constant value is not reached before $10^4$ iterations. Conversely the limit of $\sigma$ is reached very quickly ($10^2$ iterations). Moreover the oscillations displayed by $m$ and $\gamma_2$ (a and d) reveal the complicated structure of the "Chaotic zone", when the motion takes place, that is around invariant curve the presence of cantori slows down the diffusion. Like $\sigma$, the convergence of $\gamma_1$ (c) is quite fast ($\sim 10^2$ iterations), however the small value does not reflect a symmetric distribution as seen on Fig. 5 b. In this case the non zero value of $\sigma$ may be explained in the chaotic region by the presence of a dense set of hyperbolic points which may have different eigenvalues and different orientations of the eigen vectors.

## 3.2. TRANSVERSAL EXPLORATION ALONG THE X-AXIS

Since the standard mapping exhibits all the features of conservative Hamiltonian systems with two degrees of freedom, i.e. invariant curves, islands, cantori, chaotic orbits, we have plotted on Fig. 3 a, b, c, d. the values of the estimators $m$, $\sigma$, $\gamma_1$, and $\gamma_2$ after 10 000 iterations with the following initial conditions: $y_0$ fixed equal to zero and $x_0$ spacing the $x$ axis from 0 to $\pi$ with a step equal to 0.02. On Fig.3 a, we recover the well known results taking LCEs as indicators of stochasticity, i.e. zero

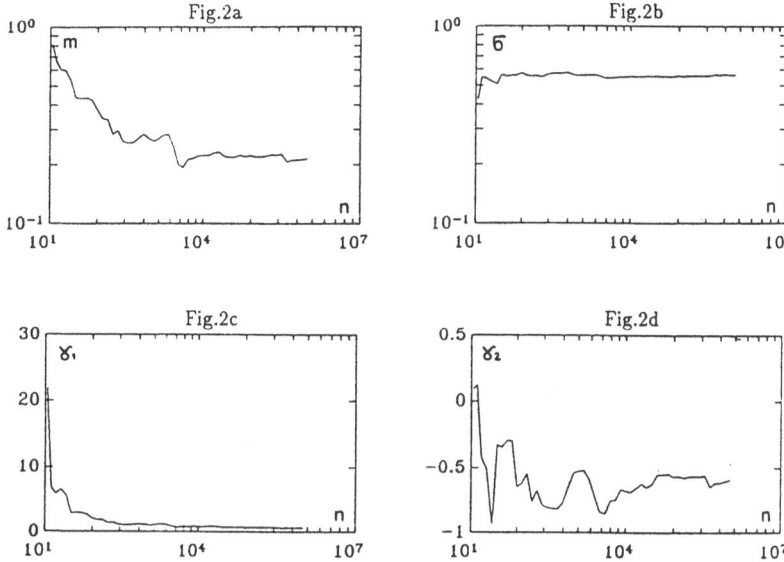

Fig. 2. Same as Figure 1 but for a chaotic orbit ($a = -1.3$, $x_0 = 2.$, $y_0 = 0.$)

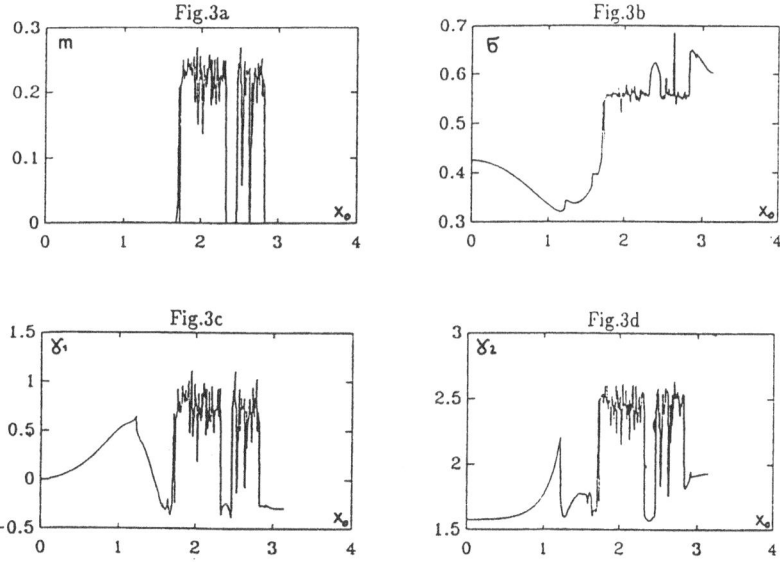

Fig. 3. (a), (b), (c), (d) show respectively the variations along the $x$ axis of the generalized Lyapunov indicators (GLI) $m$, $\sigma$, $\gamma_1$, $\gamma_2$ computed at 10 000 iterations at $y_0 = 0$ for a step 0.05 for the coarse graining.

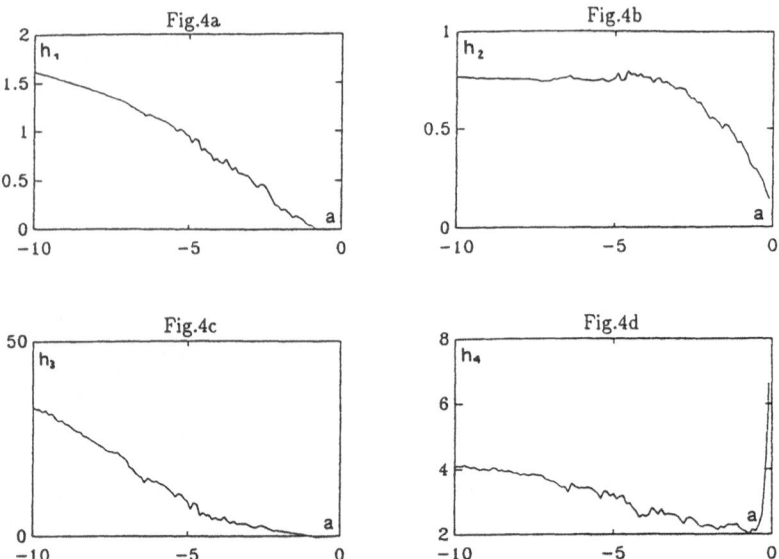

Fig. 4. Generalized Kolmogorov entropies $h_1$, $h_2$, $h_3$, $h_4$ versus the parameter axis $a$. Coarse graining step equal to $-0.2$. For each value of the parameter $a$, $M = 50$ $N = 2000$.

values indicate ordered region and important variations of the values of $m_{10000}$ even for orbits starting close to each other. The same behaviour, as far as continuity is concerned appears for the variations of $\gamma_1$ and $\gamma_2$. Let us notice the smallest and more discriminating variation of $\sigma$. In the ordered region, the continuous decrease of $\sigma$ to $\sim 0.32$ when $x$ goes to 1 may be due to both the flatness of the elliptic like invariant curve and the variation of the rotation number. More studies on simple models are needed to discriminate between these two effects.

## 4. Kolmogorov like Entropy

Pesin's formula gives the relation between the LCEs and the Kolmogorov entropy of a system. Let $\varrho(P)$ denote the sum of all positives LCEs. In our case, one has

$$\varrho(P) = \chi_1(P)$$

The formula states that the total entropy is given by

$$h = \int_M \varrho(P)\, d\mu.$$

Consequently, the local quantity $\varrho(P)$ defines a density of Kolmogorov entropy. It is related to the exponential stretch of a small volume of the phase space, in the directions corresponding to the positive LCEs and the Kolmogorov entropy gives

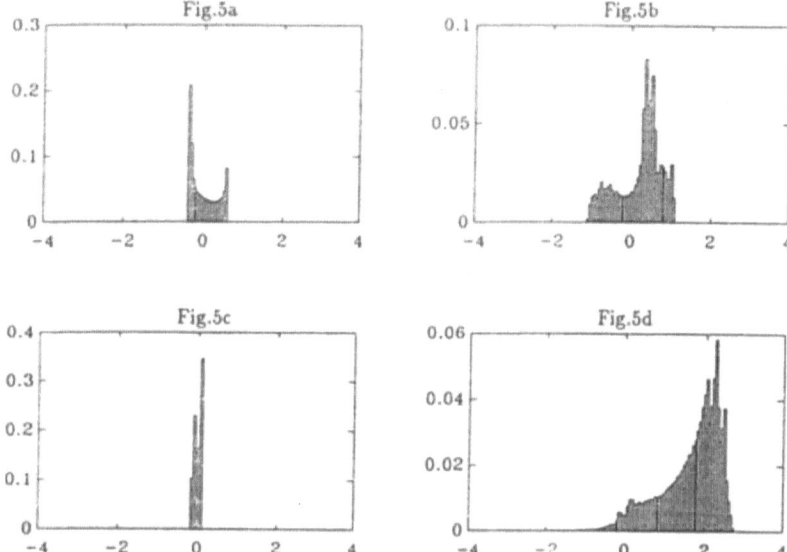

Fig. 5. Typical distributions of the local Lyapunov exponents $ln(\alpha_i)$ for different orbits of the standard mapping

a    $x_0 = 1.$    $y_0 = 0.$    $a = -1.3$(invariant curve)
b    $x_0 = 2.$    $y_0 = 0.$    $a = -1.3$ (a chaotic orbit)
c    $x_0 = 1.$    $y_0 = 0.$    $a = -0.1$ (invariant curve in the circulation case)
d    $x_0 = 1.$    $y_0 = 0.$    $a = -10.$ (strong chaotic orbit)

an average over the phase space of this density. Following the same philosophy we compute using a Monte Carlo method the quantities

$$h_k = \frac{1}{NM} \sum_{j}^{M} \sum_{i=1}^{N} ln(\alpha_{ij})$$

$j$ refers to an orbit $j$ whose initial conditions have been taken at random and $i$ refers to the number of iterations. The subscript $k$  ($k = 1, 4$) defines the quantities $h_1$, $h_2$, $h_3$ and $h_4$. Where $h_1$ is an estimation of the Kolmogorov entropy, $h_2$ is a measure of the dispersion of this entropy. While $h_3$ and $h_4$ estimate respectively the mean over the phase space of the asymmetry and the flatness parameters defined previously. The variations along the parameter axis $a$, of $h_1$, $h_2$, $h_3$ and $h_4$ are plotted respectively on Fig. 4 a, 4 b, 4 c and 4 d.

The three quantities $h_1$, $h_3$ and $h_4$ increase with the absolute value of the parameter $a$, while $h_2$ reaches a constant value for $| a | > 5$, which means that the distribution of the local Lyapunov indicators $[ln(\alpha_{ij})]$ is more or less translated towards the positive axis.

## 5. Conclusion

The first moments of the distribution of the local Lyapunov exponents, and particularly the $r.m.s.$ $\sigma$ seem to provide good hints of the global stochasticity of orbits of hamiltonian systems (or symplectic mappings). We plan to continue this preliminary study in two directions.

First, using simpler ad hoc mappings we intend to characterize, by the four first moments of the distribution, the geometry and kinematics of regular orbits, and as far as chaotic orbits are concerned we will try to measure the influence of the magnitude and orientation of hyperbolic invariant manifolds.

Second, since many distributions are far from the normal one and show, for instance, bimodal feature, the tools of artificial intelligence may provide better parameters to characterize the slope of such distributions.

## Acknowledgements

We want to thank A. Morbidelli for stimulating discussions.

## References

Benettin, G., Galgani, L., Giorgilli, A. and Strelcyn, J.M.: 1980, "Lyapunov characteristic exponents for smooth dynamical systems; a method for computing all of them", *Meccanica* ,**15**

Crisanti, A., Paladin, G. and Vulpiani, A.: 1988, "Generalized Lyapunov exponents in high-dimensional chaotic dynamics and products of large random matrices", *Journal of Statistical Physics*, **53**, N° 314.

Froeschlé, C.: 1984, "The Lyapunov characteristic exponents and applications", *Journal de Mécanique théorique et appliquée*, Numéro spécial.

Fujisaka, H.: 1983, "Statistical dynamics generated by fluctuations of local Lyapunov exponents", *Progress of Theoretical Physics*,**70**, N° 5.

Lichtenberg A.J.,Lieberman M.A.: 1983, "Regular and Stochastic Motion". Applied Mathematical Sciences Vol.**38**, Springer Verlag.

# GENERALIZED LYAPUNOV EXPONENTS INDICATORS IN HAMILTONIAN DYNAMICS: AN APPLICATION TO A DOUBLE STAR SYSTEM

E. LOHINGER*

*Institute of Astronomy, University of Vienna, Türkenschanzstr. 17, 1180 Wien, Austria*

C. FROESCHLÉ

*Observatoire de la Côte d'Azur, B.P. 229, 06304 Nice Cedex 4, France*

and

R. DVORAK

*Institute of Astronomy, University of Vienna, Türkenschanzstr. 17, 1180 Wien, Austria*

**Abstract.** The Lyapunov characteristic numbers (LCNs) which are defined as the mean value of the distribution of the local variations of the tangent vectors to the flow (=ln $\alpha_k^i$) (see Froeschlé, 1984) have been found to be sensitive indicators of stochasticity. So we computed the distribution of these local variations and determined the moments of higher order for the integrable and stochastic regions in a binary star system with $\mu = 0.5$.

**Key words:** Double star systems – Lyapunov exponents – chaotic motion

## 1. Introduction

For the study of the stochasticity of dynamical systems the theory of the Lyapunov characteristic exponents (LCEs) plays an important role. These exponents, which measure the mean exponential rate of divergence of nearby trajectories, were introduced by Lyapunov (1907) and were adapted to the theory of dynamical systems and to the ergodic theory by Oseledec (1968).
Since the numerical work of Hénon and Heiles (1964) where the divergence of nearby trajectories was used to characterize stochasticity of a phase space trajectory for the first time, much work has been done to give a precise quantitative definition of exponential divergence and thus of stochasticity (see e.g. Froeschlé (1970), Froeschlé and Scheidecker (1973), Ford (1975), Chirikov (1979)).
This time not only the LCEs or more precisely the Lyapunov characteristic numbers (LCNs) were used for the characterization of chaos, but also the higher moments of the distribution of the local variations of the tangent vectors to the flow (see Froeschlé et al. in this volume). More precisely we computed for a model described in §2 the standard deviation ($\sigma$) and the Fisher coefficients, these are the asymmetry ($\gamma_1$) and the flatness ($\gamma_2$), of the ln $\alpha_k^i$.
In §3 we describe the numerical computation and discuss the results in §4.

---

* Present address: Observatoire de la Côte d'Azur, Nice

*Celestial Mechanics and Dynamical Astronomy* **56**: 315–322, 1993.
© 1993 *Kluwer Academic Publishers.*

## 2. The Model

We used the three dimensional restricted problem of three bodies, where the two masses $m_1$ and $m_2$ (called primaries) revolve in circular orbits around their center of mass and the third body $m_3$ with infinitesimal mass moves in the gravitational field of the primaries without influencing them. As unit of mass we took the total mass of the primaries, as unit of length the distance $m_1 m_2$, and the unit of time was chosen in a way that the angular velocity of the primaries is unity. In a rotating frame of coordinates $(\xi, \eta, \zeta)$ $m_1$ and $m_2$ remain motionless in their locations on the $\xi$-axis: $(-\mu, 0, 0)$ and $(1 - \mu, 0, 0)$, where $\mu = \frac{m_2}{m_1 + m_2}$ ($= m_2$ in our case). The equations of motion for the massless body are:

$$\ddot{\xi} = 2\dot{\eta} + \xi - (1 - \mu)\frac{\xi + \mu}{r_1^3} - \mu\frac{\xi - 1 + \mu}{r_2^3}$$

$$\ddot{\eta} = -2\dot{\xi} + \eta - (1 - \mu)\frac{\eta}{r_1^3} - \mu\frac{\eta}{r_2^3}$$

$$\ddot{\zeta} = -(1 - \mu)\frac{\zeta}{r_1^3} - \mu\frac{\zeta}{r_2^3}$$

where $r_1$ and $r_2$ are the distances to the primaries:

$$r_1 = [(\xi + \mu)^2 + \eta^2 + \zeta^2]^{1/2} \text{ and } r_2 = [(\xi - 1 + \mu)^2 + \eta^2 + \zeta^2]^{1/2}.$$

For detailed information about the restricted three body problem see for example Szebehely (1967).

The initial conditions for our computations were taken from the paper of Gonczi and Froeschlé (1981), where the starting positions for the satellite-type motions are defined as:

$$\xi = 0.5 - d_0; \qquad \eta = 0; \qquad \zeta = 0$$

with the velocity components:

$$\dot{\xi} = 0.235749(d_0)^{-1/2}; \quad \dot{\eta} = -0.640312(d_0)^{-1/2} + d_0; \quad \dot{\zeta} = \dot{\xi}.$$

When increasing the parameter $d_0$ of this system, which measures the initial distance between the massless body and the mass $m_2$ ($d_0 = r_2(t = 0)$), the perturbation due to $m_1$ becomes more and more important and therefore the motion of the massless body changes from a regular to a stochastic one.

For a binary system with mass ratio $\mu = 0.5$ Gonczi and Froeschlé (1981) found this transition by means of the LCNs (Eq.1 of §3), when taking the following values for $d_0$:

$$d_0 = 0.10; \; 0.13; \; 0.22; \; 0.24; \; 0.255; \; 0.27; \; 0.31; \; 0.35; \; 0.40; \; 0.50 \, .$$

Since the LCNs are defined as the mean values and therefore the first moment of the distribution of $\ln \alpha_k^i$ (see Eq.1 of §3), we computed all the distributions for these satellite-type motions (see Fig.2-4) and examined the possibility to determine the transition to chaotic motion through the higher moments of these distributions.

# 3. The Computation of the $\ln \alpha_k^i$

The numerical computations were carried out using the Bulirsch-Stoer integration method (see Bulirsch and Stoer, 1966), which applies the Richardson extrapolation, where a large interval $H$ is divided step by step into $n$ finer subintervals $h = \frac{H}{n}$ (a conventional sequence of $n$ is: 2,4,6,8,12,...$[n_j = 2n_{j-2}]$). The integrations are done by the modified midpoint method and for the extrapolation technique rational functions are used instead of power series with its limited radius of convergence, to enable large step sizes. The Bulirsch-Stoer method is one of the best methods to obtain high-accuracy solutions to ordinary differential equations.

To compute the local variations of the tangent vectors to the flow ($\ln \alpha_k^i$, we call them local Lyapunov characteristic numbers (LLCNs)) we have to apply the definition of the LCEs of oder $k$. As the basic concepts of the Lyapunov characteristic exponents are briefly described in Froeschlé et al. (this volume) or more detailed in Benettin et al. (1980) resp. Froeschlé (1984), we only recall that part of the theory used for our calculations.

Let $(v_1 \ldots v_k)$ a system of $k$ independent vectors of $T_x(M)$ (= the tangent space of the manifold $M$ at point $x$), the LCEs of order $k(k \in [1, N])$ are defined as:

$$\chi^k(x, v_1, \ldots, v_k) = \lim_{t \to \infty} \frac{1}{t} \ln vol^k[D\phi_x^t(v_1) \ldots D\phi_x^t(v_k)]$$

$vol^k$ denotes the $k$ dimensional volume of the corresponding parallelepiped and $D\phi_x^t$ is the tangent mapping of the flow $\phi^t$ from $T_x(M)$ onto the tangent space of point $\phi^t(x)$.

This limit is shown to exist and is finite for almost all $x \in M$, thus we can write for almost all vectors $v_i(i = 1, \ldots, k)$

$$\chi^k(x, v_1, \ldots, v_k) = \sum_{i=1}^{k} \chi_i(x).$$

Numerical studies have shown that in a stochastic region the vectors become very large and the angles between them are very small so that the accuracy of the computation will be degraded. To avoid this difficulty Benettin et al. (1980) suggested to replace the evolved tangent vectors by a set of new orthonormal vectors at regular time intervals $\tau$, using the Gram-Schmidt procedure. Therefore the LCEs of a point $x$ in the $n$ dimensional phase space are defined as:

$$\chi_i(x) = \lim_{N \to \infty} \underbrace{\frac{1}{N\tau} \sum_{k=1}^{N} \ln \alpha_k^i}_{\text{LCNs}} \tag{1}$$

with $t = N\tau$ and $\alpha_k^i$ is the Euclidian norm of the component of the vector $v_k^i$, which is orthogonal to all previously orthonormalized vectors $v_k^j(j < i)$; and $i$

depends on the dimension of the phase space (in our case: $i = 1, \ldots, 6$).

Strictly speaking the whole system to integrate consists of 42 differential equations, since we need six variational equations for each vector $(v_k^i)$ in addition to the equations of motions.

As the system in consideration is an Hamiltonian one it is sufficient to examine only the first three LLCNs because of $\chi_i(x) = -\chi_{N-i+1}(x)$.

Applying this to our numerical study we took the $\ln \alpha_k^i$ (LLCNs) of each renormalization, which was done every time unit for an integration time $t = 20000$ units, i.e. 20000 revolutions of the primaries.

## 4. The Numerical Results

### 4.1. THE GENERALIZED LYAPUNOV INDICATORS

Fig.(1.a-d) show the geometrical effects of the system in consideration. As the calculations were done twice, using different values of accuracy[1] of the computation: $\varepsilon = 10^{-9}$ and $\varepsilon = 10^{-13}$, we have two curves for each LLCN. Fig.(1.a) shows the variation of the mean values which correspond to the LCNs. Here we see the same behavior as already found by Gonczi and Froeschlé (1981):

In the integrable zone (when $d_0 < 0.25$) all values remain close to zero, while the strong increase, especially of $\ln \alpha_k^1$ and $\ln \alpha_k^2$ (because of the disappearance of two isolating integrals) indicates stochasticity for $d_0 \geq 0.255$.

This figure shows clearly the strong robustness of $\ln \alpha_k^1$ and $\ln \alpha_k^2$ even for the chaotic zone whereas the variation of $\ln \alpha_k^3$ is extremely sensitive to the accuracy $\varepsilon$ of the computations. Similarly, we could only produce the correct value of the Jacobi constant during the integrations, if we chose $\varepsilon = 10^{-13}$ or smaller. In contrast to this, the higher order moments of $\ln \alpha_k^3$ seemed not to be affected by changes in $\varepsilon$ between $10^{-9}$ and $10^{-13}$ (see Fig.(1.b-d)).

The variation of sigma[2] (see Fig.(1.b))

$$\sigma = \sqrt{E[(x - m)^2]} \qquad \text{with } m = E(x)$$

is more or less constant in the chaotic zone, and its value increases with $\ln \alpha_k^1$. This is not valid for the integrable region, since here the variation reflects the geometrical properties of the manifold on which the motion takes place (see Froeschlé et. al. in this volume).

Additional to the mean $(m)$ and the $r.m.s$ $(\sigma)$ we computed the Fisher coefficients (see Fig.(1.c) and (1.d))

$$\gamma_1 = \frac{\mu_3}{\sigma^3} \qquad \text{and} \qquad \gamma_2 = \frac{\mu_4}{\sigma^4} - 3$$

---

[1] for a definition and interpretation of $\varepsilon$ see Bulirsch and Stoer (1966)

[2] here expressed in terms of expectation values

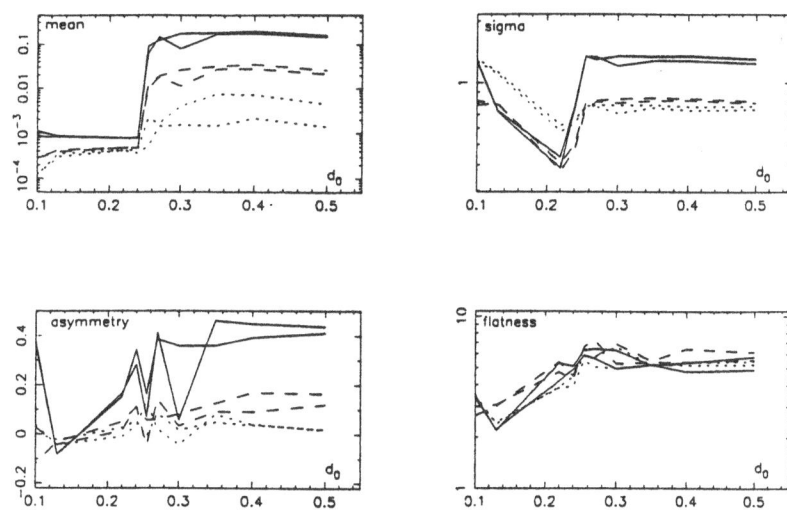

Fig. 1. a,b,c,d: show respectively the variation of the mean, the sigma, the asymmetry and the flatness for the different initial distances $d_0$. The full lines are drawn for $\ln \alpha_k^1$, the dashed lines for $\ln \alpha_k^2$ and the dotted lines for $\ln \alpha_k^3$, using $\varepsilon = 10^{-9}$ and $\varepsilon = 10^{-13}$

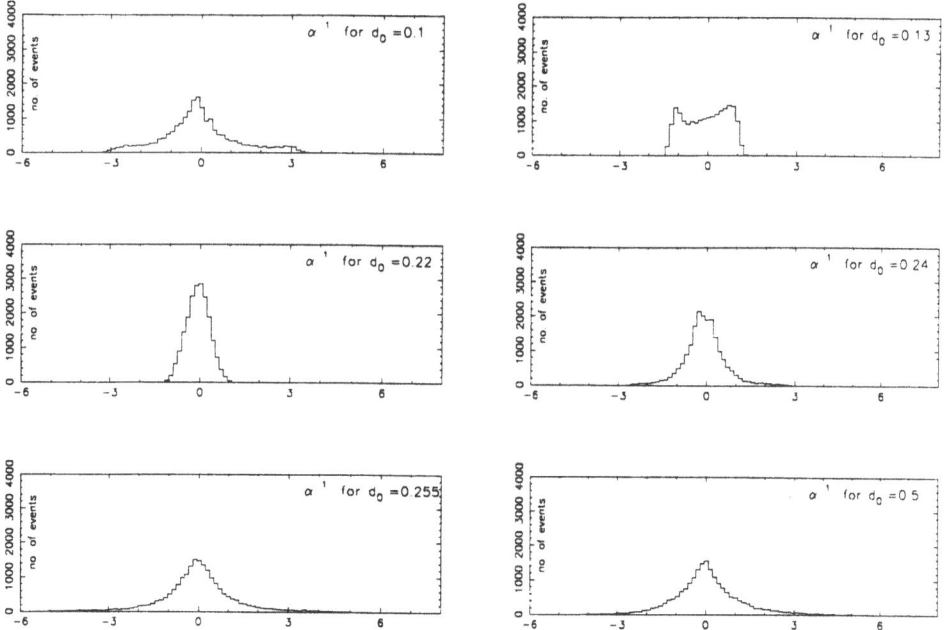

Fig. 2. The distribution of $\ln \alpha_k^1$ for the different initial distances $d_0$, using $\varepsilon = 10^{-13}$.

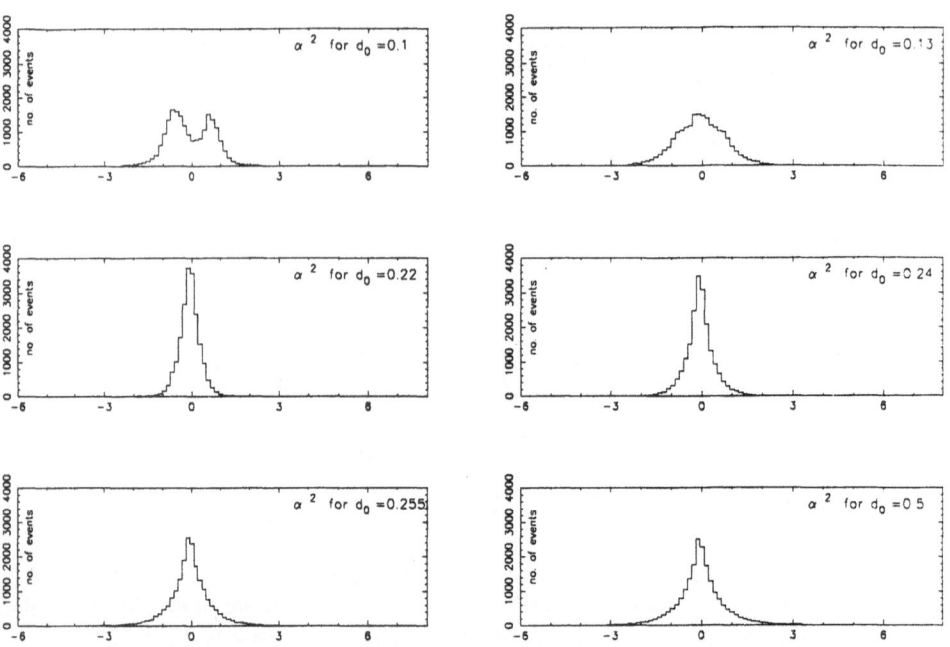

Fig. 3. The distribution of $\ln \alpha_k^2$ for the different initial distances $d_0$, using $\varepsilon = 10^{-13}$.

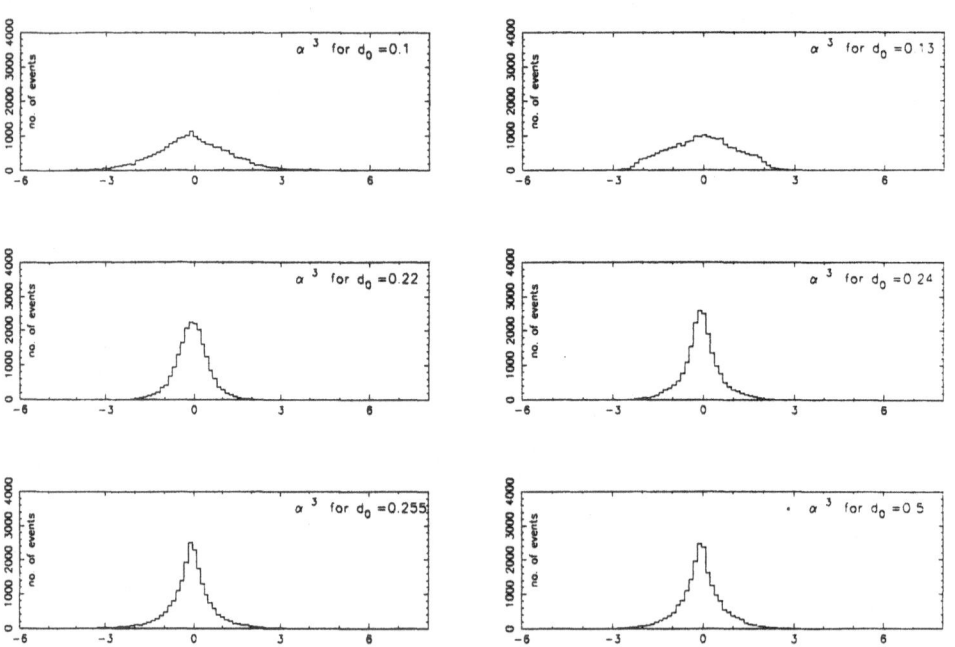

Fig. 4. The distribution of $\ln \alpha_k^3$ for the different initial distances $d_0$, using $\varepsilon = 10^{-13}$.

with $\mu_p = E[(x - m)^p]$. These coefficients measure respectively the asymmetry and the flatness with respect to the normal distribution. As the variation of $\gamma_1$ is very close to zero, except for $\ln \alpha_k^1$, one can find a distribution which is more or less symmetrical. The flatness shows the same behavior for the integrable and stochastic regions as $\sigma$ but of smaller extend. However, in all cases we are quite far from a normal distribution.

## 4.2. THE DISTRIBUTION AND CONCLUSION

The figures (2) - (4) show the distributions of $\ln \alpha_k^i$ ($i = 1, 2, 3$) for different initial distances $d_0$, which result from the computations with accuracy $\varepsilon = 10^{-13}$. The x-axis is defined by the interval [-6,8], which is divided into 100 bins. They show clearly that in the regular zone ($0.1 \leq d_0 < 0.25$) the modal value (i.e. the value on the x-axis with the highest number of events) of each distribution is to be found in different intervals of the x-axis and the shape of the distribution changes when we increase $d_0$. While in the stochastic region the modal value of each distribution is close to zero and the shape remains nearly the same for all $d_0 > 0.25$. This constancy of distribution suggests a uniform chaos for the satellite-type motions in a binary system with $\mu = 0.5$ and an initial distance $> 0.25$.

As it is well known, that the significance of a statistical analysis increases with the number of available data, we intend to continue our study of the satellite-type motions in doing the computations for about 100000 revolutions of the primaries. Furthermore, this analysis will be applied to binary systems with different mass ratios and last but not least we will use this technique for the characterization of stochasticity in cometary dynamics.

## Acknowledgements

The authors want to thank Dr. R. Gonczi for stimulating discussions and Dr. Ch. Froeschlé for her critical reading of this paper.

## References

Benettin G., Galgani L., Giorgilli A., Strelcyn J. M.: 1980,"Lyapunov characteristic exponents for smooth dynamical systems; a method for computing all of them".*Meccanica* vol. 15, part I+II, p. 9-30

Bulirsch R., Stoer J.: 1966, "Numerical Treatment of Ordinary Differential Equations by Extrapolation Methods". *Numerische Mathematik* 8, p. 1-13.

Chirikov B. V.: 1979, "An universal instability of many dimensional oscillator systems". Phys. Rep. 52, p. 263-379.

Ford J.: 1975, "The statistical mechanics of classical analytic dynamics, fundamental problems in statistical mechanics". ed. E.G.D. Cohen, Vol. III (North-Holland, Amsterdam), p. 215-255.

Froeschlé C.: 1970, "A numerical study of the stochasticity of dynamical systems with two degrees of freedom". *Astron. Astrophys.* 9, p. 15-23.

Froeschlé C., Scheidecker J. P.: 1973, "On the disappearance of isolating integrals in systems with more than two degrees of freedom". *Astrophys. and Space Sc.*, vol. **25**, p. 373-386.

Froeschlé C.: 1984, "The Lyapunov characteristic exponents and applications". *Jounal de Mécanique théorique et appliquée*, Numéro spécial, p. 101-132.

Froeschlé C., Froeschlé Ch., Lohinger E.: 1992, "Generalized Lyapunov characteristic indicators and corresponding Kolmogorov like entropy of the standard mapping", this volume.

Gonczi R., Froeschlé C.: 1981, "The Lyapunov characteristic exponents as indicators of stochasticity in the restricted three-body problem". *Celest. Mech.* **25**, p. 271-280.

Hénon M., Heiles C.: 1964, "The applicability of the third integral of motion, some numerical experiments". *Astron. Journal* **69**, p. 73-79.

Lyapunov A. M.: 1907, "Problème Général de la Stabilité du Mouvement" (transl. from Russian), *Ann. Fac. Sci. Univ. Toulouse* **9**, p. 203-475. Reproduce<d in *Ann. Math. Study*, vol. **17**, Princeton 1947.

Oseldec V. I.: 1968, "The Multiplicative Ergodic ". Theorem. The Lyapunov Characteristic Numbers of Dynamical Systems (in Russian). *Trudy Mosk. Mat. Obsch.* **19**, p. 179-210. English translation in *Trans. Mosc. Math. Soc.* **19**, p. 197, 1968.

Szebehely V.: 1967, "The Theory of Orbits". Academic Press, New York and London

# ASTEROID 522 HELGA IS CHAOTIC AND STABLE

ANDREA MILANI and ANNA M. NOBILI

*Department of Mathematics, Pisa University,*
*Via Buonarroti 2, I-56100 Pisa, Italy, and*
*Groupe E.U.R.O.P.A., Observatoire de Paris/Meudon*
*E-mail TWIN2@ICNUCEVM.BITNET*

**Key words:** chaos, stability, asteroids

A dynamical system with a positive maximum Lyapounov exponent is defined to be chaotic. It is possible to obtain an indication on the value of the maximum Lyapounov exponent by monitoring the function $\gamma(t) = \log\{|v(t)|/|v_0|\}$ where $v(t)$ is the solution of the equations of relative motion, linearized around the solution under study, with initial conditions $v(0) = v_0$ chosen at random. By definition, the maximum Lyapounov exponent is: $\lim_{t \to +\infty} \gamma(t)/t$. Any calculation being finite, only an estimate can be obtained by means of a linear fit to $\gamma(t)$ for the available time span. For *Helga*, the best fit over a $7\,Myr$ integration is with a slope of $1.45 \times 10^{-4}\,yr^{-1}$, i.e. the Lyapounov time is only about $6,900\,yr$. The reference orbit of *522 Helga*, as well as the variational equations, have been computed within quite a complete physical model, accounting for the perturbations of the outer planets Jupiter, Saturn, Uranus and Neptune; all initial conditions have been referred to the centre of mass of the inner Solar System, to decrease the secular effects of the neglected perturbations from the inner planets. The observed chaotic behaviour is therefore not an abstract property of some simplified model, but is on the contrary intrinsic to the orbit of the real asteroid. Full details are given in Milani and Nobili (1992).

The orbital elements of *522 Helga* do not show signs of large scale instabilities. The comparatively large variations in semimajor axis appear to be associated with the orbit getting in and out of the 7:12 mean motion resonance with Jupiter. The argument $7\lambda - 12\lambda_J$ (where $\lambda$ is the mean longitude of the asteroid, and $\lambda_J$ that of Jupiter) is a slow variable, completing only $\simeq 830$ revolutions in $7\,Myr$. It alternates periods of ostensible libration with periods of circulation, which is typical of a chaotic situation.

With a semimajor axis between 3.628 and 3.632 $AU$, an eccentricity between zero and 0.094 and a small inclination ($< 3.9°$), *Helga* could in principle get as close as $\simeq 0.9\,AU$ to Jupiter. Such a small close approach distance would make the apparent macroscopic stability over the $7\,Myr$ integration time span rather difficult to understand. Instead, it is found that *Helga* never gets closer to Jupiter than $1.31\,AU$, and the reason can be understood by monitoring the behaviour of the variables $e, e_J, \varpi, \varpi_J$; along the entire time span of our numerical integration, either $\varpi - \varpi_J$ is in libration around $0°$, or the eccentricities $e, e_J$ are close to their minimum values (Milani and Nobili, 1984). While $e$ has rather large variations, the proper eccentricity $e_P$ (as computed with a synthetic method, as in Schubart, 1988) has an average value of 0.039 and is remarkably stable with time (standard deviation 0.0015). The small value of *Helga*'s $e_P$, with $e_P$ either smaller than or comparable to the forced eccentricity, is the true explanation for its long term macroscopic stability. All objects in the region with an original proper eccentricity significantly larger than the forced one must have been ejected, and indeed the asteroid population in the region around *Helga* is very sparse (Milani and Nobili, 1985). It is worth stressing that the protection mechanisms effective for *Helga* have nothing to do with the 7:12 resonance, which on the other hand is responsible for the observed chaotic behaviour.

The most spectacular examples of chaos in the Solar System are those in which the chaotic domain contains a region of very close approaches (even the singularity of collision) and chaos does explain the absence of real objects. The planets themselves have been found to be chaotic, but for the major outer planets the chaotic region is very thin ($10^{-4}$ in eccentricity). On the contrary, chaos makes the eccentricity of Mercury change by 0.05 over only 40 times the Lyapounov time, which is by no means a weak instability. The orbit of Pluto has now been computed for 50 times the Lyapounov time; the chaotic character of the orbit is confirmed but still it shows no significant instability. Chaos is turning up everywhere in the Solar System, but the true meaning of this chaos is not yet understood. This work shows the first clear–cut example in which the Lyapounov exponent has very little significance: *Helga* is definitely an example of stable chaos. By stable chaos we mean that the average change of the proper elements over one Lyapounov time is very small. Although a rigorous theory for the occurrence of stable chaos is not available, it can be argued that if the resonance responsible for the onset of chaos is not the main protection mechanism, then the chaotic domain does not contain a region where the perturbations are significantly increased and macroscopic instability should not be expected. It is therefore likely that stable chaos occurs in many other problems of Solar System dynamics.

## References

Milani, A. and Nobili, A.M.: 1984, "Resonant structure of the outer asteroid belt", *Celestial Mechanics*, **34**, 343–355.

Milani, A. and Nobili, A.M.: 1985, "The depletion of the outer asteroid belt.", *Astron. Astrophys.*, **144**, 261–274.

Milani, A. and Nobili, A.M.: 1992, "An example of stable chaos in the Solar System", *Nature*, in press.

Schubart, J.: 1988,"Resonant asteroids between the main belt and Jupiter's orbit.", *Celestial Mechanics*, **43**, 309–317.

# CLASSICAL PERIODIC ORBITS
# AND QUANTUM MECHANICAL
# EIGENVALUES AND EIGENFUNCTIONS

G. CONTOPOULOS

*Department of Astronomy, University of Florida, Gainesville Fl.32611, U.S.A.*
*and Department of Astronomy, Panepistimiopolis GR-157 83 Athens, Greece*

**Abstract.** We have calculated several families of classical periodic orbits in simple Hamiltonian systems of two degrees of freedom and the corresponding quantum mechanical eigenvalues and eigenfuctions. We have found that in most cases the eigenfunctions have their maxima and minima on some simple periodic orbits. These periodic orbits are of several resonant types and can be either stable or unstable. In the latter case the quantum Poincaré surfaces of section are very different from the classical Poincaré surfaces of section.

**Key words:** Chaos – periodic orbits – eigenvalues – eigenfunctions

## 1. Introduction

The relation between classical and quantum mechanics is a subject that has always attracted great interest. However, despite the extensive literature on the subject, there are still some basic problems that are unsolved. The most important refers to the relations between classical and quantum chaos. In fact the very notion of quantum chaos is still elusive (Kleppner 1991) and many people wonder if a linear equation like Schrödinger's equation

$$-\frac{\hbar^2}{2m} \sum \frac{\partial^2 \Psi}{\partial q_i^2} + V\Psi = i\hbar \frac{\partial \Psi}{\partial t} \ , \tag{1}$$

can produce chaos at all.

On the other hand Schrödinger's equation is derived from the classical Hamiltonian

$$\frac{1}{2m} \sum p_i^2 + V = E \ , \tag{2}$$

if we replace the classical quantities $p_i$ and $E$ by the operators $\hbar \partial(.)/i\partial q_i$ and $i\hbar \partial(.)/\partial t$ , while leaving $q_i$ the same. Therefore if a classical problem is chaotic why not the corresponding quantum problem ?

Of course, the fact that Planck's constant $\hbar$ is different from zero makes all the classical details that are smaller than $\hbar$ ambiguous in quantum mechanics. This is the spirit of Heisenberg's uncertainty principle. However if the chaotic behaviour affects regions in phase space that are much larger than $\hbar$ in action one should expect chaos also in quantum mechanics. Furthermore the correspondence principle asserts that any quantum mechanical result tends to a corresponding classical one if $\hbar$ decreases and tends to zero.

*Celestial Mechanics and Dynamical Astronomy* **56**: 325–336, 1993.

But despite the correspondence principle the quantum analog of the classical "transition to chaos" is not clear. This uncertainty has led people to two very different approaches to the problem.

The first approach (Ford *et al.* 1991) goes as far as to reject the correspondence principle altogether.

The other approach (Weissman and Jortner 1982, Berry 1987, Gutzwiller 1990, 1992, and many others) emphasizes various manifestations of chaos in quantum mechanics, although the "transition to chaos" is not so clear and abrupt as in classical mechanics[1].

However most people that work in this field do not have a clear-cut answer to the problem of quantum chaos, but try to collect further evidence by elucidating certain aspects of the problem.

One particular problem refers to the connection between classical periodic orbits and quantum mechanical eigenvalues and eigenfunctions.

Gutzwiller (1971) was the first to emphasize the role of the classical periodic orbits in deriving the quantum mechanical eigenfunctions. He considered the eigenfunctions as produced by the interference of infinite waves along periodic orbits. However the most surprising result was that in most cases only *one* periodic orbit determines the basic form of the eigenfunction (Heller et al. 1980, Davis and Heller 1981, Farantos and Tennyson 1987, Davis 1988, Founariotakis *et al.* 1989). The maxima and minima of an eigenfunction are in most cases along a particular periodic orbit. This is true even in the case of unstable orbits. In classically ergodic cases, when no stable orbits exist at all, Heller (1984) found that the eigenfunctions follow some unstable orbits that he called "scars".

These phenomena were observed in models of particular 3-atomic molecules, like $HCN$, $HNC$, $O_3$, etc, and also in some mappings, like the stadion, that represent ergodic dynamical systems.

## 2. Periodic Orbits in a Simple Hamiltonian

In order to study more systematically the relations between classical periodic orbits and quantum mechanical eigenvalues and eigenfunctions, we explored a simple Hamiltonian system of two coupled harmonic oscillators with potential

$$V = \frac{1}{2}\left(\omega_1{}^2 x^2 + \omega_2{}^2 y^2\right) + \epsilon x^2 y \tag{3}$$

that has been studied extensively from the classical point of view (Contopoulos 1966, 1970).

---

[1] The abruptness of the transition to chaos in classical mechanics was first noticed by Hénon and Heiles (1964). These authors calculated the chaotic region as a function of the energy. For small energies this region is extremely small. When the energy increases beyond a critical value, the chaotic region increases abruptly, until it covers almost all the available space. The abruptness of the onset of chaos was observed in other cases also (see, e.g. Barbanis 1966) and was explained as due to an "interaction of resonances" (Contopoulos 1966, Rosenbluth *et al.* 1966, Zaslavskii and Chirikov 1972).

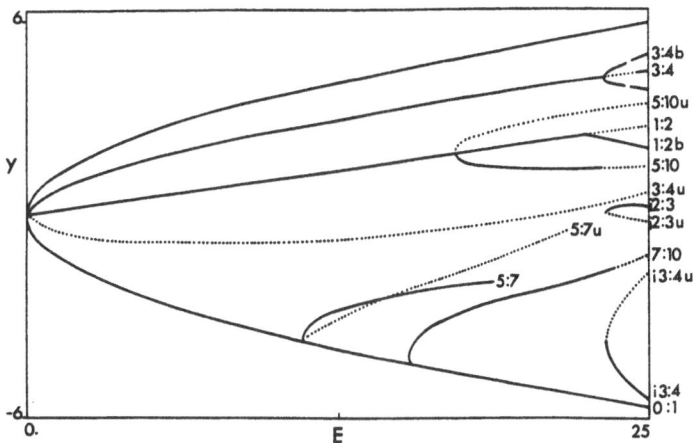

Fig. 1. Characteristics of the main families of periodic orbits (y as a function of E).(—) stable and (· · ·) unstable families. The families (5:7), (5:7u) and (7:10) bifurcate from the boundary (family (0:1)), the families (5:10), (1:2b) bifurcate from the "central" family (1:2), the family (3:4b) bifurcates from the resonant family (3:4), and the families (i2:3),(i2:3u),(i3:4),(i3:4u) are irregular.

The most important periodic orbits are those of small period. A detailed study of the dependence of such periodic orbits on the perturbation $\epsilon$ , for $\omega_1{}^2 = 0.9$, $\omega_2{}^2 = 1.6$ and fixed energy $E = 0.00765$ was made by Contopoulos (1970). We distinguish two types of orbits.

(1) Regular, that are generated from the orbits of the unperturbed case $\epsilon = 0$ by bifurcation, and

(2) Irregular that are not related to the above. Irregular orbits exist only for $\epsilon$ above a minimum value $\epsilon_{min}$, depending on the particular family studied.

In a recent paper (Founariotakis et al. 1989) we considered $\epsilon$ fixed ($\epsilon = -0.08$) and variable energy $E$. This problem is equivalent to the above with an appropriate scaling.

The various types of orbits are characterized by the numbers of oscillations along $x$ and $y$. Examples of such orbits are given in Figs. 2-5 and 7-9 below.

The characteristics of the main types of orbits are given in Fig. 1. The limiting curve that encloses all the other characteristics is the family $x = 0$, or family (0:1). The "central" family, or family (1:2), is the one that is reduced to the straight line $y = 0$ if $\epsilon = 0$ ($E \neq 0$). In our particular case we have $\omega_1/\omega_2 = 3/4$. Thus there are two resonant families, one stable (3:4), and one unstable (3.4u), that start at the origin ($y = 0$, $E = 0$). However if the ratio $\omega_1/\omega_2$ is different from 3/4 these families bifurcate either from the central family, or from the boundary.

There are several other families bifurcating from the stable families above. All these families are regular.

However there are also families that do not bifurcate from any of the regular families. Such families are generated in pairs (one stable and one unstable) at appropriate minimum values of $\epsilon$. Such are the families (i2:3), (i2:3u), the families (i3:4),(i3:4u), and others.

These results are typical of generic systems of two degrees of freedom.

In systems of three degrees of freedom we have also bifurcations of various families of periodic orbits, but their analysis is more complicated. Typical examples were given by Contopoulos and Magnenat (1985) and Contopoulos (1986a,b).

The study of periodic orbits will be used now in conjunction with the quantum mechanical eigenvalues and eigenfunctions.

## 3. Quantum Mechanical Eigenvalues and Eigenfunctions

If we set in Schrödinger's equation (1) for two degrees of freedom

$$\Psi = \phi \exp(-iEt/\hbar) , \tag{4}$$

where $E$ is the energy, we find the time independent Schrödinger's equation

$$-\frac{\hbar^2}{2m} \sum_{i=1}^{2} \frac{\partial^2 \phi}{\partial q_i{}^2} + V\phi = E . \tag{5}$$

This equation has solutions only for particular values of $E$, the eigenvalues $E_k$, that constitute the spectrum.

In order to find the spectrum in particular cases we start with an initial wave packet for $t = 0$

$$\Psi(\bar{q}, t = 0) = \sqrt{\frac{m}{\pi\hbar}} \sqrt{\omega_1 \omega_2} \times$$
$$\times \exp\left[ \sum_{i=1}^{2} \frac{i}{\hbar} p_{i0}(q_i - q_{i0}) - \frac{m\omega_i}{2\hbar}(q_i - q_{i0})^2 \right] , \tag{6}$$

where $\omega_1, \omega_2$ are the frequencies of the potential (3), and $(q_{10}, q_{20}, p_{10}, p_{20})$ are the initial conditions of a particular periodic orbit.

Such a wave function is called a minimum uncertainty wave function and represents two initial gaussian distributions along $q_1$ and $q_2$, around the point $(q_{10}, q_{20})$ in configuration space. The dispersions are

$$\Delta q_i = \sqrt{\hbar/2m\omega_i} \tag{7}$$

in $q_i$ , and

$$\Delta p_i = \sqrt{\hbar m\omega_i/2} \tag{8}$$

in $p_i$. The product of the dispersions is

$$\Delta q_i \Delta p_i = \hbar/2 . \tag{9}$$

Thus the initial gaussian wave function represents the minimum uncertainty of the initial conditions compatible with Heisenberg's principle.

This wave spreads in time, and its evolution is found by solving the time-dependent Schrödinger's equation (1) to derive the wave function $\Psi(\bar{q}, t)$ at time $t$.

Then we calculate the autocorrelation function

$$R(t) = \iint_{-\infty}^{\infty} \Psi^*(\bar{q}, t = 0)\Psi(\bar{q}, t) \, dq_1 dq_2 . \tag{10}$$

If we write the initial wave function as a linear combination of the eigenfunctions $\phi_k$ corresponding to the various eigenvalues $E_k$

$$\Psi(\bar{q}, t = 0) = \sum_k c_k \phi_k(\bar{q}) , \tag{11}$$

we find the wave function $\Psi(\bar{q}, t)$ at time $t$, in the form

$$\Psi(\bar{q}, t) = \sum_k c_k \phi_k(\bar{q}) \exp(-iE_k t/\hbar) . \tag{12}$$

Then the autocorrelation function becomes

$$R(t) = \sum_k \mid c_k \mid^2 \exp(-iE_k t/\hbar) . \tag{13}$$

The Fourier transform of $R(t)$ gives then a sum of delta functions at all $E_k$, i.e. it gives the spectrum

$$I(E) = \int_{-\infty}^{\infty} R(t) \exp(iEt/\hbar) \, dt = \sum_k \mid c_k \mid^2 \delta(E - E_k) . \tag{14}$$

We will not describe here the numerical details of calculating the spectrum. The important point is that this method allows to identify the eigenvalues $E_k$ and the amplitudes $\mid c_k \mid^2$ of the spectral lines.

In order to find the eigenfunctions $\phi_k$ corresponding to the eigenvalues $E_k$ it is more convenient to solve the time-independent Schrödinger's equation.

Finally we calculate the so-called "quantum Poincaré surfaces of section" as follows. For every eigenfuntion $\phi(x, y)$ in cartesian coordinates we calculate the function

$$\Phi(X, Y, P_x, P_y) = \iint_{-\infty}^{\infty} \phi(x, y) \times \tag{15}$$

$$\exp\left[ i\frac{P_x}{\hbar}(x - X) + i\frac{P_y}{\hbar}(y - Y) - \sigma_x(x - X)^2 - \sigma_y(y - Y)^2 \right] dx dy ,$$

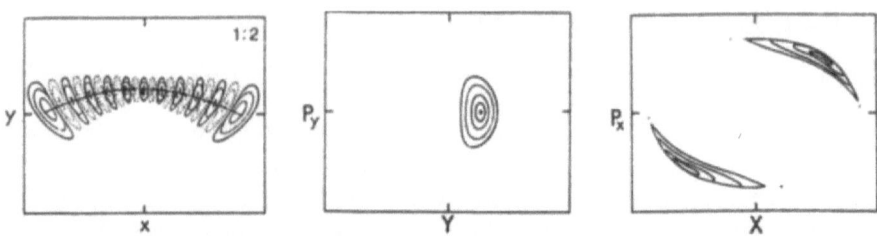

Fig. 2. (a) The eigenfunction $\phi_k(x, y)$ corresponding to the eigenvalue $E_k = 19.484$ super-imposed on the classical periodic orbit of type (1:2) for $E = E_k$. The solid lines represent positive values of $\phi$ at 30%, 50%, 70% and 85% of the maximum amplitude. The dotted lines give the corresponding levels of negative $\Psi$. (b) and (c) Quantum Poincaré surfaces of section on the planes $(Y, P_y)$ and $(X, P_x)$ respectively.

where $\sigma_x$, $\sigma_y$ are given quantities. This function allows the construction of diagrams similar to the classical Poincaré surfaces of section. If we write

$$\frac{1}{2}(P_x^2 + P_y^2) + V(X, Y) = E_k \,, \tag{16}$$

we can solve this equation for $P_x$ and insert its value in the equation

$$\Phi(X, Y, P_x, P_y) = \text{const} \,. \tag{17}$$

Then we set $X = 0$ and find curves

$$f(Y, P_y) = \text{const} \,, \tag{18}$$

for various values of the constant. These curve are like invariant curves on a classical surface of section. In this way we construct a $(Y, P_y)$ diagram containing many curves of the type (18), for a fixed value of the energy $E = E_k$. This is a quantum Poincaré surface of section.

In a similar way we can solve Eq. (16) for $P_y$ and insert its value in Eq. (17), setting then $Y = 0$, and find invariant curves of the form

$$\bar{f}(X, P_x) = \text{const} \,, \tag{19}$$

that form also quantum Poincaré surfaces of section. In the following we will give examples of quantum Poincaré surfaces of section both in the plane $(Y, P_y)$ and in the plane $(X, P_x)$.

Figure 2a gives the eigenfunction corresponding to the eigenvalue $E_k = 19.484$. Superimposed on the eigenfunction is the periodic orbit of type (1:2) for the particular value of the energy $E = E_k$. We see that the maxima and minima of the eigenfunction are almost exactly on the periodic orbit.

Figures 2b and 2c give quantum Poincaré surfaces of section corresponding to this eigenfunction, on the planes $(Y, P_y)$ and $(X, P_x)$. These are very similar to the corresponding classical Poincaré surfaces of section. They contain one set of

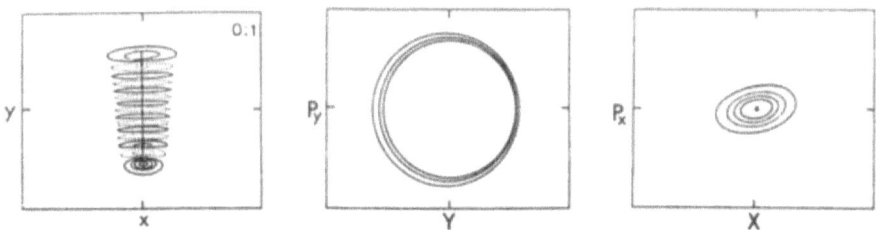

Fig. 3. As in Fig.2 for $E_k = 18.777$. The corresponding classical periodic orbit is of type (0:1) (the $y$-axis).

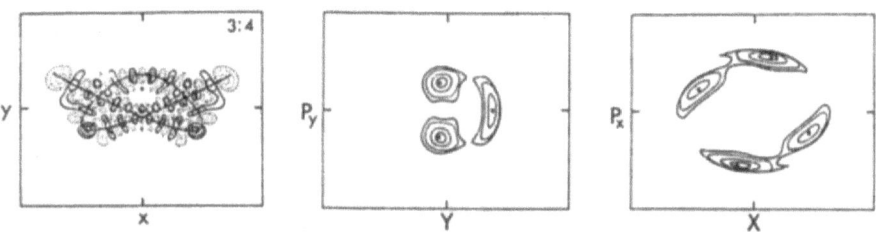

Fig. 4. As in Fig.2 for $E_k = 19.225$. The corresponding classical periodic orbit is of type (3:4).

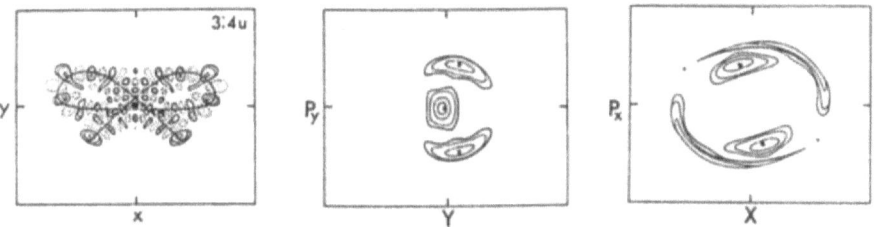

Fig. 5. As in Fig.2 for $E_k = 19.264$. The corresponding classical periodic orbit is the unstable orbit of type (3:4u).

closed curves in Fig. 2b (one island) and two sets of closed curves in Fig. 2c (two islands). The numbers of islands in the $(Y, P_y)$ and $(X, P_x)$ cases correspond to the type of resonance (1:2) of the periodic orbit.

A different form of eigenfunction is shown in Fig.3, for $E_k = 18.777$. This corresponds to a resonant orbit of type (0:1), which is exactly along the $y$-axis. The quantum Poincaré surface of section $(Y, P_y)$ consists of curves near the outer boundary, closing around the center, but not forming any islands. The quantum Poincaré surface of section $(X, P_x)$ consists of one set of islands around the center. These forms are consistent with the type of resonance in this case (0:1).

Figure 4 gives the eigenfunction and the Poincaré surfaces of section $(Y, P_y)$, $(X, P_x)$ for $E_k = 19.225$. This case corresponds to the resonance (3:4). Thus we find 3 islands in $(Y, P_y)$ and 4 islands in $(X, P_x)$. The main maxima and minima of the eigenfunction are on the classical periodic orbit (3.4). But there are some

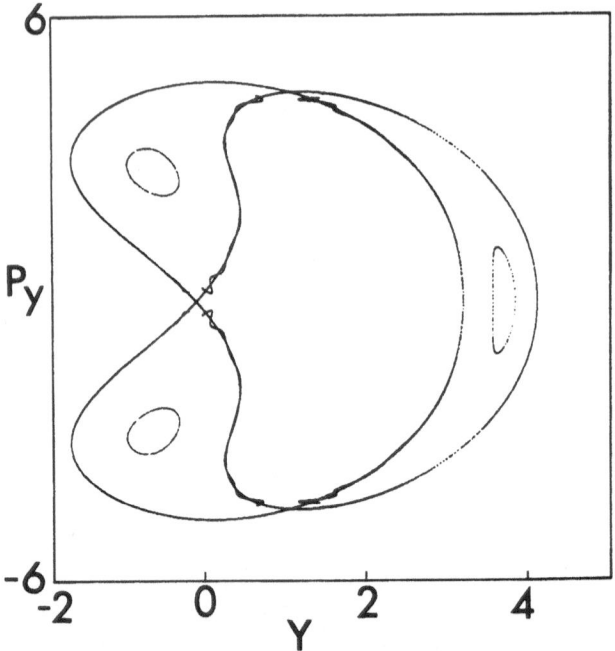

Fig. 6. The classical Poincaré surface of section $(Y, P_y)$ for $E = 19.264$ in the regions of the orbits (3:4) and (3:4u).

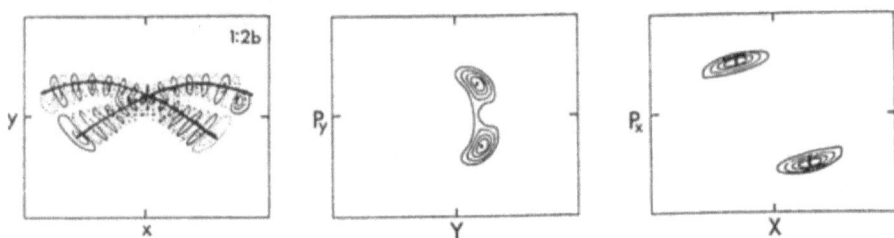

Fig. 7. As in Fig.2 for $E_k = 24.864$. There are two corresponding symmetric classical periodic orbits of type (1:2b).

secondary maxima and minima that are on both sides of this orbit. It seems that in this case the eigenfunction indicates some interference phenomena that may be due to other resonances.

Figure 5 gives the eigenfunction and the Poincaré surfaces of section $(Y, P_y)$, $(X, P_x)$ for $E_k = 19.264$. This case corresponds to the unstable periodic orbit (3:4u). It is surprising that the quantum Poincaré surfaces of section seem to indicate a stable (3:4) orbit, with 3 islands on the $(Y, P_y)$ plane and 4 islands on the $(X, P_x)$ plane. In Fig. 6 we give the classical Poincaré surface of section $(Y, P_y)$ for the orbits (3:4) and (3:4u) at the same energy $E = 19.264$. The unstable orbit is

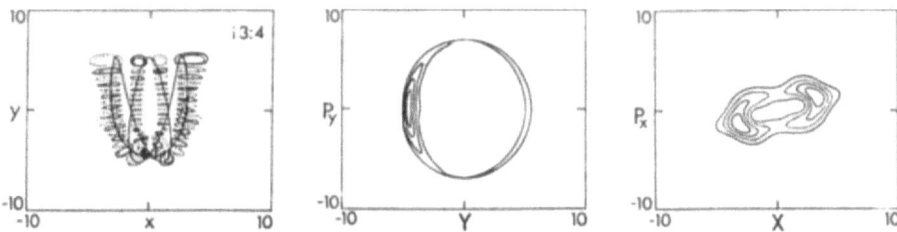

Fig. 8. As in Fig.2 for $E_k = 23.827$. The corresponding classical periodic orbit belongs to an irregular family of type (i3:4).

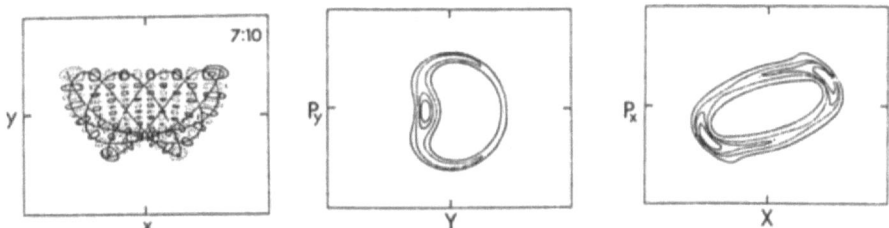

Fig. 9. As in Fig. 2 for $E_k = 20.265$. The corresponding classical periodic orbit is of type (7:10).

characterized by its asymptotic curves that surround the stable islands. It is obvious that the quantum Poincaré surface of section are quite different from the classical Poincaré surfaces of section in the case of the orbit (3:4u). The same is true for the $(X, P_x)$ plane, and for all eigenfunctions corresponding to unstable periodic orbits (scars).

Figure 7 gives the eigenfunctions and the Poincaré surfaces of section $(Y, P_y)$, $(X, P_x)$ for $E_k = 24.864$. In this case we have two different, but symmetric, periodic orbits of type (1:2b). The family (1:2b) bifurcates from the family (1:2), when the latter becomes unstable (Fig.1).

We see that the eigenfunction populates, in a symmetric way, the regions of both periodic orbits. Its maxima and minima are close to these orbits but not exactly on them. The quantum Poincaré surfaces of section form two islands in both $(Y, P_y)$ and $(X, P_x)$ planes, although the resonance is of type (1:2). This behaviour is due (1) to the fact that we have two orbits and not only one, and (2) to the fact that the axis $Y = 0$ intersects each periodic orbit only once. In fact the plane $(X, P_x)$ is not a classical Poincaré surface of section because it does not intersect all the orbits. Such a surface is called a "local surface of section". On the other hand the plane $(Y, P_y)$ is a real Poincaré surface of section, because of the symmetry of the potential with respect to the $Y$-axis, that forces all orbits to cross this axis (except for the boundary).

More generally in the case of a resonance (n:m) the number of islands on the quantum Poincaré surface of section $(Y, P_y)$ is equal to $n$ if there is only one

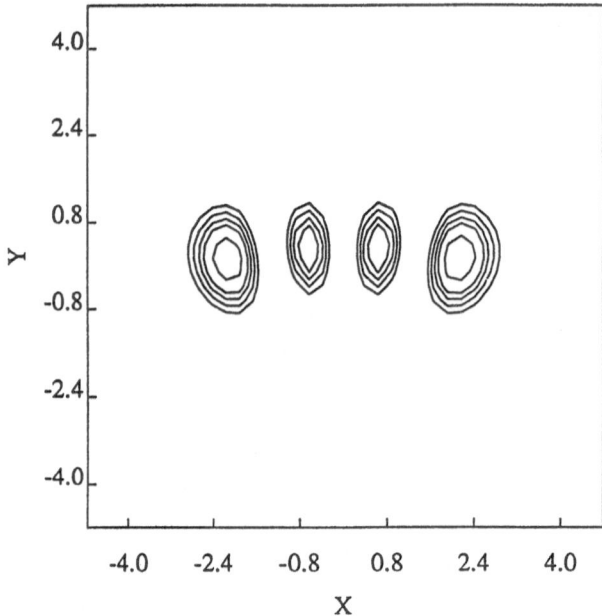

Fig. 10. Projection of the probability density $\mid \phi_k \mid^2$ on the $(x, y)$ plane in the 3-dimensional potential (20), with $\omega_1^2 = 0.9$, $\omega_2^2 = 1.6$, $\omega_3^2 = 0.4$, $\epsilon = -0.08$, $\eta = -0.01$, for an eigenvalue $E_k = 4.26$. This case corresponds to a resonant periodic orbit of type 3:6:2.

periodic orbit, but $2n$ if there are two symmetric periodic orbits. On the other hand the $(X, P_x)$ plane contains $m$ islands (or $2m$ islands in the case of two symmetric orbits) *or less* (if the surface $Y = 0$ has less intersections with the periodic orbit).

Figure 8 gives the eigenfunction and the quantum Poincaré surfaces of section $(Y, P_y), (X, P_x)$ for $E_k = 23.827$. This case corresponds to the irregular family (i3:4). There is no difference between this case and other cases corresponding to regular families. The maxima and minima of the eigenfunctions are close, but not exactly on the periodic orbit.

On the other hand the number of islands on the quantum Poincaré surfaces of section are not well defined. A first inspection gives two islands in the $(Y, P_y)$ plane and two in the $(X, P_x)$ plane. However in both cases the curves are deformed, in the sense that one may consider two secondary maxima on the left of Fig. 8b and two secondary maxima both on the left and the right of Fig. 8c. If we consider these to be the "real" islands, we have three islands in $(Y, P_y)$ and four islands in $(X, P_x)$, as found in the classical case.

However this case already emphasizes the difficulties that we encounter with higher order periodic orbits. The quantum Poincaré surfaces of section do not define clearly the same number of islands that we find in the classical case, but some islands seem to merge. This is due to the fuzziness of the quantum case that does not allow details smaller than $\hbar$ in action to be seen. For the same reason the forms of the eigenfunctions corresponding to higher order periodic orbits do not fit clearly with these orbits.

Nevertheless we found reasonable agreement between the forms of the eigenfunctions and periodic orbits up to order (5:7),(6:8),(5:10) and (7:10). But the agreement of higher order eigenfunctions with the corresponding periodic orbits is less satisfactory (Fig.9).

Similar results were found in a Morse potential (Farantos *et al.* 1989), and in more complicated potentials, representing various triatomic molecules.

These results were extended to systems of 3 degrees of freedom (Farantos and Founariotakis 1990). We are presently applying these methods to the simple 3-dimensional potential

$$V = \frac{1}{2}(\omega_1^2 x^2 + \omega_2^2 y^2 + \omega_3^2 z^2) + \epsilon x z^2 + \eta y z^2 ,\tag{20}$$

whose periodic orbits have been studied extensively in the classical case (Contopoulos and Magnenat 1985, Contopoulos 1986a,b). In this case also the eigenfunctions have their maxima and minima along particular periodic orbits. An example of such an agreement is shown in Fig. 10 (compare with Fig.2). We conclude that the close agreement between quantum mechanical eigenfunctions and classical periodic orbits is quite general.

## Acknowledgements

This research was supported by the Greek Ministry of Research and Technology (grant No. 89EΔ 86).

## References

Barbanis, B.: 1966, *Astron.J.*, **75**, 415.
Berry, M.V.: 1987, *Proc.R.Soc.Lond.*, **A413**, 183.
Contopoulos, G.: 1966, in Nahon, F. and Hénon, M. (eds) "Les Nouvelles Methodes de la Dynamique Stellaire", (CNRS, Paris) and *Bull.Astron.*, Ser. **3**, **2**, 223, 1967.
Contopoulos, G.: 1970, *Astron.J.*, **75**, 96.
Contopoulos, G.: 1986a, *Celest.Mech.*, **38**, 1.
Contopoulos, G.: 1986b, *Astron. Astrophys.*, **161**, 244.
Contopoulos, G. and Magnenat, P.: 1985, *Celest.Mech.*, **37**, 387.
Davis, M.J.: 1988, *J.Phys.Chem.*, **92**, 3124.
Davis, M.J. and Heller, E.J.: 1981, *J.Chem.Phys.*, **75**, 3916.
Farantos, S.C. and Founariotakis, M.: 1990, *Chem.Phys.*, **142**, 345.
Farantos, S.C. and Tennyson, J.: 1985, *J.Chem.Phys.*, **82**, 800.
Farantos, S.C., Founariotakis, M.: and Polymilis, C.: 1989, *Chem. Phys.*, **135**, 347.
Ford, J., Mantica, G. and Ristow, G.H.: 1991, *Physica D*, **50**, 493.
Founariotakis, M., Farantos, S.C., Contopoulos, G. and Polymilis, C.: 1989, *J.Chem.Phys.*, **91**, 1389.
Founariotakis, M., Farantos, S.C. and Tennyson, J.: 1988, *J.Chem.Phys.*, **88**, 1598.
Gutzwiller, M.C.: 1971, *J.Math.Phys.*, **12**, 343.

Gutzwiller, M.C.: 1990, *Chaos in Classical and Quantum Mechanics*, Springer Verlag.
Gutzwiller, M.C.: 1992, *Scient.Amer.* January, 26.
Heller, E.J.: 1984, *Phys.Rev.Lett.*, **53**, 1515.
Heller, E.J., Stechel, E.B. and Davis, M.J.: 1980, *J.Chem.Phys.*, **73**, 4720.
Hénon, M. and Heiles, C.: 1964, *Astron.J.*, **69**, 73.
Kleppner, D.: 1991, *Physics Today*, August, 9.
Rosenbluth, M.N., Sagdeev, R.Z., Taylor, J.B. and Zaslavskii, G.M.: 1966, *Nucl. Fusion*, **6**, 297.
Weissman, Y. and Jortner, J.: 1982, *J.Chem.Phys.*, **77**, 1486.
Zaslavskii, G.M. and Chirikov, B.V.: 1972, *Sov.Phys.Uspekhi*, **14**, 549.

# GENERALIZED LEAST-SQUARES ADJUSTMENT, A TIMELY BUT MUCH NEGLECTED TOOL

HEINRICH EICHHORN

*University of Florida, Gainesville, Florida, 32611-2015*

**Abstract.** A considerable amount of work has been done to make least-squares adjustment a more versatile tool, and it is the purpose of this article to give brief descriptions of all generalized least-squares adjustment techniques that have been developed so far. Algorithms have been established which will allow one to work with problems in which any number of (possibly correlated) observations occurs explicitly in any of the nonlinear condition equations, and in which at least a subset of the adjustment parameters may be known to be samples from a multivariate normal distribution with known covariance matrix but unknown modes. Furthermore, "filtering" algorithms have been developed which allow one to take advantage of previous results in finding revised adjustments when either the set of observations, or the set of adjustment parameters or both have been augmented over the sets on which the previous adjustment was based. The most recent generalization, in which the condition equations are not available explicitly but only through the differential equations whose solutions they are, was presented during this Colloquium. (Kallrath *et al.*, 1992)

**Key words:** Least-squares adjustment – data analysis

## 1. Introduction

The community of statistical adjusters consists of several sets: The TOLS (traditional ordinary least squares) set, the professional statisticians who practice LS, the mathematicians who expand the methods available for the solution of data adjustment problems and the scientists (and engineers) who are actively trying to find ways to arrive at the most accurate values of certain parameters. Unfortunately, the intersections of these sets are not very numerous; the intersection of all four of these sets may almost be the empty set.

The statistical adjustment of data, when practiced as point estimation, is not in the mainstream of interest of contemporary statistics. In the typical statistics text, TOLS is given short shrift as a closed subject, not worthy of the attention of first-rate research statisticians on their steep path to a successful and well-begranted academic (or other professional) career. When statisticians devote their efforts to data adjustment, they usually go in the direction of finding *robust* estimators—more about this below in Section 3.

## 2. The Practitioners of Least Squares

Statisticians understand by Ordinary Least Squares (OLS) the solution of least-squares problems in which exactly one of the uncorrelated and equally precise observations occurs in each of the linear condition equations. It is usually practiced by *experimental scientists* (*e.g.*, certain subsets of the astronomers), whose primary contributions consist either of the design and construction of scientific measuring devices or of making careful measurements toward the estimation of certain parameters. These people cannot—given the realities of scientific work—spend much time keeping up with the progress of adjustment theory and applying the results thereof to their work; this would keep them from applying the necessary time and effort to making significant contributions to new knowledge. In practice, they must estimate their parameters with the tried-and-true methods they learned as students or acquainted themselves with at some point in their careers. As a rule, these methods are not "bad", they give plausible and credible results, the procedures for their application are straightforward and ready programs for this purpose are often available. Unfortunately, these methods do normally not lead to the best available [*i.e.*, most accurate—in the technical sense as being closest to the (unavailable) truth—] estimates and therefore waste a more or less significant portion of the effort that was expended on making precise and accurate observations.

*Professional statisticians* pay little attention to least squares, they usually consider it a tiny subsection of the field of point parameter-estimation.

*Mathematicians* have devoted considerable effort to the development of techniques required to perform the computations—frequently very heavy arithmetic—in least-squares adjustments, *e.g.*, for the inversion of symmetric matrices (Cholesky, Banachiewicz). The development of techniques to solve OLS problems directly from the condition- (not constraint!) equations without establishing the normal equations (*cf.* Lawson and Hanson 1974) has occupied the minds of some superior mathematicians. This approach also appears to have the advantage, that the large condition number of the matrix of a poorly conditioned system of condition equations is not squared as it would be if the problem were conventionally solved by forming normal equations. The methods for finding solutions of systems of nonlinear equations, another field toward whose progress many mathematicians have contributed, are directly applicable to the solution of least-squares problems that are based on nonlinear condition equations. Finally, taking advantage of the sparseness of matrices such as occur frequently in least-squares problems ordinarily produces a significant gain in the accuracy of the elements of their inverses, *cf.* Branham (1990).

Unfortunately, most users of the least-squares adjustment technique regard it as a mere trade tool, which comes pre-packaged and in various degrees of sophistication. If the tool is complicated, learning how to use it may consume a lot of time and in addition be difficult, especially when—as happens not infrequently—the author of the instructions for using the tool is less than perfectly lucid. One also encounters often the—much wiser than immediately apparent—attitude that "if it ain't broken, don't fix it!", which produces a reluctance in much of the community of users to abandon a tried-and-true, even though not optimally accurate adjustment technique in favor of a new, untried and usually not yet debugged one even though it promises to deliver more accurate parameter estimates, albeit at the cost of a significantly more complicated programming effort.[1]

---

[1] I shall charitably pretend that no one would prefer a simple but slightly imperfect method over a "perfect" one just because the latter requires a much larger—by orders of magnitude—arithmetic effort. After all, we do nowadays have computers. There are, however, investigators who still cling

Those scholars, finally, who *devote a significant portion of their productive working hours to the development of improved generalized least-squares adjustment techniques* frequently do not manage to apply their results to real problems. If past experience is an indicator, the mere publication of a paper describing improved methods without numerical or analytical examples is time ill spent. Because the titles of such papers quite often give no clear and complete indication of their contents and potential significance, they are more often than not overlooked exactly by those investigators whose results would profit significantly by the application of the methods suggested in them.

Peter Brosche (1985) has very perceptively described this, and he suggests the name *"eucriny"* for that science which concerns itself with the most accurate and proper analysis of directly observed data.

## 3. Remarks on Robustness

A significant portion of contemporary research in the statistics of point estimators deals with finding "robust" estimators. These are, crudely speaking, estimators which depend but little on the distribution function of the individual data estimates, that is to say, which depend little on the probability density distribution function of the observation errors. This can be justified as follows:

An adjustment must be based on a *model function*, that is, a rigorous functional relationship between the observables and the adjustment parameters.[2] In many instances, however, the exact form of this relationship is not only not known, but there is frequently also no certainty as to exactly how many adjustment parameters the function requires. In these cases, the model function basically represents an interpolation formula with as many adjustment parameters as are necessary to produce "acceptable" residuals (whatever this means to a particular investigator). Such formulas may then be regarded as analytical approximations to the unknown exact relationships between observables and parameters. The exact physical or geometrical meaning of the adjustment parameters in such approximation functions is usually hidden. When the (inaccessible) actual observation errors follow a certain distribution, the error estimates produced by carrying out the adjustment with an analytical approximation to the (also unavailable) accurate model function will no longer obey the same distribution function as the actual residuals. Estimators which depend little on the error distribution function will then come closer to the "truth", if the form of the distribution function has little influence on them. There is, however, another side to this coin:

to techniques whose sole justification is a reduction of the arithmetic effort to where it is manageable on manual desk calculators, *e.g.*, the computing of "normal points".

[2] Until recently, this model function had to be explicitly st ated in analytical form. During this colloquium, however, Kallrath *et al.* (1992) have developed a least-squares algorithm in which the model function need not be explicitly stated in analytical form, but may be the solution of a specified and explicitly stated differential equation.

Suppose we have a set $\{l_1, l_2, \cdots, l_n\}$ of data points, all estimates of the same quantity $l$, and now we look for a principle that drives the adjustment, which means finding that estimate which has a better chance of differing from $l$ less than any other. We do this by finding that $l$ which satisfies certain conditions to which the $l_\nu - l$ are subject. We might, for example, compute $l$ thus:

$$l = \frac{1}{n} \sum_{\nu=1}^{n} l_\nu, \tag{1}$$

which satisfies

$$\sum_{\nu=1}^{n} (l_\nu - l)^2 = Min! \tag{2}$$

The $l$ found from Eq. (1) is the arithmetic mean, and we see that $dl = \frac{1}{n} dl_\nu$; a change in $l_\nu$ propagates into $l$.

We might, as one of the alternatives, choose $l$ such that

$$l_\mu < l < l_\lambda; \quad \mu = 1, \ldots, \frac{n}{2}, \lambda = \frac{n}{2} + 1, \ldots, n \qquad \text{for } n \text{ even}$$
$$l_\mu < l = l_\kappa < l_\nu; \quad \mu = 1, \ldots, \frac{n-1}{2}, \kappa = \frac{n+1}{2}, \lambda = \frac{n+3}{2}, \ldots, n \text{ for } n \text{ odd.}$$

We note in this case, in which we recognize $l$ as the *median* of the $l_\nu$, that $l$ is totally insensitive to any changes in the values of the $l_\nu$, except for $l_\kappa$ for odd $n$ and $l_{\frac{n}{2}}$ and $l_{\frac{n}{2}+1}$ for even $n$, as long as the number of $l_\nu$ with $\nu < \frac{n}{2}$ and $\nu > \frac{n}{2}$ is not altered. This is extreme robustness: The median will not change, regardless of the values of the outliers.

There is no doubt, that such estimators have their very important place, but not in the context of parameter estimation. The fundamental assumption in parameter estimation is, that the *target parameter*—the parameter we wish to estimate— actually exists and has an unambiguous numerical value. It must—obviously—be the aim of any data adjustment to find that estimate which has the best chance to be closest to this value. This suggests, that the driving principle for the adjustment should be the *principle of maximum likelihood* and **no other.**

## 4. Power Methods

We may formulate the problem thus: Given a matrix $M_{m \times n}$; $m \geq n$ and an $m$-vector $l$, find that $n$-vector $x$ which satisfies

$$\sum_{\mu=1}^{m} |\sum_{\nu=1}^{n} (m_{\mu\nu} x_\nu - l_\mu)|^k = Min! \tag{3}$$

Note, that this is not always the same as

$$|Mx - l|^k = Min!$$

The vector $x$ thus found satisfies

$$Mx - l = 0 \tag{4}$$

if and only if $R(M) = R(M|l)$; this would happen only in a freak case, because $R(M) \leq n$, (ordinarily $R(M) = n$), and $R(M|l) \leq n + 1$ (if $m \geq n + 1$); ordinarily $R(M|l) = n + 1$ and the system (4) therefore has then no solution.

To satisfy the condition (3) for a given (integer) $k$ is called *find $x$ using* $L_k$. *Ordinary Least Squares* (OLS) is thus the determination of an estimate $\hat{x}$ of $x$ in Eq. (4) such that

$$|M\hat{x} - l|^2 = (\hat{x}^T M^T - l^T)(M\hat{x} - l) = Min!,$$

which leads to the well known result

$$\hat{x} = M^+ l, \tag{5}$$

where the pseudoinverse $M^+$ of $M$, in this case $(m > n)$, is given by

$$M^+ = (M^T M)^{-1} M^T. \tag{6}$$

This intimate connection between the pseudoinverse and OLS—$L_2$— has been known for some time.

If we were interested in a robust method, we might prefer $L_1$; this, however, does not lead to an unambiguous result, *cf.* the case $m = 2$, $n = 1$ and $M = \begin{pmatrix} 1 \\ 1 \end{pmatrix}$ and $l = \begin{pmatrix} a \\ b \end{pmatrix}$. The system Eqs. (4) then yields $x = a$, $x = b$ and we look for that $\hat{x}$ which satisfies Eq. (3) for $k = 1$, that is

$$S = |\hat{x} - a| + |\hat{x} - b| = Min!$$

Now we see, that

$$S = x - a + b - x = b - a$$

for any $x$ which satisfies $a \leq x \leq b$ so that $\hat{x}$ is not uniquely determined, but $S$ is the minimum as one can see if one considers an $x$ outside the interval $[a, b]$. This lack of determinacy is alleviated as $m$ and $n$ grow, but never totally eliminated. The $L_1$ method, much more robust than the $L_2$ method—least squares—is useful for identifying observing blunders (and outliers in general), but will certainly not find the likely most accurate estimate $\hat{x}$ for $x$ when $l$ is regarded as a bias free vector. Branham (1990) gives an an algorithm for finding the $L_1$ solution.

## 5. Ordinary Least Squares (OLS)

We base the derivation of the algorithm on such assumptions that it is possible to generalize from them.

Suppose we have a matrix $A_{m \times n}$ and there is an $m$-vector $x$ of directly observable quantities. Let the $m$-vector $x_0$ be a vector of bias-free directly observed estimates of $x$, such that

$$x = x_0 + \xi. \tag{7}$$

Let the components of $\xi$ be bias-free and randomly distributed, following the Gaussian normal probability density distribution function

$$\varphi(\xi) = Ce^{-\frac{1}{2\sigma}\xi^T\xi}, \tag{8}$$

where $\sigma$ is the variance of the components of $\xi$. Let there also be a vector $a$, such that the *accurate* equations

$$Aa = x \tag{9}$$

are rigorously satisfied, meaning that

$$Aa - x_0 = \xi. \tag{10}$$

If we now want to determine that vector $\xi$ for which the probability density distribution function reaches a maximum, but in such a way that the conditions laid down by the *condition equations* Eq. (9)[3] are *rigorously* satisfied, we proceed as follows.

We are looking for that $\xi$ at which the probability density distribution function reaches a maximum, *i.e.*, $\varphi(\xi) = Max!$ This is equivalent to

$$\xi^T\xi = Min!, \tag{11}$$

subject to the conditions $Aa = x$. By using Eq. (10) in Eq. (11), $\xi$ is eliminated from Eq. (9) and we get

$$S = (Aa - x_0)^T(Aa - x_0) = a^TA^TAa - 2a^TAx_0 + x_0^Tx_0 = Min!$$

After the elimination of $\xi$, the only free parameters in this equation are the components of $a$, so that the necessary equations for the determination of that $a$, which minimizes $\xi^T\xi$ become

$$\left(\frac{\partial S}{\partial a}\right) = 2A^TA\hat{a} - 2A^Tx_0 = 0$$

or

$$\hat{a} = (A^TA)^{-1}A^Tx_0. \tag{12}$$

$\hat{a}$ is, of course, not $a$, but an estimate for it. Estimates $\hat{\xi}$ for $\xi$ follow from Eq. (10):

$$\hat{\xi} = A\hat{a} - x_0. \tag{13}$$

Eq. (12) will obviously have an unambiguous solution only for $m \geq n$, otherwise $A^TA$ would be rank deficient and therefore singular.

It can be shown, that $\sigma(A^TA)^{-1}$ is the covariance matrix of $\hat{a}$. One can also derive an expression for the covariance matrix of the components of $\hat{\xi}$; this is not important in our context and will therefore be omitted.

Eq. (12) is also the solution of the problem to minimize $|\xi| = |Aa - x_0|$, whence the name "least squares".

---

[3] Many authors call these *observation equations*. I prefer to call them *condition* equations for the following reason: The mathematical context in which they appear in the problem is clearly that of condition equations which restrict the minimum value of the function $\xi^T\xi$ of the components of the vector $\xi$.

## 6. First Group of Generalizations

Note that the minimization of $|\xi|$ by OLS encompasses only a very simple and straightforward case:

1. All components of $x_0$ are presumed to be equally precise (variance $\sigma$) and uncorrelated, which leads to the distribution function $\varphi(\xi)$ given by Eq. (8).
2. The condition equations (9) relating $a$ and $x$ are linear.
3. Exactly one component of $x$ occurs in each of the condition equations (9).

We now generalize the problem by abandoning the restrictive assumptions 1. through 3. and now allow, that

- The components of $x_0$ are correlated and not equally precise: Their covariance matrix is $\sigma$, which we (must) regard as known, at least except for a factor.
- The $n$-vector $a$ of the adjustment parameters ("unknowns" or "adjustment unknowns") and the $m$-vector $x$ of the observables are related by a system of *not necessarily linear* equations

$$F(x, a) = O, \tag{14}$$

- which implies that any component of the $p$-vector $F$ may explicitly depend on any number (including zero) of components of the vector $x$ of observables. (The simplest case is the fitting of a straight line $ax + by + 1 = 0$, when both $x$ and $y$ are observed and subject to possibly correlated observing errors.)

In this case, the probability density distribution function of $\xi$ is

$$\varphi(\xi) = C e^{-\frac{1}{2}\xi^T \sigma^{-1}\xi},$$

instead of Eq. (11) we have

$$S = \xi^T \sigma^{-1}\xi = Min! \tag{15}$$

and Eq. (14) replaces Eq. (9). In this case, the number $p$ of components of the condition equations (14) must not exceed the number $m$ of components of the vector $x$ of observables; it is thus necessary that $p \leq m$. When $p < m$, which is usually the case, we can no longer eliminate $\xi$ by expressing it in terms of $a$, as we did in the case of OLS. The problem is readily solved, however, if one considers that $S(\xi)$ is to be minimized while $\xi$ is subjected to the equations of condition (literally!)

$$F(x_0 + \xi, a) = O \tag{16}$$

as a consequence of Eqs. (7) and (14).

Following Lagrange, we introduce a $p$-vector $\Lambda$ and remember that at the solution, where Eq. (16) is satisfied, $F = O$ and the product of $F$ with any number is still zero, so that Eq. (17) will be correct (at the solution) when Eqs. (15) and (16) are fulfilled, and therefore

$$S = \xi^T \sigma^{-1}\xi - 2\Lambda^T F. \tag{17}$$

We note that we have $m + n$ free parameters in Eq. (17), which we can vary in order to find the $p$ components of $\Lambda$. Writing

$$\mathbf{X}_{p \times m} = \left(\frac{\partial \mathbf{F}}{\partial \mathbf{x}}\right) \tag{18}$$

at the solution point, we get

$$\left(\frac{\partial S}{\partial \mathbf{x}}\right) = 2\sigma^{-1}\xi - 2\mathbf{X}^T \Lambda = O, \tag{19}$$

whence, since $\sigma^{-1}$ is evidently nonsingular,

$$\xi = \sigma \mathbf{X}^T \Lambda. \tag{20}$$

Also, with

$$\mathbf{A}_{p \times n} = \left(\frac{\partial \mathbf{F}}{\partial \mathbf{a}}\right), \tag{21}$$

we get

$$\left(\frac{\partial S}{\partial \mathbf{a}}\right)_{n \times 1} = \mathbf{A}^T \Lambda = O. \tag{22}$$

Using Eq. (20) in Eq. (16), we have

$$F(\mathbf{x}_0 + \sigma \mathbf{X}^T \Lambda, \mathbf{a}) = O. \tag{23}$$

The rigorous Eqs. (22) and (23) form a system of $n + p$ equations, called "normal equations", for the $n + p$ unknown components of $\mathbf{a}$ and $\Lambda$. $\Lambda$ is, of course, of interest only because it is needed to compute $\hat{\xi}$ by Eq. (20).

These normal equations are linear exactly then, when the condition equations are linear in $\mathbf{x}$ and $\mathbf{a}$ (and contain no cross terms); in every other known case they are unmanageable. Consider that X itself is a function of $\mathbf{a}$ and $\xi$, thus also of $\Lambda$ by Eq. (20), and that $\Lambda$ occurs explicitly in Eq. (23).

It was pointed out by Eichhorn & Clary (1974) and by Jefferys (1980, 1981) that these equations must be satisfied *at the solution, i.e.* that $\hat{\mathbf{a}}$ and $\hat{\xi}$ must then have attained their correct values. A rigorous solution is plausibly impossible, thus one must proceed iteratively (or by some equivalent procedure).

To do this, we linearize the condition equations (23) for which purpose we need approximations for $\mathbf{a}$ and for $\mathbf{x}$. For the latter, the observations $\mathbf{x}_0$ may serve; but for the former, we must somehow get an approximation $\mathbf{a}_0$ to $\mathbf{a}$, such that $\mathbf{a} = \mathbf{a}_0 + \alpha$. We denote by $\mathbf{X}_0$, $\mathbf{A}_0$ and $F_0$ these matrices (and this vector) evaluated at $\mathbf{a} = \mathbf{a}_0$ and $\mathbf{x} = \mathbf{x}_0$. Thus we obtain the linearized normal equations from Eqs. (22) and (23), considering Eqs. (18) and (21)

$$\begin{aligned} \mathbf{X}_0 \sigma \mathbf{X}_0^T \Lambda + \mathbf{A}_0 \alpha + F_0 &= O + 0(2) \\ \mathbf{A}_0^T \Lambda &= O + 0(2) \end{aligned} \tag{24}$$

This linear system in the components of $\Lambda$ and of $\alpha$ has an unambiguous solution if the problem is physically and/or geometrically at all determinate. Let these solutions be $\alpha_1$ and $\Lambda_1$; from the latter, we get $\xi_1$ from Eq. (20). Now that we have these first approximations to $\alpha$ and $\xi$, we start—if necessary—the iteration process by evaluating $X, A$ and $F$ at $a_1 = a_0 + \alpha_1$ and $x_1 = x_0 + \xi_1$; this leads to a sequence

$$
\begin{aligned}
X_\nu \sigma X_\nu^T \Lambda_\nu + A_\nu \alpha_\nu + F_\nu &= O + 0(2) \\
A_\nu^T \Lambda_\nu &= O + 0(2) \\
\xi_\nu &= \sigma X_\nu^T \Lambda_\nu \\
a_{\nu+1} &= a_\nu + \alpha_\nu \\
x_{\nu+1} &= x_\nu + \xi_\nu,
\end{aligned}
\tag{25}
$$

which may be judged as having converged when the series $\sum a_\nu$ and the series $\sum \xi_\nu$ have converged and—not to forget— $F_\nu \to 0$.

## 7. The Solution of the Normal Equations

The matrix

$$
\begin{pmatrix} X\sigma X^T & A \\ A^T & O \end{pmatrix}
$$

of the linearized normal equations (24) is always nonsingular for realistic problems. The submatrix $X\sigma X^T$, however, can be nonsingular only when each individual condition equation, that is, each of the component equations of the system Eqs. (16) depends on at least one of the components of $x$ (that is, if there are no pure *parameter constraints*), otherwise it will evidently be rank deficient.

For nonsingular $X\sigma X^T$, we get

$$
\Lambda = -(X\sigma X^T)^{-1}(A\alpha + F_0)
\tag{26}
$$

which, inserted in the second of Eqs. (24), leads to an explicit solution for $\alpha$:

$$
\alpha = -[A^T(X\sigma X^T)^{-1}A]^{-1}A^T(X\sigma X^T)^{-1}F_0,
\tag{27}
$$

and $\xi$ is again computed by Eq. (20), from which we derive

$$
\xi = \sigma X^T(X\sigma X^T)^{-1}\{A[A^T(X\sigma X^T)^{-1}A]^{-1}A^T(X\sigma X^T)^{-1} - I\}F_0.
\tag{28}
$$

Even though, as just stated, Eqs. (26) – (28) become invalid when the condition equations contain pure parameter constraints, we know that the entire matrix of the system of normal equations is nonsingular as long as the problem itself is physically and/or geometrically well defined. In this case, we can accomplish the inversion by partitioning the matrix, *cf.* Brown (1955) and Jefferys (1980,1981).

A further complication arises, when *only* parameter constraints render the problem [and thus the matrix of the system Eqs. (24)] nonsingular. Appropriate partitioning of the system matrix will also in this case make the inversion possible, *cf.* Jefferys (1979), Eichhorn (1988).

The explicit computation of the covariance matrix of the parameter estimates $\hat{a}$ and the correction estimates $\hat{\xi}$ must be an integral part of a properly carried out adjustment, apart from the fact that these matrices are needed for a number of applications, as we shall see below.

## 8. Further Generalization of the Problem

So far, we have regarded the target parameters ("adjustment unknowns" in traditional astrometrese) as completely independent of each other, as well as unrestricted. We now consider the case, that the adjustment parameters consist of two groups: $c$, the "ordinary" adjustment parameters and $b$, those parameters which may, in groups, themselves be random variates with known covariance matrices. Accordingly, we write $a^T = (b^T\ c^T)$. We now assume the following structure for the components of $b$:

$$
b = \begin{pmatrix}
b_1 + \beta_{11} \\
b_1 + \beta_{12} \\
\vdots \\
\hline
b_1 + \beta_{1n_1} \\
b_2 + \beta_{21} \\
b_2 + \beta_{22} \\
\vdots \\
\hline
b_2 + \beta_{2n_2} \\
\vdots \\
\hline
b_n + \beta_{n1} \\
b_n + \beta_{n2} \\
\vdots \\
b_n + \beta_{nn_n}
\end{pmatrix}
=
\begin{pmatrix}
b_1 e_{n_1} + \beta_1 \\
b_2 e_{n_2} + \beta_2 \\
\vdots \\
b_n e_{n_n} + \beta_n
\end{pmatrix},
\tag{29}
$$

where $e_\nu$ is a $\nu$-vector whose every component is 1. We may accordingly write $b^T = b_c^T + \beta^T$. $b$ and its summands are $\sum_{\nu=1}^{n} n_\nu$-vectors, whose meaning is clear from the foregoing.

We now assume, that $\beta$ is a bias free random vector, whose components are distributed following a multivariate Gaussian distribution $\psi$ and whose covariance matrix is $\rho$:

$$
\psi(\beta) = K e^{-\frac{1}{2}\beta^T \rho^{-1}\beta}.
\tag{30}
$$

The following examples should convince that this is not just contrived: Suppose we want to estimate the individual distances of the member stars in a moving cluster from observing their radial velocities and the components of their proper motions. (These constitute the vector $x$ or $x_0$, as the interpretation goes.) Conventionally, one would solve for the components of the uniform velocity of the cluster's center of gravity—presuming that each member star moves with the same velocity—and individually for the distance of each star (or, when the material is scarce, for the distance of the cluster's center of gravity, assuming that this would be a sufficiently accurate approximation to the distance of each star). We know, however, from physics and geometry, that it would be more accurate to assume the velocity components to follow a trivariate Gaussian normal distribution about the velocity of the center of gravity, and that the stars' distances scatter—in first approximation normally—around the distance to the center. We would have $n = 4$ in this case: $b_1, b_2$ and $b_3$ are the $x$, $y$ and $z$ velocity components, respectively, of the center of gravity and $b_4$ is the distance to the cluster's center. $n_\nu$ (for $\nu = 1, 2, 3, 4$) is the number of stars involved in the analysis. The $\beta_{1\nu}, \beta_{2\nu}$ and $\beta_{3\nu}$ are the stars' velocity components with respect to that of the center of gravity, and the $\beta_{4\nu}$ are the excesses of the individual stars' distances over that of the cluster's geometrical center. We can estimate the components of $\rho$ (or at least its diagonal elements) from getting an idea of the velocity dispersion from the observed radial velocities, and of the cluster's depth (related to the variance of $\beta_{4\nu}$) from its angular diameter in conjunction with an (at least approximately) known distance. Our model for $b$, as laid down in Eq. (29) fits this situation almost exactly.

The solution of the problem is nearly trivial when one considers that the joint distribution function of our random vectors is $\varphi\psi$, which leads to the statement that the function to be minimized is

$$\xi^T \sigma^{-1} \xi + \beta^T \rho^{-1} \beta. \tag{31}$$

From then on, everything follows analogously to the previously established pattern. The solution was published some time ago (Eichhorn 1978), but this powerful tool has been largely ignored by the community. There are situations galore in which it would be applied with profit.

## 9. Filtering

This refers to the utilization of available estimates which could be improved upon by the incorporation of additional observations which were not utilized for calculating them.

In principle, one could gather *all* observations, including those already utilized for computing the existing estimates and adjust them by the standard procedures to obtain new, improved estimates. Sometimes, however, the only still available data are the original parameter vector estimate $\hat{a}$ and its covariance matrix $Q$, but not the original equations of condition. In addition, we now have equations of condition

$$F(x, a) = O, \tag{32}$$

which may or may not be the same ones used for the computation of the existing estimate $\hat{a}$, with $a$ being the vector by which we want to replace it and $x$ the observable quantities whose estimates $\hat{x}_0$ we have directly observed.

We now regard the existing previous estimate $\hat{a}$ as an observation whose co-variance matrix $(Q)$ we know and interpret $a$ as the vector of target parameters; this gives

$$\hat{a} - a = 0 \tag{33}$$

as an additional set of equations of condition. Regarding the Eqs. (32) and (33) as the new complete set of condition equations, we see that the rôles of $X$, $A$, $F_0$, $\sigma$, $a_0$ and $x_0$ are assigned as follows:

$$X : \begin{pmatrix} I & 0 \\ 0 & X \end{pmatrix} \quad A : \begin{pmatrix} -I \\ A \end{pmatrix} \quad F_0 : \begin{pmatrix} 0 \\ F_0 \end{pmatrix}$$

$$\sigma : \begin{pmatrix} Q & 0 \\ 0 & \sigma \end{pmatrix} \quad a_0 : \hat{a} \quad x_0 : \begin{pmatrix} \hat{a} \\ x_0 \end{pmatrix}.$$

Remember also the "inversion lemma" from matrix theory:

$$(A + BD^{-1}C)^{-1} \equiv A^{-1} - A^{-1}B(D - CA^{-1}B)^{-1}CA^{-1}. \tag{34}$$

This identity is going to play a key rôle in our computations.

Applying Eq. (27), we now obtain

$$\alpha = [Q^{-1} + A^T(X\sigma X^T)^{-1}A]^{-1}A^T(X\sigma X^T)^{-1}F_0. \tag{35}$$

In many applications, the order of the matrix $Q$ is larger than that of $X\sigma X^T$. If we remember, that $Q$ is (in most standard situations) proportional to the *inverse* of the matrix of the normal equations, it is clear, that $Q^{-1}$ will already be known. But it is not even needed, because according to Eq. (34),

$$[Q^{-1} + A^T(X\sigma X^T)^{-1}A]^{-1} = Q - QA^T(X\sigma X^T + AQA^T)^{-1}AQ, \tag{36}$$

which requires only the inversion of a matrix of the order of the number of new observations. This is the basic idea of Kalman filtering which gives, accordingly, the same results that would have been obtained by a regular least squares adjustment. Note that the covariance matrix $Q$ of the previously existing estimates need not be nonsingular, and that this method can be applied successively, incorporating the "new" observations in small groups.

An even more general situation arises, when new observations $x$, which could have been (but were not) utilized in the original estimation of a vector $a$ of parameters had they been available for this purpose, require in addition the first estimation of a vector $b$ of *additional* parameters. Eq. (32) must then be replaced by

$$F(x; a, b) = O, \tag{37}$$

where $b$ is the vector of the now essential, additional target parameters. This situation occurs not infrequently in some problems of astrometry and of geodesy which require what is called a "block adjustment". The correspondences in this case are as follows:

$$X : \begin{pmatrix} I & 0 \\ 0 & X \end{pmatrix} \quad A : \begin{pmatrix} -I & 0 \\ A & B \end{pmatrix} \quad F_0 : \begin{pmatrix} O \\ F_0 \end{pmatrix}$$

$$\sigma : \begin{pmatrix} Q & 0 \\ 0 & \sigma \end{pmatrix} \quad a_0 : \begin{pmatrix} a_0 \\ b_0 \end{pmatrix} \quad x_0 : \begin{pmatrix} \hat{a} \\ x_0 \end{pmatrix} \quad \alpha : \begin{pmatrix} \alpha \\ \beta \end{pmatrix},$$

with $Q = \left( \frac{\partial F}{\partial b} \right)$.

By strictly following the algorithm outlined above, we obtain a system in $(\alpha^T \; \beta^T)$ whose matrix is

$$\begin{pmatrix} Q^{-1} + A^T(X\sigma X^T)^{-1}A & A^T(X\sigma X^T)^{-1}B \\ B^T(X\sigma X^T)^{-1}A & B^T(X\sigma X^T)^{-1}B \end{pmatrix}.$$

Eichhorn (1989a) has shown that this matrix, too, can be inverted without having to invert any matrix of an order larger than that of $X\sigma X^T$. This is done by applying the formulas for the inversion of a matrix partitioned into a $2 \times 2$ matrix of matrices with quadratic matrices on the diagonal, cf. e.g., Eichhorn (1988). This is a generalization of the Kalman filter.

## 10. Condition and Efficiency

The matrices which must be inverted to solve the normal equations of a least-squares problem are normally symmetric and positive definite.

The *condition number* $\alpha$ is in essence the factor by which the inaccuracy in the elements of a matrix is multiplied in the elements of the inverse. For positive definite matrices $\alpha = \lambda_{max}/\lambda_{min}$, the ratio of the matrix' largest to its smallest eigenvalue.

The *efficiency* $e$ of a vector $a$ whose covariance matrix is $Q$ was defined by Eichhorn (1989b) as

$$e = \sqrt[n]{\frac{|Q|}{\prod_{\nu=1}^{n} q_{\nu\nu}}}, \tag{38}$$

which is the $n$-th root of the ratio of the product of the eigenvalues of $Q$—the product of the variances of the components when $a$ is decoupled—divided by the product of the variances of the components of $a$, *i.e.*, that of the terms on the main diagonal of $\mathbf{Q}$.

Both a large condition number and a low efficiency indicate that the system of equations, which was solved for $a$ does not produce a very accurate estimate for $a$, but for different reasons.

Low efficiency signals large correlations between the components of $a$, which originate invariably from off-diagonal terms in $\mathbf{Q}$. The covariance matrix of a vector with efficiency 1 will therefore be diagonal.

$\mathbf{Q}_{n \times n}$ is symmetric and positive definite and therefore can be written in the form

$$\mathbf{Q} = \mathbf{XDX}^T, \tag{39}$$

where $\mathbf{D} = diag(\lambda_n, \cdots, \lambda_1)$ (The eigenvalues $\lambda_\nu$ of $\mathbf{Q}$ are ordered such, that $\lambda_n \geq \lambda_{n-1} \geq \cdots \geq \lambda_1$) and $\mathbf{X}$ is the orthogonal matrix of its normalized eigenvectors (belonging to the $\lambda_\nu$ as in the sequence indicated above).

Since covariance matrices transform according to the rule

$$\mathbf{Q}_{bb} = \left(\frac{\partial b}{\partial a}\right) \mathbf{Q}_{aa} \left(\frac{\partial b}{\partial a}\right)^T \tag{40}$$

for $b = b(a)$, the covariance matrix of the orthogonal transform $b$

$$b = \mathbf{X}^T a \tag{41}$$

of $a$ is the diagonal matrix $\mathbf{D}$ of the eigenvalues of $\mathbf{Q}$. Note that $\mathbf{Q}$ and $\mathbf{D}$ have the same eigenvalues from which follows, that the matrices of the systems from which $a$ and $b$, respectively, were determined, are also equally well conditioned: $a$ is determined from the system

$$\mathbf{XDX}^T a = l \tag{42}$$

and $b$, because $a = \mathbf{X}b$, from

$$\mathbf{XD}b = l. \tag{43}$$

The matrices of the systems of Eqs. (42) and (43) have the same condition number, because $\mathbf{X}$ is orthogonal and thus perfectly conditioned, and therefore we might rewrite Eq. (43) as

$$\mathbf{D}b = \mathbf{X}^T l \tag{44}$$

without change of conditioning.

Conditioning changes for the vector $c$, however, which we might call the vector of *normalized parameters*, whose covariance matrix $\mathbf{Q}_{cc} = \mathbf{I}_n$. It is evidently given by

$$c = \mathbf{W}^{-1}b = \mathbf{W}^{-1}\mathbf{X}^T a, \tag{45}$$

where $\mathbf{WW} = \mathbf{D}$. Considering Eq. (43), it is the solution of the system

$$\mathbf{XDW}c = \mathbf{X}^T l, \tag{46}$$

or, without change of conditioning

$$\mathbf{DW}c = \mathbf{X}^T \mathbf{X}^T l, \tag{47}$$

whose condition number $(\lambda_n / \lambda_1)^{\frac{3}{2}}$ is larger than $\lambda_n / \lambda_1$, the condition number of $\mathbf{D}$. *Transforming to "normalized" unknowns whose covariance matrix is the identity matrix does therefore not improve the conditioning.*

While the transformation to fully efficient, *i.e.*, noncorrelated or decoupled parameters is very clear-cut, a clean analysis of the conditioning is not all that obvious, because of the necessity to invert $\mathbf{X}\sigma\mathbf{X}^T$ as well.

## Acknowledgements

I am indebted to Mr. Karl Wodnar (Vienna University Observatory) for a critical reading of the manuscript.

## References

Branham, R.L.: 1990, *Scientific data analysis*. New York &c. (Springer)
Brosche, P.: 1985, *Naturwissenschaften*, **772**, 668
Brown, D.C.: 1955, *Ballistic Research Laboratories Report No. 937*, Aberdeen Proving Grounds, Md.
Eichhorn, H.: 1978, *Mon. Not. Roy. astron. Soc.*, **182**, 335
Eichhorn, H.: 1988, *Astrophys. Journ.*, **334**, 465
Eichhorn, H.: 1989a, *Anz. d. Oesterr. Ak. d. Wiss.*, **126**, 89
Eichhorn, H.: 1989b, *Bull. Astr. Inst. Czechosl.*, **40**, 394
Eichhorn, H. & Clary, W.G.: 1974, *Mon. Not. Roy. astron. Soc.*, **166**, 425
Jefferys, W. H. 1979, *Astron. Journ.*, **84**, 175
Jefferys, W. H.: 1980, *Astron. Journ.*, **85**, 177
Jefferys, W. H.: 1981, *Astron. Journ.*, **86**, 149
Kallrath, J., Schlöder, J. & Bock, H. G.: 1992, *Celest. Mech.*, this issue
Lawson, C. L., & Hanson, R. J.: 1974. *Solving Least Squares Problems*, Englewood Cliffs, N.J. (Prentice Hall)

# LEAST SQUARES PARAMETER ESTIMATION IN CHAOTIC DIFFERENTIAL EQUATIONS

JOSEF KALLRATH*

*BASF-AG, ZXT/C, Kaiser-Wilhelm-Str.52, D-6700 Ludwigshafen, Federal Republic of Germany*

and

JOHANNES P. SCHLÖDER and HANS GEORG BOCK

*IWR and Institut für Angewandte Mathematik, Universität Heidelberg, Im Neuenheimer Feld 368, D-6900 Heidelberg, Federal Republic of Germany*

**Abstract.** A recent least squares algorithm, which is designed to adapt implicit models to given sets of data, especially models given by differential equations or dynamical systems, is reviewed and used to fit the Hénon-Heiles differential equations to chaotic data sets.

This numerical approach for estimating parameters in differential equation models, called the *boundary value problem approach*, is based on discretizing the differential equations like a boundary value problem, *e.g.* by a multiple shooting or collocation method, and solving the resulting constrained least squares problem with a structure exploiting generalized Gauss-Newton-Method (Bock,1981).

Dynamical systems like the Hénon-Heiles system which can have initial values and parameters that lead to positive Lyapunov exponents or phase space filling Poincaré maps give rise to chaotic time series. Various scenarios representing ideal and noisy data generated from the Hénon-Heiles system in the chaotic region are analyzed *w.r.t.* initial conditions, parameters and Lyapunov exponents. The original initial conditions and parameters are recovered with a given accuracy. The Lyapunov spectrum is then computed directly from the identified differential equations and compared to the spectrum of the "true" dynamics.

**Key words:** least squares techniques – numerical parameter estimation – boundary value problem approach – dynamical systems – Hénon-Heiles system – Lyapunov spectrum

## 1. Introduction

The method of least squares, introduced by Gauss (1809) to determine the orbits of Ceres and Pallas, is still of great significance for the analysis of observations, in astronomy as well as in experimental sciences in general. Since the time of Gauss, numerical methods for solving several types of least squares problems have been developed and improved, and there is still active research in that area. For a review on the methods of least squares as known and used in astronomy we refer to Eichhorn (1992).

A common basic feature and limitation of least squares methods used in astronomy, which is however seldom explicitly noted, is that they require an explicit model which is fitted to the data. Most models have a well founded base in physics, and are often described in terms of differential equations, which are often not solvable in a closed analytical form. Therefore, it seems desirable to develop solution algorithms that require only the statement of the model equations as a side condition, which implicitly determines the solution. The demand for such techniques

* presently at IWR, Universität Heidelberg, Im Neuenheimer Feld 368, D-6900 Heidelberg, Germany

*Celestial Mechanics and Dynamical Astronomy* **56**: 353–371, 1993.
© 1993 *Kluwer Academic Publishers.*

becomes even stronger when considering the rapidly increasing field of non-linear dynamics in physics and astronomy, non-linear reaction kinetics in chemistry, and non-linear systems describing ecosystems in biology or environmental sciences.

In this paper, a *boundary value problem approach* is reviewed and used to fit a four dimensional, quadratic system of Hamiltonian differential equations with four unknown initial values $x_{10}$, $x_{20}$, $x_{30}$ and $x_{40}$ and three unknown parameters $a, b$ and $c$ to chaotic time series. For $a = 1$, $b = 1$ and $c = -1$ the equations (see section 5 for details)

$$\dot{x}_1 = x_3$$

$$\dot{x}_2 = x_4$$

$$\dot{x}_3 = -ax_1 - 2x_1 x_2$$

$$\dot{x}_4 = -bx_2 - x_1{}^2 - cx_2{}^2 \tag{1.1}$$

are identical to the system investigated by Hénon and Heiles (1964) describing the motion of a star within the potential of a cylindrical galaxy.

Usually, experimental data from dynamical systems in the chaotic region are analyzed with time series analysis techniques (Eckmann and Ruelle, 1985), and the goal is to identify basic characteristic quantities of the system like Lyapunov exponents $\lambda_i$, or in dissipative systems the dimension of the attractor. These proce-dures, however, have the disadvantage that in order to obtain the $\lambda_i$ from time series one may easily need a few thousand data points according to Zeng *et al.*(1991) or Holzfuss and Parlitz (1991).

While the $\lambda_i$ are certainly a useful tool to distinguish between quasi-periodic or chaotic orbits, it still seems to be a greater advantage if the underlying dynamical system of the time series could be derived from the data. Therefore, one objective of this paper is to demonstrate that it is actually possible to fit the solution of a differential equation system to a chaotic time series and estimate the parameters, and that a few hundred data points are sufficient to reach that goal. In addition, we use the differential equations and the identified parameters to compute the Lyapunov spectrum by the algorithm of Bennetin *et al.* (1980).

Since a dynamical system at first sight has the structure of an initial value problem, an obvious approach is to integrate the equations with some initial values and parameters, to insert the trajectories into the least squares function, and to correct the unknowns by an iterative procedure till a minimum is reached. However, the numerical error propagation involved in the numerical integration of dynamical systems with a strong sensitivity to changes in initial values, *e.g.* those with positive Lyapunov exponents $\lambda_i$, prohibits the link of an initial value problem solver to a least squares solver in this fashion.

On the other hand, however, one may expect that the trajectories are very sensitive to perturbations of the parameters which is another way of saying that the problem of parameter estimation itself may be especially well-posed.

The problem remains to find *adequate numerical methods* that are able to cope with parameter-dependent instabilities in differential equations and are thus capable to recover parameters from given noisy chaotic trajectories.

It will be shown that on numerical methods based on the *boundary value problem* (BVP) *approach* as described in section 4 are indeed able to solve such problems (section 5) because of their stability properties. The suitability of this class of methods for the identification of unstable and chaotic processes was explained in Bock and Schlöder (1986), a thorough investigation to chaotic data from dissipative systems is given in Baake *et al.* (1992).

## 2. The Structure of the Problem

Parameter estimation means to optimally adapt models to given data by determining their parameter values such that the deviation of model and data is minimized in a suitable norm. In this paper we concentrate on models given by ordinary differential equations (ODE)

$$\dot{x} = f(t, x, p; signQ(t, x, p)), \quad x \epsilon R^{n_d}, \quad p \epsilon R^{n_p} \qquad (2.1)$$

where $t$ is the independent variable, $x \epsilon R^{n_d}$ are the states, and $p \epsilon R^{n_p}$ is the vector of unknown parameters. Depending on the vector of signs of the state and parameter dependent switching functions $Q$ the right hand side $f$ is defined piecewise smooth. At a root $\tau$ of a component of $Q$, jumps of the states

$$x(\tau^+) = x(\tau^-) + s(\tau^-, x(\tau^-), p) \qquad (2.2)$$

are admitted, and the right hand side $f$ may change discontinuously

$$\dot{x} = \begin{cases} f_i \quad (t, x, p), & t \leq \tau \\ f_{i+1}(t, x, p), & t > \tau \end{cases} \qquad (2.3)$$

Additional requirements on the solution of the ODE (2.1), like periodicity, initial or boundary conditions, range restrictions to the parameters can be formulated in a vector of (component wise) equality and inequality conditions

$$r_2[t_1, x(t_1), ..., t_k, x(t_k), p] = 0 \; or \geq 0 \qquad (2.4)$$

All in all we arrive at a fairly general boundary value problem with jumps (2.1) and switching condition (2.2) and (2.3).

More general dynamics (implicit differential equations, differential algebraic systems, discretized partial differential equations etc.) can often be reduced in principle to the above form. However, efficient and stable algorithms have to take into account their special structures. For a treatment of the differential algebraic case see Bock *et al.* (1988).

The boundary value problem is linked to experimental data via minimization of a least squares objective function

$$\ell_2(\boldsymbol{x}, \boldsymbol{p}) := \|r_1[t_1, \boldsymbol{x}(t_1), ..., t_k, \boldsymbol{x}(t_k), \boldsymbol{p}]\|_2^2 = \min_{\boldsymbol{x}, \boldsymbol{p}} \qquad (2.5)$$

This includes the probably most common case, where one component of the vector function $r_1$ is given by

$$r_{1j}[t_1, \boldsymbol{x}(t_1), ..., t_k, \boldsymbol{x}(t_k), \boldsymbol{p}] = [\eta_{ij} - g_{ij}(t_i, \boldsymbol{x}(t_i), \boldsymbol{p})]/\sigma_{ij} \qquad (2.6)$$

leading to

$$\ell_2(\boldsymbol{x}, \boldsymbol{p}) := \sum_{i,j} \sigma_{ij}^{-2} \cdot [\eta_{ij} - g_{ij}(t_i, \boldsymbol{x}(t_i), \boldsymbol{p})]^2 \qquad (2.7)$$

Here, $\eta_{ij}$ is the observed value which is related to states and parameters at time $t_i$ by

$$\eta_{ij} = g_{ij}(t_i, \boldsymbol{x}(t_i), \boldsymbol{p}) + \epsilon_{ij} \qquad (2.8)$$

The numbers $\epsilon_{ij}$ are the measurement errors, $\sigma_{ij}^2$ are weights that have to be adequately chosen due to statistical considerations, $e.g.$ as the measurement error variances. In the case of independent Gaussian $N(0, \sigma_{ij}^2)$ errors and known variances $\sigma_{ij}^2$ (up to a common factor $\beta^2$) the solution of the least squares problem is a maximum likelihood estimate.

Note, that also correlated measurements may be considered by (2.5) using an according formulation of the vector $r_1$.

Note further, that $r_1$ may be a nonlinear function of states and parameters and observation times which themselves may be disturbed by measurement errors. Neither all components of the states nor dense data are required (see sect.5). A procedure for the solution of such a problem should of course deliver an optimal fit that satisfies all restrictions. In addition, it is indispensable in practical problems to deliver also information about the statistical quality of the parameters and the states ($e.g.$ in terms of a $variance - covariance$ matrix of these quantities).

## 3. The Initial Value Problem Approach

The obvious approach to estimate parameters in ODE, which is also implemented in most commercial packages, is to guess parameters and initial values for the trajectories, compute a solution of an initial value problem (IVP) (2.1) and iterate the parameters and initial values in order to improve the fit. However, this procedure often exhibits poor performance or even does not work at all depending on the (usually parameter dependent) stability behavior of the IVPs that may lead $e.g.$ to exploding trajectories or solutions that run into steady states with extremely poor sensitivities $w.r.t.$ parameter variations. In the case of dynamical systems with positive Lyapunov exponents this approach is not advisable since error propagation dominates the system.

On the other hand the parameter estimation problem itself may be very well posed. Since small perturbations in the parameters may lead to strong perturbations of the trajectories, given data of a trajectory can be expected to determine the parameters very well. The next section describes an approach that optimally exploits the inverse structure of parameter estimation problems and is thus able to cope with the disastrous error propagation in chaotic systems.

## 4. The Boundary Value Problem Approach

In the following we sketch a versatile realization of an alternative method — the *boundary value problem approach* introduced in Bock (1981). It consists of a multiple shooting method for the discretization of the boundary value problem side condition and a generalized Gauss-Newton-Method for the solution of the resulting structured nonlinear constrained least squares problem. A detailed description and analysis of this family of methods is included in Bock (1987).

Depending on the stability behaviour of the ODE and the availability of information about the process (measured data, qualitative knowledge about the problem, etc.) the user chooses a suitable grid $\mathcal{T}_m$ of $m$ multiple shooting nodes $\tau_j$ ($m - 1$ subintervals $\mathcal{I}_j$)

$$\mathcal{T}_m : \tau_1 < \tau_2 < ... < \tau_m, \quad \Delta\tau_j := \tau_{j+1} - \tau_j, \quad 1 \le j \le m - 1 \qquad (4.1)$$

covering the interval where measurements are given ($[\tau_1, \tau_m] \supseteq [t_0, t_f]$). At each grid point new variables $\{s_j := x(\tau_j)\}$ are introduced and $m - 1$ initial value problems

$$\dot{x} = f(t, x, p), \quad x(\tau_j) = s_j, \quad t\epsilon[\tau_j, \tau_{j+1}] \qquad (4.2)$$

are considered on the subintervals. The $m - 1$ vectors of initial values $s_j$, the value $s_m$ at the end point, the parameter vector $p$ and the observation times $t_i$ are summarized in an augmented vector

$$z^t := (s_1^t, ..., s_m^t, p^t, t_1, ..., t_k) \qquad (4.3)$$

The least squares function (2.5) is then regarded as a function of this vector

$$\ell_2(z) = \|\hat{r}_1[x(t_1, z), ..., x(t_k, z), z] \qquad (4.4)$$

or in the special case (2.7) $\ell_2(z) = \sum_j \sigma_{ij}^{-2} \cdot [\eta_{ij} - \hat{g}_{ij}(x(t_i, z), z)]^2$, and the ODE (2.1) is replaced by the matching conditions

$$x(\tau_{j+1}; s_j, p) - s_{j+1} = 0, \quad 1 \le j \le m - 1 \qquad (4.5)$$

which ensure continuity of the final trajectory. Note, however, that the initial trajectories provided with initial guesses $z_i$ and the trajectories in the course of

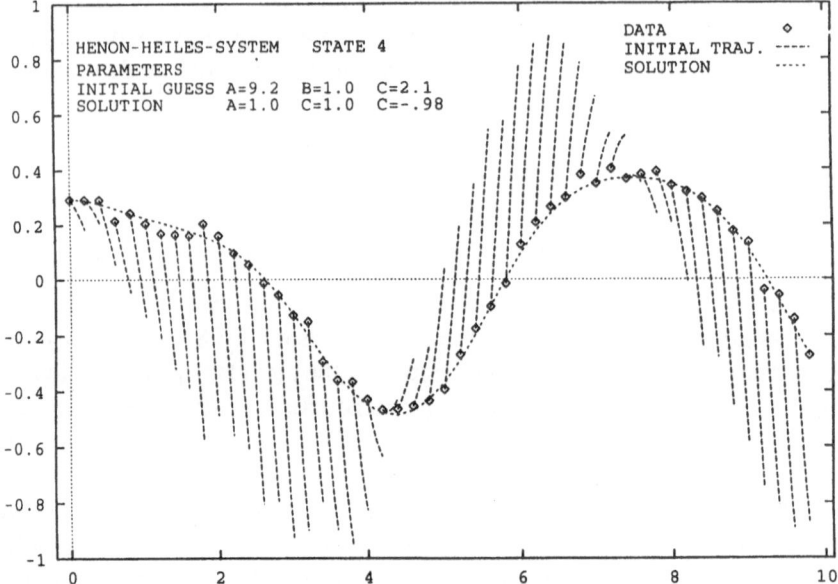

Fig. 1. Data, solution and initial trajectory for $x_4$ in the Hénon-Heiles system generated with initial values $x_{10} = x_{20} = 0$, $x_{30} = 0.3$ and $x_{40} = -0.41$ and parameters $(a, b, c) = (+9.2, +1, +2.1)$ as initial guesses for the fitting routine. This figure should demonstrate that the initial trajectories may be discontinuous and deviate significantly from the solution.

the iterative procedure are allowed to be discontinuous (see Fig.1). By choosing the multiple shooting intervals $\Delta\tau_j := \tau_{j+1} - \tau_j$ sufficiently small, existence of a (discontinuous) initial trajectory can be guaranteed under mild conditions, and error propagation can be controlled and limited.

Formally, the least squares problem described so far is a constrained optimization problem of the type

$$\min_z \{\|F_1(z)\|_2^2 \mid F_2(z) = 0 \; or \; \geq 0 \; \epsilon \; \mathbf{R}^{n_c}\} \tag{4.6}$$

where $n_c$ is the number of constraints (continuity constraints (4.5) and additional constraints (2.4)). This usually large constrained structured non-linear problem is solved by a damped generalized Gauss-Newton method, which is described here for the equality constrained case (see Bock (1987), where also the treatment of inequalities is described). Starting with an initial guess $z_0$ the variables are iterated via

$$z_{k+1} = z_k + \alpha_k \cdot \Delta z_k \tag{4.7}$$

with a damping constant $\alpha_k$, $0 < \alpha_{min} \leq \alpha_k \leq 1$. The increment $\Delta z_k$ is the solution of the problem linearized at $z_k$

$$\min_{\Delta z_k}\{ \| J_1(z_k)\Delta z_k + F_1(z_k)\|_2^2 | J_2(z_k)\Delta z_k + F_2(z_k) = 0, J_i(z_k) := \partial_z F_i(z_k)\}$$
$$(4.8)$$

Under appropriate assumptions $w.r.t.$ the regularity of the Jacobians $J_i$ there exists a unique solution $\Delta z_k$ of the linear problem and a unique linear mapping $J_k^+$ obeying the relations

$$\Delta z_k = -J_k^+ F(z_k), \quad J_k^+ J_k J_k^+ = J_k^+, \quad J_k^t := [J_1(z_k)^t, J_2(z_k)^t] \quad (4.9)$$

The solution $\Delta z_k$ of the linear problem, or formally the generalized inverse $J_k^+$ (Bock 1981) of $J_k$, follows uniquely from the Kuhn-Tucker conditions

$$J_1^t J_1 \Delta z_k - J_2^t \lambda_c + J_1^t F_1 = 0, \quad J_2 \Delta z_k + F_2 = 0 \qquad (4.10)$$

where $\lambda_c \epsilon R^{n_c}$ is a vector of Lagrange multipliers. For the numerical solution $\Delta z_k$ of the linear constrained problem (4.8) several structure exploiting methods have been developed that compute special factorizations of $J_1$ and $J_2$ and thus implicitly, but not explicitly the generalized inverse $J^+$. There even exist methods (Schlöder,1988) that generate and decompose the Jacobians simultaneously, and that require an amount of computing time for the generation of the derivatives that is proportional only to the number of degrees of freedom and not to the number of variables.

The availability of the Jacobians $J_1$ and $J_2$ allows rank checks in every iteration and automatic detection of violations of the regularity assumptions. In that case automatic regularization and computation of a relaxed solution is possible.

The iteration (4.7) can be forced to converge globally to a stationary point of the problem if the damping factors $\alpha_k$ are chosen appropriately. In the treatment of a large number of practical problems strategies based on "natural level functions" have proven to be very successful (see Appendix A). In the region of local convergence of the full step method, the algorithm converges linearly to a solution that is stable to statistical variations in the observations. An iterate $z_k$ is accepted as solution $z^*$ of the nonlinear constrained problem, if a scaled norm of the increments $\Delta z_k$ is below a user specified tolerance $(10^{-5}, say)$.

As the Jacobians and their decompositions are available in each iteration, covariance and correlation matrices are easily computable for the full variable vector $z$. In large scale problems this is usually not desirable since sparsity is destroyed. Therefore very fast algorithms have been developed that compute only the diagonal elements of the covariance matrix. For details of the algorithms, the treatment of more general classes of problems and the properties of the optimization package PARFIT we refer to Bock (1987).

The integrator, which is also used for the computations of the Lyapunov spectrum, is based on an extrapolation method by Bulirsch and Stoer (1966). In addition,

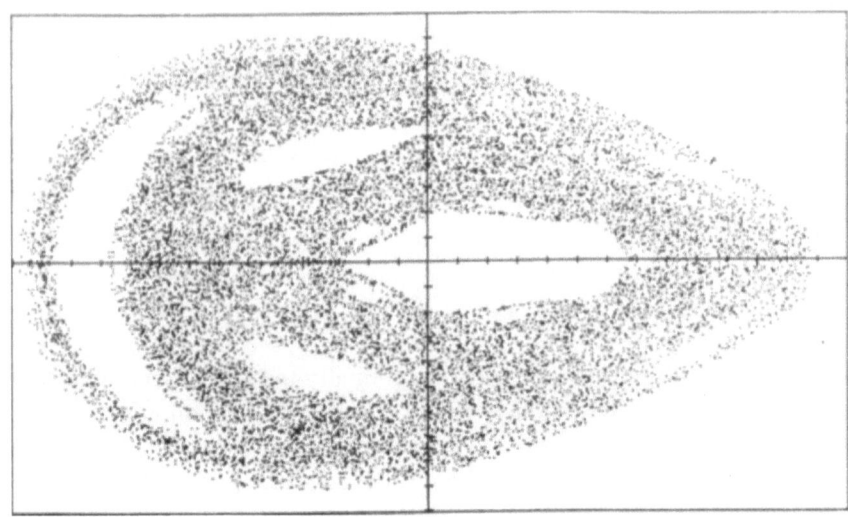

Fig. 2. Poincaré map $x_4$ versus $x_2$ in the Hénon-Heiles system generated with initial values $(x_{10}, x_{20}, x_{30}, x_{40}) = (0, 0, +0.3, -0.41)$ and parameters $(a, b, c) = (1, 1, -1)$. The plane of section is defined by $x_1 = 0$ and $x_3 \geq 0$. The ranges of the plot are $-0.5 \leq x_2 \leq +0.7$ and $-0.5 \leq x_4 \leq +0.5$.

it allows the user to compute simultaneously the sensitivity matrices $G$ or $H$ which are the most costly part in computing the Jacobians $J_i$

$$G(t; t_0, \boldsymbol{x}_0, \boldsymbol{p}) := \frac{\partial}{\partial \boldsymbol{x}_0} \boldsymbol{x}(t; t_0, \boldsymbol{x}_0, \boldsymbol{p})$$

$$H(t; t_0, \boldsymbol{x}_0, \boldsymbol{p}) := \frac{\partial}{\partial \boldsymbol{p}} \boldsymbol{x}(t; t_0, \boldsymbol{x}_0, \boldsymbol{p}) \tag{4.11}$$

via the so-called *internal numerical differentiation* (Appendix B) as introduced by Bock (1981), which does not require the often cumbersome and error prone formulation of the variational differential equations

$$\frac{dG}{dt} = \boldsymbol{f}_x(t, \boldsymbol{x}, \boldsymbol{p}) \cdot G, \quad \frac{dH}{dt} = \boldsymbol{f}_x(t, \boldsymbol{x}, \boldsymbol{p}) \cdot H + \boldsymbol{f}_p(t, \boldsymbol{x}, \boldsymbol{p}) \tag{4.12}$$

by the user.

## 5. Results for the Hénon-Heiles System

In this section, we consider the Hamiltonian

$$H(x, y, \dot{x}, \dot{y}) = \frac{1}{2}(\dot{x}^2 + \dot{y}^2) + U(x, y) \tag{5.1}$$

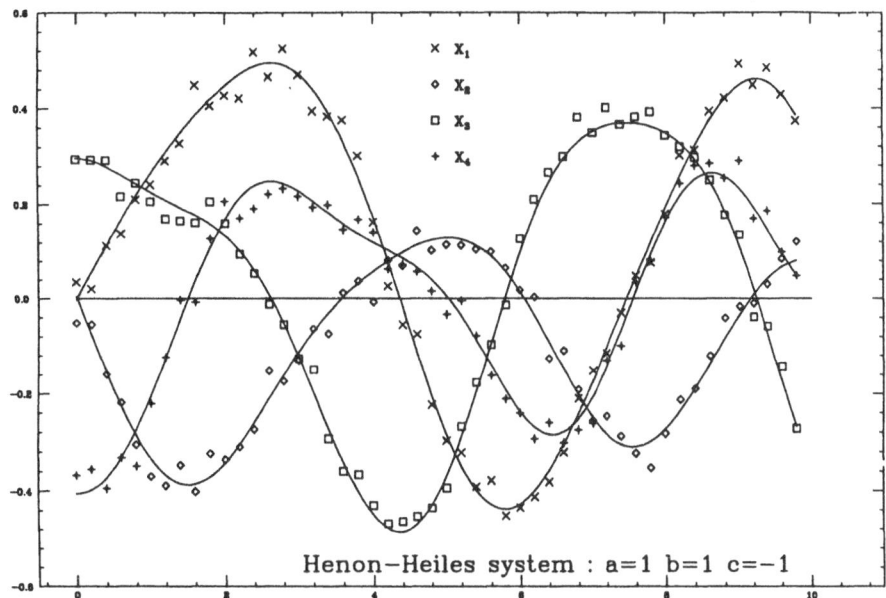

Fig. 3. Trajectories in the Hénon-Heiles system generated with initial values $x_{10} = x_{20} = 0$, $x_{30} = 0.3$ and $x_{40} = -0.41$ and parameters $(a, b, c) = (1, 1, -1)$. Gaussian random numbers with standard deviation $\sigma = 0.05$ are added to the trajectories generated by integration of the dynamical system.

and the potential function

$$U(x, y) = \frac{1}{2}\left(ax^2 + by^2 + 2x^2y + \frac{2}{3}cy^3\right) \qquad (5.2)$$

From that we derive the equations of motion

$$\ddot{x} = -U_x = -ax - 2xy \qquad (5.3)$$

$$\ddot{y} = -U_y = -by - x^2 - cy^2 \qquad (5.4)$$

or — via ( $x_1 = x$, $x_2 = y$, $x_3 = \dot{x}$, $x_4 = \dot{y}$) — the equivalent first-order system $\dot{x} = f(x)$

$$\dot{x}_1 = x_3$$

$$\dot{x}_2 = x_4$$

$$\dot{x}_3 = -ax_1 - 2x_1x_2$$

$$\dot{x}_4 = -bx_2 - x_1^2 - cx_2^2 \qquad (5.5)$$

For the sake of computing the Lyapunov spectrum in a later section we also write down the linearized or variational differential equations along a given nominal trajectory $x$

$$\dot{\xi}_1 = \xi_3$$

$$\dot{\xi}_2 = \xi_4$$

$$\dot{\xi}_3 = f_{31}\xi_1 + f_{32}\xi_2, \quad f_{31} = -(a + 2x_2), \quad f_{32} = -2x_1$$

$$\dot{\xi}_4 = f_{41}\xi_1 + f_{42}\xi_2, \quad f_{41} = -2x_1, \quad f_{42} = -(b + 2cx_2) \qquad (5.6)$$

Choosing the parameters $p = (a, b, c) = (+1, +1, -1)$ we get the system investigated by Hénon and Heiles (1964). For energies between $E^- = 1/12$ and $E^+ = 1/8$ one observes a transition from regular behavior characterized by a family of curves which almost completely fill the available space in the $(x_2, x_4)$-Poincaré surface of section $(x_1 = 0, \dot{x}_3 \geq 0)$ to ergodicity. As shown in Figure 2 one *ergodic* trajectory [e.g. initial values $x_0 = (x_{10}, x_{20}, x_{30}, x_{40}) = (0, 0, 0.3, -0.41)$] may cover large parts of the permissible area

$$U(0, x_2) + \frac{1}{2} \cdot x_4^2 \leq E = 0.12905 \qquad (5.7)$$

in the $x_1 = 0$ plane.

In order to set up interesting scenarios for our parameter fitting algorithm we have varied the parameters $(a, b, c)$, have inspected the Poincaré's surfaces of section and have chosen the initial conditions $x_0$ such that they are placed in the chaotic region.

The generation of *simulated data* is performed in two steps starting with the choice of $(x, p)_s$. By integration of (5.5) we generate a simulated trajectory. Since we expect to have a strong error propagation due to the chaoticity we choose a very stringent local error bound for the integrator ($\epsilon_{int} = 10^{-13}$) and restrict the integration interval such that the global error remains significantly below the intended accuracy of the estimation problem ($\epsilon = 10^{-5}$). To simulate noisy data we add Gaussian random numbers of standard deviation $\sigma$ within a $3\sigma$ strip to our fiducial trajectory. Alternatively, we might use a generalization of the boundary value problem approach to multiple experiments (Schlöder, 1988), in which the time range is split and the according data are treated as separate time series.

In the different scenarios summarized in Tables I to III we vary the parameters $p_s$ and initial conditions $x_0$ for the integration as well as the noise $\sigma$ added to the simulated trajectories. Figure 3 shows the trajectories in the Hénon-Heiles system generated with initial values $x_0 = (0, 0, 0.3, -0.41)$, parameters $p = (1, 1, -1)$ and Gaussian random numbers with standard deviation $\sigma = 0.05$ added to generate the *simulated* data.

## TABLE I

This table defines the setup for generating trajectories with energy $E_s = 0.125$, i.e. initial values $x_{0s} = (0.00, 0.00, \sqrt{0.1275} \approx 0.35707, -0.35)$ and parameters $(a, b, c)_s = (1, 1, -1)$, and contains different scenarios with 400 'exact' data points in the first part. Given some initial guesses $a_i$, $b_i$ and $c_i$, the original parameters are recovered with an absolute error less than $5 \cdot 10^{-5}$. The second part of the table contains simulated data. In both cases we used 100 multiple shooting nodes. The length of the data series is $t_{end} = (n-1) \cdot \Delta t$. The estimated parameters and their statistical errors derived from the covariance matrix are listed as well as the initial guesses. The estimated errors can be transformed to confidence intervals by multiplying them with the Fisher factor $\gamma = 3.75$. All Lyapunov exponents have been computed but we write down only $\lambda_1 = 0.044$, $\lambda_2$ and $\lambda_3$ are of opposite sign with absolute values of the order of $10^{-7}$ while we find $\lambda_4 = -\lambda_1$ with five digits accuracy.

|         |            | S11   | S12   | S13   | S14   | S15   | S16   | S17     | S18     | S19     |
|---------|------------|-------|-------|-------|-------|-------|-------|---------|---------|---------|
| Interv.: | $\Delta t$ | 0.4   | 0.4   | 0.4   | 0.2   | 0.2   | 0.2   | 0.2     | 0.2     | 0.2     |
| Nodes:  | $n$        | 100   | 100   | 100   | 100   | 100   | 100   | 100     | 100     | 100     |
| St. dev: | $\sigma$  |       |       |       |       |       |       | 0.01    | 0.03    | 0.05    |
| Guess:  | $a_i$      | 5     | 5     | 10    | 10    | 10    | 20    | 10      | 10      | 10      |
|         | $b_i$      | 5     | 5     | 10    | 10    | 10    | 20    | 10      | 10      | 10      |
|         | $c_i$      | -2    | 2     | 2     | 2     | 10    | 10    | 2       | 2       | 2       |
|         | $E_i$      | $E_s$ | $E_s$ | $E_s$ | $E_s$ | $E_s$ | $E_s$ | 0.124   | 0.122   | 0.130   |
| Iter.:  | $k$        | 6     | 6     | 6     | 6     | 6     | 16    | 6       | 11      | 14      |
| Estim.: | $x_{10}$   |       |       |       |       |       |       | -0.0003 | -0.0017 | -0.0036 |
|         | $x_{20}$   |       |       |       |       |       |       | 0.0006  | 0.0031  | 0.0067  |
|         | $x_{30}$   |       |       |       |       |       |       | 0.3574  | 0.3585  | 0.3603  |
|         | $x_{40}$   |       |       |       |       |       |       | -0.3496 | -0.3479 | -0.3455 |
| Estim.: | $a$        |       |       |       |       |       |       | 0.9999  | 0.9995  | 0.9989  |
|         | $b$        |       |       |       |       |       |       | 0.9995  | 0.9969  | 0.9926  |
|         | $c$        |       |       |       |       |       |       | -1.0062 | -1.0335 | -1.0764 |
| Estim.: | $E$        |       |       |       |       |       |       | 0.1250  | 0.1248  | 0.1246  |
| Estim.: | $\Delta x_{10}$ |  |       |       |       |       |       | 7.6D-4  | 3.9D-3  | 8.4D-3  |
|         | $\Delta x_{20}$ |  |       |       |       |       |       | 7.6D-4  | 3.9D-3  | 8.1D-3  |
|         | $\Delta x_{30}$ |  |       |       |       |       |       | 4.1D-4  | 2.1D-3  | 4.5D-3  |
|         | $\Delta x_{40}$ |  |       |       |       |       |       | 4.6D-4  | 2.4D-3  | 5.1D-3  |
|         | $\Delta a$ |       |       |       |       |       |       | 3.2D-4  | 1.7D-3  | 3.6D-3  |
|         | $\Delta b$ |       |       |       |       |       |       | 2.5D-3  | 1.3D-2  | 2.7D-2  |
|         | $\Delta c$ |       |       |       |       |       |       | 1.2D-2  | 6.3D-2  | 1.4D-1  |

## TABLE II

This table defines the setup for generating trajectories with energy $E_s = 0.12905$, *i.e.* initial values $x_{0s} = (0, 0, 0.3, -0.41)$ and parameters $(a, b, c)_s = (1, 1, -1)$, and contains different scenarios for generating *simulated data*. The estimated parameters and their statistical errors derived from the covariance matrix (Fisher factor $\gamma = 3.75$ to convert to confidence intervals) are listed as well as the initial guesses. As in Table 1 we only write down $\lambda_1 = 0.053$.

|            |             | S21     | S22     | S23     | S24     | S25     | S26     |
|------------|-------------|---------|---------|---------|---------|---------|---------|
| Intervals: | $\Delta t$  | 0.2     | 0.2     | 0.2     | 0.2     | 0.2     | 0.2     |
| Nodes:     | $n$         | 50      | 50      | 100     | 100     | 100     | 100     |
| Range:     | $t_{end}$   | 9.8     | 9.8     | 19.8    | 19.8    | 19.8    | 19.8    |
| Points:    | $n$         | 200     | 200     | 400     | 400     | 400     | 400     |
| St. dev.:  | $\sigma$    | 0.050   | 0.020   | 0.020   | 0.050   | 0.025   | 0.050   |
| Guess:     | $a_i$       | +9.2    | +1.2    | +1.2    | +1.2    | +10.0   | +10.0   |
|            | $b_i$       | +1.0    | +1.0    | +1.0    | +1.0    | +10.0   | +10.0   |
|            | $c_i$       | +2.1    | -2.0    | -2.0    | -2.0    | +20.0   | +20.0   |
|            | $E_i$       | 0.124   | 0.124   | 0.124   | 0.111   | 0.127   | 0.148   |
| Iter.:     | $k$         | 7       | 5       | 5       | 6       | 7       | 14      |
| Estim.:    | $x_{10}$    | -0.0036 | -0.0009 | -0.0013 | -0.0049 | -0.0018 | -0.0049 |
|            | $x_{20}$    | 0.0036  | 0.0009  | 0.0016  | 0.0062  | 0.0022  | 0.0062  |
|            | $x_{30}$    | 0.2963  | 0.2991  | 0.2997  | 0.2988  | 0.2996  | 0.2988  |
|            | $x_{40}$    | -0.4060 | -0.4090 | -0.4081 | -0.4024 | -0.4073 | -0.4024 |
| Estim.:    | $a$         | 0.9958  | 0.9989  | 0.9986  | 0.9944  | 0.9980  | 0.9944  |
|            | $b$         | 1.0047  | 1.0011  | 1.0007  | 1.0030  | 1.0010  | 1.0030  |
|            | $c$         | -0.9815 | -0.9957 | -1.0079 | -1.0317 | -1.0111 | -1.0317 |
| Estim.:    | $E$         | 0.1263  | 0.1284  | 0.1282  | 0.1256  | 0.1278  | 0.1256  |
| Estim.:    | $\Delta x_{10}$ | 9.5D-3  | 2.4D-3  | 1.6D-3  | 6.3D-3  | 2.3D-3  | 6.3D-3  |
|            | $\Delta x_{20}$ | 8.7D-3  | 2.2D-3  | 1.7D-3  | 6.6D-3  | 2.4D-3  | 6.6D-3  |
|            | $\Delta x_{30}$ | 3.9D-3  | 9.8D-4  | 7.6D-4  | 3.0D-3  | 1.1D-3  | 3.0D-3  |
|            | $\Delta x_{40}$ | 4.8D-3  | 1.2D-3  | 8.4D-4  | 3.3D-3  | 1.2D-3  | 3.3D-3  |
|            | $\Delta a$  | 8.4D-3  | 2.0D-3  | 8.4D-3  | 3.3D-3  | 1.2D-3  | 3.3D-3  |
|            | $\Delta b$  | 3.4D-2  | 9.1D-3  | 1.1D-3  | 4.5D-3  | 1.5D-3  | 4.5D-3  |
|            | $\Delta c$  | 1.7D-1  | 4.2D-2  | 7.5D-3  | 3.0D-3  | 1.1D-2  | 3.0D-3  |

In order to demonstrate the robustness of the algorithm in Figure 4 we present trajectories obtained with initial values $x_o = (0, 0, 0.3, -0.41)$ and parameters $p_i = (9.2, 1, 2.1)$ as initial guesses for the fitting routine. As listed in Table II, the original parameters $p = (1, 1, -1)$ are well recovered although the initial trajectories correspond to escape orbits and are far off from the solution.

Tables I to III contain three different scenarios, and in each of them we have tried to identify the parameters under different circumstances, *i.e.* different distances between the initial parameters and solution parameters, different noise and available data for fitting, and different lengths $t_{end}$ of the time series.

The parameter estimations listed in Table I have been derived from *exact* data, *i.e.* from trajectories not disturbed by noise. These experiments demonstrate the

TABLE III

This table defines the setup for generating trajectories with energy $E_s = 0.1849$, *i.e.* initial values $x_{0s} = (0, 0, 0.43, 0.43)$ and parameters $(a, b, c)_s = 1.3 \cdot (1, 1, -1)$, and contains different scenarios for re-estimating the parameters from *simulated data*. The initial parameter guess is $p_i = (5, 5, 5)$ corresponding to a guess energy of $E_i = 0.2027$. The estimated parameters and their statistical errors are listed as a function of the data points considered for fitting. If a curve has weight 0 then it is not considered in the least-squares problem. As in Table I we only write down $\lambda_1 = 0.037$.

| | | S31 | S32 | S33 | S34 | S35 | S36 |
|---|---|---|---|---|---|---|---|
| Intervals: | $\Delta t$ | 0.2 | 0.2 | 0.2 | 0.2 | 0.4 | 0.4 |
| Nodes: | $n$ | 100 | 100 | 100 | 100 | 100 | 100 |
| Range: | $t_{end}$ | 19.8 | 19.8 | 19.8 | 19.8 | 39.6 | 39.6 |
| St. dev.: | $\sigma$ | 0.05 | 0.05 | 0.05 | 0.05 | 0.05 | 0.07 |
| Weight: | $w$ | 1111 | 1110 | 1100 | 1000 | 1000 | 1111 |
| Iter.: | $k$ | 5 | 5 | 5 | 11 | 7 | 6 |
| Estim.: | $x_{10}$ | 0.0017 | 0.0041 | -0.0030 | -0.0144 | -0.0066 | 0.0083 |
| | $x_{20}$ | 0.0110 | 0.0054 | 0.0042 | 0.0362 | 0.0509 | -0.0080 |
| | $x_{30}$ | 0.4023 | 0.4324 | 0.4338 | 0.4433 | 0.4322 | 0.4329 |
| | $x_{40}$ | 0.4834 | 0.4367 | 0.4371 | 0.4272 | 0.4276 | 0.4292 |
| Estim.: | $a$ | 1.3010 | 1.2965 | 1.3024 | 1.3618 | 1.2984 | 1.2997 |
| | $b$ | 1.2905 | 1.3000 | 1.3019 | 1.2021 | 1.3150 | 1.2937 |
| | $c$ | -1.2683 | -1.2848 | -1.2882 | -0.9491 | -1.3957 | -1.2669 |
| Estim.: | $E$ | 0.1868 | 0.1889 | 0.1896 | 0.1900 | 0.1854 | 0.1859 |
| Estim.: | $\Delta x_{10}$ | 6.0D−3 | 6.3D−3 | 9.0D−3 | 1.7D−2 | 1.2D−2 | 8.5D−3 |
| | $\Delta x_{20}$ | 6.9D−3 | 9.6D−3 | 1.0D−2 | 4.0D−2 | 3.5D−2 | 1.2D−2 |
| | $\Delta x_{30}$ | 3.3D−3 | 3.6D−3 | 4.8D−3 | 1.2D−2 | 6.3D−3 | 6.0D−3 |
| | $\Delta x_{40}$ | 3.0D−3 | 4.2D−3 | 4.2D−3 | 1.8D−2 | 9.2D−3 | 5.5D−3 |
| | $\Delta a$ | 6.0D−3 | 6.9D−3 | 8.4D−3 | 6.5D−2 | 4.3D−3 | 3.0D−3 |
| | $\Delta b$ | 9.D−3 | 1.2D−2 | 1.2D−2 | 8.9D−2 | 1.3D−2 | 6.0D−3 |
| | $\Delta c$ | 3.0D−2 | 3.9D−2 | 3.9D−2 | 3.7D−1 | 7.2D−2 | 3.1D−2 |
| Fisher: | $\gamma$ | 3.75 | 3.75 | 3.75 | 3.90 | 3.90 | 3.75 |

large region of convergence of the method. Table II contains scenarios defined by time series with 200 or 400 data points perturbed with Gaussian random numbers varying between $\sigma = 0.02$ and $0.05$. We choose the standard deviation $\sigma$ according to

$$\sigma \approx 0.1 \cdot \max(|x_i(t)|) \tag{5.8}$$

and use only those perturbed data points which lie within a $3\sigma$ strip. The error limits given in the tables are derived from the diagonal elements of the covariance matrix. In order to get confidence intervals based on a 95% confidence level, these values have to be multiplied with the Fisher factor

$$\gamma := \sqrt{\ell_1 \cdot F(1 - \alpha, \ell_1, \ell_2)}, \quad \ell_2 = n - \ell_1, \quad \alpha = 0.95 \tag{5.9}$$

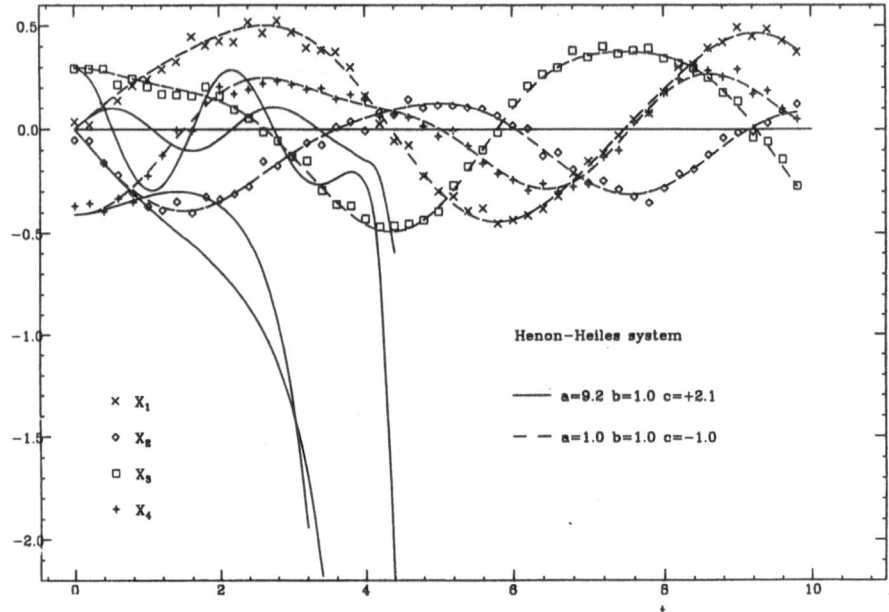

Fig. 4. Trajectories in the Hénon-Heiles system generated with initial values $x_{10} = x_{20} = 0$, $x_{30} = 0.3$ and $x_{40} = -0.41$ and parameters $(a, b, c) = (+9.2, +1, +2.1)$ as initial guesses for the fitting routine. As listed in Table I the original parameters $(a, b, c) = (1, 1, -1)$ are well recovered although the trajectories corresponding to the initial parameter guesses are far of from the solution.

where $\ell_1 = 7$ gives the degree of freedom, and $F(1 - \alpha, \ell_1, \ell_2)$ is the $F$-distribution (Abramowitz and Stegun, 1970) for the quantil $\alpha$. The Fisher factor is also listed in the tables. As expected the parameters are determined more accurately if the number $n$ of data points or $t_{end}$ is increased.

The scenarios in Table III belong to a different set of initial values and parameters corresponding to an energy of $E = 0.1849$ slightly below the escape energy $E_{esc} \approx 0.21$ derived from the contour lines of the potential. Starting with some moderately perturbed parameter guesses we decrease $n$ by disregarding successively the $x_4$, $x_3$ and $x_2$ data. For this selective procedure we observe, of course, that the sum of squared residuals also decreases, while the errors of the parameters increase. Scenario S34 contains only the $x_1$ data and the error of parameter $c$ increases to more than 10%. In scenario S35 the error is reduced again by increasing $\Delta t$ to 0.4, or equivalently $t_{end}$ to 39.6 which gives a time series of doubled length. This experiment shows that in order to identify the parameters of the Henon-Heiles system, two curves, for instance the $x_1$ and $x_2$ curves, prove to be sufficient, and even one curve is sufficient if it contains enough data points. This result agrees well with similar experiences of Baake $et$ $al.$ (1991) in the dissipative Lorenz and Rössler systems. Eventually, in S36 we contaminated all curves with Gaussian errors ($\sigma = 0.07$) which is again 10% of the amplitude of the $x_i$-curves.

The Lyapunov spectrum $\lambda = \{\lambda_1, \lambda_2, \lambda_3, \lambda_4\}$ also contained in Tables I to III has been computed with the code listed in Wolf *et al.* (1985). This algorithm is based on the work of Bennetin *et al.* (1980). For a continuous dynamical system in an $n$-dimensional phase space the long-term evolution of an infinitesimal $n$-sphere of initial conditions is obtained by generation of appropriate directional derivatives, *e.g.* via the integration of (5.6), or by using internal numerical differentiation techniques (sect. 4). This sphere will become a $n$-ellipsoid with principal axes $r_i(t)$ which will lead to the $i^{th}$ one-dimensional Lyapunov exponent $\hat{\lambda}_i$ (Wolf *et al.*, 1985)

$$\hat{\lambda}_i := \lim_{t \to \infty} \frac{1}{t} \cdot log_2 \frac{r_i(t)}{r_i(0)}, \quad \lambda_i = \ln(2) \cdot \hat{\lambda}_i \approx 0.693 \cdot \hat{\lambda}_i \qquad (5.10)$$

In order to get a numerically stable procedure Bennetin *et al.* (1980) suggested that at regular time intervals $\Delta t_{orth}$ the evolved tangent vectors associated with the $n$-sphere mentioned above are replaced by a set of new orthonormal vectors, using the Gram-Schmidt procedure, which avoids small angles between the vectors defining the $n$-ellipsoid.

We would like to stress that the $\lambda_i$ are ordered with $\lambda_1$ being the largest exponent, and that although they are related to the expanding or contracting nature of different directions in phase space it is not possible to assign a well-defined direction with a given exponent. Since the Hénon-Heiles differential equations are derived from a Hamiltonian system with $f = 2$ degrees of freedom we use some additional facts on the Lyapunov spectra of such systems (Froeschlé 1984) such as symmetry with respect to zero, which in our particular case lead to $\lambda_1 \geq 0$, $\lambda_2 = \lambda_3 = 0$ and $\lambda_4 = -\lambda_1$ and $\lambda_1 + \lambda_2 + \lambda_3 + \lambda_4 = 0$ corresponding to the Liouville theorem. While the exponents may be interpreted geometrically as average exponential rates of divergence or convergence of nearby orbits in phase space or growth of an infinitesimal volume element, the $log_2$ function in definition (5.10) allows also to measure the rate at which the system dynamics creates or destroys information in $bits/orbit$. The complete Lyapunov spectra $\lambda$ for the analyzed scenarios were calculated with an integration time of $t_{int} = 10^6$ leading to spectra with $\lambda_1 \approx 5 \cdot 10^{-2}$ while those exponents $\lambda_2$ and $\lambda_3$ to be expected to vanish reached values between $10^{-7}$ and $10^{-5}$ [for comparison in the case $E = 0.125$ see Bennettin et al. (1976)]. For regular orbits, identified as closed curves in the Poincaré's surfaces of section, we found that all Lyapunov exponents tended to zero as expected, *i.e.* their absolute values fell below $10^{-7}$.

Concerning the interpretation of Lyapunov exponents derived from fitted ODE we would like to add the following remarks: The tables contain only the spectra $\lambda$ for the generated trajectories and their associated parameters $p_s$, and not those for the identified parameters $\tilde{p}$ since for even slightly perturbed parameters $\tilde{p}$, $\tilde{\lambda}_1$ differed by about 20 to 30% from $\lambda_1$. The reason is that slightly different initial values or parameters may lead to a significantly different energy surface. Similar to the strong dependence of the Lyapunov exponents on the distance $d_0$ between the third object and the second primary in the restricted three body problem investigated by Gonczi and Froeschlé (1981), in our case we have a strong sensitivity of the Lyapunov exponents $w.r.t.$ energy $E$. For the Hénon-Heiles system this is in particular true when $E \approx 1/8$. We are forced to adopt the following point of view: The Lyapunov spectra derived from fitted ODE can be used to give a qualitative picture of ergodicity. They are still a quantitative tool to indicate chaos. In particular, this is important in dynamical systems with more than 2 degrees of freedom since in those cases the method of Poincaré's surfaces of section is inconvenient to use. And, finally, it is still an efficient method to determine the number of isolating integrals.

The major advantage, however, is that there is no explicit limitation as in a finite time series. Once the ODE have been identified the computation of the $\lambda_i$ can be performed to any desired degree of accuracy. For some celestial mechanics problems, for instance that of the stochasticity of the orbit of Halley's comet (Froeschlé and Gonczi, 1988), there are typical integration times of $10^6$ years required to reach convergence. In most cases, however, there is no a priori condition determining $t_{int}$ available. Therefore the only chance is to integrate for a long period until convergence.

## 6. Conclusions

The *boundary value problem approach* recovers the parameters of the Hénon-Heiles system in a variety of scenarios differing in initial values and parameters of the simulated trajectories, noise of the Gaussian random errors and distance of the initial guesses of the parameters from the solution. From the identified dynamical systems the Lyapunov exponents are derived and are found to obey the conditions for Hamiltonian systems. In order to identify the parameters of the Hénon-Heiles system two components, $e.g.$ the $x_1$ and $x_2$ curves, prove to be sufficient, and even one component is sufficient if it contains enough data points. This is a very advantageous property, since in real experiments measurements, $e.g.$ of instance population densities in biological systems, concentrations in chemical and pharmacokinetical systems, or observations in astronomy or celestial mechanics seldom yield data for the whole phase space.

## Appendix

## A. Damping Strategies

Full step (Gauss-)Newton methods usually require unrealistically good initial guesses to guarantee convergence. To alleviate this problem step size strategies have been developed. According to an appropriately chosen level function the damping factor $\alpha_k \epsilon (\alpha_{min}, 1]$ is chosen in every iteration such that the value of the level function is reduced for the new iterate.

Depending on the choice of the level function which serves as a measure of the distance of the actual iterate to the solution different iterates are computed. For parameter estimation algorithms the following two level functions are important (Bock, 1981)

$$T_1(z) = \frac{1}{2} \cdot \|F_1(z)\|_2^2 + \sum_{i=1}^{n_c} \beta^i \cdot |F_2^i(z)| \tag{A1}$$

where the weights $\beta_i$ satisfy $\beta^i > |\lambda_c(z_k)|$ for all $(\Delta z_k, \lambda_c(z_k))$, $z_k \epsilon \mathbf{D} := \{z|T(z) \le T(z_k)\}$. The vector $\lambda_c(z_k)$ is the Lagrange parameter of the problem linearized at $z_k$, $\Delta z_k$ the solution of the linearized problem. The damping factor $\alpha_k$ is determined by

$$T_1(z_k + \alpha_k \cdot \Delta z_k) = \min_{\alpha_k} \tag{A2}$$

It can be shown under mild conditions that then the Gauss-Newton-Method starting with an arbitrary initial guess converges to a stationary point.

Unfortunately, the practical benefits of this theoretically elegant result are limited: Even in mildly nonlinear ill-conditioned problems the direction of steepest descent of $T_1$ and the search direction $\Delta z_k$ are nearly orthogonal. Intolerable small step sizes are computed.

As a remedy to that so called "natural level functions" (cf. Deuflhard, 1974)

$$T_n^k(z) = \|J^+(z_k)F(z)\|_2^2 \tag{A3}$$

have been developed. These iteratively reweighted functions adapt to the local geometry of the problem. Together with special simplified line searches they allow significantly larger step sizes than by using the level function $T_1$, which are, in particular independent of the condition number. In the treatment of practical problems they have proven to be very effective.

## B. Internal Numerical Differentiation

Consider the solution $y(t)$ of an IVP

$$\dot{y} = f(t, y, p), \quad y(t_0) = y_0 \tag{B1}$$

as a function of the initial values and parameters $y(t) := Y(t, y_0, p)$. Provided differentiability of this function, the question of reliable and user friendly computation of $\partial Y/\partial y_0$ and $\partial Y/\partial p$ arises. In PARFIT an approximation of these quantities is generated by numerical differentiation of the approximating scheme $\hat{Y}(t, y_0, p)$ which is established by an integration procedure in the course of the computation of an approximation of $y(t) =: Y(t, y_0, p)$. $\hat{Y}$ is constituted by the chosen underlying integration method (extrapolation, Runge Kutta, multistep, etc.) and the step size and order strategies. The derivatives $\partial \hat{Y}/\partial y_0$ and $\partial \hat{Y}/\partial p$ are computed by finite differences or by adjoint schemes. This requires no preparation of the user and is implemented in such a way that even discontinuous dynamics are tractable. Details and refined variants can be found in Bock (1987) and Schlöder (1988).

## Acknowledgements

J. Kallrath gratefully acknowledges many encouraging discussions with V. Schulz and M. Steinbach (Institut für Mathematik, University of Augsburg) and their assistance concerning the application of PARFIT. Thanks are also directed to R. Dvorak (Institut für Astronomie, Vienna) for hospitality at Vienna Observatory and fruitful discussions on celestial mechanics, and to E. Baake (MPI-EB,Tübingen) for critical reading of the manuscript. We would also like to thank the careful referee (H. Eichhorn, Gainesville, University of Florida) for a couple of remarks and comments which helped to improve this paper. This work was sponsored by *Deutsche Forschungsgemeinschaft* under "Forschungsschwerpunkt Anwendungsbezogene Optimierung und Steuerung".

## References

Abramowitz, M. and Stegun, I.A. : 1970, *Handbook of Mathematical Functions*, Dover Publication, 9th printing.
Baake, E., Baake, M., Bock, H.G. and Briggs, K.M. : 1992, 'Fitting Ordinary Differential Equations to Chaotic Data', *Phys.Rev.* A45.
Bennetin, G., Galgani, L. and Strelcyn, J.M.: 1976, 'Kolmogorov entropy and numerical experiments', *Phys.Rev.* A14, 2338-2345.
Bennetin, G., Galgani, L., Giorgilli, A. and Strelcyn, J.M.: 1980, 'Lyapunov Characteristic Exponents for Smooth Dynamical Systems — A Method for Computing all of them', *Meccania* (March), 9-20, 21-30.
Bock, H.G. : 1981. in: Ebert, K.H.,Deuflhard, P. & Jäger, W. (Eds.) *Modelling of Chemical Reaction Systems*, Springer Series in Chemical Physics, Springer, Heidelberg.
Bock, H.G. : 1987. 'Randwertproblemmethoden zur Parameteridentifizierung in Systemen nichtlinearer Differentialgleichungen', *Bonner Mathematische Schriften* 183, Bonn.
Bock, H.G. and Schlöder, J.P. : 1986, 'Recent progress in the development of algorithm and software for large-scale parameter estimation problems in chemical reaction systems', in: Kotobh, P.(Ed.) *Automatic Control in Petrol , Petrochemical and Desalination Industries*, IFAC Congress, Pergamon, Oxford.
Bock, H.G., Eich, E. and Schlöder, J.P. : 1988, 'Numerical Solution of Constrained Least Squares Boundary Value Problems in Differential Algebraic Equations', in: Strehmel(Ed.) *Numerical Treatment of Differential Equations*, BG Teubner, Leipzig.
Bulirsch, R. and Stoer, J. : 1966, 'Numerical Treatment of Ordinary Differential Equations by Extrapolations Methods', *Numerische Mathematik* 8, 1–13.
Eckmann, J.-P. and Ruelle, D.: 1985, 'Ergodic Theory of chaos and strange attractors', *Rev.Mod.Phys.* 57, 617–657.
Eichhorn, H. : 1992, 'Generalized Least Squares Adjustment, a Timely but Much Neglected Tool', *Celestial Mechanics and Dynamical Astronomy*, this issue.
Deuflhard, P.: 1974, 'A modified Newton Methode for the Solution of Ill-conditioned Systems of Nonlinear Equations with Applications to Multiple Shooting', *Numerische Mathematik* 22, 289–311.

Froeschlé, C. : 1984, 'The Lyapunov Characteristic Exponents — Applications to Celestial Mechanics', *Celest. Mech.* **34**, 95–115.

Froeschlé, C. and Gonczi, R. : 1988, 'On the Stochasticity of Halley like Comets', *Celest. Mech.* **43**, 325-330.

Gauss, C.F. : 1809, '*Theoria Motus Corporum Coelestium in Sectionibus Conicus Solem Ambientium*', Perthes, F. and Besser, J. H., Hamburg, [reprinted in Carl Friedrich Gauss Werke, Vol. VII, Königliche Gesellschaft der Wissenschaften zu Göttingen, Göttingen 1871.]

Gonczi, R. and Froeschlé, C. : 1981, 'The Lyapunov Characteristic Exponents as Indicators of Stochasticity in the Restricted Three-Body problem', *Celest. Mech.* **25**, 271-280.

Hénon, M. and Heiles, C. : 1964, 'The Applicability of the Third Integral of Motion: Some Numerical Experiments', *Astron.J.* **69**, 73-79.

Holzfuss, J. and Parlitz, U. : 1991, 'Lyapunov exponents from time series', in: Arnold, L., Cranel, H. and Eckmann, J.-P., *Lyapunov Exponents*, Proceedings, Oberwolfach 1990, 263–271.

Schlöder, J.P. : 1988, 'Numerische Methoden zur Behandlung hochdimensionaler Aufgaben der Parameteridentifizierung', *Bonner Mathematische Schriften*, **187**, Bonn.

Wolf ,A., Swift, J.B.,Swinney, H.L. and Vastano, J.A. : 1985, 'Determining Lyapunov Exponents from a Time Series', *Physica* **16D**, 285–317.

Zeng, X., Eykholt, R. and Pielke, R.A. : 1991, 'Estimating the Lyapunov-Exponents from Short Time Series of Low Precision', *Physical Review Letters*, **66**, 3229–3232.

# THE ARRANGEMENT IN MEAN ELEMENTS SPACE OF THE
# PERIODIC ORBITS CLOSE TO THAT OF THE MOON

G.B. VALSECCHI

*I.A.S. – C.N.R., Reparto di Planetologia, viale dell' Università 11, 00185 Roma, Italy*
*E-mail: GIOVANNI@IRMIAS (EARN/BITNET)*

E. PEROZZI

*Telespazio s.p.a., via Tiburtina 965, 00156 Roma, Italy*

A.E. ROY

*Department of Physics and Astronomy, University of Glasgow, Glasgow G12 8QQ, U.K.*

and

B.A. STEVES

*Department of Mathematics, Glasgow Polytechnic, Cowcaddens Rd, Glasgow, U.K.*

**Abstract.** In a simplified model of the Earth-Moon-Sun system based on the restricted circular 3-dimensional 3-body problem, it is possible to find numerically a set of 8 periodic orbits whose time evolutions closely resemble that of the Moon's orbit. These orbits have a period of 223 synodic months (i.e. the period of the Saros cycle known for more than two millennia as a means of predicting eclipses), and are characterized by a secular rotation of the argument of perigee $\omega$. Periodic orbits of longer durations exhibiting this last feature are very abundant in Earth-Moon-Sun dynamical models. Their arrangement in the space of the mean orbital elements $\bar{e}$-$\bar{i}$ for various values of the lunar mean motion is presented.

**Key words:** Periodic orbits – motion of the Moon – Saros cycle

## 1. Introduction

The analysis by Perozzi et al. (1991) of the motion of the Moon over one Saros cycle has led to the discovery of a set of 8 symmetric periodic orbits in the restricted circular 3-dimensional 3-body problem that are remarkably close to the Moon's orbit (Roy et al., 1991; Valsecchi et al., 1991, 1992). The 8 orbits have the same duration of 223 synodic lunar months, but are composed of different pairs of mirror configurations (Roy and Ovenden, 1955). In the restricted circular 3-dimensional 3-body problem there are 16 different ways in which a mirror configuration may be obtained. Therefore there must exist 8 distinct periodic orbits with the same basic periodicities since a symmetric periodic orbit must contain two different mirror configurations, (Valsecchi et al., 1991, 1992).

In this paper we explore the space of the mean lunar orbital elements $\bar{e}$-$\bar{i}$ (eccentricity and inclination) using various values of the lunar mean motion, in order to discover the manner in which periodic orbits similar to the Saros are arranged.

To this purpose, in Section 2, we discuss the various periods involved in the the formation of periodic orbits similar to the Saros, taking from Delaunay's theory the expressions necessary to compute these periods. In Section 3, we present the

*Celestial Mechanics and Dynamical Astronomy* 56: 373–380, 1993.
© 1993 *Kluwer Academic Publishers.*

arrangement of the periodic orbits in $\bar{e}\text{-}\bar{i}$ space and discuss its main features. Conclusions are drawn in Section 4.

## 2. The Basic Periods

The periods involved in the Saros, and thus in the associated set of periodic orbits, are the synodic, anomalistic and nodical (also called draconitic) lunar months.

The synodic month $T_{\Delta\lambda}$ is the period of revolution of the angular variable $\Delta\lambda = \lambda - \lambda'$, where $\lambda$ and $\lambda'$ are the mean longitudes of the Moon and Sun respectively. The anomalistic month $T_M$ is the period of revolution of the mean lunar anomaly $M$, and the nodical month $T_\theta$ is the period of revolution of $\theta = \omega + M$, where $\omega$ is the argument of the lunar perigee.

The observed mean durations of these three months are $T_{\Delta\lambda} = 29.530589$ d, $T_M = 27.554550$ d and $T_\theta = 27.212221$ d. In the Saros they are related by

$$223T_{\Delta\lambda} \simeq 239T_M \simeq 242T_\theta. \tag{1}$$

Since the difference of duration between $T_M$ and $T_\theta$ is due to the secular rotation of $\omega$, we can also write

$$223T_{\Delta\lambda} \simeq 239T_M \simeq 3T_\omega \tag{2}$$

where $T_\omega$ is the period of revolution of $\omega$. In the periodic orbits associated with the Saros relation (2) holds exactly.

We may ask whether there are other sets of periodic orbits characterized by a relationship of the form (3), i.e.

$$N_{\Delta\lambda}T_{\Delta\lambda} = N_M T_M = N_\omega T_\omega \tag{3}$$

$N_{\Delta\lambda}$, $N_M$ and $N_\omega$ being suitable integers.

Let us start by keeping $\bar{n}$, the *observed mean* mean motion of the Moon, and $n'$, the mean motion of the Sun, fixed at their present values, so that

$$T_{\Delta\lambda} = \frac{2\pi}{\bar{n} - n'} \tag{4}$$

is also fixed at its present value; $\bar{n}$ differs from the *keplerian* mean motion $n$ by a small term (Brouwer and Clemence, 1961):

$$\bar{n} = n(1 - \nu^2) \tag{5}$$

where $\nu = n'/\bar{n}$.

The mean durations of $T_M$ and $T_\omega$ are given by

$$T_M = \frac{2\pi}{\bar{n} - \dot{\omega}} \tag{6}$$

$$T_\omega = \frac{2\pi}{\dot{\omega} - \dot{\Omega}} \tag{7}$$

as functions of the secular rotation rates of the longitudes of the lunar apse $\dot{\bar{\omega}}$ and node $\dot{\Omega}$.

We can obtain $\dot{\bar{\omega}}$ and $\dot{\Omega}$ as functions of $\nu$, $\bar{e}$, $\bar{i}$ using the formulae given by Delaunay (1872)

$$\dot{\Omega} = -\bar{n}$$
$$\left[ \left( \frac{3}{4} - \frac{3}{2}\gamma^2 + \frac{3}{2}\bar{e}^2 + \frac{9}{8}e'^2 \right. \right.$$
$$\left. + \frac{51}{8}\gamma^2\bar{e}^2 - \frac{9}{4}\gamma^2 e'^2 - \frac{21}{64}\bar{e}^4 + \frac{9}{4}\bar{e}^2 e'^2 + \frac{45}{32}e'^4 \right) \nu^2$$
$$- \left( \frac{9}{32} - \frac{27}{16}\gamma^2 - \frac{189}{32}\bar{e}^2 + \frac{23}{32}e'^2 \right.$$
$$\left. + \frac{27}{16}\gamma^4 + \frac{567}{16}\gamma^2\bar{e}^2 - \frac{99}{16}\gamma^2 e'^2 - \frac{675}{256}\bar{e}^4 - \frac{349}{16}\bar{e}^2 e'^2 \right) \nu^3$$
$$- \left( \frac{273}{128} - \frac{843}{128}\gamma^2 - \frac{2739}{128}\bar{e}^2 + \frac{3261}{256}e'^2 \right) \nu^4$$
$$- \left( \frac{9797}{2048} - \frac{7185}{1024}\gamma^2 - \frac{165411}{2048}\bar{e}^2 + \frac{73423}{1024}e'^2 \right) \nu^5$$
$$- \frac{199273}{24576}\nu^6 - \frac{6657733}{589824}\nu^7$$
$$\left. + \left( \frac{45}{32}\nu^2 + \frac{1935}{512}\nu^3 \right) \alpha^2 \right] \tag{8}$$

$$\dot{\bar{\omega}} = \bar{n}$$
$$\left[ \left( \frac{3}{4} - 6\gamma^2 - \frac{3}{8}\bar{e}^2 + \frac{9}{8}e'^2 \right. \right.$$
$$\left. - \frac{45}{4}\gamma^4 + \frac{69}{8}\gamma^2\bar{e}^2 - 9\gamma^2 e'^2 - \frac{3}{32}\bar{e}^4 - \frac{9}{16}\bar{e}^2 e'^2 + \frac{45}{32}e'^4 \right) \nu^2$$
$$+ \left( \frac{225}{32} - \frac{189}{8}\gamma^2 - \frac{675}{64}\bar{e}^2 + \frac{825}{32}e'^2 \right.$$
$$\left. + \frac{1107}{16}\gamma^4 + \frac{81}{32}\gamma^2\bar{e}^2 - \frac{349}{4}\gamma^2 e'^2 - \frac{2475}{64}\bar{e}^2 e'^2 \right) \nu^3$$
$$+ \left( \frac{4071}{128} - \frac{3963}{32}\gamma^2 - \frac{31605}{512}\bar{e}^2 + \frac{61179}{256}e'^2 \right) \nu^4$$
$$+ \left( \frac{265493}{2048} - \frac{335403}{512}\gamma^2 - \frac{1483665}{4096}\bar{e}^2 + \frac{1767849}{1024}e'^2 \right) \nu^5$$
$$+ \left( \frac{12822631}{24576} - \frac{25291729}{16384}\bar{e}^2 \right) \nu^6$$
$$+ \left( \frac{1273925965}{589824} + \frac{352038885}{1179648}\bar{e}^2 \right) \nu^7$$

$$+ \frac{71028685589}{7077888} \nu^8 + \frac{32145882707741}{679477248} \nu^9$$
$$+ \left( \frac{45}{32} \nu^2 + \frac{7425}{512} \nu^3 \right) \alpha^2 \Bigg] \tag{9}$$

in which $\gamma = \sin \frac{\bar{i}}{2}$ and

$$\alpha = \frac{\bar{a}}{a'} = \sqrt[3]{\frac{\mu \nu^2 (1 - \nu^2)^2}{1 + \mu}} \tag{10}$$

$\mu$ being the ratio of the mass of the Earth-Moon system to that of the Sun, $\bar{a}$ being the mean semimajor axis of the lunar orbit and $a'$, $e'$ being the semimajor axis and eccentricity of the geocentric orbit of the Sun.

## 3. Periodic Orbits in the $\bar{e}$-$\bar{i}$ Space

Using the formulae of Section 2 we have made a search for pairs $\bar{e}$-$\bar{i}$ such that (3) is satisfied. The procedure uses a 2-dimensional root finder taken from Press et al. (1986). We have set $e' = 0$ in (8) and (9), so that the pairs of $\bar{e}$-$\bar{i}$ found correspond to periodic orbits of the restricted *circular* problem. The value of $\nu$ is taken equal to the observed one.

Figure 1 shows the orbits found in the region $0 \leq \bar{e} \leq 0.1$, $0° \leq \bar{i} \leq 10°$, for $N_\omega \leq 10$ (upper left), $N_\omega \leq 20$ (upper right), $N_\omega \leq 40$ (lower left), and $N_\omega \leq 80$ (lower right). The point located at $\bar{e} = 0.0549$, $\bar{i} = 5.133°$, easily identifiable in the lower plots because it breaks the regularity of the patterns, gives the actual orbit of the Moon for comparison.

Figure 2 shows all the orbits found for $N_\omega \leq 115$ on a larger scale, to allow a better estimate of the details. The figure resembles a city map, with "squares", large "avenues" and an intricate (and to a certain degree self-similar) network of narrower "streets". The "squares" are situated at the crossings of "avenues", and have at their centres a point representing a periodic orbit whose $N_\omega$ is much smaller than those of the surrounding orbits. The "avenues" and the "streets" have at their centres "traffic lines", and get narrower (and the "squares" likewise get smaller) as we allow larger and larger values of $N_\omega$.

The point representing the periodic orbits associated with the Saros (i.e. the point for $N_{\Delta\lambda} = 223$, $N_M = 239$, $N_\omega = 3$) is located near the centre of Fig. 2 at the centre of the most conspicuous "square" in the figure, the point's coordinates being $\bar{e} = 0.0623$ and $\bar{i} = 5.57°$ (it is tempting to name it "Saros square"!). Interestingly the point representing the actual mean elements of the Moon is rather close to "Saros square", as is readily seen in Fig. 1.

Each of the points represented in Figs. 1 and 2 actually corresponds to a set of 8 periodic orbits, all of which have the same periods and mean elements, as already discussed in the Introduction; sample checking of some of the points found has provided verification. This is done numerically by finding the sets of

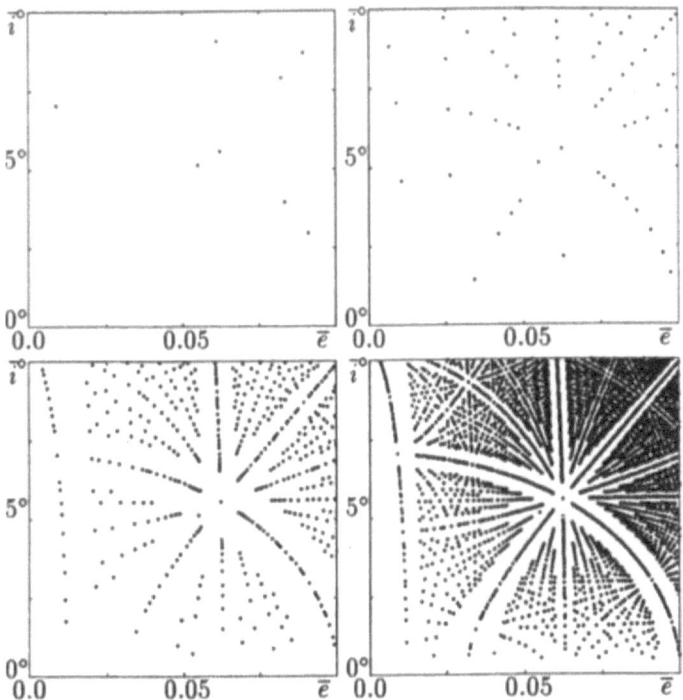

Fig. 1. Mean eccentricities and inclinations of periodic orbits of the restricted circular 3-dimensional 3-body problem close to the orbit of the Moon, for a number of revolutions of the argument of perigee $\omega$ of up to 10 (upper left), 20 (upper right), 40 (lower left) and 80 (lower right).

*osculating* elements which describe the initial conditions of the periodic orbits and is the same method used to find the periodic orbits associated with the Saros. It has been possible to find the initial conditions for periodic orbits having up to $N_{\Delta\lambda} = 8534$, $N_M = 9146$, and $N_\omega = 115$. In particular, the set of periodic orbits with these coefficients is associated with the long-term lunar cycle first attributed to Hipparchus (Steves, 1990).

The structure of periodic orbits shown in Figs. 1 and 2 has been computed for the current value of $\nu$, the ratio of the mean motions of the Sun and the Moon. Using the same procedure it is possible to obtain such a structure for other values of $\nu$, in order to understand the effects of changing $\overline{n}$ through the action of the tides raised on the Earth by the Moon.

Figure 3 shows the results for 4 different values of $\nu$: $\nu = 0.999\nu_0$, $\nu = 0.9995\nu_0$, $\nu = 1.0005\nu_0$ and $\nu = 1.001\nu_0$, where $\nu_0$ is the current value of $\nu$. The first two values have been met by the lunar orbit in its past evolution, while the other two are likely to be met in the future.

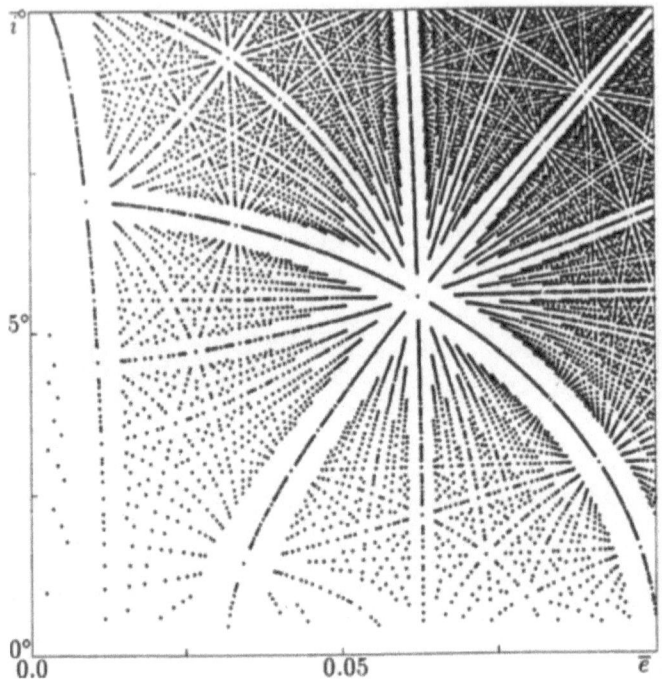

Fig. 2. Same as Figure 1 but for up to 115 revolutions of the argument of perigee $\omega$.

To understand the overall trend we concentrate on the position in the plots of the "Saros square". This feature is rather easily identifiable in the diagram for $\nu = 0.9995\nu_0$, at $\bar{e} = 0.0871$, $\bar{i} = 5.41°$, and for $\nu = 1.0005\nu_0$, at $\bar{e} = 0.0137$, $\bar{i} = 5.72°$, respectively to the right and to the left of the present position (see Fig. 2 for comparison). For $\nu = 0.999\nu_0$ it has still to "enter" from the right, and for $\nu = 1.001\nu_0$ it has just been "absorbed" from the left.

Due to the different dependencies of $\dot{\Omega}$ and $\dot{\bar{\omega}}$ (i.e.(8) and (9)) on $\bar{e}$ and $\bar{i}$, increasing values of $\nu$ cause the inclination of the periodic orbits to increase slowly while the eccentricity decreases at a much faster rate. In the plots it is also noticeable that, at constant $\nu$, there is a general decrease of the density of periodic orbits for decreasing $\bar{i}$ and especially for decreasing $\bar{e}$.

It should be noted that if a substantial fraction of the periodic orbits shown in Fig. 2 turns out to be unstable, then the figure would be an Arnold's web relative to the lunar problem, i.e. it would show points that pertain to regions of the $\bar{e}$-$\bar{i}$ space through which Arnold diffusion is possible (we are grateful to A. Milani, who called our attention to this).

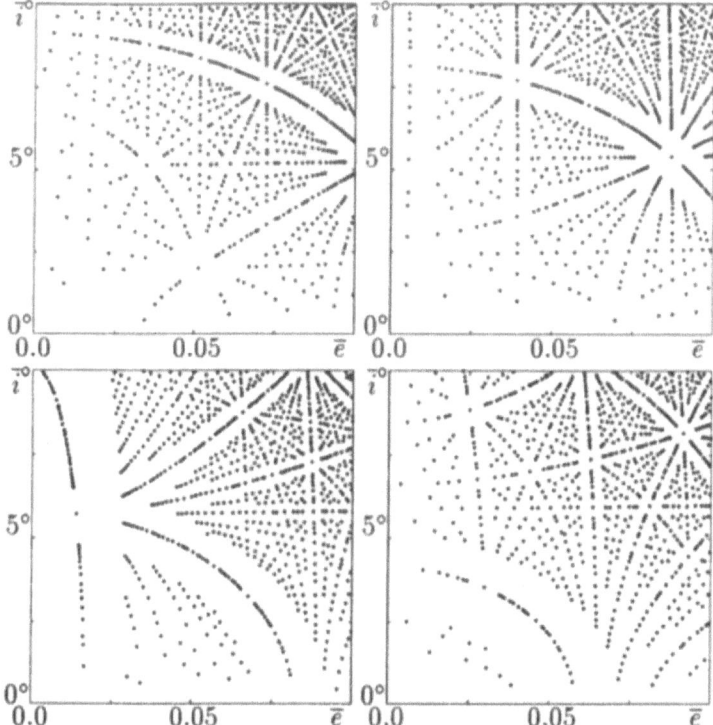

Fig. 3. Same as Figure 1 for a number of revolutions of the argument of perigee $\omega$ up to 50, and for $\nu = 0.999\nu_0$ (upper left), $\nu = 0.9995\nu_0$ (upper right), $\nu = 1.0005\nu_0$ (lower left) and $\nu = 1.001\nu_0$ (lower right), where $\nu_0$ is the current value of $\nu$.

## 4. Conclusions

Given $\bar{n}$, Delaunay's expressions for the secular motions of $\Omega$ and $\tilde{\omega}$ combined with (3), enable the computations of the mean eccentricities and inclinations of periodic orbits in the restricted circular 3-dimensional 3-body Earth-Moon-Sun model.

These orbits are arranged in a rather peculiar pattern in $\bar{e}$-$\bar{i}$ space, with certain areas displaying a lower density of points. As the period of the periodic orbits studied increases, these areas of lower density get filled with points at a slower rate than other regions of the $\bar{e}$-$\bar{i}$ space.

For different values of the lunar mean motion, the network of points moves in a direction almost parallel to the $\bar{e}$-axis, with the points representing orbits with given ratios of the periods disappearing at $\bar{e} = 0$ as one increases $\bar{n}$.

The current values of the mean eccentricity and inclination of the lunar orbit are rather close to one of the most conspicuous areas devoid of points in $\bar{e}$-$\bar{i}$ space. This agrees with the finding that the lunar orbit is well approximated by the set of periodic orbits (Roy et al., 1991; Valsecchi et al., 1991, 1992) occupying the centre of the area of avoidance.

## Acknowledgements

We are grateful to M. Carpino, A. Celletti, Cl. Froeschlé, M. Hénon, A. Milani, A. Morbidelli and A.M. Nobili for discussions, suggestions and encouragement.

## References

Brouwer D., Clemence, G.M.: 1961, *Methods of Celestial Mechanics*, Academic, New York

Delaunay Ch.: 1872, 'Note sur les mouvements du périgée et du noeud de la Lune', *Compt. rend. hebdom. Acad. Sci.* **74**, 17-21

Perozzi E., Roy A.E., Steves B.A., Valsecchi G.B.: 1991, 'Significant high number commensurabilities in the main lunar problem. I: the Saros as a near periodicity of the Moon's orbit', *Celest. Mech. & Dynam. Astron.* **52**, 241-261

Press W.H., Flannery B.P., Teukolsky S.A., Vetterling W.T.: 1986, *Numerical Recipes*, Cambridge Univ. Press, Cambridge

Roy A.E., Ovenden M.W.: 1955, 'On the occurrence of commensurable mean motions in the solar system. II. The mirror theorem', *Mon. Not. R. Acad. Sci.* **115**, 296-309

Roy A.E., Steves B.A., Valsecchi G.B., Perozzi E.: 1991, in A.E. Roy, ed(s)., *Predictability, Stability and Chaos in N-Body Dynamical Systems*, Plenum, London, 273-282

Steves B.A.: 1990, *Finite-Time Stability Criteria for Sun-Perturbed Planetary Satellites*, PhD Thesis, Univ. Glasgow

Valsecchi G.B., Perozzi E., Roy A.E., Steves B.A.: 1991, *Periodic orbits close to that of the Moon. A postscript to a discovery by the ancient Chaldeans*, I.A.S. Internal Report n. 2

Valsecchi G.B., Perozzi E., Roy A.E., Steves B.A., 1992, 'Periodic orbits close to that of the Moon', submitted to *Astron. Astrophys.*

# RESONANCE TRAPPING OF CIRCUMSTELLAR DUST PARTICLES
# BY AN ALLEGED PLANET

H. SCHOLL

*Observatoire de Nice, Dept. Cassini, B.P. 229, F-06304 NICE Cedex 4, France*

F. ROQUES

*Observatoire de Meudon, DAEC, F-92195 Meudon, France*

and

B. SICARDY

*Université Paris VI and Observatoire de Meudon, DAEC, F-92195 Meudon, France*

**Abstract.** The dynamical evolution of dust particles forming a circumstellar disk around $\beta$ Pictoris is followed by numerical simulations on a Connection Machine. The disk appears to be cleared inside a radius of about 20 AU. We integrate simultaneously the orbits of 8,000 dust particles subjected to Poynting-Robertson drag and perturbed by one alleged planet. The simulations show that a planet revolving about $\beta$ Pictoris at a mean distance of 20 AU with a mass of at least $2 * 10^{-5}$ central stellar mass can confine the disk by outer resonance trapping. The azimuthal density distribution of particles which shows very strong variations. appears to be stationary in a frame rotating with the planet.

**Key words:** Resonance trapping – Poynting-Robertson effect – circumstellar disks

## 1. Introduction.

The combination of IRAS and ground-based imagery indicates that the $\beta$-Pictoris circumstellar disk is mainly composed of dust particles and that it extends out to at least 250 AU (Smith and Terrile, 1984, Artymowicz et al. 1989) from the central star. In its inner part the disk appears to be cleared inside a radius of about 20 AU. Since observations yield radii of dust particles in the micron range, dust particles are expected to spiral towards the central star due to the Poynting-Robertson effect (e.g. Burns et al. 1979). Semi-major axis and eccentricity are decreasing while the inclination of a particle's orbit is not effected. No inner boundary of the disk would form since the particles appear to fall towards the central star. Which mechanism confines the disk ?

We conjecture that the inner boundary of the circumstellar disk is due to the presence of a planet. In case of a completely thin disk (the inclinations of particle orbits are nearly zero), encounters between the planet and particles which cross the planetary orbit may clear the region interior to the planetary orbit. Observations, however, indicate a thickness of the disk corresponding to maximum orbital inclinations of about 8°. This diminishes the collision probability for the dust particles : no observable clearing region would appear. A comparatively large number of particles would cross the planetary orbit without a catastrophic encounter.

Is resonance trapping of particles a possible mechanism to confine the disk? It is well known that dust particles entering a Kirkwood gap (inner resonance with Jupiter) while spiraling towards the Sun are not trapped. (Gonczi et al. 1982, Torbett and Smoluchowski 1982). In these inner resonance regions, where a particle

*Celestial Mechanics and Dynamical Astronomy* **56**: 381–393, 1993.
© 1993 *Kluwer Academic Publishers.*

revolves faster than the Jupiter about the Sun, the loss of momentum and energy of
a particle's orbit due to the Poynting-Robertson effect cannot be counterbalanced
by a corresponding transfer from the planetary to the particle's orbit. In an outer
resonance where a particle revolves slower than the planet, however, gain and loss
of orbital momentum as well as of orbital energy may be kept in equilibrium. The
time scale for this equilibrium state depends on the values for the planetary mass,
for the eccentricity of the planetary orbit, for the particle's radii and depends on the
order of the resonance. These values can be determined by numerical simulations
and by theory. The simulations are expected to yield additional information about
the radial and azimuthal density distribution of the particles in the disk which are
trapped in resonances, a result which may stimulate searches for an alleged planet.

The Connection Machine CM-2, a massively parallel computer, appears to be an
ideal machine to study collective dynamical phenomena of many particles like disk
confinement or accumulation of material due to resonance trapping. Thousands of
orbits can be integrated in parallel and the large amount of resulting data can be
easily visualized. In the following sections we describe first the physical model,
the implementation of our model on a Connection Machine CM-2, and in the
last section we present our first results concerning density distribution and disk
confinement.

## 2. The Numerical Model

The physical model is basically the three-dimensional restricted three-body problem, central star - alleged planet - dust particle, modified by radiation pressure and
Poynting-Robertson effect.

The corresponding differential equations for the motion of a particle are:

$$\frac{d^2 \mathbf{r}_i}{dt^2} = -k^2 m (1 - \beta) \frac{\mathbf{r}_i}{|\mathbf{r}_i|^3} \tag{1}$$

$$- k^2 m_p \left( \frac{\mathbf{r}_i - \mathbf{r}_p}{|\mathbf{r}_i - \mathbf{r}_p|^3} + \frac{\mathbf{r}_p}{|\mathbf{r}_p|^3} \right) \tag{2}$$

$$- \frac{k^2 m \beta}{c} \frac{\mathbf{v}_i}{|\mathbf{r}_i|^2} - \frac{k^2 m \beta}{c} \frac{\mathbf{r}_i (\mathbf{r}_i * \mathbf{v}_i)}{|\mathbf{r}_i|^4} \tag{3}$$

The symbols t, k, c, m, and $m_p$ denote respectively time, Gauss constant,
velocity of light, mass of the central star, and mass of the planet. $\mathbf{r}_i$ and $\mathbf{v}_i$ denote
respectively the position and the velocity vector of the i-th particle. $\mathbf{r}_p$ is the
position vector of the planet. The forces acting on the i-th particle are (1) due to
the central star, (2) due to the planet and (3) due to the Poynting-Robertson effect.
The second term in (3) was omitted since it contributes usually very little to the
Poynting-Robertson effect. We have carried out a few tests including this term
which had no significant effect on our results.

The parameter $\beta$ is defined by:

$$\beta = \frac{\alpha c}{m G} \tag{4}$$

$$\alpha = \frac{3 E}{16 \pi c^2 s \rho}, \tag{5}$$

where m is the mass of the central star, G is the gravitational constant, E is the luminosity of the star, c is the velocity of light, s is the particle's radius, and $\rho$ is the particle's mean density.

The parameter $\beta$ can be simply expressed by

$$\beta = \frac{radiation\ force}{gravitational\ force} \tag{6}$$

The latter relation implies that the value for $\beta$ is independent of the particle's distance from the central star since both gravitational and radiation force follow the same distance law.

Radiation pressure and Poynting-Robertson effect are consequences of the interaction between dust grains and photons emitted from the star (Burns et al. 1979). Photons carry energy and momentum. A dust grain which is hit by photons feels a pressure due to the momentum carried by photons. This is called the radiation pressure. As a result, the dust grains feel the gravitational attraction exerted by the central star reduced by a factor of $(1 - \beta)$. Parameter $\beta$ ranges from 0 to 1. For particles of micron size, $\beta$ has values of the order of 0.1.

A dust grain absorbs the photons which are emitted from the central star anisotropically. Since the dust grain is in thermal equilibrium, the absorbed energy is reradiated isotropically. The reradiated photons carry momentum. As a consequence of the anisotropic gain and the isotropic loss of momentum, the particles' orbits loose momentum; the particles spiral slowly into the central star. This is the Poynting-Robertson effect. Analytical expressions are known for the two-body problem central star - dust particle which describe the timely variations of eccentricity and semi-major axis of a particle's orbit (Wyatt and Whipple 1950). In particular, these analytical expressions allow an estimation for the time interval to spiral into the central star. In the three-body problem, however, these estimations can be totally wrong due to the trapping of dust particles in resonances.

## 3. The Parameters of the Model and Starting Values

Since $\beta$ Pictoris is classified as a A5-star, we put its mass equal to 1.5 solar masses. The planetary mass is given in units of the mass of $\beta$ Pictoris. The semimajor axis of the planetary orbit is set equal to 20 AU.

The parameters of the model are : planetary mass $m$, the eccentricity of the planetary orbit $e_P$, the radiation pressure parameter $\beta$, and the initial osculating elements for the particle orbits.

The different sets of starting osculating elements represent different scenarios and stages for the dynamical evolution of the circumstellar disk. In the following we concentrate on the simple scenario that all particles have originated far from the central star and spiral towards the central star without any mutual collisions.

While spiraling towards the central star, the eccentricities of the particle orbits decrease. The inclinations are not effected. This is true as long as the particles do not cross a resonance where significant changes in eccentricity and inclination may occur. Preliminary calculations with a planetary mass of $10^{-4}$ showed that the first important resonance encountered by the particles is the outer 2/1 resonance located near 28 AU. The exact location depends on the value for $\beta$. For larger planetary masses the first important resonance is located at larger distances from the central star.

We choose starting semi-major axes at random between 30 and 40 AU. At this distance, the eccentricities can expected to have decreased to nearly zero while the inclinations are expected not to have changed. We assume an uniform distribution of orbital inclinations between 0° and 8°. All the angle variables are selected at random between 0° and 360°.

## 4. Implementation of the Numerical Model on a Connection Machine.

### 4.1. ARCHITECTURE OF THE CONNECTION MACHINE

Which integrator is suitable for our problem on the CM-2 ? This is not a trivial question since the massive parallelism of the Connection Machine may not be favourable for schemes known to be fast on traditional sequential computers. The Connection Machine CM-2 is a SIMD (single instruction multiple data) machine; all active processors execute the same instruction simultaneously on different data (eg. Hillis 1985, 1987). From the point of view of a FORTRAN scientist, the CM-2 can be considered as a sample of micro-computers which all execute the same instruction of a code simultaneously and which can exchange data on a very rapid network. The speed of a code running on a CM-2 is mainly determined by (1) the intrinsic speed of the floating point accelerators implemented in the micro-computers, (2) by the code dependent need of transporting data between the microcomputers, and (3) by the number of simultaneously active processors which also depends on the user's code.

We used a so-called CM-2/8k with 256 micro-computers (each associated with one floating point accelerator) and with 1 Megabyte memory each. Data in the memory of each micro-computer is handled by 32 simple data processors. This gives a total of 8192 (8k) data processors which can handle 8192 data elements in parallel. A total of 256 floating point operations are carried out in parallel.

From the point of view of data storing, the memory of the CM-2/8k can be considered as being divided up into 8192 sections each controlled by a data processor mentioned above. A vector appearing in a code is stored in such a way that one component of the vector is stored in one memory section. Scalars like the number $\pi$ are stored separately outside of the Connection Machine with a special fast link to the memory sections. The virtual architecture of the Connection Machine allows the storage of vectors with $2^n * 8192$ elements, $n$ being a positive integer. One processor controls in such a case $2^n$ sections.

It is one of the arts of programming a Connection Machine to associate the parallel variables of the problem with memory sections. The aim of this association is to minimize communication costs among sections during the execution of the code. For instance, the addition of two components of a vector $B(1234) = A(5) + A(211)$ requires the transport of $A(5)$ to the memory section where $A(211)$ is stored. After the arrival of $A(5)$ the addition $A(5) + A(211)$ is carried out and the result has to be transported to the section where $B(1234)$ is stored. On the other hand, the computation of $B(1234) = A(1234) + C(1234)$ does not need any transport between memory sections since all the three components A(1234), B(1234) and C(1234) are stored in the same memory section.

In our problem it is natural to associate all parameters which describe the state of one particle with the same memory section: the components of the position and the velocity vector of one particle and the parameter $\beta$ which is related to the particle's size are stored in one section. Since no interactions among the particles are taken into account, no communications among the memory sections are necessary. The problem is the storage of the planetary coordinates. The corresponding position vector $r_P$ is a priori not a parallel variable since it contains only two components; the planetary orbit is taken as the plane of reference in our model. There are two possibilities to store $r_P$: either replicate it in each memory section and make it a parallel variable or store it in the memory for scalar variables. Our choice of the first possibility is directly related to the implementation of the integration scheme on the CM-2. As a consequence, we integrate the restricted three-body problem 8192 times in parallel.

## 4.2. CHOICE OF A SUITABLE INTEGRATOR

It is obvious that an integrator with variable step size is preferable since the eccentricities of particles entering a resonance may be pumped up very strongly. Choosing an integrator with variable step size on a massively parallel machine (this is also true for other parallel machines) may have a very unpleasant consequence: it is the particle with the smallest step size which slows down the remaining particles. Hence, the advantage of using a scheme with variable step size on a sequential computer is not necessarily an advantage for parallel computers.

For our problem there is a way out of this dilemma, since the particles do not interact. Each particle can progress on its own proper time since we calculate the restricted three-body problem 8192 times independently in parallel. This is possible if the statements of the integration scheme are independent on the value of the time step.

The dilemna of using an integrator with variable step size on a SIMD machine might be also overcome by regularizing the equations of motion (1-3) (suggestion by V. Szebehely). Then, an integrator with fixed step size can be used unless eccentricities reach values of nearly 1.

There are three well known types of integration schemes with variable step size: Bulirsch-Stoer, Adams (predictor-corrector) and Runge-Kutta.

Besides the test for step size, the Bulirsch-Stoer scheme (Bulirsch and Stoer 1966) makes another test, namely for the number of extrapolation points. Such tests should be avoided on SIMD machines since they slow down the computation: In a sample of 8192 particles we can expect always at least one with a negative test. This means that all the other particles have to wait and their corresponding processors are inactive while an additional computation has to be carried out for the particles with the negative test.

A similar statement holds for Krogh's DVDQ integrator (Krogh 1970) which is a particularly fast predictor-corrector scheme with variable order and variable step size. It is the test for the order which slows down this scheme on a SIMD machine.

Runge-Kutta schemes (e.g. Press et al. 1989) carry out only one test which slows down the code on a SIMD machine, namely the test for step size. We choose the classical 4-th order Runge-Kutta integrator, because our CM-2 has floating point accelerators with only 32 bits (single precision). This allows the integration of orbits over 10000 planetary revolutions, a sufficient time scale for our purpose.

We like to illustrate how the particles advance on their proper time. Suppose that particles A and B are on the same proper time, $t_A = t_B$ and that they have the same values for their respective step sizes $h_A = h_B$. The Runge-Kutta scheme requires four evaluations of the equations of motion in the interval $[t_A, t_A + h_A]$ for particle A and $[t_B, t_B + h_B]$ for particle B. This can be done in parallel. For the time test the same calculations are carried out but with half the step size and the results for the corresponding calculations with the full and with half the step size are compared. This is done in parallel. Suppose that the test is positive for particle A and negative for particle B: This requires a different computation for A which has to reduce its step size and for B which tries a larger step size. These are different algorithms which can not carried out in parallel on a SIMD machine. Sequential computation is necessary. It may occur that only one processor is active for reducing the step size. This is the only part in the code which does not guarantee full parallelism.

After the determination of the new step sizes $h_A$ and $h_B$, the evaluation of the equations of motion in the interval $[t_A, t_A + h_A]$ for particle A and $[t_B, t_B + h_B]$ for particle B can be again carried out in parallel. Particle A and B have now different proper time.

Statistically, the whole cloud of 8192 particles evolved in all our models at about

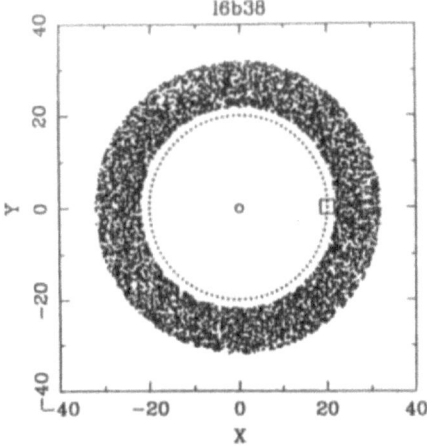
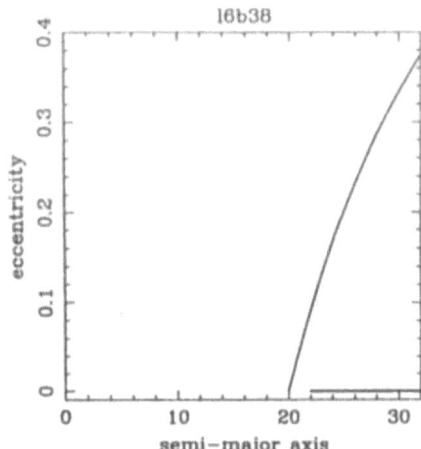

Fig. 1. To the left, starting configuration in a cartesian coordinate system for the circular 3-D model with a planetary mass of $10^{-6}$ central stellar mass. The position of the planet on its orbit (dashed circle) is marked by a square. The central star is marked by a small circle. To the right, starting configuration of the model of the left in a semi-major axis vs. eccentricity diagram. Particles form a straight line between 22 and 32 AU. The curved line corresponds to a pericentric distance of 20 AU and indicates planet-crossing.

the same rate. A run was stopped when 80 percent of the sample was integrated over 10000 planetary revolutions.

A model with 8192 particles needs between 4 and 12 hours CPU time on a CM-2/8k depending on the choice of values for the model parameters.

## 5. Major Results

We describe in the following the outcome of three runs which show the essential results. In all the three runs, $\beta$ is set equal to 0.3. Our major conclusions are independent for $\beta$ ranging from 0.1 to 0.3. In the first example, we show that a planetary mass of $10^{-6}$ central stellar mass is too small to trap particles in resonances. Also possible close encounters between the planet and particles moving over the planetary orbit do not form a visible boundary of the disk. No disk confinement occurs. The planet revolves on a circular orbit at a distance of 20 $AU$ about $\beta$ Pictoris. The starting inclinations of the particles vary between 0° and 8°. Starting eccentricities are all zero.

Two different coordinate systems are used to present the results: a cartesian system defined by the planetary orbital plane, and an orbital element system, semi-major axis vs. eccentricity. In the first system, the positions of the particles are projected on the cartesian system. Figures 1a and b show the particles in both coordinate systems just before the particles start to move over the planetary orbit

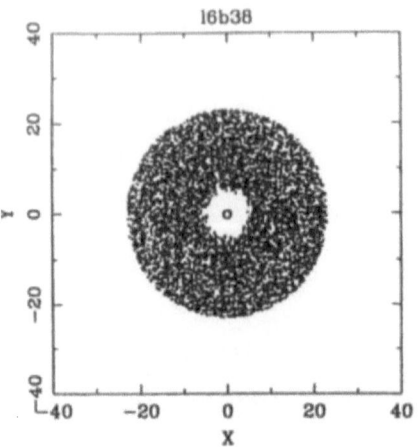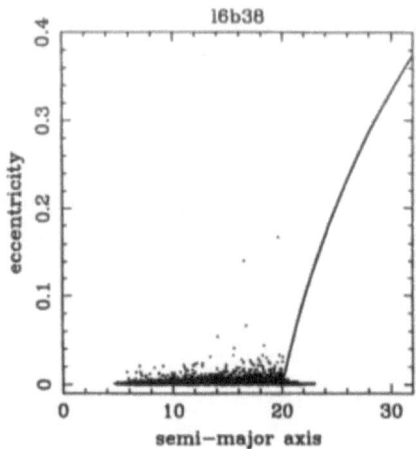

Fig. 2. Particle distribution 6500 planetary revolutions after figure 1a. Most particles have moved over the planet's orbit without a catastrophic encounter. A planet with $10^{-6}$ central stellar mass has no observable effect. The eccentricities of most particles are only slightly changed which demonstrates the negligible effect of the planet on the orbits of the particles.

due to the Poynting-Robertson effect. The dashed line in figure 1a represents the planetary orbit with a semi-major axis of 20 AU. The position of the planet is marked by a square and the central star is marked by a small circle at the origin of the coordinate system.

In figure 1b, the curved dashed line corresponds to a pericentric distance of 20 AU and, hence, indicates crossing of the planetary orbit. The orbits of particles situated below this dashed line do not cross the planetary orbit because their corresponding eccentricities are too small. Particles situated above this line between 20 and 40 AU are planet-crossers. We like to point out that planet-crossing does not necessarily mean a close encounter with the planet. In particular in resonance regions, planet-crossers may avoid close encounters due to the libration of their respective critical resonance argument. On the other hand, non-resonant particles situated above this line can be expected to encounter the planet.

The particles in figure 1b form a straight line between 22 and 32 AU. Due to the Poynting-Robertson effect, particles move from the right to the left in this diagram and may move over the planet's orbit with the possibility of close encounters with the planet. It is natural to assume that such encounters may clear the region interior to 20 AU and that the main mechanism to form an inner boundary of the disk is based on the ejection of particles by encounters with the planet. This is, however, a wrong conjecture according to our results.

Figures 2 shows the particles 6500 planetary revolutions after figure 1. Most of the particles have moved over the planetary orbit without a catastrophic encounter. The presence of a planet with a mass of $10^{-6}$ would have no observable effect.

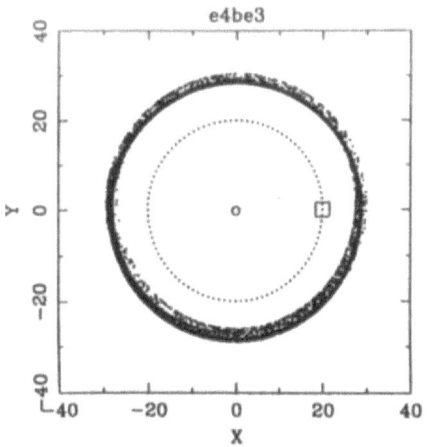
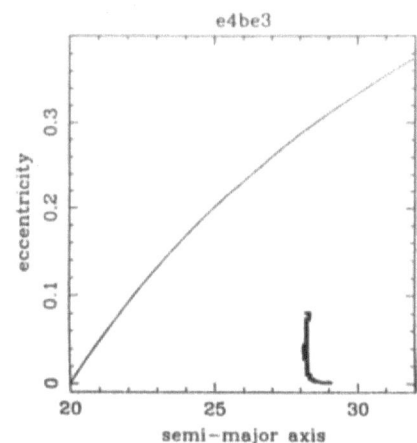

Fig. 3. Evolved state of a circular 2-D model with a planetary mass of $10^{-4}$ central stellar mass. Most particles are trapped in the 2/1 resonance and pump up eccentricities.

Even in a purely planar model (all particle inclinations are zero), the number of catastrophic encounters is not sufficient to clear the region interior to the planetary orbit according to our simulations.

In the second run, a planetary mass of $10^{-4}$ is used and the planetary orbit is circular. Since resonance trapping occurs for such a planetary mass, the starting semi-major axes of the particles are put outside of the first important resonance, the 2/1 resonance. Particles start between 32 and 33 $AU$ on circular orbits with an inclination of 0° (2-D model).

Figure 3 show the state of the system when part of the particles are trapped in the 2/1 resonance at 28.188 $AU$; this value depends on the parameter $\beta$. The remaining particles have not yet arrived in the 2/1 resonance. During resonance trapping, the eccentricities of the particles are pumped up. A semi-analytical description for eccentricity pumping is given in a different paper in this volume (Lazarro et al.).

After a total of 7500 planetary revolutions, all the particles are trapped in the 2/1 resonance (figure 4). This has a strong effect on the azimuthal distribution of particles near the planetary orbit. At an angle of about 90° behind the planet, particles appear to accumulate. The whole pretzel-like pattern (figure 4a) revolves with the speed of the planet. Since particle speed is different from the planetary speed, one has the impression of a trailing wave sweeping through the disk. Of course, this is only an apparent wave since no interactions among particles are taken into account in our model. The trailing wave may be an observable effect and may be the basis to detect a planet. We concede that our figure shows the best case, i.e. a planar model where all particles have the same value for $\beta$. In a more realistic model with differing values for $\beta$, the trapping would occur at different distances

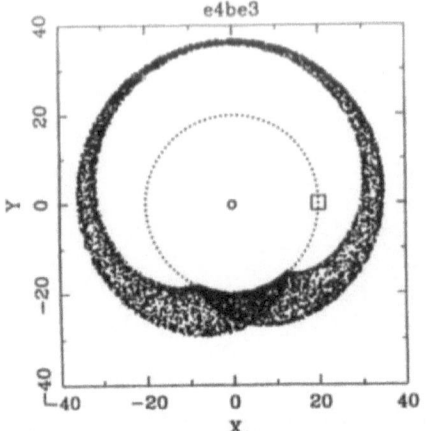
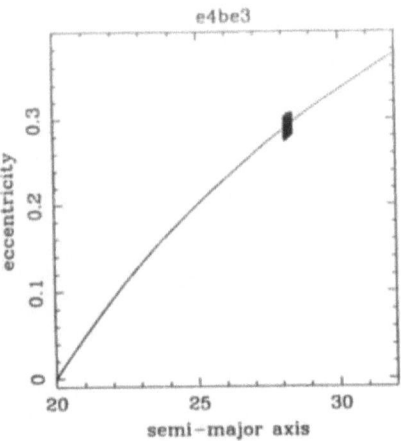

Fig. 4. A pretzel like pattern appears 3000 planetary revolutions after figures 3a and b. The azimuthal density distribution of the particles is not uniform. Planetary motion is counter-clockwise. All the particles are trapped in the 2/1 resonance.

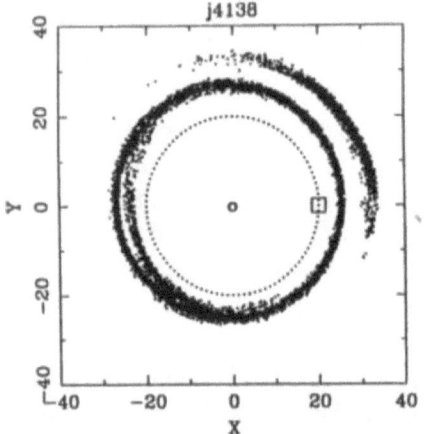
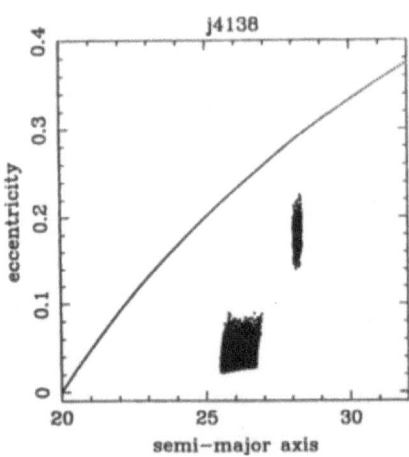

Fig. 5. Evolved state of a circular 2-D model with an eccentricity of 0.01 for the planetary orbit and a planetary mass of $10^{-4}$ central stellar mass. Two arcs appear. One arc is clearly visible in the upper right part of the diagram. The second arc is situated in the lower left part opposite to the first arc. Part of the particles have crossed the 2/1 resonance without being trapped. The particles trapped in the 2/1 resonance form the two arcs in figure 5a

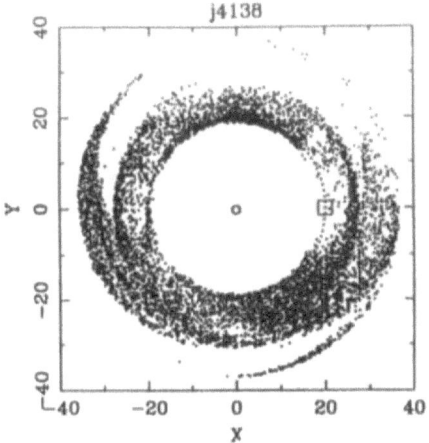
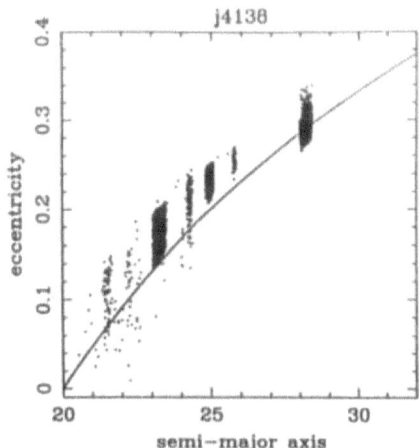

Fig. 6. A complex structure appears 3000 planetary revolutions after figs. 5a,b. Near the planet a minimum in the azimuthal density distribution of particles appears. Maxima of the density distribution are behind and ahead of the planet. Planetary motion is counter-clockwise. The large majority of particles is trapped in resonances. The spectrum of major outer resonances appears. During the trapping, eccentricities are pumped up.

from the central star, since the position of a resonance depends on $\beta$ (effect of radiation pressure). We found that the value of about 90° is independent of $\beta$ and, hence, our main result concerning the possibility of detecting a planet holds for a 2-D model.

Figure 5 shows the results for a 3-D model with a planet on an eccentric orbit, $e_P = 0.01$. Starting values are like in the previous model except that the inclinations of the particle orbits are distributed uniformly between 0° and 8°. In figure 5, part of the particles are trapped in the 2/1 resonance while others have crossed this resonance without being trapped. Two arcs appear in figure 5a. One arc is clearly visible in the upper right part of the diagram. The second arc is situated in the lower left part opposite to the first arc. Such a feature is not totally unexpected since the formation of arcs at resonances is a well known phenomenon (Sicardy 1991).

After 8000 planetary revolutions (figure 6), the large majority of the particles is trapped in mean motion resonances. Figure 6a shows the resonance spectrum. Most particles are planet-crossers but without encountering the planet. Resonances provide a protection against encounters. Particles are trapped in particular in the resonance 2/1 (28.2 AU), 3/2 (23.3 AU), 5/3 (25.0 AU), 4/7 (25.8 AU), and 4/3 (21.5 AU). It is interesting to note that the large majority of particles is trapped in resonances. Only a small percentage of particles crosses all the resonances and reaches the orbit of the planet. The outer resonances act like a system of barriers.

The concentration of particles behind the planet is striking like in the previous circular model. This result is independent of $\beta$.

Like in the planar circular model, also in the 3-D model with a planet on an eccentric orbit, an inner region is cleared where particle density drops off very sharply. Also this result is independent of $\beta$.

## 6. Conclusions

We have shown that a planet with a mass of $10^{-4}$ central stellar mass confines a disk of micron sized dust particles. Simulations searching a critical value for the planetary mass show that the confinement works for masses larger than $2 * 10^{-5}$ central stellar mass. The principal mechanism is trapping of dust particles in outer resonances. A consequence of this resonance trapping is a non-uniform azimuthal distribution of disk material which appears to be static in a reference frame rotating with the planet. This result is valid for a large range of particle sizes confirmed also by the semi-analytical theory of Lazarro et al. (this volume).

Of course, our model can be criticized on the basis that we neglect collisions among the dust particles as well as the total force of the circumstellar disk acting on the planet. One of the major difficulties to include collisions in a simulation is the lack of observational data which allows an estimation of the particle size distribution in the $\beta$ Pictoris circumstellar disk. Collisions are expected to decrease a particle's radius which means an increase of $\beta$. As soon as $\beta$ reaches a critical value, the radiation force exceeds the gravitational force acting on a particle and the particle is blown out of the system. Its binding energy becomes positive. Such a particle is known as a $\beta$-meteoroid. The first effect, increase of $\beta$ does not effect our major conclusions since we found that they are independent of $\beta$. The second effect, the disappearance of particles may effect our conclusions. The relevant parameters are the trapping time scale and the destruction time scale to become a $\beta$-meteorid. If the destruction time scale is some orders of magnitude smaller than the trapping time scale, no particular density distribution as described above would appear.

## Acknowledgements

The calculations were carried out on the Connection Machine CM-2 at INRIA, Sophia Antipolis. This work was supported by the Programme National de Planétologie.

## References

Artymowicz, P., Burrows, C. and Paresce F.: 1989, "The Structure of the beta Pictoris Circumstellar Disk from Combined IRAS and Coronographic Observations.", *Astrophysical Journal*, **337**, 494-513

Bulirsch, R. and Stoer J.: 1966, "Numerical Treatment of Ordinary Differential Equations by Extrapolation Methods.", *Numer. Math.*, **8**,1-13

Burns, J.A., Lamy, P. L. and Soter S.: 1979, "Radiation Forces on Small Particles in the Solar System.", *Icarus*, **40**,1-40

Gonczi, R., Froeschlé, Ch. and Froschlé, C.: 1982, "Poynting-Robertson drag and orbital resonances", *Icarus*, **51**, 633-654

Hillis, D;: 1985, *The Connection Machine*, MIT Press Cambridge, Mass.

Hillis, D.: 1987,"The Connection Machine.", *Scientific American*, **256**, 108-115

Krogh, F. T.: 1970,*JPL Technical Utilization Document, No.CP-238*

Lazarro, D., Sicardy, B., Roques, F. and Scholl, H.: 1992, "Orbital Resonances and Poynting-Robertson Drag Confining Dust Particles in $\beta$-Pic Disk", this volume

Press, W.H., Flannery, B.P., Teukolsky S.A. and Vetterling W.T.: 1989, "Numerical Recipes", Cambridge Univ. Press

Sicardy, B: 1991,"Numerical Exploration of Planetary Arc Dynamics", *Icarus*, **89**, 197-219

Smith, B A. and Terrile, J.: 1984, "A Circumstellar Disk around $\beta$ Pictoris", *Nature*, **226**, 1421-1424

Torbett, M. and Smoluchowski R.: 1982, "Motion of the Jovian commensurability resonance and the character of the celestial mechanics in the asteroid zone", *Astron. Astrophys.*, **110**, 43-49

Wyatt, S.P.jr. and Whipple, F.L.: 1950, "The Poynting-Robertson Effect on Meteor Orbits", *Astrophys. Journ.*, **111**, 134-141

# ORBITAL RESONANCES AND POYNTING-ROBERTSON
# DRAG CONFINING DUST PARTICLES IN $\beta$-PIC DISK

DANIELA LAZZARO

*Observatório Nacional, DAF, BR-20921 Rio de Janeiro, Brazil and*
*Observatoire de Meudon, LAM/DAEC, F-92195 Meudon, France*

BRUNO SICARDY and FRANCOISE ROQUES

*Observatoire de Meudon, LAM/DAEC, F-92195 Meudon, France*

and

HANS SCHOLL

*Observatoire de la Côte d'Azur, F-06304 Nice, France*

The observations of the $\beta$-Pictoris circumstellar disk show a cleared zone inside a radius of nearly 20 A.U. (Smith & Terrile, 1984). In order to explain this important feature, the existence of a planet inside the cleared zone has been proposed. It is suggested that the combined effects of resonances and Poynting-Robertson drag would be responsible for confining the dust particles (Roques *et al.*, 1991).

The set of equations which analytically describe the above model, in a reference system with origin at the star's center, is the Hamiltonian system of the restricted three body problem plus an added non conservative drag force proportional to the velocity ($\mathbf{F} = -\alpha d\mathbf{r}/dt$), where $\mathbf{r}$ is the position of the dust particle. The coefficient $\alpha$ depends mainly on the particle size and composition. When $\alpha$ is zero, the equations reduce to the classical equations of the perturbed motion in a conservative system and their solutions in the presence of resonances are well studied (Henrard & Lemaitre, 1983; Ferraz-Mello, 1985).

The above equations which include the effects of the gravitational attraction of the star, the perturbation due to the hypothetical planet and the dissipative term due to the P-R drag in its simplified form, are further canonically transformed in more suitables variables (Brower & Hori, 1966; Ferraz-Mello, 1992). Only secular and resonant terms were retrieved in the perturbation function and the equations were averaged over the short period terms.

The system thus obtained was numerically integrated and the results compared to those obtained by the integration of the complete equations of motion. Both integrations show that a planet larger than nearly $10 M_\oplus$ can efficiently confine dust particles at the outer resonances.

Many tests where performed in order to determine to what extent the initial orbital elements of the dust particle influence the trapping or escape from a resonance. As an example, in figure 1 it is shown how the dynamical evolution of the particle depends on the planet mass (m).

In order to study the probability of trapping or escape of dust particles near a resonance an extensive analysis has been performed with respect to the evolution of the phase space of the system, considering an "osculating conservative phase

*Celestial Mechanics and Dynamical Astronomy* **56**: 395–396, 1993.
© 1993 *Kluwer Academic Publishers.*

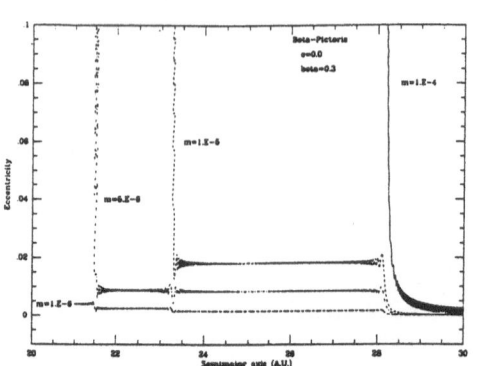

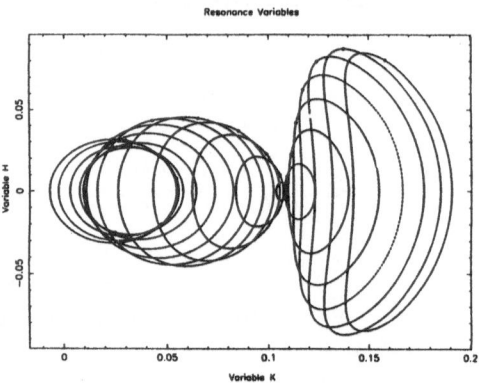

Fig. 1. Orbital elements.                    Fig. 2. Osculating phase space.

space" for different instants of the non-conservative integration. Figure 2 shows an example of this evolution in the case of the crossing, and subsequent trapping, in the 4:3 resonance of a dust particle. The variables H and K are proportional, respectively, to the sinus and cosinus of the critical angle defining the resonance. The darker points indicate the particle position in the non-conservative integration. It can be seen that the particle begins in external circulations (first curve in the left) and evolves to librations. We hope that further analysis will lead us to a suitable parametrization defining the possible behaviour of a dust particle near a resonance in the presence of a strong dissipation.

## References

Brouwer, D., Hori, G.: 1961, "Theoretical evaluation of atmospheric drag effects in the motion of an artificial satellite.", *Astron. J.*, **66**, 193–210.

Ferraz-Mello, S.: 1985, "Resonance in regular variables.", *Celest. Mech.*, **35**, 209–220.

Ferraz-Mello, S.: 1992, "Averaging the elliptic asteroidal problem with a Stokes drag.", In *Physics and Dynamics of the Small Bodies of the Solar System* (D. Benest and Cl. Froeschlé, Eds.), Ed. Frontières, (in press).

Henrard, J., Lemaitre, A.: 1983, "A second fundamental model for resonance.", *Celest. Mech.*, **30**, 197–218.

Roques, F., Scholl, H., Sicardy, B., Lazzaro, D.: 1991, "A planet in the inner clearing zone of $\beta$-Pic disk?.", *Bull. Am. Astron. Soc.*, **23**, 1233.

Smith, B., Terrile, R.J.: 1984, "Circumstellar disk around $\beta$ Pictoris.", *Science*, **226**, 1421–1424.